Geophysical Monograph Series

Including
IUGG Volumes
Maurice Ewing Volumes
Mineral Physics Volumes

Geophysical Monograph Series

143 **Mt. Etna: Volcano Laboratory** *Alessandro Bonaccorso, Sonia Calvari, Mauro Coltelli, Ciro Del Negro, and Susanna Falsaperla (Eds.)*

144 **The Subseafloor Biosphere at Mid-Ocean Ridges** *William S. D. Wilcock, Edward F. DeLong, Deborah S. Kelley, John A. Baross, and S. Craig Cary (Eds.)*

145 **Timescales of the Paleomagnetic Field** *James E. T. Channell, Dennis V. Kent, William Lowrie and Joseph G. Meert (Eds.)*

146 **The Extreme Proterozoic: Geology, Geochemistry, and Climate** *Gregory S. Jenkins, Mark A. S. McMenamin, Christopher P. McKay, and Linda Sohl (Eds.)*

147 **Earth's Climate: The Ocean–Atmosphere Interaction** *Chunzai Wang, Shang-Ping Xie, and James A. Carton (Eds.)*

148 **Mid-Ocean Ridges: Hydrothermal Interactions Between the Lithosphere and Oceans** *Christopher R. German, Jian Lin, and Lindsay M. Parson (Eds.)*

149 **Continent-Ocean Interactions Within East Asian Marginal Seas** *Peter Clift, Wolfgang Kuhnt, Pinxian Wang, and Dennis Hayes (Eds.)*

150 **The State of the Planet: Frontiers and Challenges in Geophysics** *Robert Stephen John Sparks, and Christopher John Hawkesworth (Eds.)*

151 **The Cenozoic Southern Ocean: Tectonics, Sedimentation, and Climate Change Between Australia and Antarctica** *Neville Exon, James P. Kennett, and Mitchell Malone (Eds.)*

152 **Sea Salt Aerosol Production: Mechanisms, Methods, Measurements, and Models** *Ernie R. Lewis and Stephen E. Schwartz*

153 **Ecosystems and Land Use Change** *Ruth S. DeFries, Gregory P. Anser, and Richard A. Houghton (Eds.)*

154 **The Rocky Mountain Region—An Evolving Lithosphere: Tectonics, Geochemistry, and Geophysics** *Karl E. Karlstrom and G. Randy Keller (Eds.)*

155 **The Inner Magnetosphere: Physics and Modeling** *Tuija I. Pulkkinen, Nikolai A. Tsyganenko, and Reiner H. W. Friedel (Eds.)*

156 **Particle Acceleration in Astrophysical Plasmas: Geospace and Beyond** *Dennis Gallagher, James Horwitz, Joseph Perez, Robert Preece, and John Quenby (Eds.)*

157 **Seismic Earth: Array Analysis of Broadband Seismograms** *Alan Levander and Guust Nolet (Eds.)*

158 **The Nordic Seas: An Integrated Perspective** *Helge Drange, Trond Dokken, Tore Furevik, Rüdiger Gerdes, and Wolfgang Berger (Eds.)*

159 **Inner Magnetosphere Interactions: New Perspectives From Imaging** *James Burch, Michael Schulz, and Harlan Spence (Eds.)*

160 **Earth's Deep Mantle: Structure, Composition, and Evolution** *Robert D. van der Hilst, Jay D. Bass, Jan Matas, and Jeannot Trampert (Eds.)*

161 **Circulation in the Gulf of Mexico: Observations and Models** *Wilton Sturges and Alexis Lugo-Fernandez (Eds.)*

162 **Dynamics of Fluids and Transport Through Fractured Rock** *Boris Faybishenko, Paul A. Witherspoon, and John Gale (Eds.)*

163 **Remote Sensing of Northern Hydrology: Measuring Environmental Change** *Claude R. Duguay and Alain Pietroniro (Eds.)*

164 **Archean Geodynamics and Environments** *Keith Benn, Jean-Claude Mareschal, and Kent C. Condie (Eds.)*

165 **Solar Eruptions and Energetic Particles** *Natchimuthukonar Gopalswamy, Richard Mewaldt, and Jarmo Torsti (Eds.)*

166 **Back-Arc Spreading Systems: Geological, Biological, Chemical, and Physical Interactions** *David M. Christie, Charles Fisher, Sang-Mook Lee, and Sharon Givens (Eds.)*

167 **Recurrent Magnetic Storms: Corotating Solar Wind Streams** *Bruce Tsurutani, Robert McPherron, Walter Gonzalez, Gang Lu, José H. A. Sobral, and Natchimuthukonar Gopalswamy (Eds.)*

168 **Earth's Deep Water Cycle** *Steven D. Jacobsen and Suzan van der Lee (Eds.)*

169 **Magnetospheric ULF Waves: Synthesis and New Directions** *Kazue Takahashi, Peter J. Chi, Richard E. Denton, and Robert L. Lysal (Eds.)*

170 **Earthquakes: Radiated Energy and the Physics of Faulting** *Rachel Abercrombie, Art McGarr, Hiroo Kanamori, and Giulio Di Toro (Eds.)*

171 **Subsurface Hydrology: Data Integration for Properties and Processes** *David W. Hyndman, Frederick D. Day-Lewis, and Kamini Singha (Eds.)*

172 **Volcanism and Subduction: The Kamchatka Region** *John Eichelberger, Evgenii Gordeev, Minoru Kasahara, Pavel Izbekov, and Johnathan Lees (Eds.)*

173 **Ocean Circulation: Mechanisms and Impacts—Past and Future Changes of Meridional Overturning** *Andreas Schmittner, John C. H. Chiang, and Sidney R. Hemming (Eds.)*

174 **Post-Perovskite: The Last Mantle Phase Transition** *Kei Hirose, John Brodholt, Thorne Lay, and David Yuen (Eds.)*

175 **A Continental Plate Boundary: Tectonics at South Island, New Zealand** *David Okaya, Tim Stem, and Fred Davey (Eds.)*

176 **Exploring Venus as a Terrestrial Planet** *Larry W. Esposito, Ellen R. Stofan, and Thomas E. Cravens (Eds.)*

177 **Ocean Modeling in an Eddying Regime** *Matthew Hecht and Hiroyasu Hasumi (Eds.)*

Geophysical Monograph 178

Magma to Microbe: Modeling Hydrothermal Processes at Ocean Spreading Centers

Robert P. Lowell
Jeffrey S. Seewald
Anna Metaxas
Michael R. Perfit
Editors

American Geophysical Union
Washington, DC

Library of Congress Cataloging-in-Publication Data

Magma to microbe: modeling hydrothermal processes at ocean spreading centers / Robert P. Lowell ... [et al.], editors.
p. cm. — (Geophysical monograph ; 178)
ISBN 978-0-87590-443-6
1. Hydrothermal circulation (Oceanography)—Mathematical models. 2. Seawater—Thermodynamics—Mathematical models. 3. Hydrothermal vents—Microbiology. 4. Mid-ocean ridges. 5. Sea-floor spreading. I. Lowell, Robert P.
GC228.5.M34 2008
551.1′36—dc22

2008035797

ISBN: 978-0-87590-443-6
ISSN: 0065-8448

Cover Photo: Image taken at Endeavor Segment by the Canadian remotely operated vehicle ROPOS during a Keck-NSERC funded cruise to Juan de Fuca Ridge.

CONTENTS

PREFACE

Oceanic spreading centers are among the most intriguing of Earth's features. As seafloor spreading occurs, magma rising from the mantle and traveling through a hot crystal mush zone accumulates in thin lens-shaped magma bodies below mid-ocean ridge axes. Gabbroic rocks form from the crystallizing magmas at depth while dikes emerge from the subaxial magma lens and erupt on the seafloor as basaltic lava flows that create the upper volcanic layer of the oceanic crust. During this process of crustal formation, seawater circulates downward through faults and fissures, where it is heated to temperatures greater than 400°C by the hot, newly created crust and subsequently undergoes a number of chemical reactions. These hydrothermal fluids alter and exchange elements with the surrounding igneous rock and buoyantly ascend to the seafloor though faults and fissures along the ridge axes. Upon exiting the seafloor, the metal-laden fluids rapidly mix with seawater, resulting in cooling and precipitation of metal-sulfide compounds that form hydrothermal vents associated with discrete vent chimneys and in some cases, large metal sulfide deposits. The chemical constituents of the hydrothermal fluids discharging at the seafloor provide nutrients to diverse microbial and macrofaunal ecosystems. The existence of these ecosystems has provided new insights into the evolution of life on Earth. Oceanic spreading centers are thus an important natural laboratory for studying interconnected processes of crustal formation, heat, mass and chemical transfer, and biological processes. Hydrothermal processes at oceanic spreading centers represent an important bridge between the heat transported by magma from the mantle into the crust and the existence of biological ecosystems on the seafloor.

Ridge crest hydrothermal systems have been the subject of intense study since the discovery of high-temperature black-smoker vents on the East Pacific Rise in the late 1970s. Over the past 3 decades, considerable progress has been made in understanding the complex interrelationships among the geological, geochemical and biological processes that occur at mid-ocean ridges. This progress has resulted in part from National Science Foundation support for the RIDGE program established in 1989, its successor RIDGE 2000, and support from the international community through the InterRIDGE program. To date, RIDGE 2000 has focused its attention on three sites for integrated study: the Endeavour Segment of the Juan de Fuca Ridge, with particular focus on the Main Endeavour Field; the 8-11°N region of the East Pacific Rise, with particular focus near 9°50′ N; and the Eastern Lau Spreading Center, a back-arc spreading center in the southwest Pacific Ocean.

A primary goal of RIDGE 2000 research at these integrated study sites is to "develop focused, quantitative, whole system models through coordinated, integrated and interdisciplinary experiments...." To achieve this goal, it is necessary not only to make careful observations of magmatic, tectonic, biological, and hydrothermal activity but also to develop robust, interdisciplinary, integrated mathematical and numerical models. To foster the development of such models, the RIDGE 2000 office, with support from the National Science Foundation, organized a Ridge Theoretical Institute (RTI) and Workshop at Mammoth Lakes, California from June 25-30, 2006. This volume is an outgrowth of that RTI. By providing an up-to-date review of existing models and modeling approaches in the study of hydrothermal processes at oceanic spreading centers, we hope that this volume will yield a fuller, more integrated understanding of these complex interdisciplinary processes. We expect it will lead in turn to innovative seafloor and laboratory experiments that will test and verify the accuracy of models and serve as a springboard for the development of a new generation of integrated, interdisciplinary models. The focus on modeling distinguishes this monograph from other AGU monographs dealing with seafloor hydrothermal processes, which also stemmed from previous RTIs and their InterRIDGE counterparts.

Chuck Fisher, the first Chair of the RIDGE 2000 Program from October, 2001 to October, 2005, initiated the idea of a RTI with a modeling emphasis. We gratefully acknowledge his encouragement and support. The process of developing this RTI was carried to fruition by Donna Blackman, the current RIDGE 2000 Program Chair. The logistical support of the RIDGE 2000 office at the Scripps Institution of Ocea-

Magma to Microbe: Modeling Hydrothermal Processes at Ocean Spreading Centers
Geophysical Monograph Series 178

10.1029/178GM01

nography through the hard work of Katie Phillips and others enabled the smooth functioning of the RTI and workshop. We thank all the RTI speakers, attendees, and workshop participants for their informative talks and vigorous discussions. We thank the contributors to this volume for submitting their manuscripts in timely fashion and the peer reviewers for the high-quality and timeliness of their reviews. Lastly, we thank Dawn Seigler at AGU for her assistance throughout the process and to the production staff at AGU for seeing this volume to completion.

Robert P. Lowell
Virginia Tech

Jeffrey S. Seewald
Woods Hole Oceanographic Institution

Anna Metaxas
Dalhousie University

Michael R. Perfit
University of Florida

Modeling Hydrothermal Processes at Ocean Spreading Centers: Magma to Microbe—An Overview

Robert P. Lowell

Department of Geosciences, Virginia Polytechnic Institute and State University, Blacksburg, Virginia, USA

Jeffrey S. Seewald

Department of Marine Chemistry and Geochemistry, Woods Hole Oceanographic Institution, Woods Hole, Massachusetts, USA

Anna Metaxas

Department of Oceanography, Dalhousie University, Halifax, Nova Scotia, Canada

Michael R. Perfit

Department of Geological Sciences, University of Florida, Gainesville, Florida, USA

Hydrothermal processes at oceanic spreading centers encompass a number of highly interconnected processes ranging from the transport of mantle melts beneath spreading centers to the evolution of ocean chemistry and Earth's climate. This volume, which stems from a RIDGE Theoretical Institute held at Mammoth Lakes, California in June 2006, contains papers that address the complex connections among magmatic heat supply, crustal formation, seismicity, and hydrothermal circulation as well as the complex linkages among hydrothermal circulation, vent chemistry, carbon cycling, and microbial and macrofaunal ecosystems. The last paper in this volume explores the connection between hydrothermal venting and the chemical evolution of the oceans during the Phanerozoic. From reading these papers, one should recognize the wide variety of modeling approaches used and the uneven state of model development within various subdisciplines. Models of hydrothermal circulation and vent chemistry tend to be more quantitative, whereas models of carbon cycling and biological processes tend to be more conceptual. Although many of the complex linkages among the subdisciplines are understood at a conceptual level, considerable effort must be undertaken to develop integrated quantitative models of hydrothermal processes at oceanic spreading centers.

Magma to Microbe: Modeling Hydrothermal Processes at Ocean Spreading Centers
Geophysical Monograph Series 178

10.1029/178GM02

1. INTRODUCTION

Seafloor hydrothermal systems represent an important component of Earth's thermal engine whereby magmatic heat transported from the mantle beneath oceanic spreading centers is advected to the ocean by seawater-derived vent fluids. On a global basis these systems account for 25% of Earth's heat loss and 33% of the heat loss through oceanic lithosphere [*Williams and Von Herzen*, 1974; *Sclater et al.*, 1980; *Stein and Stein*, 1994]. Hydrothermal systems also play a major role in mass transfer between the oceanic crust and overlying ocean [e.g., *Wolery and Sleep*, 1976; *Edmond et al.*, 1979; *Thompson*, 1983]. Hydrothermal sites that discharge high temperature (~350°C) metal-laden fluids along the more than 60000 km ocean ridge system are of particular interest, in part because the chemical constituents transported by hydrothermal fluids support chemosynthetic microbial and macro-biological ecosystems [e.g., *Jannasch and Wirsen*, 1979; *Jannasch*, 1983, 1995; *Hessler et al.*, 1988; *Tunnicliffe,* 1991]. The discovery of chemosynthetic ecosystems at seafloor hydrothermal vents has led to a new awareness of life in extreme environments on Earth [e.g., *Baross and Hoffman*, 1985; *Wilcock et al.* 2004] and to suggestions of the possibility of life elsewhere in the solar system [e.g., *McCollom*, 1999].

High temperature hydrothermal venting at oceanic spreading centers has been studied for about three decades. Presently, approximately 20% of the ridge length has been surveyed, and thus far about 150 vent sites have been identified. A similar number has been inferred from chemical, optical, and thermal anomalies in the overlying water column [*Baker and German*, 2004]. The total number of active hydrothermal fields, whether known or inferred, represents just a fraction of the ~ 1000 or more sites that are predicted to exist [*Baker and German*, 2004; *Lowell and DuBose*, 2005].

Submarine hydrothermal activity requires a heat source and circulation pathways through permeable crustal rocks to transfer heat. In oceanic crust, shallow, relatively thin, largely molten magma bodies extending ~1-4 kilometers across the ridge axis at subseafloor depths of a few km are thought to act as the principal heat source [e.g., *Lowell and Germanovich*, 2004; *Lowell et al.*, this volume]. The main permeable pathways are likely a result of cracks, fissures and faults, which are generated and maintained by magmatic intrusions, tectonism, thermoelastic processes, and earthquakes [e.g., *Curewitz and Karson*, 1997; *Macdonald*, 1998; *Tolstoy*, this volume]. Diking events provide local, short-term heat sources and also generate new permeability [e.g., *Lowell and Germanovich*, 1995; *Germanovich et al.*, 2000]. As seawater enters the crust and is heated, mineral dissolution and precipitation processes occur, which may alter the permeability and porosity of crustal rocks [e.g., *Lowell et al.*, 1993; *Lowell and Yao*, 2002; *Alt-Epping and Diamond*, this volume] and thus alter circulation pathways. High-temperature chemical reactions modify fluid composition, and as heated hydrothermal seawater ascends from depth to the shallow crust, mixing with cooler seawater circulating in extrusive volcanic rocks may lead to additional mineral precipitation, again altering the permeability and porosity. Chemical disequilibrium in mixing environments represents an abundant source of energy that supports the growth of microbes in the shallow crust and macro-biological ecosystems on the seafloor, which may also affect subsurface porosity and permeability. Consequently, there are linkages and feedbacks between biological activity and fluid flow [*Huber and Holden.*, this volume; *McCollom*, this volume; *Schrenk et al.*, this volume; *Shea et al.*, this volume]. High-temperature hydrothermal processes at oceanic spreading centers are therefore characterized by a complex interplay among magmatic, tectonic, hydrogeological, and biogeochemical transport processes. Mathematical and numerical modeling allows integration of these processes into a quantitative framework.

In order to understand the state of development of mathematical and numerical modeling of hydrothermal processes at oceanic spreading centers and to encourage the development of integrated, whole-system models, RIDGE 2000 organized a Ridge Theoretical Institute (RTI) and Workshop at Mammoth Lakes, California from June 25–30, 2006. This volume is an outgrowth of that RTI. The focus on modeling distinguishes this monograph from other AGU monographs dealing with seafloor hydrothermal processes, which also stemmed from previous RTIs and their InterRidge counterparts [*Humphris et al.*, 1995; *Buck et al.*, 1998; *Wilcock et al.*, 2004; *German et al.*, 2004; *Christie et al.*, 2006].

This volume provides an up-to-date view of existing models and modeling approaches in the study of hydrothermal processes at oceanic spreading centers. We hope that it will provide a fuller, more integrated understanding of these complex interdisciplinary processes, leading in turn to innovative seafloor and laboratory experiments that will verify the accuracy of models. We also anticipate that it will serve as a springboard for the development of new, integrated interdisciplinary models.

The complexity of multiple, interconnected, non-linear, and time-dependent processes that occur over a broad range of temporal and spatial scales at oceanic spreading centers presents a daunting task for modelers. Yet models are essential and, in a sense, ubiquitous. Conceptual models are a necessary prerequisite for the interpretation of experimental data and field observations. Such models also serve as starting points for the development of quantitative models that elucidate the importance of various processes and guide the interpretation of data in appropriate physical, chemical, biological, and geological contexts. The development of integrated, quantitative numerical models of hydrothermal circulation is an important goal of ocean ridges studies.

The papers in this volume address a number of interconnected processes ranging from magmatic to biogeochemical processes. Processes fundamentally associated with the deeper parts of the system are, in general, abiotic while those occurring in the shallower parts of the system may involve an important biological component. The fundamental bridge between these parts of the system is hydrothermal circulation, which transfers heat, energy, and biogeochemically important constituents from depth to the shallow crust, and ultimately the ocean. Lastly, the volume ends with a brief discussion of the effects of hydrothermal processes on ocean chemistry over geologic time.

2. MAGMA TRANSPORT AND EVOLUTION OF THE OCEANIC CRUST

Adiabatic rise of upper mantle peridotite beneath spreading centers due to convection and plate separation results in partial melting and the generation of mid-ocean ridge basalt (MORB) magma at depths beginning around 100 km. Melting is affected by the decompression of hot, buoyant peridotite that reaches the melting point (solidus) for mantle material as it rises to shallow depths beneath the ridges [*Keleman et al.*, 1997a; *Asimow et al.* 2001]. Melting continues as the mantle rises and MORB melts continue to form and accumulate in a broad region (10s to 100s of kilometers), but melts are ultimately focused so that they feed relatively narrow regions (a few kilometers) along the axis of spreading ridges [e.g., *Langmuir et al.*,1992; *Grove et al.*, 1992; *Shen and Forsyth*, 1995]. These accumulated primary melts rarely make it to the seafloor to erupt as MORB lavas without modification by a variety of physical and chemical processes. In fact, studies of ophiolites, deep drill cores, and exposures of deep crust in "tectonic windows", together with seismic studies, indicate that the greatest volume (but not all [see *Herzberg*, 2004]) of these mantle melts reside and cool at depths of ~ 1 to 10 km below the seafloor where they ultimately crystallize to form the plutonic foundations of the oceanic crust.

We know little about what occurs in this environment from direct observation or in situ sampling of the oceanic crust because the deeper parts of the crust are rarely exposed and difficult to reach by drilling. Instead, we have relied on inferences made from combined field studies of ophiolites [e.g., *Nicolas et al.*, 1996] and marine seismic studies which long ago defined a thin (< 0.5 km) upper volcanic layer (seismic Layer 2A) from the underlying plutonic sections (Layer 2B – sheeted dikes and Layer 3 - gabbroic layer) that comprise the bulk (> 5–6 km) of the oceanic crust [e.g., *Hooft et al.*, 1997; *Dunn and Toomey*, 2000; *Carbotte and Scheirer*, 2004]. Seismic and compliance studies have indicated that beneath many ridges (mostly of fast to intermediate spreading rate) there is a melt-rich zone, the axial magma chamber (AMC) or magma lens, approximately 1–4 km wide and ~30-100 m thick, that lies on top of a wider (5–7 km) and thicker zone of crystal mush (Low Velocity Zone) [*Sinton and Detrick*, 1992; *Crawford and Webb*, 2002]. Presumably, the AMC is where magma is stored before it is transported upward by dikes along faults, fractures and fissures to the seafloor. This is consistent with observations in the neovolcanic zone associated with a narrow axial summit collapse trough or axial valley where fissure-fed eruptions are most commonly located [*Fornari et al.*, 1998; *Soule et al.*, 2005; *Perfit and Chadwick*, 1998], volcanogenic earthquakes are focused [*Tolstoy*, this volume] and hydrothermal vents are concentrated [*Fornari et al.*, 2004]. Although there is evidence for what would be considered "off-axis" volcanism on some ridges, the focus of seismic, hydrothermal and magmatic activity is predominantly in the neovolcanic zone where volcanism is centered. *Maclennan* [this volume] estimates that at the northern EPR ~90% of heat released from cooling and crystallization occurs within 2 km of ridge axes, although globally 85% of the total magmatic heat is released within 10–15 km of all spreading rate ridges.

Regardless of how well this integrated model of the structure of oceanic crust appears, we do not fully understand how the middle and lower crust forms, particularly at slow spreading centers where magmatism is likely to be intermittent and tectonism is the dominant process affecting crustal structure. The model of a large, steady-state magma chamber has given way to one that invokes the presence of discrete sill-like intrusions with dimensions similar to the seismically imaged AMC, but the temporal and spatial distributions of the intrusions are unconstrained. A problem with this conceptual model is that it is unclear how a ~5 km thick horizontally layered middle and lower crust could be formed from a melt lens or multiple sills intruded in the very narrow (few kilometers wide) sub-axial region (AMC). If melt-rich lenses occur near the top of this volume, then much of the crust may be accreted by substantial downward and lateral flow. Currently, there are two general conceptual models for the formation of the middle to lower crust at fast-spreading mid-ocean ridges. In the "gabbro glacier" models, magma is fed from the mantle to the AMC where it fractionally crystallizes and the crystal residue subsides to form the lower oceanic crust. In the "sheeted sill" or "stacked pluton" models [e.g., *Kelemen et al.*, 1997b; *Natland and Dick*, 2001], magmas crystallize in thin lenses that exist throughout the crust – the AMC is the uppermost of these. In gabbro glacier-type models, the AMC acts as a mixing pot for homogenizing and fractionating mantle derived magmas and the latent heat of crystallization is largely removed from the top of the gabbroic layer, thus providing the heat for hydrothermal circulation through the upper crust [see *Coogan et al.*, 2002a].

In the sheeted sill-type models, hydrothermal circulation throughout the lower crust is required to extract the latent heat of crystallization and the AMC plays a secondary role in magma differentiation.

It is in the subaxial magma bodies or "chambers" – most likely regions with varying proportions of melt and crystals [*Singh et al.*, 1998; *Crawford and Webb*, 2002] – where physical and chemical processes, such as fractional crystallization, magma mixing and crustal assimilation, act to differentiate original melt compositions into a spectrum of MORB compositions [*Sinton and Detrick*, 1992; *Klein*, 2003] and to form a wide variety of slowly cooled gabbroic and ultramafic rocks [e.g. *Natland and Dick*, 1996, 2001]. Many of these plutonic rocks, as well as some rare gabbroic xenoliths in lavas, have macroscopic and microscopic textures that indicate they formed from the accumulation of minerals that crystallized from a wide range of magma compositions - from primitive MgO-rich basalts to FeO and TiO_2 enriched ferrobasalts and even highly evolved, Si-rich melts [e.g. *Natland et al.*, 1991; *Natland and Dick*, 2001; *Ridley et al.*, 2006]. The geochemical range of MORB can largely be generated by fractional crystallization of a primitive, high-MgO magma. The compositions and temperatures of these primary basaltic melts are still uncertain; consequently, the amount of crystallization, cooling, and heat release required to make typical MORB are difficult to determine precisely [*Langmuir et al.*, 1992; *Perfit et al.*, 1996; *Herzberg*, 2004]. *Maclennan* [this volume] discusses the difficulties in determining the heat supplied to ridges by magmatism without full knowledge of melt composition, depths of crystallization and phase proportions during crystallization. Regardless, it is clear that mantle-derived magmas must undergo significant crystallization as they cool below 1300°C to form the oceanic crust. Calculations by *Maclennan* [this volume] indicate that most of the heat supplied to MOR is from the latent heat of crystallization, specific heat and subsolidus cooling.

The magmatic heat supplied to ridges during the formation of oceanic crust is a primary driver of hydrothermal circulation. This is most evident at fast and intermediate spreading ridges, where the most vigorous hydrothermal venting is closely associated with mostly-liquid magma lenses [e.g., *Singh et al.*, 1999; *Canales et al.*, 2006]. The details of these connections are not fully understood, as quantitative models of coupled magmatic and hydrothermal processes have yet to be developed in sufficient detail. A few studies have discussed relationships between chemically evolved MORB, hydrothermal circulation and sulfide deposits [e.g. *Perfit et al.*, 1998]. *Lowell et al.* [this volume] suggest that heat transfer from a stationary convecting and crystallizing magma chamber is not sufficient to maintain stable hydrothermal heat output on decadal time scales. They argue that rapid rates of magma replenishment are necessary. Thus, knowledge of rates of supply, replenishment, crystallization and cooling of magmas residing in the crust is critical to our understanding of hydrothermal systems.

It is also clear that Fe-enrichments and associated sulfide saturation in differentiated lavas and gabbros [*Natland et al.*, 1991; *Perfit et al.*, 1998] can provide the raw materials to hydrothermal fluids for sulfide chimneys and stockworks, but our knowledge of the composition and alteration of deeper crust in areas where hydrothermal-rock reactions and mineralization occur is still very poor. *Pester et al.* [this volume] and *Cruse et al.* [this volume] show the importance of rock composition and temperature on the evolution of vent fluid chemistry from two very different mid-ocean ridge environments. *Alt-Epping and Diamond* [this volume] reviews the current state of seawater-basalt reactive transport modeling.

Clues to the depth, composition and physical conditions of hydrothermal circulation can be obtained from studies of crustal metamorphism that ranges from low-grade "alteration" in the zeolite and greenschist facies in the shallow crust to pervasive greenschist, amphibolite and even granulite facies at greater depths [*Gregory and Taylor*, 1981; *Dick et al.*, 1991; *Stakes et al.*, 1991; *Robinson et al.*, 1991; *Gillis*, 1995; *Alt*, 1995; *Gillis et al.*, 2003; *Coogan et al.*, 2002b; *Python et al.*, 2007]. Phase chemical studies (amphibole, chlorite, pyroxene), fluid inclusion work (quartz veins), and isotopic investigations of altered oceanic crustal rocks indicate metamorphism occurs deep within the oceanic gabbroic section where temperatures can be as high as 750°C [*Coogan et al.*, 2002b; *Coogan*, 2008; *Gillis et al.*, 2003]. Petrologic and isotopic evidence from the Oman ophiolite even suggest that hydrothermal fluid circulation extends through the entire thickness of the crust and into the uppermost mantle at temperatures from 500°C to over 900°C [*Bosch et al.*, 2004; *Python et al.*, 2007].

The extent to which metamorphism occurs, and the processes that control the transport of fluids both into and out of the oceanic crust, remain to be determined [*Coogan*, 2008]. Some of the metamorphism observed in ophiolites may be occurring off-axis. The transition between high-temperature circulation at the ridge axis and the low-temperature circulation that continues transport heat from the lithosphere to ages of ~ 50 m.y. is not well known [e.g. *Stein and Stein*, 1994].

Finally, magmatic volatiles such as CO_2 and CH_4 provide an important carbon source of microbial ecosystems, which in turn form the basis for macrofaunal vent communities [*Kelley et al.*, 2004]. Thus, the transport of magmatic volatiles along with other nutrients in hydrothermal fluids provides a link between crustal formation and biological processes. Presently, our understanding of these complex connections is, for the most part, conceptual and based on empirical evidence alone [*McCollom*, this volume; *Huber and Holden*, this volume; *Schrenk et al.*, this volume; *Shea et al.*, this volume].

3. GEOCHEMICAL MODELS FOR THE ORIGIN AND EVOLUTION OF HYDROTHERMAL FLUIDS

The composition of submarine hydrothermal fluids reflects the integrated effects of numerous physical, chemical, and in some cases biological processes, as seawater circulates though the oceanic lithosphere at elevated temperatures and pressures. Collectively, field, laboratory, and theoretical modeling studies have demonstrated that many aspects of vent fluid chemistry are buffered by fluid-mineral equilibria, which in turn, are strongly influenced by temperature, pressure, rock mineralogy, and water/rock ratio [*Seyfried and Ding*, 1995]. The chemistry of vent fluids is also affected by phase separation and segregation that leads to complex mixing relationships between vapors, brines, non-phase separated hydrothermal fluids, and seawater during ascent from deep-seated reaction zones to the seafloor. All of these processes are influenced by magmatic and tectonic events, which may contribute magmatic volatiles and alter both the heat source and fluid circulation pathways.

Since the initial discovery of high temperature venting at oceanic spreading centers, equilibrium thermodynamic modeling has been widely utilized as a means to constrain factors that regulate vent fluid chemistry and develop conceptual models for the evolution of hot-spring fluids (*Bowers et al.*, 1985, 1988; *Berndt et al.*, 1989; *Seyfried et al.*, 1991; *Wetzel and Shock*, 2000). These studies have examined processes occurring on a variety of scales that include the reaction path of seawater as it circulates through basaltic crust [e.g. *Bowers et al.*, 1985; *McCollom and Shock*, 1998] as well as more spatially focused studies that examine equilibrium processes in high temperature reaction zones [e.g. *Berndt et al.*, 1989; *Seyfried et al.*, 1991; *Cruse et al.*, this volume; *Pester et al.* this volume]. Mixing environments associated with hydrothermal upflow and the formation of sulfide chimneys [e.g. *Janecky and Seyfried*, 1984; *Janecky and Shanks*, 1988; *Tivey*, 2004; *Cruse et al.*, this volume; *Pester et al.*, this volume] have also been investigated. The effectiveness of such models is readily apparent when the chemistry of hydrothermal fluids is interpreted within a thermodynamic framework. For example, recent work at 86°W on the Galapagos Spreading Center (GSC) has emphasized the importance of phase separation and mixing in deep-seated reaction zones in regulating vent fluid chemistry [*Pester et al.*, this volume]. Similar processes also regulate fluid composition at the sediment-covered Middle Valley hydrothermal system on the northern Juan de Fuca Ridge [*Cruse et al.*, this volume]. At Middle Valley, however, reaction path modeling reveals rapid re-equilibration of aqueous H_2 and H_2S with metal sulfides present in the upflow zone, suggesting that the chemistry of high temperature vent fluids may not always reflect conditions in the deepest and hottest portions of the crust [*Cruse et al.*, this volume].

Recognition that fluid-rock reactions influence subsurface permeability and porosity in addition to fluid and rock chemistry has led to the development of geochemical models that consider the feedback between simple chemical interaction and subsurface fluid flow [*Sleep*, 1991; *Lowell et al.*, 1993; 2003; *Fontaine et al.*, 2001; *Martin and Lowell*, 2000; *Lowell and Yao*, 2002]. Development of more complex whole system models is just beginning and faces some important limitations when applied to submarine hydrothermal systems [*Alt-Epping and Diamond*, this volume]. In particular, reactive transport models that account for multiphase fluid flow induced by phase separation need to be developed. Although some progress is being made on modeling phase separation and transport in NaCl-H_2O fluids [*Kissling*, 2005a,b; *Geiger et al.*, 2006a,b] applications to seafloor hydrothermal systems are sparse [*Lewis and Lowell*, 2004; *Lewis*, 2007; *Lowell et al.*, this volume].

There is a strong spatial association of biological communities with areas of diffuse venting. Consequently, the evolution of fluid composition during mixing of high temperature fluids with cool seawater in shallow regions of the crust has direct implications for metabolic strategies employed by vent organisms. In particular, the composition and temperature of diffuse fluids directly constrain the amount of chemical energy that is available to support vent ecosystems [e.g. *McCollom and Shock*, 1997; *Shock and Holland*, 2004; *Tivey* 2004; *McCollom*, 2007]. Accordingly, substantial spatial variability is expected in shallow mixing zones characterized by steep chemical and thermal gradients. The composition of vent ecosystems may also reflect the availability of organic and inorganic carbon sources that are influenced by cycling of organic matter in shallow regions of the crust and carbon fluxes from depth [*McCollom*, this volume]. Chemosynthetic vent organisms have the potential to influence the chemistry of low temperature vent fluids because they survive by catalyzing thermodynamically favorable reactions. Examination of metabolically active gases (CH_4, H_2, and H_2S) suggests that microbial activity may significantly influence fluid composition in diffuse flow environments [*Von Damm and Lilley*, 2004; *Butterfield et al.*, 2004; *Pester et al.*, this volume]. Assessing the role of biological activity in regulating the chemistry of diffuse fluids, however, will require additional information on the rates of abiotic and biotic processes that may affect the abundance of metabolically relevant species, the residence times of fluids in low temperature environments, and estimates of biomass.

4. MODELING BIOLOGICAL PROCESSES FROM INDIVIDUALS TO VENT COMMUNITIES

The discovery of biological ecosystems associated with hydrothermal discharge on the seafloor revolutionized scientific thinking about biological processes in the deep ocean and the origin of life on Earth [e.g., *Baross and Hoffman*, 1985; *Wilcock et al.*, 2004] and other planetary bodies [e.g., *McCollom*, 1999]. Complex systems of chemosynthetic microbial organisms thrive in extreme environments, both within the shallow crust and within sulfide edifices, at temperatures up to 120°C [*Holden and Daniel*, 2004], as well as in symbiotic associations with macrofauna [*Childress and Fisher*, 1992]. Most species of macrofauna that occur in areas of hydrothermal discharge are only found in these habitats [*Tunnicliffe et al.*, 1998], and have developed adaptations to tolerate the prevailing physicochemical conditions. Consequently, the spatial distribution of these organisms is primarily dependent upon the flux of hydrothermal end-member fluids [*Hessler et al.*, 1988]. At present, mathematical models of biological processes within the microbial biosphere [*Huber and Holden*, this volume; *Schrenk et al.*, this volume] including carbon cycling [*McCollom*, this volume], and in macrofaunal communities [*Shea et al.*, this volume] are in their infancy.

Carbon cycling in hydrothermal systems at oceanic spreading centers is a complex process that involves the transfer of different inorganic and organic carbon compounds across several abiotic and biotic reservoirs. Because carbon is the major component of biomass, the links between biological processes, particularly within the microbial communities of the oceanic crust, and the carbon cycle must be quantified. This will in turn enable an understanding of the connection between biological and geological processes. Presently, only conceptual models of the carbon cycle within the crust are possible [*McCollom*, this volume]. To move beyond this stage, additional observational and experimental data are required, particularly on the distribution and activities of microbial communities and their interaction with the geochemical environment [*McCollom*, this volume].

Within the crust, much of the microbial activity appears to be associated with diffuse flow regions near black smoker vents, where it impacts the chemistry of diffuse flow fluids [e.g., *Von Damm and Lilley*, 2004; *Huber and Holden*, this volume]. In addition, microbial activity may affect subsurface porosity and permeability, and hence the transport of nutrients both within the microbial community, and to overlying macrofaunal communities [*Huber and Holden*, this volume]. Conceptual models of microbial discharge following magmatic events ("snowblowers") suggest, however, that this effect is likely to be small [*Lowell et al.*, this volume; *Crowell et al.*, A model for the production of sulfur floc and "snowblower" events at mid-ocean ridges, submitted to *Geochemistry Geophysics Geosystems*, 2008].

Mathematical models that describe the interactions between microbes and hydrothermal systems (*Crowell et al.*, 2008) currently do not exist. To develop such models, estimates of fluid flux at the vent field scale, as well as rates of microbial growth under realistic conditions reproduced in the laboratory setting, are required [*Huber and Holden*, this volume]. Additionally, the steep thermal and chemical gradients within hydrothermal sulfide structures can provide an important natural laboratory for the study of the seafloor microbial biosphere [*Schrenk et al.*, this volume]. However, the large spatial and temporal variability in the physical and chemical characters of these structures can limit our ability to model the associated microbial ecosystems [*Schrenk et al.*, this volume].

The dynamics of macrofaunal populations and communities are driven by dispersal of propagules between vents, recruitment and growth of individuals at a particular vent location, and interactions within and between species throughout the lifetime of the system. While our knowledge of some of these processes has been increasing through observations and experiments, only a few conceptual, and even fewer numerical models have been constructed to describe different macrofaunal population and community dynamics [*Shea et al.*, this volume]. For example, in two published studies (and one in process at the EPR), particle tracking has been embedded in water circulation models to provide a crude estimate of larval dispersal (*Marsh et al.*, 2001; *Garcia Berdeal et al.*, 2006). Although mostly unexplored (except for those based on genetics), mathematical models can be developed to better understand mechanisms of dispersal, growth of both individuals and populations of species, and community succession. Mathematical models originally developed for terrestrial or other marine systems (e.g. those linking demography and dispersal) can be adapted to seafloor vent communities, guide further data collection, and enable a mechanistic understanding of the processes regulating these communities [*Shea et al.*, this volume].

5. HYDROTHERMAL CIRCULATION

The occurrence of heat flow values near mid-ocean ridges that are much lower than predicted by conductive lithospheric cooling models led to the early recognition that hydrothermal heat transfer was likely occurring near ridge crests [*Elder*, 1965; *Talwani et al.*, 1971]. The earliest hydrothermal models discussed mining of heat from hot crustal rocks and the development of hot springs on the seafloor [*Bodvarsson and Lowell*, 1972; *Lowell*, 1975]. *Lowell and Rona* [1976] used thermal anomalies in the water column in the TAG region of the Mid-Atlantic Ridge to suggest that the heat output from oceanic hot springs could approach that of continental geothermal systems. *Ribando et al.* [1976] developed the first numerical model of cellular convection in a porous medium, as applied to hydrothermal circulation in the oceanic crust. Finally, *Lister* [1974] proposed the novel idea of fluid circulation in downward propagating thermal contraction cracks as a mechanism for extracting heat from crustal rocks. Since this pioneering work, the idea of a "cracking front" and the role of thermal stresses in modifying crustal permeability have become part of the basic conceptual framework for understanding seafloor hydrothermal processes [e.g., *Germanovich and Lowell*, 1992; *Lowell and Germanovich*, 1994; *Wilcock and Delaney*, 1996; *Seyfried and Ding*, 1995; *Wilcock and Fisher*, 2004; *Tolstoy*, this volume]. It is interesting to note that the three basic approaches to modeling hydrothermal processes at oceanic spreading centers (single-pass or pipe models, cellular convection models, and downward crack propagation models) were all in place before the first high-temperature hydrothermal systems were discovered.

The discovery of warm springs at the Galapagos Spreading Center (GSC) [*Corliss et al.*, 1979; *Edmond et al.*, 1979a,b] and high-temperature black-smoker venting at the EPR [*Spiess et al.*, 1980; *Hekinian et al.*, 1983a,b] brought forth a new era in model development. Models showed that high-temperature venting could not be driven by extracting heat from hot crustal rocks [*Strens and Cann*, 1982; *Lowell and Rona*, 1985], but required the heat supplied by cooling, crystallizing magma [*Cann and Strens*, 1982; *Lowell and Rona*, 1985; *Lowell and Burnell*, 1991; *Lowell and Germanovich*, 1994].

In the past decade, advances in mathematical and numerical modeling of hydrothermal processes have evolved along several different fronts. The discovery of event plumes on the Juan de Fuca Ridge [*Baker et al.*, 1987, 1995] has prompted a number of different modeling approaches [*Lowell and Germanovich*, 1995; *Cann and Strens*, 1989; *Wilcock*, 1997; *Palmer et al.*, 1998; *Germanovich et al.*, 2000]. The question of how hydrothermal discharge becomes focused into black smoker vents has led to models in which the permeability distribution evolves in time, either as a result of thermal stresses [e.g., *Germanovich and Lowell*, 1992; *Germanovich et al.*, 2001] or mineral precipitation [*Lowell et al.*, 1993, 2003; *Fontaine et al.*, 2001]. *Lowell et al.* [2007] discuss the importance of mixing between high temperature hydrothermal fluids and seawater circulating in highly permeable extrusives as a controlling factor on vent temperature.

Quantitative modeling of hydrothermal processes at oceanic spreading centers is now entering a stage that emphasizes linkages among complex phenomena and the development of integrated models of specific hydrothermal systems. For example, *Wilcock* [2004] and *Ramondenc et al.* [this volume] investigate the hydrothermal response to earthquakes. *Alt Epping and Diamond* [this volume] review the state of the art in reactive transport modeling. *Lowell et al.* [this volume] discuss recent advances in the study of multiphase, multi-component processes, such as phase separation, coupled magma-hydrothermal processes, and biogenic floc production. These reviews show that such integrative quantitative models are in their infancy and that many exciting advances may be expected over the next decade.

6. HYDROTHERMAL PROCESSES OVER GEOLOGIC TIME

Although this monograph is focused on models that highlight the interconnectedness among hydrothermal, magmatic, tectonic, geochemical and biological processes at present day oceanic spreading centers, the impacts of hydrothermal processes extend to the global ocean and throughout geologic time [e.g., *Kadko et al.*, 1995]. Because seafloor hydrothermal environments might have served as sites for the origin of life on Earth [*Corliss et al.*, 1981; *Baross and Hoffman*, 1985], considerable attention has been paid to hydrothermal processes on early Earth. During the Precambrian, more vigorous mantle convection coupled with faster seafloor spreading rates and younger subduction ages [*Abbott and Hoffman*, 1984] suggests that hydrothermal activity may have been prevalent throughout the ocean basins. *Lowell and Keller* [2003] estimate that high-temperature hydrothermal heat loss and chemical fluxes during the Archean may have been ~ 10 times the present value.

The composition of the Precambrian oceans was likely different from today. As a result of higher crustal production rates, ocean chemistry was predominantly mantle-buffered [e.g., *Veizer et al.*, 1982]. There was little continental material available for weathering and transport to the oceans, and the increased vigor of hydrothermal activity suggests that hydrothermal chemical fluxes had a greater impact on ocean geochemistry than they do today. The Fe flux from Archean hydrothermal systems into a generally anoxic deep ocean is thought to have been a major source of Fe for banded iron formations (BIF) [*Isley*, 1995; *Isley and Abbot*, 1999]. Moreover, the absence of sulfate in the Precambrian ocean [e.g., *Grotzinger and Kasting*, 1993] would lead to an increase in the Fe/H_2S ratio in Archean vent fluids, also promoting the availability of hydrothermally-sourced Fe for BIF [*Kump and Seyfried*, 2005]. As a result of this difference in hydrothermal vent chemistry, chemosynthetic processes at Archean vents may have been based on H_2 oxidation rather the H_2S oxidation, which prevails in modern systems [*Kump and Seyfried*, 2005].

As Earth's tectonic engine has slowed since the Precambrian, hydrothermal fluxes have gradually diminished. With the rise of atmospheric oxygen, growth of continents, and decrease in hydrothermal fluxes, ocean composition gradually evolved to be more like the present. But even during the Phanerozoic, it appears that temporal changes in hydrothermal fluxes may have had an impact on ocean chemistry [e.g., *Kadko et al.*, 1995; *Kump*, this volume]. Although the CO_2 flux from oceanic spreading centers to the ocean is small compared to annual fluxes from respiration or remineralization of continental shelf sediments [*Holser et al.*, 1988], the CO_2 flux may be important at larger time scales. Changes in spreading rates yield changes in CO_2 flux from ridges [e.g., *Berner et al.*, 1983]. *Owen and Rea* [1985] and *Rea et al.* [1990] argue that major plate reorganization at ~ 50 Ma led to increased hydrothermal activity, which they linked ultimately to a doubling of atmospheric CO_2 and global warming. These suggested linkages, however, are a subject of debate. *Kump* [this volume] discusses the inconsistencies between model predictions on the evolution of seawater Mg^{2+} and Ca^{2+} [e.g., *Berner et al.*, 1983] and the record of Phanerozoic changes provided by fluid inclusions in evaporites [e.g., *Horita et al.*, 2002; *Lowenstein et al.*, 2003]. In coming years, better quantitative understanding of the relationships among such factors as seafloor production rates, hydrothermal circulation, and ocean chemistry will lead to improved models of ocean chemistry over geologic time.

7. SUMMARY AND NEW DIRECTIONS

This volume constitutes a set of papers that highlight recent observational results from seismology, vent geochemistry and biology and reviews advances in modeling hydrothermal processes at oceanic spreading centers. Although these papers show that mathematical and numerical modeling has progressed considerably over the past decades, there is still much to be done. We expect that integrated models of multiphase multi-component, including phase separation and magma-hydrothermal interactions, reactive transport, and modeling microbial and macrofaunal ecosystems will significantly increase our understanding of these complex, interconnected processes. The biological models, in particular, are still largely conceptual and would benefit from more quantitative approaches in the collection of both microbiological and macrobiological, data. In addition, quantitative models linking biological processes with fluid circulation and the geochemical environment are sorely needed.

Acknowledgments. We thank the reviewers Tom McCollom and Bill Seyfried for their helpful comments on the original version of this manuscript.

REFERENCES

Abbott, D., and S. Hoffman (1984), Archaean plate tectonics revisted-Part1: Heat flow, spreading rate, and the age of subducting lithosphere, and their effects on the origin and evolution of continents, *Tectonics*, *3*, 429–448.

Alt, J. C. (1995), Subseafloor processes in mid-ocean ridge hydrothermal systems, in *Seafloor Hydrothermal Systems: Physical, Chemical, Biological, and Geological Interactions, Geophys. Monogr. Ser.*, vol. 91, edited by S. E. Humphris, R. A. Zierenberg, L. S. Mullineaux, and R. E. Thomson, pp. 85–114, Am. Geophys. Union, Washington, D. C.

Alt-Epping, P., and L. W. Diamond (2008), Reactive transport and numerical modeling of seafloor hydrothermal systems: A review (this volume).

Asimow, P. D., M. M. Hirschmann, and E. M. Stolper (2001), Calculations of peridotite partial melting from thermodynamic models of minerals and melts. IV. Adiabatic decompression and the compositions and mean properties of mid-ocean ridge basalts, *J. Petrology*, *42*, 963–998.

Baker, E. T., G. J. Massoth, and R. A. Feely (1987), Cataclysmic hydrothermal venting on the Juan de Fuca Ridge, *Nature*, *329*, 149–151.

Baker, E. T., G. J. Massoth, R. A. Feely, R. W. Embley, R. E. Thomson, and B. J. Burd (1995), Hydrothermal event plumes from the CoAxial eruption site, Juan de Fuca Ridge, *Geophys. Res. Lett.*, *22*, 147–150.

Baker, E. T., and C. R. German (2004), On the global distribution of hydrothermal vent fields, in *Mid-Ocean Ridges: Hydrothermal Interactions Between the Lithosphere and Oceans, Geophys. Monogr. Ser.*, vol. 148, edited by C. R. German, J. Lin, and L. M. Parson, pp. 245–266, Am. Geophys. Union, Washington, D. C.

Baross, J. A. and S. E. Hoffman (1985), Submarine hydrothermal vents and associated gradient environments as sites for the origin and evolution of life, *Origins of Life*, *15*, 327–345.

Berndt, M. E., W. E. Seyfried, Jr., and D. R. Janecky (1989), Plagioclase and epidote buffering of cation ratios in mid-ocean ridge hydrothermal fluids: Experimental results in the near supercritical region, *Geochim. Cosmochim. Acta*, *53*, 2283–2300.

Berner, R. A., et al. (1983), The carbonate-silicate geochemical cycle and its effect on atmospheric carbon dioxide over the past 100 million years, *Amer. J. Sci.*, *283*, 641–683.

Bodvarsson, G., and R. P. Lowell (1972), Ocean-floor heat flow and the circulation of interstitial waters, *J. Geophys. Res.*, *77*, 4472–4475.

Bosch, D., Jamais, M., Boudier, F., Nicolas, A., Dautria, J.-M., and Agrinier, P. (2004), Deep and high-temperature hydrothermal circulation in the Oman ophiolite—Petrological and isotopic evidence, *J. Petrology*, *45*, 1181–1208.

Bowers, T. S., A. C. Campbell, C. I. Measures, A. J. Spivack, M. Khadem, and J. M. Edmond (1988) Chemical controls on the composition of vent fluids at 13°–11°N and 21°N, East Pacific Rise. *J. Geophys. Res. 93*, 4522–4536.

Bowers, T. S., K. L. Von Damm, and J. M. Edmond (1985), Chemical evolution of mid-ocean ridge hot springs, *Geochim. Cosmochim. Acta*, *49*, 2239–2252.

Buck, W. R., P. T. Delaney, J. A. Karson, and Y. Lagabrielle, editors (1998), *Faulting and Magmatism at Mid-Ocean Ridges, Geophys. Monogr.*, vol. 106, 348 pp., Am. Geophys. Union, Washington, D. C.

Butterfield, D. A. K. K. Roe, M. D. Lilley, J. A. Huber, J. A. Baross, R. W. Embley, and G. J. Massoth (2004), Mixing, reaction and microbial acitivty in the sub-seafloor revealed by temporal and spatial variation in diffuse flow vents at Axial Volcano, in *The Subseafloor Biosphere at Mid-Ocean Ridges, Geophys. Monogr.*, vol. 144, edited by W. S. D. Wilcock, E. F. DeLong, D. S. Kelley, J. A. Baross, and S. C. Cary, pp. 269–290, Am. Geophys. Union, Washington, D. C.

Canales, J. P., S. C. Singh, R. S. Detrick, S. M. Carbotte, A. Harding, G. M. Kent, J. B. Diebold, J. Babcock, and M. R. Nedimović (2006), Seismic evidence for variation in axial magma chamber properties along the southern Juan de Fuca Ridge, *Earth Planet. Sci. Lett.*, *246*, 353–366.

Cann, J. R., and M. R. Strens (1982), Black smokers fuelled by freezing magma, *Nature*, *298*, 147–149.

Cann, J. R., and M. R. Strens (1989), Modeling periodic megaplume emissions by black smoker systems, *J. Geophys. Res.*, *94*, 12,227–12,238.

Carbotte, S. M., and D. S. Scheirer (2004), Variability of ocean crustal structure created along the global mid-ocean ridge, in *Hydrogeology of the Oceanic Lithosphere*, edited by E. E. Davis and H. Elderfield, pp. 59–107, Cambridge University Press, Cambridge.

Childress, J. J., and C. R. Fisher (1992) The biology of hydrothermal vent animals: Physiology, biochemistry and autotrophic symbioses, *Oceanogr. Mar. Biol. Annu. Rev.*, *30*, 337–441.

Christie, D. M., C. R. Fisher, S.-M. Lee, and S. Givens (Eds.) (2006), *Back-Arc Spreading Systems*, *Geophys. Monogr.*, vol. 166, 303 pp., Am. Geophys. Union, Washington, D. C.

Coogan, L. A., G. Thompson, and C. J. MacLeod, (2002a), A textural and geochemical investigation of high level gabbros from the Oman ophiolite: Implications for the role of the axial magma chamber at fast spreading ridges, *Lithos*, *63*, 67–82.

Coogan, L. A., K. M. Gillis, C. J. MacLeod, G. Thompson, and R. Hekinian (2002b), Petrology and geochemistry of the lower ocean crust formed at the East Pacific Rise and exposed at Hess Deep: A synthesis and new results, *Geochem. Geophys. Geosyst.*, Special Issue: The Oman ophiolite and ocean ridge processes, doi:10.1029/2001GC000230.

Coogan, L. A. (2008), Reconciling temperatures of metamorphism, fluid fluxes, and heat transport in the upper crust at intermediate to fast spreading mid-ocean ridges, *Geochem. Geophys. Geosyst.*, *9*, Q02013, doi:10.1029/2007GC001787.

Corliss, J. B., et al. (1979), Submarine thermal springs on the Galapagos Rift, *Science*, *203*, 1073–1083.

Corliss, J. B., J. A. Baross, and S. E. Hoffman (1981), An hypothesis concerning the relationship between submarine hot springs and the origin of life on Earth, *Oceanol. Acta*, *4*, 59–69.

Crawford, W. C., and S. C. Webb (2002), Variations in the distribution of magma in the lower crust and at the Moho beneath the East Pacific Rise at 9°–10°N, *Earth Planet. Sci. Lett.*, *203*, 117–130.

Crowell, B. N., R. P. Lowell, and K. L. Von Damm (2008), A model for the production of sulfur floc and "snowblower" events at mid-ocean ridges, *Geochem. Geophys. Geosyst.*, in press.

Cruse, A. M., J. S. Seewald, P. M. Saccocia, and R. Zierenberg (2008), Hydrothermal fluid composition at Middle Valley, northern Juan de Fuca Ridge: Temporal and spatial variability (this volume).

Curewitz, D., and J. A. Karson (1997), Structural settings of hydrothermal outflow: Fracture permeability maintained by fault propagation and interaction, *J. Volcanol. Geotherm. Res.*, *79*, 149–168.

Dick, H. J. B., P. S. Meyer, S. Bloomer, S. Kirby, D. Stakes, and C. Mawer (1991), Lithostratigraphic evolution of an in-situ section of oceanic layer 3, in *Proc. Ocean Drill. Program Sci. Results*, *118*, edited by R. Von Herzen, P. T. Robinson, et al., Ocean Drilling Program, College Station, TX, pp. 439–538. doi:10.2973/odp.proc.sr.118.128.

Dunn, R. A., and D. R. Toomey (2000), Three-dimensional seismic structure and physical properties of the crust and shallow mantle beneath the East Pacific Rise at 9°30′N, *J. Geophys. Res.*, *105*, 23,537–23,555.

Edmond, J. M., C. Measures, R. E. McDuff, L. H. Chan, R. Collier, B. Grant, L. I. Gordon, and J. B. Corliss (1979), Ridge crest hydrothermal activity and the balances of the major and minor elements in the ocean: The Galapagos data, *Earth Planet. Sci. Lett.*, *46*, 1–18.

Elder, J. W. (1965), Physical processes in geothermal areas, in *Terrestrial Heat Flow*, *Geophys. Monogr. Ser.*, vol. 8, edited by W. H. K. Lee, pp. 211–239, Am. Geophys. Union, Washington, D. C.

Fontaine, F. J., M. Rabinowicz, and J. Boulegue (2001), Permeability changes due to mineral diagenesis in fractured crust: Implications for hydrothermal circulation at mid-ocean ridges, *Earth Planet. Sci. Lett.*, *184*, 407–423.

Fornari, D. J., R. M. Haymon, M. R. Perfit, T. K .P Gregg, and M. H. Edwards (1998), Axial summit trough of the East Pacific Rise 9°–10°N: Geological characteristics and evolution of the axial zone on fast spreading mid-ocean ridges, *J. Geophys. Res.*, *103*, 9827–9855.

Fornari, D. J., M. Tivey, H. Schouten, M. R. Perfit, et al. (2004), Submarine lava flow emplacement at the East Pacific Rise 9°50′N: Implications for uppermost ocean crust stratigraphy and hydrothermal fluid circulation, in *Mid-Ocean Ridges: Hydrothermal Interactions Between the Lithosphere and the Oceans*, *Geophys. Monogr.*, vol. 148, edited by C. R. German, J. Lin, and L. M. Parson, pp. 187–217, Am. Geophys. Union, Washington, D. C.

Garcia Berdeal, I., S. L. Hautala, L. N. Thomas, and H. P. Johnson (2006), Vertical structure of time-dependent currents in a mid-ocean ridge axial valley, *Deep-Sea Res., I*, 53,367–53,386.

Geiger, S., T. Driesner, C. A. Heinrich, and S. K. Matthai (2006a), Multiphase thermohaline convection in the Earth's crust: I. A new finite element-finite volume solution technique combined with a new equation of state for NaCl–H_2O, *Transp. Porous Media*, *63*, 399–434.

Geiger, S., T. Driesner, C. A. Heinrich, and S. K. Matthai (2006b), Multiphase thermohaline convection in the Earth's crust: II. Benchmarking and application of a finite element-finite volume solution technique combined with a new equation of state for NaCl–H_2O, *Transp. Porous Media*, *63*, 435–461.

German, C. R., J. Lin, and L. M. Parson, (Eds.) (2004), *Mid-Ocean Ridges: Hydrothermal Interactions Between the Lithosphere and Ocean*, *Geophys. Monogr. Ser.*, vol. 148, 318 pp., Am. Geophys. Union, Washington, D. C.

Germanovich, L. N., and R. P. Lowell (1992), Percolation theory, thermoelasticity, and discrete hydrothermal venting in the earth's crust, *Science*, 255, 1564–1567.

Germanovich, L. N., R. P. Lowell, and D. K. Astakhov (2000), Stress dependent permeability and the formation of seafloor event plumes, *J. Geophys. Res.*, *105*, 8341–8354.

Germanovich, L. N., R. P. Lowell, and D. K. Astakhov (2001), Bifurcations in seafloor hydrothermal flow, *J. Geophys. Res.*, *106*, 473–496.

Gregory, R. T., and H. P. Taylor (1981), An oxygen isotope profile in a section of Cretaceous oceanic crust, Samail ophiolite, Oman: evidence for {delta}18O buffering of the oceans by deep (>5 km) seawater–hydrothermal circulation at mid-ocean ridges, *J. Geophys. Res. 86*, 2737–2755.

Grotzinger, J. P., and J. F. Kasting (1993), New constraints on Precambrian ocean composition, *J. Geology*, *101*, 235–243.

Haymon, R. M., et al. (1993), Volcanic eruption of the mid-ocean ridge along East Pacific Rise crest at 9°45′–52′N: Direct submersible observations of seafloor phenomena associated with an eruption event in April 1991, *Earth Planet. Sci. Lett.*, *119*, 85–101.

Hekinian, R., et al. (1983a), East Pacific Rise near 13°N: Geology of new hydrothermal fields, *Science*, *219*, 1321–1324.

Hekinian, R., et al. (1983b), Intense hydrothermal activity at the rise axis of the East Pacific Rise near 13°N: Submersible witnesses the growth of a sulfide chimney, *Mar. Geophys. Res.*, *6*, 1–14.

Herzberg, C. (2004), Partial crystallization of mid-ocean ridge basalts in the crust and mantle, *J. Petrol.*, *45*, 2389–2405.

Hessler, R. R., W. M. Smithey, M. A. Boudrias, C. H. Keller, R. A. Lutz, and J. J. Childress (1988), Spatial and temporal variation of giant clams, tube worms and mussels at deep-sea hydrothermal vents, *Bull. Biol. Soc. Wash.*, *6*, 411–428.

Holden, J. F., and R. M. Daniel (2004), Upper temperature limit of life based on hyperthermophile culture experiments and field observations, in *The Subsurface Biosphere at Mid-Ocean Ridges*, *Geophys. Monogr.*, vol. 144, edited by W. S. D. Wilcock, E. F. DeLong, D. S. Kelley, J. A. Baross, and C. Cary, pp. 13–24, American Geophysical Union, Washington, D. C.

Holser, W. T, M. Schidlowski, F. T. MacKenzie, and J. B. Maynard (1988), Biogeochemical cycles of carbon and sulfur, in *Chemical Cycles in the Evolution of the Earth*, pp. 105–174, edited by B. Gregor et al., John Wiley, New York.

Hooft, E. E. E., R. S. Detrick, and G. M. Kent (1997), Seismic structure and indicators of magma budget along the southern East Pacific Rise, *J. Geophys Res.*, *102*, 27,319–37,340.

Huber, J. and J. F. Holden (2008), Modeling the impact of diffuse vent microorganisms along mid-ocean ridges and flanks (this volume).

Humphris, S. E., R. A. Zierenberg, L. S. Mullineaux, and R. E. Thomson (Eds.) (1995), *Seafloor Hydrothermal Systems: Physical, Chemical, Biological, and Geological Interactions*, *Geophys. Monogr. Ser.*, vol. 91, 466 pp., American Geophys. Union, Washington, D. C.

Isley, A. (1995), Hydrothermal plumes and the delivery of iron to banded iron formations, *J. Geol.*, *103*, 169–185.

Isley, A. E. and D. H. Abbott (1999), Plume-related mafic volcanism and the deposition of banded iron formations, *J. Geophys. Res.*, *104*, 15,461–15,477.

Janecky, D. R. (1982), Serpentinization of peridotite within the oceanic crust: Experimental and theoretical investigations of seawater-peridotite interaction at 200°C and 300°C, 500 bars, PhD thesis, University of Minnesota, Minneapolis, MN.

Janecky, D. R., and W. E. Seyfried, Jr. (1984), Formation of massive sulfide deposits on oceanic ridge crests, incremental reaction models for mixing between hydrothermal solution and seawater, *Geochim. Cosmochim. Acta*, *48*, 2723–2738.

Janecky, D. R., and W. C. Shanks, III (1988), Computational modeling of chemical and sulfur isotopic reaction processes in seafloor hydrothermal systems, chimneys, massive sulfides, and subjacent alteration zones, *Can. Mineral.*, *26*, 805–825.

Jannasch, H. W. (1983), Microbial processes at deep-sea hydrothermal vents, in *Hydrothermal Processes at Sea Floor Spreading Centers*, edited by P. A. Rona, K. Bostrom, L. Laubier, and K. L. Smith, pp. 677–709, Plenum Press, New York.

Jannasch, H. W. (1995), Microbial interactions with hydrothermal fluids, in *Seafloor Hydrothermal Systems*, *Geophys. Monogr.*, vol. 91, edited by S. E. Humphris, R. A. Zierenberg, L. S. Mullineaux, and R. E. Thomson, pp. 273–296, Am. Geophys. Union, Washington, D. C.

Jannasch, H. W., and C. O. Wirsen (1979), Chemosynthetic primary production at East Pacific sea floor spreading centers, *Bioscience*, *29*, 592–598.

Kadko, D., J. Baross, and J. Alt (1995), The magnitude and global implications of hydrothermal flux, in *Seafloor Hydrothermal Systems*, *Geophys. Monogr.*, vol. 91, edited by S. E. Humphris, R. A. Zierenberg, L. S. Mullineaux, and R. E. Thomson, pp. 446–466, Am. Geophys. Union, Washington, D. C.

Kelemen, P. B., G. Hirth, N. Shimizu, M. Spiegelman, and H. J. B. Dick (1997a), A review of melt migration processes in the adiabatically upwelling mantle beneath oceanic spreading ridges, *Philos. Trans. R. Soc. Lond.*, *A355*, 283–318.

Kelemen, P. B., K. Koga, and N. Shimizu (1997b), Geochemistry of gabbro sills in the crust–mantle transition zone of the Oman ophiolite: Implications for the origin of the oceanic lower crust, *Earth Planet. Sci. Lett.*, *146*, 475–488.

Kelley, D. S., M. D. Lilley, and G. L. Früh-Green (2004), Volatiles in submarine environments: Food for life, in *The Subseafloor Biosphere at Mid-Ocean Ridges*, *Geophys. Monogr.*, vol. 144, edited by W. S. D. Wilcock, E. F. DeLong, D. S. Kelley, J. A. Baross, and S. C. Cary, pp. 167–190, Am. Geophys. Union, Washington, D. C.

Kissling, W. (2005a), Transport of three-phase hyper-saline brines in porous media: Examples, *Transp. Porous Media*, *60*, 141–157.

Kissling, W. (2005b), Transport of three-phase hyper-saline brines in porous media: Theory and code implementation, *Transp. Porous Media*, *61*, 25–44.

Klein, E. M. (2003), Geochemistry of the igneous ocean crust, in *Treatise on Geochemistry*, 3, edited by R. Rudnick, pp. 433–463, Elsevier.

Korenaga, J., and P. B. Kelemen (1998), Melt migration through the oceanic lower crust; a constraint from melt percolation modeling with finite solid diffusion, *Earth Planet. Sci. Lett.*, *156*, 1–11.

Kump, L. R. (2008), The role of seafloor hydrothermal systems in the evolution of seawater composition during the Phanerozoic (this volume).

Kump, L. R., and W. E. Seyfried Jr. (2005), Hydrothermal Fe fluxes during the Precambrian: Effect of low oceanic sulfate concentrations and low hydrostatic pressure on the composition of black smokers, *Earth Planet. Sci. Lett.*, *235*, 654–662.

Langmuir, C. H., E. M. Klein, and T. Plank (1992), Petrological systematics of Mid-Ocean Ridge basalts: Constraints on melt generation beneath ocean ridges, in *Mantle Flow and Melt Generation at Mid-Ocean Ridges*, *Geophys. Monogr. Ser.*, vol. 71, edited by J. P-Morgan, D. K. Blackman, and J. M. Sinton, pp. 183–280, Washington, D. C.

Lewis, K. C. (2007), Numerical modeling of two-phase flow in the sodium chloride-water system with applications to seafloor hydrothermal systems, Ph.D. thesis, Georgia Institute of Technology, Atlanta, GA.

Lewis, K. C., and R. P. Lowell (2004), Mathematical modeling of phase separation of seawater near an igneous dike, *Geofluids*, *4*, 197–209.

Lister, C. R. B. (1974), On the penetration of water into hot rock, *Geophys. J. R. Astron. Soc.*, *39*, 465–509.

Lowell, R. P. (1975), Circulation in fractures, hot springs, and convective heat transport on mid-ocean ridge crests, *Geophys. J. R. Astron. Soc.*, *40*, 351–365.

Lowell, R. P., and D. K. Burnell (1991), A numerical model for magma-hydrothermal boundary layer heat transfer in the oceanic crust, *Earth Planet. Sci. Lett.*, *104*, 59–69.

Lowell, R. P., B. W. Crowell, K. C. Lewis, and L. Liu (2008), Modeling multiphase, multi-component processes at oceanic spreading centers (this volume).

Lowell, R. P., and M. DuBose (2005), Hydrothermal activity on Europa, *Geophys. Res. Lett.*, *32*, L05202, doi:10.1029/2005GL022375.

Lowell, R. P., and L. N. Germanovich (1994), On the temporal evolution of high-temperature hydrothermal systems at ocean ridge crests, *J. Geophys. Res.*, *99*, 565–575.

Lowell, R. P., and L. N. Germanovich (1995), Dike injection and the formation of megaplumes at ocean ridges, *Science*, *267*, 1804–1807.

Lowell, R. P., and L. N. Germanovich (2004), Seafloor hydrothermal processes: Results from scale analysis and single-pass models, in *Mid-Ocean Ridges: Hydrothermal Interactions Between the Lithosphere and Oceans*, *Geophys. Monogr.*, vol. 148, edited by C. R. German, J. Lin, and L. M. Parson, pp. 219–244, Am. Geophys. Union, Washington, D. C.

Lowell, R. P., S. R. Gosnell, and Y. Yang (2007), Numerical simulations of single-pass hydrothermal convection at mid-ocean ridges: Effects of the extrusive layer and temperature-dependent permeability, *Geochem, Geophys. Geosyst.*, *8*, Q10011, doi:10.1029/2007GC001653.

Lowell, R. P., and S. M. Keller (2003), High-temperature seafloor hydrothermal circulation over geologic time and Archean banded iron formations, *Geophys. Res. Lett.*, *30*(7), 1391, doi:10.1029/2002GL016536.

Lowell, R. P., and P. A. Rona (1976), On the interpretation of near-bottom water temperature anomalies, *Earth Planet. Sci. Lett.*, *32*, 18–24.

Lowell, R. P., and P. A. Rona (1985), Hydrothermal models for the generation of massive sulfide ore deposits, *J. Geophys. Res.*, *90*, 8769–8783.

Lowell, R. P., P. Van Cappellen, and L. N. Germanovich (1993), Silica precipitation in fractures and the evolution of permeability in hydrothermal upflow zones, *Science*, *260*, 192–194.

Lowell, R. P., and Y. Yao (2002), Anhydrite precipitation and the extent of hydrothermal recharge zones at ocean ridge crests, *J. Geophys. Res.*, *107*(B9), 2183, doi:10.1029/2001JB001289.

Lowell, R. P., Y. Yao, and L. N. Germanovich (2003), Anhydrite precipitation and the relationship between focused and diffuse flow in seafloor hydrothermal systems, *J. Geophys. Res.*, *108*(B9), 2424, doi:10.1029/2002JB002371.

Lowenstein, T. K., et al. (2003), Secular variation in seawater chemistry and the origin of calcium chloride basinal brines, *Geology*, *31*, 857–860.

Macdonald, K. C. (1998), Linkages between faulting, volcanism, hydrothermal activity, and segmentation on fast spreading centers, in *Faulting and Magmatism at Mid-Ocean Ridges*, *Geophys. Monogr. Ser.*, vol. 106, edited by W. R. Buck, P. T. Delaney, J. A. Karson, and Y. Lagabrielle, pp. 27–58, Am. Geophys. Union, Washington, D. C.

Maclennan, J. (2008), The supply of heat to mid-ocean ridges by crystallization and cooling of mantle melts (this volume).

Marsh, A. G., L. S. Mullineaux, C. M. Young, and D. T. Manahan (2001), Larval dispersal potential of the tubeworm *Riftia pachyptila* at deep-sea hydrothermal vents, *Nature*, *411*, 77–80.

Martin, J., and R. P. Lowell (2000), Precipitation of quartz during high-temperature fracture-controlled hydrothermal upflow at ocean ridges: Equilibrium vs. linear kinetics, *J. Geophys. Res.*, *105*, 869–882.

McCollum, T. M. (1999), Methanogensis as a potential source of chemical energy for primary biomass production by autotrophic organisms in hydrothermal systems on Europe, *J. Geophys. Res.*, *104*, 30,729–30,742.

McCollom, T. M. (2007), Geochemical constraints on sources of metabolic energy for chemolithoautrophy in ultramafic-hosted deep-sea hydrothermal systems, *Astrobiology*, *7*, 933–950.

McCollom, T. M. (2008), Observational, experimental, and theoretical constraints on carbon cycling in mid-ocean ridge hydrothermal systems (this volume).

McCollom, T. M., and E. L. Shock (1997), Geochemical constraints on chemolithoautotrophic metabolism by microorganisms in seafloor hydrothermal systems, *Geochim. Cosmochim. Acta*, *61*, 4375–4391.

McCollom, T. M., and E. L. Shock (1998), Fluid-rock interactions in the lower oceanic crust: Thermodynamic models of hydrothermal alteration, *J. Geophys. Res.*, *103*, 547–575.

Natland, J. H., P. S. Meyer, H. J. B Dick, and S. H. Bloomer (1991), Magmatic oxides and sulfides in gabbroic rocks from Hole 735B and the later development of the liquid line of descent, in *Proc. Ocean Drill. Program Sci. Results*, *118*, edited by R. Von Herzen, P. T. Robinson, et al., pp. 75–111, Ocean Drilling Program, College Station, TX, doi:10.2973/odp.proc.sr.118.163.

Natland, J. H., and H. J. B Dick (1996), Melt migration through high-level gabbroic cumulates of the East Pacific Rise at Hess Deep: Inferences from rock textures and mineral compositions, in *Proc. Ocean Drill. Program Sci. Results*, *147*, edited by C. Mével, K. Gillis, J. Allan, and P. Meyer, pp. 21–58, Ocean Drilling Program, College Station, TX.

Natland, J. H., and H. J. B. Dick (2001), Formation of the lower ocean crust and the crystallization of gabbroic cumulates at a very slowly spreading ridge, *J. Volcanol. Geotherm. Res.*, *110*, 191–233.

Nicolas, A. (1989), *Structure of Ophiolites and Dynamics of Oceanic Lithosphere*, Kluwer, Amsterdam, 367 pp.

Nicolas, A., F. Boudier, and B. Ildefonse (1996), Variable crustal thickness in the Oman ophiolite: Implications for oceanic crust, *J. Geophys. Res.*, *101*, 17,941– 17,950.

Owen, R. M., and D. K. Rea (1985), Sealfoor hydrothermal activity links climate to tectonics: Carbon dioxide greenhouse, *Science*, *277*, 166–169.

Palmer, M. R., and G. G. J. Ernst (1998), Generation of hydrothermal megaplumes by cooling of pillow basalts at mid-ocean ridges, *Nature*, *393*, 643–647.

Perfit, M. R., et al. (1996), Recent volcanism in the Siqueiros transform fault: Picritic basalts and implications for MORB magma genesis, *Earth Planet. Sci. Lett.*, *141*, 91–108.

Perfit, M. R., and W. W. Chadwick (1998), Magmatism at mid-ocean ridges: Constraints from volcanological and geochemical investigations, in: *Faulting and Magmatism at Mid-Ocean Ridges*, *Geophys. Monogr. Ser.*, vol. 106, edited by W. R. Buck, P. T. Delaney, J. A. Karson, and Y. Lagabrielle, pp. 59–115, Am. Geophys. Union, Washington, D. C.

Perfit, M. R., W. I. Ridley, and I. Jonasson (1998), Geologic, petrologic and geochemical relationships between magmatism and massive sulfide mineralization along the eastern Galapagos Spreading Center, *Rev. Econ. Geol.*, *8*, 75–100.

Pester, N. J., D. A. Butterfield, D. I. Foustoukos, K. K. Roe, K. Ding, T. M. Shank, and W. E. Seyfried, Jr. (2008), Chemistry of diffuse flow vent fields on the Galapagos Rift (86°W): Temporal variability and subseafloor phase equilibria controls (this volume).

Python, M., G. Ceuleeneer, Y. Ishida, J.-A. Barrat, and S. Arai (2007), Oman diopsidites: A new lithology diagnostic of very high temperature hydrothermal circulation in mantle peridotite below oceanic spreading centres, *Earth Planet. Sci. Lett.*, *255*, 289–305.

Ramondenc, P., L. N. Germanovich, and R. P. Lowell (2008), Modeling the hydrothermal response to earthquakes: Application to the 1995 event at 9°50′N East Pacific Rise (this volume).

Rea, D. K., J. C. Zachos, R. M. Owen, and P. D. Gingrich (1990), Global change at the Paleocene-Eocene boundary: Climatic and evolutionary consequences of tectonic events, *Paleogeogr. Paleoclimatol. Paleoecol.*, *79*, 117–128.

Ribando, R. J., K. E. Torrance, and D. L. Turcotte (1976), Numerical models for hydrothermal circulation in the oceanic crust, *J. Geophys. Res.*, *81*, 3007–3012.

Robinson, P. T., H. J. B. Dick, and R. P. Von Herzen (1991), Metamorphism and alteration in oceanic layer 3: Hole 735B, in *Proc. Ocean Drill. Program Sci. Results*, *118*, edited by R. Von Herzen, P. T. Robinson, et al., pp. 541–552, Ocean Drilling Program, College Station, TX, doi:10.2973/odp.proc.sr.118.122.

Sclater, J. G., C. Jaupart, and D. Galson (1980), The heat flow through oceanic and continental crust and the heat loss of the Earth, *Rev. Geophys.*, *18*, 269–311.

Schrenk, M., J. F. Holden, and J. A. Baross (2008), Microbial interactions and processes with minerals and fluids in deep-sea hydrothermal sulfide deposit (this volume).

Seyfried, Jr., W. E., and K. Ding (1995), Phase equilibria in subseafloor hydrothermal systems: A review of the role of redox, temperature, pH and dissolved Cl^- on the chemistry of hot spring fluids at mid-ocean ridges, in *Seafloor Hydrothermal Systems*, *Geophys. Monogr.*, vol. 91, edited by S. E. Humphris, R. A. Zierenberg, L. S. Mullineaux, and R. E. Thomson, pp. 248–272, Am. Geophys. Union, Washington, D. C.

Seyfried, Jr., W. E., K. Ding, and M. E. Berndt (1991), Phase equilibria constraints on the chemistry of hot spring fluids at mid-ocean ridges, *Geochim. Cosmochim. Acta*, *55*, 3559–3580.

Shea, K., A. Metaxas, C. R. Young, and C. R. Fisher (2008), Processes and interactions in macrofaunal assemblages at hydrothermal vents: A modeling perspective (this volume).

Shock, E. L., and M. E. Holland (2004), Geochemical energy sources that transport the subseafloor biosphere, in *The Subseafloor Biosphere at Mid-Ocean Ridges*, *Geophys. Monogr.*, vol. 144, edited by W. S. D. Wilcock, E. F. DeLong, D. S. Kelley, J. A. Baross, and S. C. Cary, pp. 153–165, Am. Geophys. Union, Washington, D. C.

Sinton, J. M., and R. S. Detrick (1992), Mid-ocean ridge magma chambers, *J. Geophys. Res.*, *97*, 197–216.

Singh, S. C., G. M. Kent, J. S. Collier, A. J. Harding, and J. A. Orcutt (1998), Melt to mush variations in crustal magma properties along the ridge crest at the southern East Pacific Rise, *Nature*, *394*, 874–878.

Sleep, N. H. (1991), Hydrothermal circulation, anhydrite precipitation and thermal structure at ridge axes, *J. Geophys. Res.*, *96*, 2375–2387.

Soule, S. A., D. J., Fornari, M. R. Perfit, M. A. Tivey, W. I. Ridley, and H. Schouten (2005), Channelized lava flows at the East Pacific Rise crest 9°–10°N: The importance of off-axis lava transport in developing crustal architecture, *Geochem. Geophys. Geosyst.*, *6*, Q08005, doi:10.1029/2005GC000912.

Spiess, F. N., et al. (1980), East Pacific Rise: Hot springs and geophysical experiments, *Science*, *207*, 1421–1433.

Stakes, D., C. Mével, M. Cannat, and T. Chaput (1991), Metamorphic stratigraphy of Hole 735B, in *Proc. Ocean Drill. Program Sci. Results*, *118*, edited by R. Von Herzen, P. T. Robinson, et al., pp. 153–180, Ocean Drilling Program, College Station, TX, doi:10.2973/odp.proc.sr.118.127.

Stein, C. A., and S. Stein (1994), Constraints on hydrothermal heat flux through the oceanic lithosphere from global heat flow, *J. Geophys. Res.*, *99*, 3081–3095.

Strens, M. R., and J. R. Cann (1982), A model of hydrothermal circulation in fault zones at mid-ocean ridge crests, *Geophys. J. R. Astron. Soc.*, 71, 225–240.

Talwani, M., C. C. Windisch, and M. G. Langseth (1971), Reykjanes Ridge crest: A detailed geophysical study, *J. Geophys. Res.*, *76*, 473–517.

Thompson, G. (1983), Basalt-seawater interaction, in *Hydrothermal Processes at Seafloor Spreading Centers*, edited by P. A. Rona, K. Boström, L. Laubier, and K. L. Smith, *NATO Conf. Ser. IV, Mar. Sci.*, *12*, 225–278.

Tivey, M. K. (2004), Environmental conditions within active seafloor vent structures: Sensitivity to vent fluid composition and fluid flow, in *The Subseafloor Biosphere at Mid-Ocean Ridges*, *Geophys. Monogr.*, vol. 144, edited by W. S. D. Wilcock, E. F. DeLong, D. S. Kelley, J. A. Baross, and S. C. Cary, pp. 137–152, Am. Geophys. Union, Washington, D. C.

Tolstoy, M. (2008) Seismological constraints on magmatic and hydrothermal processes at mid-ocean ridges (this volume).

Tunnicliffe, V. (1991), The biology of hydrothermal vents: Ecology and evolution, *Oceanogr. Mar. Biol. Annu. Rev.*, *29*, 319–407.

Tunnicliffe, V., A. G. MacArthur, and D. McHugh (1998), A biogeographical perspective of the deep-sea hydrothermal vent fauna, *Adv. Mar. Biol.*, *34*, 353–442.

Veizer, J., W. Compston, J. Hoefs, and H. Nielsen (1982), Mantle buffering of the early oceans, *Naturwissenschaften*, *69*, 173–180.

Von Damm, K. L. (1995), Controls on the chemistry and temporal variability of seafloor hydrothermal fluids, in *Seafloor Hydrothermal Systems*, *Geophys. Monogr.*, vol. 91, edited by S. E. Humphris, R. A. Zierenberg, L. S. Mullineaux, and R. E. Thomson, pp. 222–247, Am. Geophys. Union, Washington, D. C.

Von Damm, K. L., and M. D. Lilley (2004), Diffuse flow hydrothermal fluids from 9°50′N East Pacific Rise: Origin, evolution, and biogeochemical controls, in *The Subseafloor Biosphere at Mid-Ocean Ridges*, *Geophys. Monogr.*, vol. 144, edited by W. S. D. Wilcock, E. F. DeLong, D. S. Kelley, J. A. Baross, and S. C. Cary, pp. 245–268, Am. Geophys. Union, Washington, D. C.

Wetzel, L. R., and E. L. Shock (2000) Distinguishing ultramafic-, from basalt-hosted, submarine hydrothermal systems by comparing calculated vent fluid compositions, *J. Geophys. Res.*, *105*(B4), 8319–8340.

Wilcock, W. S. D. (2004), Physical response of mid-ocean ridge hydrothermal systems to local earthquakes, *Geochem. Geophys. Geosyst.*, *5*(11) Q11009, doi:10.1029/2004GC000701.

Wilcock, W. S. D. (1997), A model for the formation of transient event plumes above mid-ocean ridge hydrothermal systems, *J. Geophys. Res.*, *102*, 12,109–12,121.

Wilcock, W. S. D., and J. R. Delaney (1996), Mid-ocean ridge sulfide deposits: Evidence for heat extraction from magma chambers or cracking fronts, *Earth Planet. Sci. Lett.*, *145*, 49–64.

Wilcock, W. S. D., E. F. DeLong, D. S. Kelley, J. A. Baross, and S. C. Cary (Eds.) (2004), *The Subseafloor Biosphere at Mid-Ocean Ridges*, *Geophys. Monogr. Ser.*, vol. 144, 399 pp., Am. Geophys. Union, Washington, D. C.

Williams, D. L., and R. P. Von Herzen (1974), Heat loss from the Earth: New estimate, *Geology*, *2*, 327.

Wolery, T. J., and N. H. Sleep (1976), Hydrothermal circulation and geochemical flux at mid-ocean ridges, *J. Geol.*, *84*, 249–275.

R. P. Lowell (corresponding author) Department of Geosciences, Virginia Polytechnic Institute and State University, Blacksburg, Virginia 24061, USA. (rlowell@vt.edu)

J. S. Seewald, Department of Marine Chemistry and Geochemistry, Woods Hole Oceanographic Institution, Woods Hole, Massachusetts 02543, USA.

A. Metaxas, Department of Oceanography, Dalhousie University, Halifax, Nova Scotia, Canada B3H 4J1.

M. R. Perfit, Department of Geological Sciences, University of Florida, Gainesville, Florida 32611-2120, USA.

Modeling Multiphase, Multicomponent Processes at Oceanic Spreading Centers

Robert P. Lowell

Department of Geosciences, Virginia Polytechnic Institute and State University, Blacksburg, Virginia, USA

Brendan W. Crowell,[1] Kayla C. Lewis,[2] and Lei Liu[3]

School of Earth and Atmospheric Sciences, Georgia Institute of Technology, Atlanta, Georgia, USA

Hydrothermal systems at oceanic spreading centers are characterized by complex interrelationships among magmatic heat sources that generate buoyancy-driven fluid flow, geochemical and biogeochemical processes that affect fluid composition, and tectonic stresses that generate permeable fluid pathways. These stresses, along with mineral dissolution and precipitation, affect the porosity and permeability of crustal rocks and thereby exert feedbacks on the circulation. Vent chemistry shows that hydrothermal fluid at ocean ridges undergoes phase separation, which affects both the dynamics of flow and mineral species carried in solution. To understand hydrothermal processes at oceanic spreading centers, mathematical and numerical models of multiphase, multicomponent processes are needed. In this paper, we first present a steady-state *basic model* for high-temperature circulation using the single-pass model approach. We then present some idealized multiphase, multicomponent models that address: (a) magma crystallization and replenishment, (b) phase separation, and (c) mineral precipitation related to fluid mixing and biogeochemical processes. We show that magma chamber convection is insufficient to maintain steady-state hydrothermal heat output, thus showing the importance of magma replenishment. We show that phase separation in a single-pass system leads naturally to vent salinities that change from sub-seawater to super-seawater values over time. Finally, we show that mixing between hot hydrothermal fluid and seawater in the shallow crust can explain in

[1]Now at University of California, San Diego, Scripps Institute of Oceanography, San Diego, California, USA.

[2]Now at University of Chicago, Department of Geophysical Sciences, Chicago, Illinois, USA.

[3]The co-authors all contributed equally to this paper and thus are listed in alphabetical order.

Magma to Microbe: Modeling Hydrothermal Processes at Ocean Spreading Centers
Geophysical Monograph Series 178

10.1029/178GM03

part the relationship between focused high-temperature venting and nearby diffuse flow and that the generation of biogenic floc in the shallow crust appears to have a generally minor effect on crustal porosity over a magmatic cycle.

1. INTRODUCTION

Hydrothermal systems essentially consist of a heat source and a fluid circulation system. For high-temperature hydrothermal systems at oceanic spreading centers, the heat comes primarily from the emplacement of magma bodies in the shallow crust. The circulating fluid is mainly seawater, with some input of magmatic volatiles. In the simplest scenario, seawater enters the crust through faults and fractures and descends along permeable pathways to near the top of the magma body where it reaches temperatures exceeding 400°C. The heated fluid then ascends through another highly permeable fault or fracture zone and emerges at the seafloor as mineral-rich "black smoker" vents. *Kelley et al.* [2002] and *German and Von Damm* [2003] provide recent reviews of seafloor hydrothermal processes. Figure 1 shows schematically the layered structure of the crust at oceanic spreading centers extending at fast to intermediate rates, and hypothetical fluid pathways in the crust. The magma distribution, as shown in Figure 1, consists mainly of a thin mostly liquid melt lens atop a broad partially molten region extending to the base of the crust.

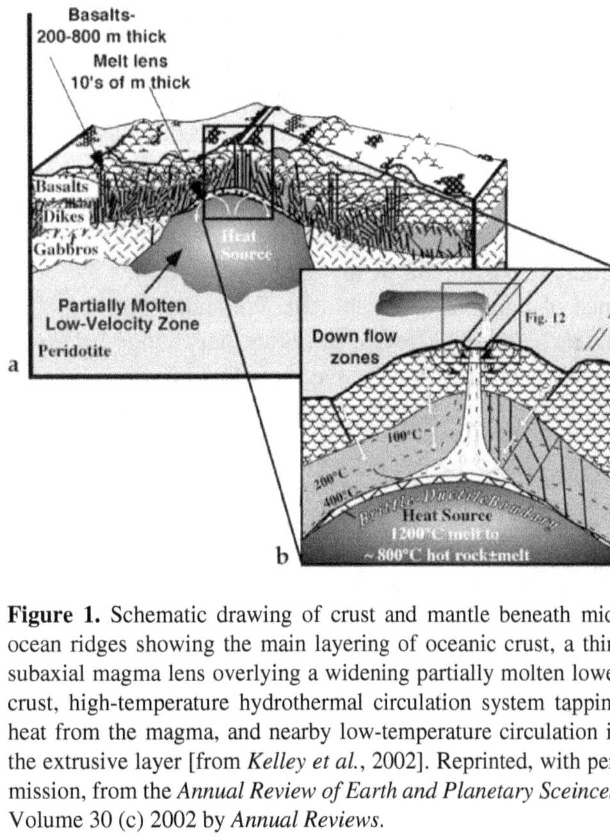

Figure 1. Schematic drawing of crust and mantle beneath mid-ocean ridges showing the main layering of oceanic crust, a thin, subaxial magma lens overlying a widening partially molten lower crust, high-temperature hydrothermal circulation system tapping heat from the magma, and nearby low-temperature circulation in the extrusive layer [from *Kelley et al.*, 2002]. Reprinted, with permission, from the *Annual Review of Earth and Planetary Sceinces*, Volume 30 (c) 2002 by *Annual Reviews*.

At slow-spreading ridges, liquid magma may be present intermittently, the partially molten zone may be narrower [*Sinton and Detrick*, 1992], and circulation may extend to greater depths in the crust. At the trans-Atlantic geotraverse (TAG) hydrothermal field on the Mid-Atlantic Ridge (MAR), the distribution of seismicity suggests that hydrothermal circulation penetrates into the deep crust along detachment faults [*deMartin et al.*, 2007]. Seismic refraction data do not record the presence of a shallow or midcrustal magma chamber, thus suggesting that the TAG hydrothermal system is driven by a magma body at or below the base of the crust [*Canales et al.*, 2007]. Recently, an extensive magma body has been detected beneath the Lucky Strike hydrothermal field on the MAR [*Singh et al.*, 2006]. Even at the ultraslow-spreading Gakkel Ridge in the Arctic Ocean, hydrothermal activity appears to be linked to recent volcanism [*Michael et al.*, 2003; *Baker et al.*, 2004]. Thus it is likely that magmatic heat sources drive high-temperature hydrothermal circulation at oceanic spreading centers regardless of the spreading rate.

The Lost City vent field on 1.5-million-year-old lithosphere west of the MAR is an end-member hydrothermal system that may be driven in part by heat released from the serpentinization of peridotite [*Kelley et al.*, 2001]. Limited hydrothermal modeling has been performed on these types of systems [*Lowell and Rona*, 2002; *Emmanuel and Berkowitz*, 2006]. Geochemical data suggest that serpentinization may influence the vent fluid chemistry at high-temperature systems such as Rainbow and Logatchev on the MAR, which probably also have magmatic heat inputs [*Seyfried et al.*, 2004]. Detailed discussion of hydrothermal systems hosted in ultramafic rocks is outside the scope of this paper.

This paper focuses on ridges spreading at fast and intermediate rates. In these systems, high-temperature hydrothermal circulation extracts heat mainly from the mostly liquid convecting melt lens. As this occurs, the magma lens, which is near its liquidus, begins to cool and crystallize, resulting in a decline in heat output from magma to the hydrothermal system. Magma may be transported from the underlying mushy zone to maintain the magma pressure and provide additional heat to the magma lens. Heat from the convecting magma lens is transported to the hydrothermal system by conduction across an impermeable boundary layer. A detailed understanding of the magma-hydrothermal system, therefore, must include how magma cooling, crystallization, and magma replenishment affect the heat transport to the hydrothermal circulation system.

Diking is another factor in the coupling between magmatic and hydrothermal processes. Pressurization of the magma lens by input from below could lead to dike propagation from near the tips of the magma lens [e.g., *Fialko*, 2001; *Sim*, 2004]; whereas hydrothermal cooling of the magma may lead to dike propagation at the center of the magma lens [*Sim*, 2004]. Diking, in turn, may lead to event plumes [e.g., *Baker et al.*, 1987, 1995; *Lowell and Germanovich*, 1995; *Germanovich et al.*, 2000] and to transient effects in the hydrothermal output [e.g., *Von Damm et al.*, 1995, 1997; *Von Damm*, 2004].

For the sake of convenience, the hydrothermal circulation system can be divided into two parts: (a) a shallow part that is essentially confined to the high porosity, high permeability extrusive layer and (b) a deeper part that penetrates the lower porosity sheeted dikes to near the top of the magma lens. In the shallow part of the system, fluid temperatures are generally low, except for the focused high-temperature upflow channels. Shallow circulation can occur both near the ridge crest and far off axis, resulting in the low-temperature alteration of the crust over many millions of years [see *Alt*, 1995, and reference therein]. Near the ridge axis, seawater circulating in the extrusive section may mix with deep seated, high-temperature hydrothermal fluid ascending toward the seafloor. This mixing likely leads to the precipitation of minerals in the shallow crust [see *Alt*, 1995, and references therein]. The effects of mineral precipitation on crustal permeability patterns may result in the observed complex interplay between focused and diffuse flow [e.g., *Lowell et al.*, 2003, 2007] as well as the distribution and evolution of seafloor biological communities [*Shea et al.*, this volume].

Microbial processes may affect the both the physics and chemistry of fluid circulation in the shallow crust. In recharge zones, biological process such as sulfate reduction may affect the fluid composition and microbial growth may lead to changes in the porosity and permeability of the recharge zone. In addition, as heated, hydrothermally altered fluid ascends into the shallow crust, it tends to drive circulation of cool seawater in the extrusive layer [*Lowell et al.*, 2007]. Typically, there is some mixing between the seawater in the extrusives and the high-temperature hydrothermal fluid. This mixture emerges as patches of relatively low-temperature diffuse flow that occurs in close proximity to the focused black smoker vents.

In shallow circulation systems with temperatures less than ≈120°C [*Holden and Daniel*, 2004], the hydrothermal-seawater fluid mixture is used by a variety of microbial species that constitute a vast, relatively unknown microbial ecosystem in the shallow oceanic crust [e.g., *Jannasch*, 1983, 1995]. Microbial growth may lead to the formation of biofilms, waste products, or biomineralization effects that could affect the porosity and permeability of the shallow crust [e.g., *Rogers et al.*, 2003; *Edwards et al.*, 2004]. The interplay between microbial processes and the fluid dynamics of the shallow crust has, so far, received little attention.

Recharge fluids that penetrate deep into the system undergo further reactions. As the temperature reaches ≈150°C, anhydrite will begin to precipitate from seawater [*Bischoff and Seyfried*, 1978]. As the temperature approaches 400°C or higher, a number of water-rock reactions occur that transform the initial seawater in the hydrothermal fluid. For example, the Mg and SO_4 initially contained in seawater are lost to the rock; and the hydrothermal fluid becomes enriched in Ca, Si, and a number of trace metals (see review by *Von Damm* [1995]). Because these geochemical processes affect the volume of rock occupied by the altered minerals and seawater, the porosity and permeability of the circulation pathways may vary in time and space. These changes can be modeled in terms of multiphase, multicomponent processes in which the solid phase remains stationary.

Moreover, the chlorinity of hydrothermal vent fluids entering the ocean is rarely equal to that of seawater. Measured values range from less than 10% of seawater to approximately twice that of seawater; and in systems for which time series measurements of vent fluid chlorinity are available, it is found that the chlorinity often changes over time [e.g., *Von Damm*, 1995, 2004]. The best explanation for these features of hydrothermal vent fluid chlorinities is that the fluid undergoes phase separation at the high temperatures and pressures occurring beneath the seafloor. Because seawater can be approximated as a NaCl-H_2O solution, phase separation results in the fractionation of salt between a relatively low salinity vapor phase and a relatively high salinity brine phase [e.g., *Bischoff and Pitzer*, 1989; *Bischoff*, 1991]. In some cases, halite may precipitate from the solution [e.g., *Lewis and Lowell*, 2004]. The partitioning of salt between the fluid phases and the precipitation of halite affects not only the dynamics of the hydrothermal system but also its chemistry. *Butterfield et al.* [1997] have found evidence for halite precipitation at the Co-Axial segment of the Juan de Fuca Ridge (JDFR), whereas low Br/Cl ratios in the 1991 vent fluids from the East Pacific Rise (EPR) indicate halite dissolution [*von Damm*, 2000].

An understanding of hydrothermal processes at oceanic spreading centers requires that seawater and hydrothermal fluid resulting from biogeochemical processes and water-rock reactions in the crust be treated as multicomponent fluids. This requirement adds considerable complexity to the description of these systems. Moreover, the processes that are occurring throughout the system involve multiple phases. These include solid, liquid, and gas phases in the subsurface magma bodies, the dissolution and precipitation of solid mineral phases as a result of biogeochemical processes and water-rock reactions, and the phase separation of hydrothermal fluid itself.

Modeling of hydrothermal processes at oceanic spreading centers as multicomponent, multiphase fluids is in its infancy. In this paper, we first develop a simple model of high-temperature hydrothermal circulation that provides insight into some of the key physical aspects of these systems. We call this the "basic model," which is described in section 2. We then introduce multicomponent, multiphase complexities and give examples of how some of these processes can be investigated in the context of still relatively simple models. We conclude by highlighting future directions for multicomponent, multiphase modeling of seafloor hydrothermal systems.

2. THE BASIC MODEL

Although high-temperature hydrothermal circulation at oceanic spreading centers is a very complex phenomenon, some basic features of the process can be described, and to some extent quantified, using relatively simple mathematical models. One of the most fundamental simplifications is to consider hydrothermal circulation and heat transfer in terms of a single-pass or pipe model. In this model, the circulation system is made up of one-dimensional recharge and discharge flow paths, along with a heating region near the base of the system in which conductive heat transfer between the hydrothermal fluid and the underlying magma body occurs (Figure 2). The exact locations and extent of hydrothermal recharge zones are unknown, but deep recharge may occur both on and off axis, and be partly controlled by faults and fissures. *Lowell and Yao* [2002] argue that the effective area of hydrothermal recharge must be extensive to prevent clogging of flow paths by anhydrite. Some areas that have been sealed by anhydrite precipitation may be reopened as a result of subsequent cooling and dissolution of anhydrite or as a result of stresses that may reopen these channels or generate new ones. These stresses may be thermal [*Germanovich and Lowell*, 1992; *Germanovich et al.*, 2001] or tectonic [e.g., *Sohn et al.* 1999; *Tolstoy*, this volume]. Discharge zones are likely controlled by faults and fractures as well [e.g., *Gente*, 1986; *Karson and Rona*, 1990].

By using this highly idealized flow geometry and heat transfer regime, the mathematical description of hydrothermal circulation becomes greatly simplified. Key elements of the physical processes remain in place, however, and when the model is used in conjunction with observational data, the main results appear to be surprisingly robust. As shown below, this model yields estimates of crustal permeability and the thickness of the basal conductive boundary layer that are consistent with estimates derived from more complex mod-

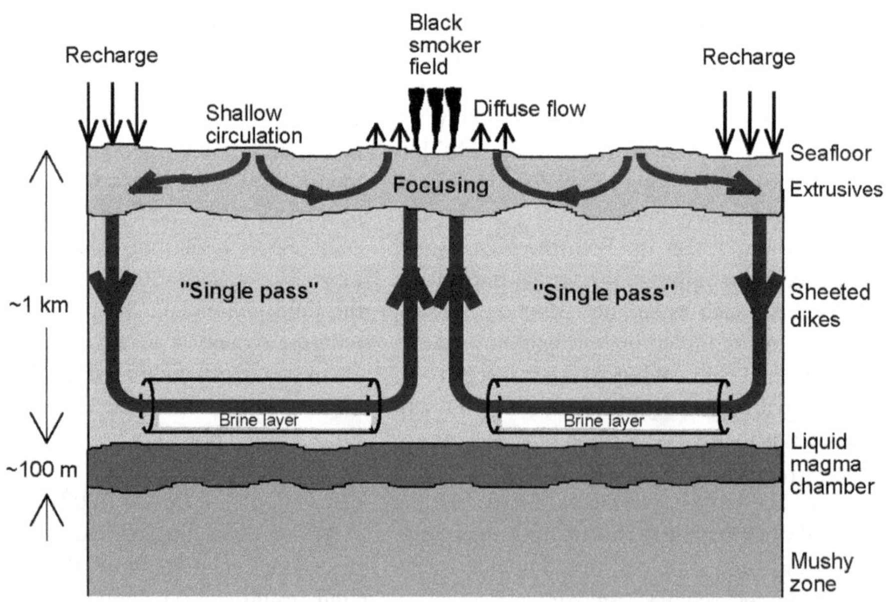

Figure 2. Cartoon of a single-pass hydrothermal circulation cell at a mid-ocean ridge. The single-pass refers to the deep circulation system in which fluid circulates downward into the ocean crust, flows more or less horizontally near the top of the subaxial magma chamber at the base of the sheeted dikes, and ascends back to the surface. Heat conducted from the magma is taken up by the fluid in the horizontal limb, where high-temperature water-rock reactions and phase separation occur. Phase separation is denoted by the brine at base of the circulation cell. Focused high-temperature flow is thought to occur in the deep single-pass limb; diffuse flow may occur as a result of mixing of the deep circulation with shallower circulation in the extrusives [from *Germanovich et al.,* 2000].

eling approaches. Consequently, single-pass models have had a long history of use in hydrothermal modeling, dating from early models of warm springs and continental geothermal systems [e.g., *Elder*, 1981] as well as from the earliest models of seafloor hydrothermal processes [e.g. *Bodvarsson and Lowell*, 1972; *Lowell*, 1975; *Strens and Cann,* 1982] to the present [e.g., *Lowell and Germanovich*, 1994; *Pascoe and Cann*, 1995; *Lowell et al.*, 2003, 2007; *Alt-Epping and Diamond*, this volume; *Ramondenc et al*., this volume]. Reviews of the single-pass modeling approach, and a comparison of this approach with cellular convection models in porous boxes, may be found in the work of *Lowell* [1991] and *Lowell and Germanovich* [2004].

As a result of its conceptual accessibility and mathematical simplicity, we use the single-pass approach as the framework for describing a "basic model" of seafloor hydrothermal systems. We will then advance from this state to explore some of the multiphase, multicomponent complications that can be added with relative ease. The goal of this section is to highlight a number of the key issues mentioned in section 1 and to point the way to the development of more detailed numerical modeling approaches.

2.1. Constraints on the Basic Model

The basic single-pass model as developed here is actually an inverse model in which we use observational constraints such as hydrothermal temperature, heat output, and surface area of the vent field to determine subsurface permeability, which at present cannot be measured directly. In addition, by estimating the area of heat extraction from observations of vent field spacing and cross-axis width of the subaxial melt lens, we can estimate the thickness of the conductive boundary layer between the base of the hydrothermal system and the top of the melt lens. Although ~300 ridge crest vent fields are known [*Baker and German*, 2004], data to constrain permeability and boundary layer thickness are only available for a small fraction of them. Table 1 lists the pertinent thermal data and observed vent field areas to constrain single-pass models. In addition to the data in Table 1, an estimate of the heat uptake area A_m is needed to determine the magma-hydrothermal conductive boundary layer thickness. This area can be estimated from the cross-axis width of the magma lens, obtained from seismic data, and the spacing of vent fields along the ridge. The depth to the subaxial magma chamber plays a smaller role.

For many vent fields, data on the depth and extent of subsurface magma are not yet available, but the cross-axis extent of the magma lens ranges between approximately 500 m and 4 km, and the depth of the mostly liquid magma lens typically between 1 and 3 km [e.g., *Detrick et al.*, 1987, 1993; *Collier and Sinha*, 1990; *Navin et al.*, 1998; *Sinha et al.*, 1998; *Crawford et al.*, 1999; *Singh et al.*, 2006; *Van Ark et al.*, 2007]. The vent field spacing at fast to intermediate spreading ridges typically ranges between a few hundred meters and a few kilometers [*Gente et al.*, 1986; *Delaney et al.*, 1992; *Kelley et al.*, 2002; *Wilcock and Fisher*, 2004]. Thus the heat uptake area $A_m \sim 10^6$ to 10^7 m^2. The principal observational constraints used to develop a generic model of seafloor hydrothermal systems are given in Table 2.

2.2. Basic Model Parameters

In addition to the observational constraints given in the previous section, quantitative model calculations also require a number of physical parameters. Although there is some variability and uncertainty in these parameters for any particular hydrothermal field, the resulting uncertainty in the model results is small compared to the range of model output values (i.e., permeability and boundary layer thickness). Moreover, the simplifications in the model do not appear to introduce significant error in the results. Hence we believe that the results of the model are robust to the degree of accuracy portrayed. Table 3 presents the symbols used to define the pertinent parameters in the equations below, as well as the values used in the model.

2.3. Steady-State Conservation Equations and Results of the Basic Model

The basic single-pass model employs a simple expression for conservation of mass, namely, that a total mass flux Q (kg/s) enters the recharge zone from the ocean and leaves the crust through the discharge zone (Figure 2). Conservation of momentum employs an integrated version of Darcy's law (analogous to Ohm's law of electric circuits):

$$Q = \frac{\hat{H}}{R} \tag{1}$$

where $\hat{H}$ is the driving force stemming from the integrated buoyancy difference between the hot discharge zone with a temperature $T_h(z)$ and the cold recharge zone with a temperature $T_r(z)$, and R is the total flow resistance along the circulation path. That is,

$$\hat{H} = \rho_{f0} g \int_0^H \alpha(z)\,(T_h(z) - T(z))\,dz \tag{2}$$

and

$$R = \int_0^L \frac{\nu}{kA} ds \tag{3}$$

Table 1. Thermal data from seafloor hydrothermal systems.[a]

Reference	Location	Vent Field Area (A_d), m^2	Vent Fluid Flow Rate (m/s)	T_h, °C	Vent Heat Flux, MW	Integrated Heat Flux (H_t), GW	Basis for Estimation/ Remarks
Rona and Trivett [1992]	Axial Volcano (JdFR)	1.E+04	0.2–0.9	108–326	2.4–6.4		*Discrete*: direct measurements
					15–75		*Diffuse:* standard plume model
Baker et al. [1990]	Axial Volcano (JdFR)	2.E+05				0.8	Water column survey
Rosenberg et al. [1988]	Endeavour (JdFR)	4.E+04		up to 400		1–5	Radon (^{222}Rn) measurements
Thomson et al. [1992]	Endeavour (JdFR)	8.E+07				6–18	Water column survey
Schultz et al. [1992]	Endeavour (JdFR)	2.E+01	0.07–0.15	7–13	53.5–62.9		*Diffuse:* Electromagnetic flow meter
Veirs et al. [2006]	Endeavour (JdFR)	2.E+05			8–42		Horizontal heat flux from plume survey
Bemis et al. [1993]	Endeavour (JdFR)					0.07–0.24	Simple plume theory with buoyant plumes
	South Cleft (JdFR)					0.02–0.07	Simple plume theory with buoyant plumes
Ginster et al. [1994]	Endeavour (JdFR)		0.6–6.2	296–374	3.6–87.3	0.29–0.44	Direct measurements and sum over no. of smokers
	South Cleft (JdFR)		1.3–3.5	235–350	7.9–10.4	0.04–0.06	Direct measurements and sum over no. of smokers
Baker and Massoth [1986]	South Cleft (JdFR)					0.29–0.87	Deep-tow survey and plume model
Baker and Massoth [1987]	Endeavour (JdFR)					0.6–2.8	Deep-tow survey and plume model
	South Cleft (JdFR)					0.23–0.92	Deep-tow survey and plume model
Baker et al. [1993]	North Cleft (JdFR)			104–324	161–319	0.48–0.96	*Discrete*: Linear plume theory and mixing model
					324–642		*Diffuse*: Linear plume theory and mixing model
Baker [1994]	North Cleft (JdFR)					0.38–0.94	Deep-tow survey and nonbuoyant plume theory
Gendron et al. [1994]	North Cleft (JdFR)					1.4–4	Radon (^{222}Rn) measurements
Baker et al. [1998]	Co-Axial (JdFR)					0.25–25	Deep-tow survey and plume model

Stein and Fisher [2001]	Middle Valley (JdFR)	3.E+08		180–276	0.13		*Advective*: individual buoyant plume measurements
						0.27	*Conductive + advective* (conductive may also contain a *diffuse* flow component)
Lonsdale and Becker [1985]	Southern Trough (GB)		1–2	270–314	86–201		*Discrete*: assuming 10 chimneys of 80 cm^2 orifice
Fisher and Becker [1991]	Southern Trough (GB)	3.E+07				0.01–0.25	*Conductive + advective* (conductive may also contain a *diffuse* flow component)
Ramondenc et al. [2006]	9°50'N (EPR)	1.E+05	0.1–0.3	345–388	40		*Discrete:* Extrapolation of direct measurements
			0.04	10	300–1.6 × 10^4		*Diffuse:* Extrapolation of direct measurements
McConachy et al. [1986]	11°N (EPR)	3.E+03	0.4–1.2	347	3.0–25		Direct measurements and orifices 3 cm in diameter
Little et al. [1987]	11°N (EPR)	3.E+03			2.9–4.5		Nonlinear plume theory
Macdonald et al. [1980]	21°N (EPR)			344–356	0.2–0.3		Direct measurements and orifices 30 cm in diameter
Converse et al. [1984]	21°N (EPR)		0.7–2.4	275–350	140–300		Flow meter measurements
Rudnicki and Elderfield [1992]	TAG (MAR)			360–364		0.5–0.9	Solution to entrainment in buoyant plume
Rona et al. [1993]	TAG (MAR)	3.E+04		365	200–250		Simple plume theory with buoyant plumes
Murton et al. [1999]	Broken Spur (MAR)					0.25–0.3	Plume theory
Hashimoto et al. [2001]	Kairei (CIR)	3.E+03		360			
Rudnicki and German [2002]	Kairei (CIR)	3.E+03		360		0.07–0.12	Standard plume theory

[a] "Vent Field Area" corresponds to the area investigated in the specific study. The actual area of the vent field may be somewhat different. "Vent Fluid Flow Rate" corresponds to the flow rate estimated at the exit of discrete venting. "Vent Heat Flux" refers to measurements made at one or more discrete vent sites, whereas "Integrated Heat Flux" refers to estimates based on measurements made in the water column overlying the vent field and indicates hydrothermal heat output on a vent field scale. If the flow type (discrete or diffuse) is not indicated in the last column, this means the total (combined) flow.

JdFR, Juan de Fuca Ridge; GB, Guaymas Basin; EPR, East Pacific Rise; MAR, Mid-Atlantic Ridge; CIR, Central Indian Ridge.

Table 2. Observational constraints—the generic single-pass model.

Symbol	Definition	Value
A_m	Heat uptake (magma) area	~10^6–10^7 m^2
A_d	Vent field area	~10^3–10^4 m^2
H_t	Hydrothermal heat output	~10^8–10^9 W
T_h	Hydrothermal temperature	≈400°C
T_m	Magma temperature	≈1200°C

where α is the thermal expansion coefficient of the fluid, H is the height of the layer, L is the length of the flow path, A its cross-sectional area, ν is the kinematic viscosity of the fluid, and k is the permeability, respectively. For simplicity, we will assume that the recharge temperature $T_r = 0$ and the discharge temperature $T_h(z) = T_h$ (a constant) and that the flow resistance is dominated by the discharge limb. Letting k_h, ν_h, and A_h represent the average values of permeability, kinematic viscosity, and cross-sectional area of the discharge zone, respectively, equation (1) can be written:

$$Q = \frac{\rho_{f0} g \alpha_h k_h T_h A_h}{\nu_h} \tag{4}$$

Conservation of energy comes from a simple heat balance between heat conducted across a boundary layer of thickness

Table 3. Symbols, definitions, and values (parameters and vent field observables) used to constrain mathematical models.

Symbol	Definition	Value
a	Thermal diffusivity	10^{-6} m^2
A	Cross-sectional area of fluid flow path	variable
c_f	High-temperature specific heat of fluid	6×10^3 J (kg°C)$^{-1}$
D	Thickness of convecting magma lens	100 m
Fm	Turbulent heat flux from convecting magma	
g	Acceleration due to gravity	10 m s^{-2}
H	Height of convecting layer	10^3 m
$\hat{H}$	Total driving head	
k	Rock permeability	variable
k_{ext}	Permeability of extrusives	variable
h	depth of extrusive layer	variable
Nu	Nusselt number	
Q	Total mass flux	
R	Integrated flow resistance	
Ra	Rayleigh number	
t	Time	
T	Temperature	
T_s	Solidus temperature of magma	1070°C
T_L	Liquidus temperature of magma	1200°C
T_m	Bulk temperature of convecting magma	
z	Vertical Cartesian coordinate	
Greek symbols		
α	Thermal expansion coefficient of fluid	~10^{-3} °C^{-1}
α_m	Thermal expansion coefficient of magma	~10^{-5} °C^{-1}
$\chi(T_m)$	Crystallinity of magma as a function of temperature	
δ	Conductive thermal boundary layer between magma and base of hydrothermal system	
δ_m	Conductive boundary layer at top of magma chamber	
λ_r	Thermal conductivity of rock	2.0 W (m °C)$^{-1}$
λ_m	Thermal conductivity magma	2.0 W (m °C)$^{-1}$
ν	Kinematic viscosity of fluid	~10^{-7} m^2 s^{-1}
ν_{m0}	Kinematic viscosity of crystal-free basalt magma	~10^{-1} m^2 s^{-1}
ρ_f	Fluid density	10^3 kg m^{-3}
Subscripts		
0	Reference value	
h	Discharge zone	
m	Magma	
r	Recharge zone	

δ between the top of the magma body and the base of the hydrothermal system. For simplicity, we assume that all the heat transported across the boundary layer is carried upward by the hydrothermal system without loss. Then conservation of energy can be written simply as

$$\frac{\lambda_r(T_m - T_h/2)A_m}{\delta} = c_f Q T_h \sim H_t \quad (5)$$

where λ_r is the thermal conductivity of rock, T_m is the temperature of magma, A_m is the area of heat uptake at the base of the hydrothermal system, c_f is the specific heat of the hydrothermal fluid, and H_t is the total heat output, respectively. The observed heat outputs at high-temperature ridge crest hydrothermal systems typically range between 10^8 and 10^9 W (see Tables 1 and 2).

By using observed values of T_h, H_t, A_m, and A_h in an average or generic sense together with the parameters in Table 3, equations (4) and (5) can be solved upon inspection for the unknown conductive boundary layer thickness δ and bulk permeability of the discharge zone *k*. The results are: $2 \leq \delta \leq$ 20 m.

$$10^{-13} \leq k_h \leq 10^{-11} \text{ m}^2 \quad (6)$$

In the rare case of low H_t and large A_m, $\delta \sim 200$ m. The results of the "basic model" do not change significantly if more complex single-pass models are used [e.g., *Lowell and Burnell*, 1991; *Lowell and Germanovich*, 1994]. For example, the assumed value of T_r in the basic model only affects the buoyancy of the system, and hence the flow rate. Using a value of T_r = 200°C rather than 0°C reduces the buoyancy drive by only 25%, and does not change the range of k_h or δ given by equation (6) [*Lowell and Germanovich*, 1994]. On the other hand, the basic model assumes that heat transfer from magma to the hydrothermal system occurs without loss and that hydrothermal heat is carried to the seafloor without loss. Inclusion of these factors would result in smaller values of δ than those given by equation (6).

Obviously, assuming that T_r = 0°C in the entire recharge zone precludes the possibility of chemical reactions there. Chemical reactions and other features such as mixing between hydrothermal fluid and seawater can be included in single-pass models by enhancements to the "basic model" described above [e.g., *Lowell and Yao*, 2002; *Lowell et al.*, 2003, 2007]. These effects are discussed further in ss-section 3. *Ramondenc et al.* [this volume] use a two-loop single-pass model to discuss the response of hydrothermal systems to earthquakes. *Alt-Epping and Diamond* [this volume] use a single-pass model to investigate water-rock reactions.

The "basic model" developed here provides insight into important features of high-temperature hydrothermal circulation at the ocean ridge crest. In particular, the results show that the permeability, particularly in the discharge zone, must be high and that the conductive boundary layer between magma and hydrothermal system must be kept thin. These features stem from the fact that both the temperature and the heat output of the system are high. The results in equation (6) show that heat output and vent temperature together provide significant constraints on the behavior of hydrothermal activity in the oceanic crust. The values given in equation (6) represent generic results based on "typical observational data" from Table 2 and the parameter values in Table 3. Specific values for individual hydrothermal systems can be determined by using the system-specific data in Table 1 and more accurate values for system-specific parameters from Table 3.

Although the basic model provides insight into some of the general features of high-temperature hydrothermal circulation, many important aspects require more complex models, most of which require models of multiphase, multicomponent fluid flow. These aspects include magmatic processes such as magma crystallization and cooling, magma chamber replenishment, and eruption dynamics, phase separation of seawater, water-rock reactive transport, mineral dissolution and precipitation, and biogeochemical processes. These processes are discussed in the next section.

3. MULTIPHASE, MULTICOMPONENT MODELS

3.1. Magmatic Processes

The structure of the oceanic crust is intimately tied to magmatic processes (Figure 1). The uppermost layer of extrusives is underlain by sheeted dikes. The subaxial, mostly liquid magma is a thin lens-like layer above a partially molted mushy zone. The mushy zone extends to the base of the gabbroic crust and broadens somewhat with depth [e.g., *Sinton and Detrick*, 1992; *Dunn et al*, 2000; *Sinha and Evans*, 2004; *Maclennan*, this volume]. Magmatic processes that build the crust include magma flow through the mushy zone, the formation and evolution of the largely liquid shallow melt lens, convection and crystallization, fractionation of melt and the exsolution of gases in the melt lens, the generation of dikes, eruptions, and flow of lavas on the seafloor. These magmatic processes both influence and are influenced by high-temperature hydrothermal circulation within the axial zone as well as by near-axis convection that may occur on the flanks of the mushy zone [*Mottl*, 2003; *Cherkaoui et al.*, 2003].

Mathematical and numerical models of complex multiphase, multicomponent, coupled magma-hydrothermal processes are not yet available, but progress is being made in a number of areas. For example, the initiation of dikes from the subaxial magma lens depends on coupled magmatic and hydrothermal processes. An increase in magma pressure resulting from the influx of fresh magma to the magma lens generates tensile stresses near the tips of the magma lens, whereas hydrothermal cooling and crystallization of the melt lens lead to tensile stresses near the center of the lens. Consequently, dikes resulting from pressurization of the shallow magma lens will propagate from the tips of the lens, whereas dikes resulting from hydrothermal cooling and crystallization of the melt lens will propagate from the center of the lens [*Sim*, 2004]. In each case, dikes propagate nearly vertically and those associated with magma pressurization tend to reach the seafloor [*Sim*, 2004]. Central dikes are not likely to reach the seafloor, even if degassing of the magma is included [*Ramondenc*, 2007]. Possible connections between dike emplacement, earthquakes, and the response of hydrothermal venting are discussed further by *Ramondenc et al.* [this volume] and Germanovich et al. (Seismic and hydrothermal evidence for a diking event on the East Pacific Rise crest at 9°50′N, East Pacific Rise, submitted to *Journal of Geophysical Research*, 2008). We discuss below new results that relate hydrothermal heat output to magma convection, crystallization, and replenishment; however, fully coupled models of magmatic processes and hydrothermal circulation are not yet available.

The results derived for the basic model assume a steady-state situation, but basaltic magma at ≈1200°C is at its liquidus [*Sinton and Detrick*, 1992; *Maclennan*, this volume], so extracting heat from the magma chamber will result in crystallization and cooling. One way to treat magma crystallization is to simply assume that magma freezes from the top downwards, resulting in a thickening of the conductive boundary layer [e.g., *Lowell and Rona*, 1985; *Lowell and Burnell*, 1991; *Lowell and Germanovich*, 1994]. In this case, $\delta \sim (at)^{1/2}$ and $H_t \sim t^{-1/2}$ (see equation (5)). Inserting equation (4) into (5) then leads to the result $T_h \sim t^{-1/4}$. Such a rapid decay in hydrothermal vent temperature and heat output has not been observed in long-lived systems that are presumably driven by basal magmatic heat sources (such as the systems in Table 1). Therefore, the model of a thickening boundary layer and magma freezing from the top is oversimplified. Mechanisms that maintain the thin boundary layer must be invoked.

A thin boundary layer may be maintained, in part, as a result of downward crack propagation and creation of new permeability in the growing boundary layer [*Lister*, 1974, 1982]. This idea is conceptually appealing and has often been invoked in the context of hydrothermal heat transfer [e.g., *Mével and Cannat*, 1991; *Seyfried and Ding*, 1995; *Wilcock and Delaney*, 1996; *Sohn et al.*, 1998, 1999; *Johnson et al.*, 2000; *Wilcock and Fisher*, 2004; *Tolstoy*, this volume]. The model developed by *Lister* [1974] is physically flawed, however, because it neglects horizontal compressive stresses resulting from the lithostatic load, even as cracks propagate to depths of kilometers in the crust. Studies of lower crustal fracturing in the Oman ophiolite and the Hess Deep also call into question many details of the Lister model [*Manning and MacLeod*, 1996; *Manning et al.*, 2000].

Another way to consider downward crack propagation is to associate it with diking events [*Bodvarsson*, 1982; *Lowell and Germanovich*, 1994]. Although this mechanism is mechanically sound, diking events are irregular in time and space. Consequently, if the generation of permeability near the top of the subaxial magma chamber were associated with diking events, hydrothermal temperature and heat output would tend to fluctuate significantly with time. Moreover, in systems that appear to be driven mainly by diking events such as that associated with the Co-Axial eruption on the Juan de Fuca Ridge, heat flux actually decays more rapidly than $t^{-1/2}$ [*Baker et al.*, 1998]. This decay rate is explained in part by lateral heat loss from fluid ascending near the dike margins [*Baker et al.*, 2004].

Crack propagation and the thickness of the thermal boundary layer have implications for vent fluid chemistry as well as for heat transfer. Incompatible trace elements such as Li, B, and K are readily leached from basalt at high temperatures [e.g., *Seyfried et al.*, 1984]; and the high concentrations of these elements in vent fluids [e.g., *Von Damm et al.*, 1985] are consistent with low water-rock ratios in the reaction zone. Moreover, the relatively constant concentrations of these elements in vent fluids over time argues for the continued availability of relatively unaltered basalt [*Seyfried et al.*, 1991], suggesting downward migration of hydrothermal fluids into newly solidified magma [*Seyfried and Ding*, 1995].

It is difficult to use *Lister*'s [1974, 1982] model directly to explain these chemical data, however, because this model concerns the downward propagation of large-scale macrocracks separated by ~0.1–1 m. For such widely separated cracks, the fluid has little direct access to rock and hence the extraction of trace elements is problematical. On the other hand, if the heat uptake-reaction zone suggested by the single-pass model (Figure 2) is a few tens of meters thick and its area is $\sim 10^6$ to 10^7 m^2, then its mass is $\sim 10^{11}$–10^{12} kg, assuming a rock density of 3×10^3 kg/m^3. From equation (5), a hydrothermal system transporting 500 MW of heat has a mass flow rate Q

~ 200 kg/s. After 30 years, the mass of fluid that has passed through the system is ~2×10^{11} kg. If the bulk porosity of the reaction zone is 1%, the apparent water-rock ratio would be ~10–100; however, the chemical reactivity depends critically on the exposed rock surface area, which may be quite high for densely cracked rocks even if the porosity is low. For example, a set of parallel cracks 10^{-4} m wide and spaced 10^{-2} m apart would give rise to a porosity of 1% and a permeability of 10^{-11} m^2. In this case, the fluid would effectively react with the entire rock volume of the reaction zone, and the effective water-rock ratio would approach 1. High-temperature microcracking at the grain scale may occur in the solidified roof of the magma chamber [*Manning et al.*, 2000]. Fluids entering these cracks may be able to extract trace elements and transport them to the hydrothermal system even though heat is transported mainly by conduction. Finally, the reaction zone may expand laterally as a result of magma replenishment. This possibility is discussed further below.

In a convecting magma chamber, solidification does not necessarily occur at the roof. Because crystals are denser than magma, they may either be entrained within the convecting interior or settle toward the base of the magma. In either case, the location of the solid upper boundary between the convecting magma and the hydrothermal system may not change; however, heat transfer between the convecting magma and the overlying hydrothermal system must be investigated. Moreover, the magma chamber may grow as new magma is transported from the underlying mushy zone. The additional heat supply from magma replenishment may help maintain heat transport from the convecting magma to the hydrothermal system [e.g., *Lowell and Germanovich*, 1994; *Humphris and Cann*, 2000].

To investigate the relationship between magma convection and hydrothermal heat output, we construct a simple heat balance model for a convecting cooling magma body of thickness $D \sim 100$ m following the approach of *Huppert and Sparks* [1988]. For convecting magma with average temperature T_m initially at its liquidus T_L and upper boundary at its solidus T_s, the Rayleigh number is

$$Ra = \frac{\alpha_m g (T_m - T_s) D^3}{a_m \nu_m} \tag{7}$$

Assuming $T_m - T_s = 100°C$, $D = 100$ m, and other parameters for basaltic magma (Table 3), $Ra \sim 10^{11}$, which is much greater than the critical Rayleigh number $Ra_c \sim 10^3$. If $D \sim 10$ m and $\Delta T \sim 1°C$, then $Ra \sim 10^6$, which is still much greater than Ra_c. Therefore, the magma heat flux F_m may be found from the dimensionless Nusselt number Nu by using the classical relationship between Nu and Ra for a vigorously convecting fluid layer:

$$F_m = \frac{\lambda_m (T_m - T_s)}{D} Nu$$

$$Nu \sim Ra^{1/3} \tag{8}$$

The total heat output from the magma chamber is $F_m A_m$. This heat output is linked directly to that of the hydrothermal system by equating the magma heat output to the heat output expressed by equation (5).

As heat is lost through the upper surface, the magma crystallizes and cools; the Rayleigh number decreases, and F_m decreases with time [*Huppert and Sparks*, 1988; *Liu*, 2007; Liu and Lowell, Hydrothermal heat output from a convecting, crystallizing, and replenished magma chamber, submitted to *Journal of Geophysical Research*, 2008, hereinafter referred to as Liu and Lowell, submitted manuscript, 2008]. As an example, we assume that vigorous convection maintains a uniform crystal suspension in the convecting magma (Figure 3). Analogous models in which crystals settle to the bottom of the magma lens are discussed by *Liu* [2207] and Liu and Lowell (submitted manuscript, 2008). If crystals remain suspended, the Ra decreases primarily because the kinematic viscosity ν_m increases with increasing crystallinity $\chi(T_m)$ according to the formula [*Marsh*, 1981]

$$\nu_m = \nu_{m0}[1 - 1.67\chi(T_m)]^{-2.5} \tag{9}$$

According to equation (9), the kinematic viscosity of magma approaches infinity when the crystal content reaches 60%. At this point, convection stops. Following *Huppert and Sparks* [1988], we assume the crystallinity of the magma is related to the magma temperature by the relationship $\chi(T_m) = 7200/T_m - 6$.

Figure 4 shows total magmatic (hydrothermal) heat output and hydrothermal temperature as a function of time t as magma crystallizes and cools. Figure 4(a) shows that heat output from the magma chamber, and hence hydrothermal heat output, decreases from 10^9 Wto 10^7 W within about a 10-year period. The convection system ceases at 18 years because magma crystal content reaches 60%. Figure 4b shows that hydrothermal temperature also decreases rapidly. As in the basic model, these results are not consistent with observations, suggesting that magma convection alone is not enough to maintain high heat output and observed temperatures from seafloor hydrothermal systems.

We then investigated whether magma replenishment and growth of the magma lens would help maintain steady heat output on decadal time scales. For simplicity, we assumed magma at its liquidus T_L entered the magma chamber

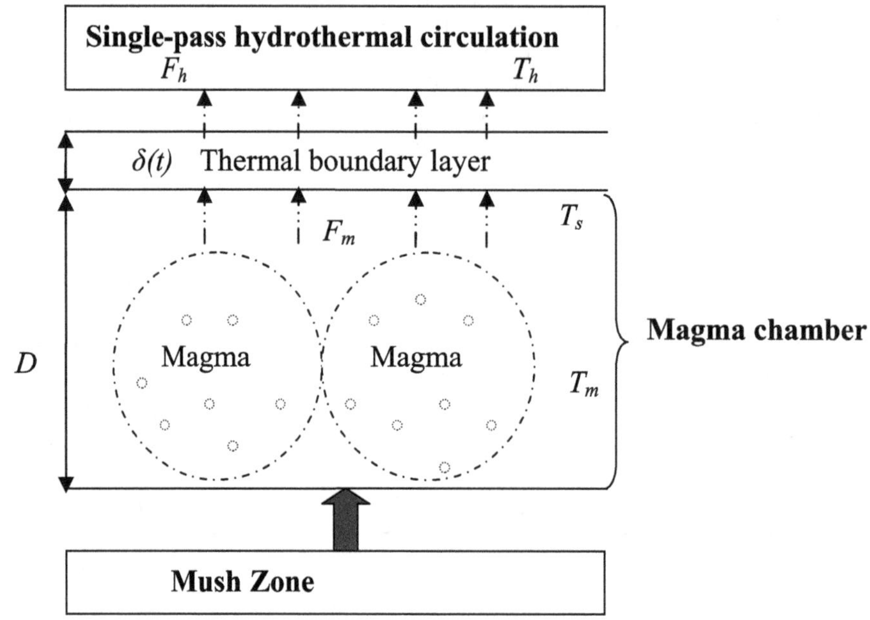

Figure 3. Schematic of a coupled magma-hydrothermal system underlain by a partially molten region. We consider a turbulently convecting, crystallizing magma chamber of temperature T_m height D cooled from above by hydrothermal circulation at temperature T_h. The magma is initially at its liquidus and the roof is at the solidus T_s. Magmatic heat flux $F_m(t)$ is conducted across a boundary layer $\delta(t)$ to the hydrothermal system. Magma heat output is related to hydrothermal heat output by the relation $F_m(t)A_m = F_h(t)A_h$ [from *Liu*, 2007].

through the basal area A_m at a constant rate u_0. We assumed the thickness D of the magma chamber remained constant, so the addition of magma resulted in an increase in the lateral extent of the magma lens and its surface area A_m. We did not consider mechanical processes associated with magma inflation or the possibility of magma eruption. We stopped replenishment after 30 years. Mathematical details are discussed more fully by *Liu* [2007] and Liu and Lowell (submitted manuscript, 2008).

Figure 5 shows magmatic heat output versus time when magma replenishment is included. The results show that if the rate of replenishment is rapid enough ($u_0 = 10^{-6}$ m/s), the heat output increases with time, whereas for slow rates of magma replenishment ($u_0 = 10^{-9}$ m/s), the heat output still decayed rapidly below observed levels. For $u_0 = 10^{-7}$ and 10^{-8} m/s, the heat output stabilized within the observed range for decades. During this time, the magma volume increases by approximately a factor of 2. As the magma chamber grows laterally, and hydrothermal fluid flows across the top of the newly emplaced magma, the hydrothermal system undergoes high-temperature reaction with fresh rock. Lateral extension of the magma chamber may thus provide a mechanism for exposing fresh rock to high-temperature hydrothermal alteration rather than downward propagation of a cracking front.

The velocity $u_0 = 10^{-7}$ m/s corresponds to an initial magma flow rate of 0.1 m^3/s, which increases to twice that value after 20 years. This rate of magma input is about 1 order of magnitude greater than that needed for steady-state production of 6 km of crust at a total spreading rate of 10 cm/yr, but about 1 order of magnitude less than magma inputs into Kilauea volcano over a similar period [*Dzurizin et al.*, 1984; *Humphris and Cann*, 2000].

The models presented here suggest that magma convection alone may not be sufficient to maintain stable high hydrothermal temperature and heat output for decadal time scales. Modest rates of magma replenishment may provide the heat required to maintain steady hydrothermal output on decadal time scales, however. These results need to be tested further by developing more realistic multiphase, multicomponent models of magma chamber crystallization and replenishment and coupling these to hydrothermal processes. Such models would include two- and three-dimensional models of magma convection and incorporate temperature- and crystallinity-dependent viscosity [e.g., *McBirney and Murase*, 1984; *Spera*, 2000], crystal nucleation and growth near the upper boundary layer [e.g., *Brandeis and Jaupart*, 1986; *Worster et al.*, 1990], crystal settling and flow through cumulate mush [e.g., *Martin and Nokes*, 1989; *Jellinek and Kerr*, 1999, 2001], and magma degassing [e.g., *Huppert and Woods*, 2002; *Ramondenc*, 2007].

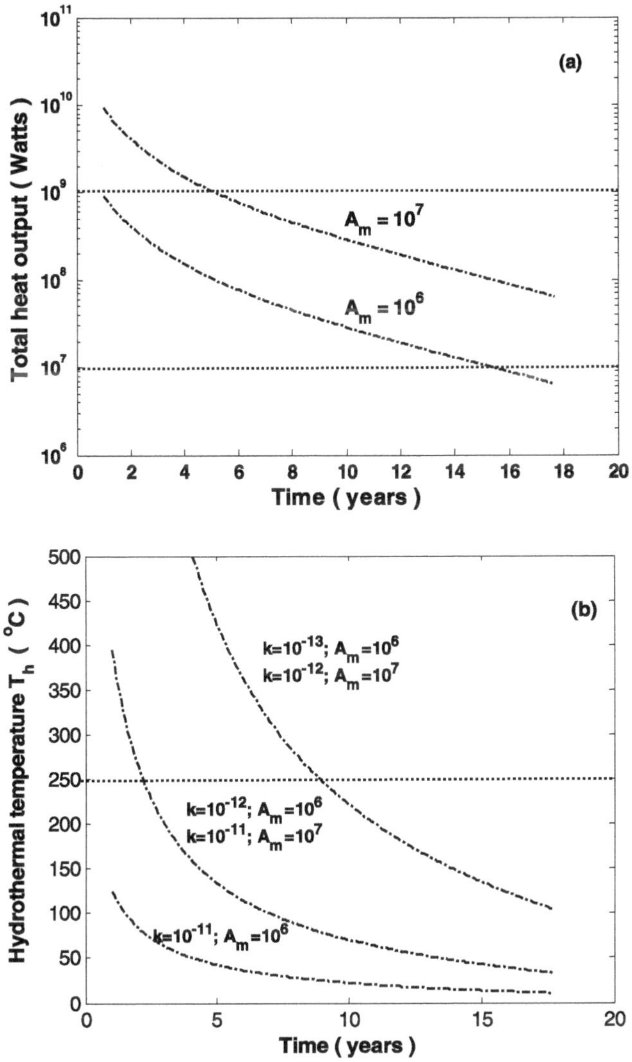

Figure 4. (a) Total heat output from the convecting magma for two different magma surface areas, for the model shown in Figure 3. T_L = 1200°C and T_s = 1070°C. The horizontal lines at 10^9 and 10^7 W mark the range of observed hydrothermal heat outputs. (b) Resulting hydrothermal temperature versus time for different values of permeability and magma area, assuming $A_h = 10^4$ m². The horizontal line in (b) denotes black smoker temperatures (from Liu and Lowell, submitted manuscript, 2008).

Models of processes within a convecting magma chamber also need to be coupled with hydrothermal convection and dike emplacement models [*Sim*, 2004], as well as with mechanical and physical models of magma transport within the lower crust [e.g., *McKenzie*, 1984; *Spiegelman*, 1993a, 1993b; *Henstock et al.*, 1993; *Kelemen et al.*, 1997; *Kelemen and Aharonov*; 1998; *Maclennan et al.*, 2002]. Magma transport within the lower crust needs to be linked to the evolution and replenishment of the melt lens. Finally, coupled models of magma-hydrothermal processes need to be consistent with geophysical data on crustal structure and petrological data. For example, seismic data from the southern EPR show regions that contain a thin relatively crystal free magma lens and nearby regions in which the lens is much more crystalline [*Singh et al.*, 1999]. These authors point out that hydrothermal activity is enhanced in the neighborhood of crystal-free magma, suggesting a close link between hydrothermal activity and inputs of fresh magma into the shallow crust. They further argue that the presence of a solid floor indicates cooling and crystallization of melt in the magma lens. Although most mid-ocean ridge basalts contain very few phenocrysts [e.g., *Bryan*, 1983], lava compositions reflect both fractionation and mixing in shallow magma chambers in differing degrees, with fractionation effects dominating at fast-spreading ridges, and mixing effects dominating at slow-spreading ridges [*Sinton and Detrick*, 1992]. Detailed analyses of lavas from 9°31′N on the EPR indicate that rapid changes in magma chemistry occur as a result of both crystallization and replenishment [*Perfit et al.*, 1994]. In the context of the magma replenishment model described here, low phenocryst contents of erupted lavas would suggest that eruptions occur in conjunction with replenishment events. Quantitative models that account for the full range of complex processes that occur in magma-hydrothermal systems are likely to emerge slowly.

3.2. Phase Separation

Since the inception of studies of vent fluid chemistry at mid-ocean ridge hydrothermal sites, it has been noted that

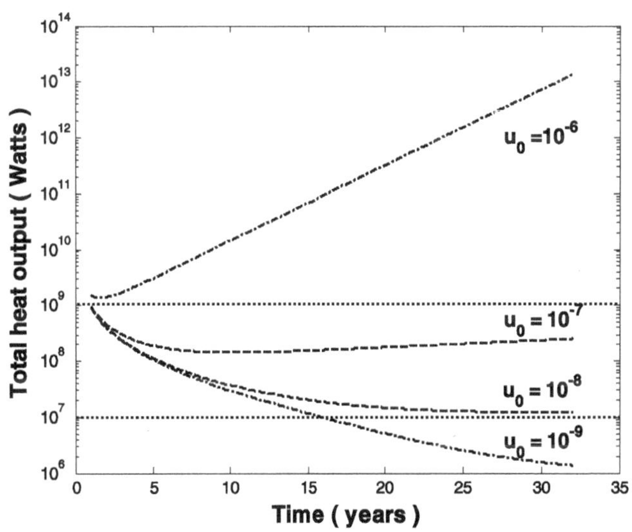

Figure 5. Total magmatic (hydrothermal) heat output assuming magma replenishment at different constant velocities u_0 (m/s). Replenishment stopped after 30 years [from *Liu*, 2007].

the salinity of these fluids is rarely the same as that of the surrounding seawater (Figure 6). Figure 6 shows that the salinity, expressed as wt % NaCl, ranges from nearly twice that of seawater (e.g., at Monolith vent on the JDFR) to less than 10% of the seawater value (e.g., at A vent on the EPR after the 1991 eruption). In some cases, the salinity has remained above or below seawater values for many years (e.g., 21°N EPR); whereas in other cases, repeated sampling over more than a decade shows marked fluctuations in salinity values over time (e.g., Figure 6a).

It is generally recognized that these differences are the result of phase separation as fluids heated near subsurface magma are driven into the two-phase regime [e.g., *Von Damm,* 1995; *Seyfried and Ding,* 1995; *Seyfried et al.,* 2003]. When phase separation of a NaCl-H_2O fluid occurs, a low-salinity vapor phase is formed in equilibrium with high-salinity brine (Figure 7). This process affects both the physics and chemistry of hydrothermal processes at oceanic spreading centers. Because the brine is denser than vapor, phase separation leads to segregation of the phases. Brine tends to settle toward the base of the system where it affects heat transfer between the hydrothermal system and the underlying magma [*Bischoff and Rosenbauer*, 1989; *Lowell and Germanovich*, 1997].

The generation and the fate of brines in seafloor hydrothermal systems are not well understood. At EPR 21°N, fluids with salinities less than seawater have been venting for decades with no evidence of brine discharge (Figure 6b). In the region near 9°50′N on the EPR that was affected by the 1991 magmatic eruption, mainly low-salinity fluids have been venting for the past decade (Figure 6a); however, F vent, located at 9°17′N, changed from vapor to brine within a

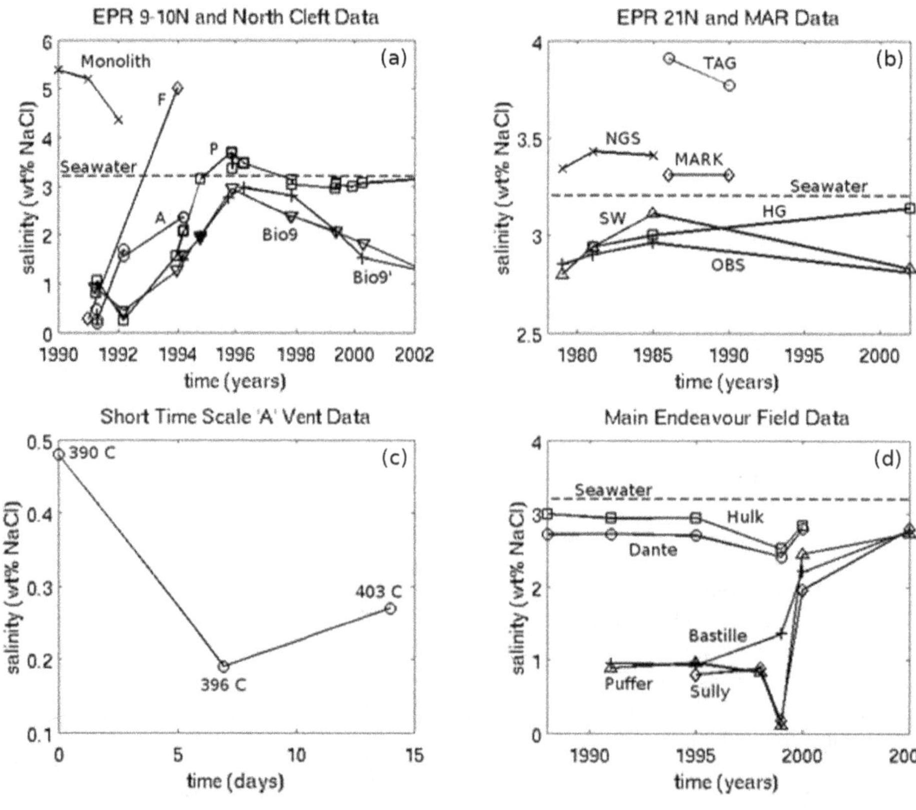

Figure 6. Summary of current vent fluid salinity data. (a) Monolith is a North Cleft site on the JDF; the others are from EPR 9–10°N. (b) TAG and MARK refer to MAR sites; the others refer to EPR at 21°N. (c) A 2-week time series of vent salinity and temperature at "A" vent at 9°N EPR. (d) Sites from the Main Endeavor Vent Field, JDF. "A" vent data are from *Von Damm et al.* [1995], F vent data are from *Von Damm et al.* [1997], and data for the other vents at EPR 9–10°N are from *Von Damm* [2004]; EPR 21°N data are from *Campbell et al.* [1988b] and *Von Damm et al.* [2002]; JDF Data are from *Lilley et al.* [2003], except for the year 1998 and 2005, which come from *Butterfield et al.* [1994] and Seyfried (personal communication), respectively; North Cleft and MAR data are from *Butterfield and Massoth* [1994] and *Edmond et al.* [1995], respectively. [Updated from *Lewis and Lowell*, 2004]

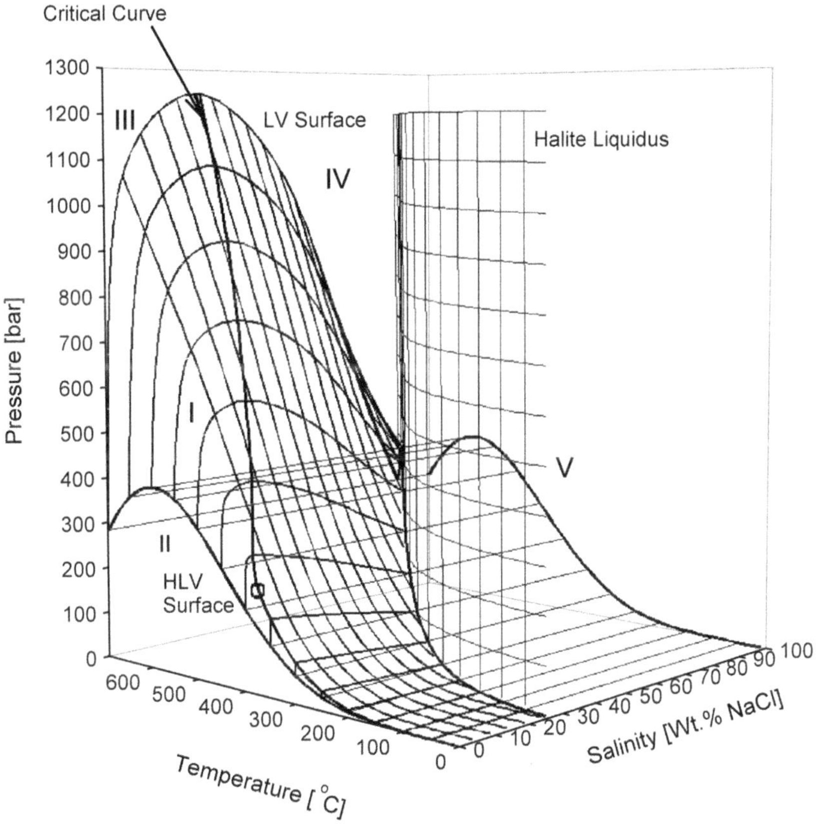

Figure 7. The *P-T-X* phase diagram of the binary NaCl-H_2O system. Roman numerals I through V represent different regions of phase space as described in the text. LV refers to the liquid-vapor equilibrium surface and HLV refers to the three-phase halite-liquid vapor equilibrium surface. The open circle is the critical point of pure water [from *Geiger et al.*, 2006a]. With kind permission from Springer Science+Business Media: *Transport in Porous Media*, Multiphase thermohaline convection in the Earth's crust: I. A new finite element-finite volume solution technique combined with a new equation of state for NaCl-H_2O, v. 63, 2006, p. 403, Geiger, S., T. Driesner, C. A. Heinrich, and S. K Matthai, Figure 1.

few years [*Von Damm et al.*, 1997]. *Von Damm* [2004] has suggested that brines generated in the 9°50′N region of the EPR migrate southward to regions of lower topography, and where high-salinity venting is observed [*Von Damm*, 2000]. *Fontiane and Wilcock* [2006] suggest that brine formed beneath the Main Endeavour Vent field may migrate along the ridge axis to adjacent vent fields. If phase separation occurs at shallow depths, perhaps as a result of dike emplacement, brines may not settle to the base of the system and may be flushed from the system relatively quickly when the dike cools [*Lewis and Lowell*, 2004]. Such a flushing event may have been missed in the time series data from EPR 9°50′N; or it may be reflected in the gradual increase in vent salinities observed, including the higher-than-seawater salinities recorded at P vent (Figure 6a).

Brine transport from the base of the hydrothermal system is controversial. *Bischoff and Rosenbauer* [1989] argue that the brine layer convects beneath the seawater circulation cell and that chemical exchange between the brine layer and the overlying fluid is consistent with vent salinity data. *Lowell and Germanovich* [1997] suggest that the brine layer may be depleted by diffusion or by dynamical mixing. *Schoofs and Hanson* [2000] develop numerical models of brine depletion for a single-phase fluid, but this model does not address how brine might have formed in the first place. *Fontaine and Wilcock* [2006] argue that the brine layer is stably stratified and does not convect. They further argue that brine may be immobilized in lower permeability regions of the crust, although they also argue that even brines with salinities of 20% may rise buoyantly in the discharge limb of a convection cell. The resolution of these various viewpoints awaits the development of numerical models of phase separation and transport. As discussed below, these models are just beginning to emerge.

Phase separation also affects the fluid chemistry. The Cl ion concentration in the fluid strongly affects the solubility of other minerals; hence the vapor and brine components of the phase separated fluids are likely to have substantially different major and trace element compositions [*Seyfried and Ding*, 1995]. At most vent fields, the *P-T* conditions of fluids at the seafloor place them in the liquid phase regime, suggesting that the measured chlorinity values represent mixing among a low-salinity vapor phase, a high-salinity brine, and seawater [e.g., *Von Damm and Bischoff*, 1987; *Kelley et al.*, 1993].

The ubiquitous evidence for phase separation in vent fluids at oceanic spreading centers and the effects of phase separation on the dynamics and chemistry of the fluids speak to the importance of developing mathematical and numerical models to investigate these processes. Although numerical simulators for two-phase flow in pure water systems have been available for a long time [e.g. *Faust and Mercer,* 1979; *Preuss,* 1991; *Hayba and Ingebritsen*, 1994; *Xu and Lowell*, 1998], appropriate simulators for NaCl-H_2O systems are just now becoming available. The reasons for this lag in development stem largely from the complexity of the NaCl-H_2O phase diagram (Figure 7).

The phase diagram in *P-T-X* space shows five distinct regions of phase space: (I) the two-phase liquid-vapor region in which salt-rich brine and salt-poor vapor coexist, (II) a two-phase region of solid salt-pure vapor equilibrium, (III) a single-phase region of low-salinity vapor-like fluid that exists at high pressures, (IV) a single-phase region of higher salinity liquid-like fluid, and (V) a two-phase region of halite-liquid equilibrium. In addition, there is a surface along which the three phases (liquid-vapor-halite) are in equilibrium. Single-phase regions lie above the liquid-vapor surface.

Experimental work [*Sourirajan and Kennedy*, 1962; *Rogers and Pitzer*, 1982; *Bischoff and Pitzer*, 1989; *Bischoff*, 1991] has provided information on volumetric properties, and phase equilibria for certain regions of the NaCl-H_2O system. These data have provided a framework for developing theoretical equations of state [e.g., *Archer*, 1992; *Bischoff and Rosenbauer*, 1994; *Anderko and Pitzer*, 1993]. Equations of state formulations are often valid only over limited ranges of pressure and temperature. Even more problematical, however, is the fact that the formulations often do not join smoothly at the boundaries between their ranges of applicability. Anderko and Pitzer's approach, based on free energy considerations, is inconvenient to use for calculating fluid properties in terms of state variables (such as *P-T-X*) used in numerical simulators. *Palliser and McKibben* [1998a, 1998b, 1998c] employed correlation functions to calculate fluid properties; however, their method used different functions above and below the critical point and results in nonphysical artifacts where the two functions join. As a result, it is difficult to use the Palliser and McKibben formulation across the range of parameter space encountered in many applications, including hydrothermal systems at oceanic spreading centers. The earliest work on multiphase NaCl-H_2O hydrothermal systems has included models of the behavior of NaCl-H_2O heat pipes [*Bai et al.*, 2003] and approximate solutions for phase separation near a subseafloor igneous dike [*Lewis and Lowell*, 2004]. Both of these papers used equation of state formulations with limited ranges of applicability in *P-T-X* space.

Several groups [*Xu et al.*, 2001; *Berndt*, 2001; *Driesner*, 2001] simultaneously recognized the need to develop more comprehensive equations of state and ways to calculate the thermodynamic variables needed for numerical computations. *Driesner and Heinrich* [2003, 2007] and *Driesner* [2007] developed their own correlation equations based on available experimental data, and their equations of state have been incorporated into a new numerical code that combines finite element and finite volume methods [*Geiger et al.*, 2006a, 2006b]. The well-known numerical code TOUGH2 [*Preuss*, 1991] has been modified to incorporate saline fluid. The new code, NaCl-TOUGH2, uses the Palliser and McKibben equations of state [*Kissling*, 2005*b*]. Application to the Taupo volcanic zone in New Zealand and other examples are given by *Kissling* [2005a]. Finally, *Lewis* [2007] has developed a finite control volume code called FISHES (Fully Implicit Seafloor Hydrothermal Event Simulator) for simulating two-phase flow in a NaCl-H_2O fluid for temperatures between 0° and 800°C, pressures between 8.5 and 100 MPa, and salinities between 0 and 100%. These ranges incorporate the regions of *P-T-X* space encountered in seafloor hydrothermal systems. Thermodynamic properties are calculated using linear interpolation from lookup tables that were constructed from the formulations of *Tanger and Pitzer* [1989], *Archer* [1992], and *Anderko and Pitzer* [1993]. Salinities on the vapor-halite surface were calculated using the correlation equations of *Palliser and McKibbin* [1998a].

As an example of two-phase flow and phase separation in a seafloor hydrothermal system using FISHES, we show simulation results based on the single-pass geometry and parameters shown in Figure 8. Additional details are given by *Lewis* [2007] and Lewis and Lowell (Numerical modeling of two-phase flow in the NaCl-H_2O system I: Introduction of a numerical method and benchmarking, *Journal of Geophysical Research*, 2008). With the model geometry and parameters shown, we emplace a linear temperature gradient at the base of the model ranging from 420°C at the left-hand boundary to 300°C at the right-hand boundary. This temperature distribution ensures that a small two-phase region develops near the lower left-hand corner of the model. The

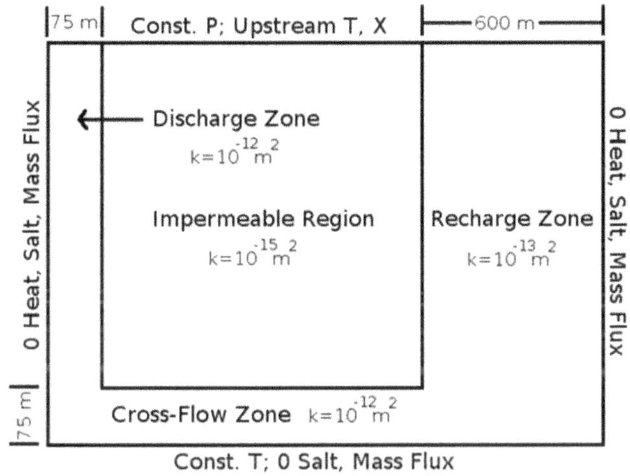

Figure 8. Basic geometry for a single-pass model simulation of two-phase flow in a ridge crest hydrothermal system. There is no salt flux through the side boundaries, which are also thermally insulated and impermeable. Mass and salt fluxes are also zero through the bottom boundary and a temperature distribution is imposed there. The top boundary is at constant pressure, and upstream weighting is used to determine values of temperature and salinity of fluid that crosses the boundary.

two-phase region is allowed to develop for 3 years, and then the temperature is reduced from 420° to 390°C in the left hand corner so that a one-phase system returns. The total simulation time is 10 years. Figure 9 shows the circulation pattern and isotherms for when the two-phase region is developing. Figure 10 highlights the extent of the two-phase region in the lower left-hand corner of the system. Figure 11 shows the evolution of vent salinity and temperature over the 10-year simulation time.

As the two-phase zone develops, saline brine settles in the lower left-hand corner of the system and vapor rises buoyantly in the discharge zone. The salinity of the vapor leaving the base of the system is approximately 2.5 wt % NaCl. As the vapor rises to the seafloor, however, it mixes with seawater and vents as a single-phase fluid. Figure 11 shows that as a result of phase separation, vent salinity is less than seawater; but as a result of mixing, it is greater than the salinity of the initial vapor phase. Figure 11 also shows that the brine that is initially stored at the base of the system ascends to the surface after the system returns to the single-phase state. The brine is also mixed with single-phase seawater, and after a few years, fluids with salinity slightly greater than seawater discharge at the seafloor. After the stored brine is flushed

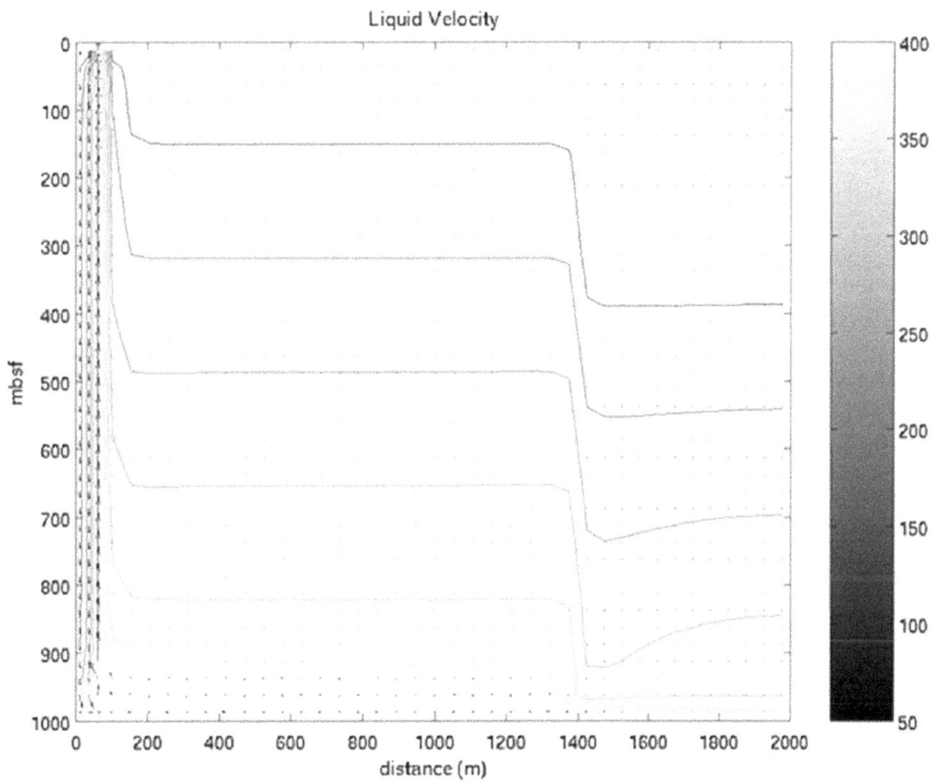

Figure 9. Flow patterns and isotherms for the single-pass model geometry and parameters shown in Figure 8 [from *Lewis*, 2007].

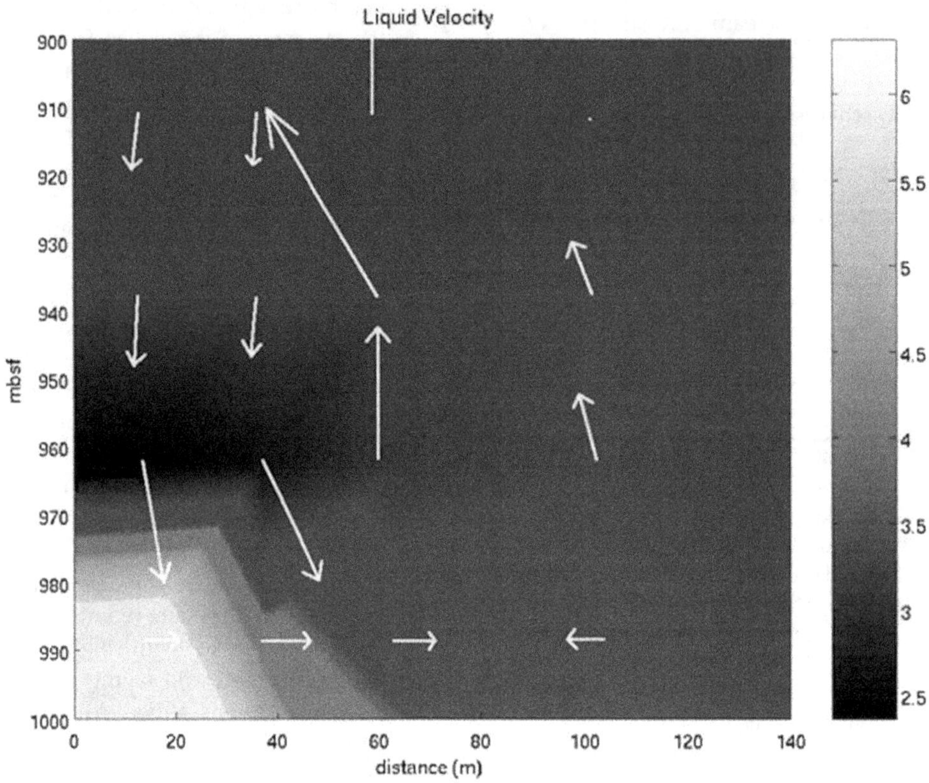

Figure 10. Expanded view of the lower left-hand corner of Figure 9 showing the extent and salinity distribution of the two-phase zone at the base of the system [from *Lewis*, 2007].

from the system, the vent salinity returns to that of normal seawater.

Although not correct in detail, the results shown in Figure 11 resemble the salinity data recorded at P vent between 1994 and 2002 (Figure 6). Temperatures at P vent also decreased between 1994 and 1996, the period during which brine was being discharged [*von Damm*, 2004]. The subsequent rise in temperature and decrease in salinity suggest that the two-phase regime has again been reactivated at depth. More detailed parametric analyses are needed to model the evolution of salinity, temperature, and heat output at hydrothermal vents. For example, *Lewis* [2007] shows that as the two-phase regime becomes laterally more extensive, the vent salinity decreases, even if the maximum temperature at the base of the system stays the same.

Numerical codes for multiphase processes in NaCl-H_2O hydrothermal systems have been developed only recently; so there has been little modeling of phase separation in hydrothermal systems at oceanic spreading centers at present. Consequently, there is a wide array of problems awaiting the implementation of numerical models. Such problems include employing the most basic models of phase separation to determine the fate of brine in hydrothermal systems, either related to phase separation at depth near the subaxial magma lens, or at shallower depths as hydrothermal circulation responds to dike emplacement. Models are also needed to address the evolution of vent salinity after a magmatic event (e.g., the salinity profiles in Figure 6) on time scales of weeks as well as decades. Such models would test the hypothesis that vent salinity near the EPR Integrated Study Site between 9 and 10°N shows evidence that the hydrothermal system is shoaling [*Von Damm*, 2004] and provide insight into the evolution of vent salinity at the Main Endeavour Field on the JDF after the magma-tectonic event of 1999 [*Lilley et al.*, 2003]. Models are also needed to explain the general N-S gradient of vent salinity and temperature along the Main Endeavour Field [*Butterfiled et al.*, 1994].

3.3. Geochemical and Biogeochemical Processes

Geochemical and biogeochemical processes at oceanic spreading centers can most easily be described in the con-

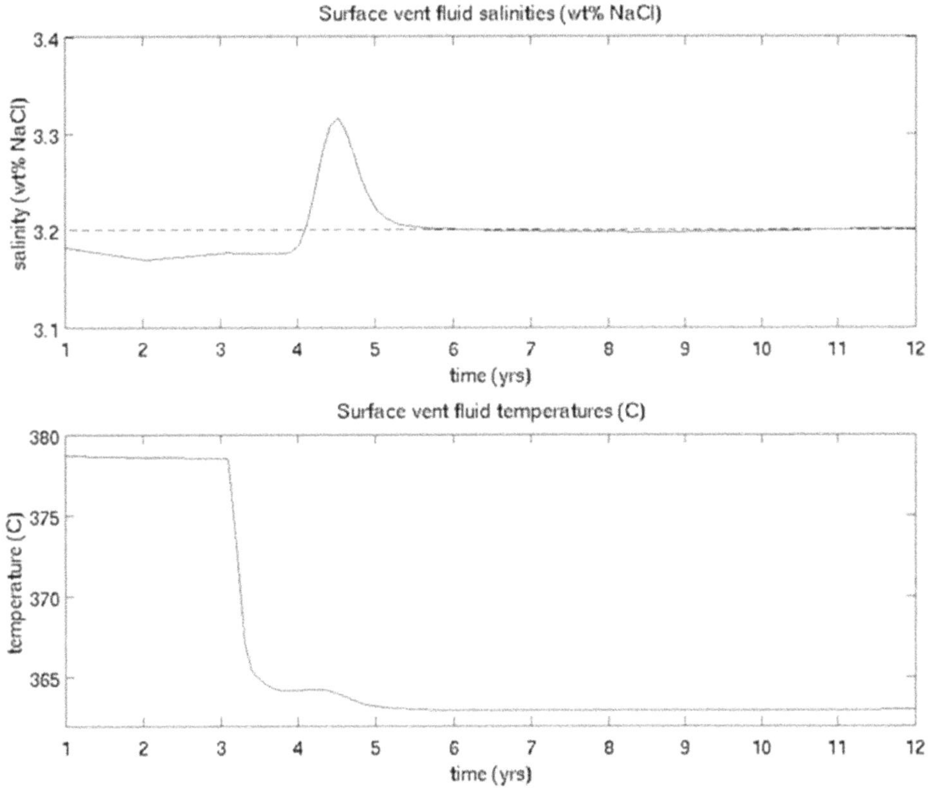

Figure 11. Vent fluid temperatures and salinities corresponding to an initial two-phase simulation (first 3 years) followed by purely single-phase flow in which brine is flushed from deep within the system. The model geometry and parameters are shown in Figure 8. The dashed line shows the equivalent normal seawater salinity (3.2 wt % NaCl) [from *Lewis*, 2007].

text of shallow and deep parts of the hydrothermal circulation system. Within these two regimes chemical reactions may occur that change the porosity, and hence the permeability structure of the crust. Conceptually, these reductions or increases in porosity that result from the precipitation or dissolution of a solid phase can be viewed as a multiphase process in which the solid phase is immobile. This approach is not the standard manner for investigating reactive transport problems, but the method has been used in problems related to the formation and dissolution of gas hydrates in seafloor sediments [*Xu*, 2004].

The shallow part of the circulation system is confined to the high-porosity, high-permeability extrusive layer, which ranges in thickness between ~ 200 and 800 m in thickness (Figure 1). Much of the seawater entering the crust likely circulates within the extrusives. Over many millions of years, this circulation results in low-temperature alteration of the crust. The reduction in porosity by vein filling minerals changes the seismic velocity characteristic of the shallow crust as it moves off axis [*Houtz and Ewing*, 1976; *Wilkens et al.*, 1991; *Jacobsen*, 1992]. It is highly likely that microbial processes are also occurring in the extrusives. Biogeochemical reactions may alter the chemistry of the circulating seawater [e.g., *Von Damm and Lilley*, 2004] as well as the physical characteristics of the crust, but the importance of these processes has not been quantified. Except for the regions of the crust beneath focused high-temperature venting, fluid temperatures in the shallow part of the system are generally lower than ~100°C.

As seawater penetrates into the underlying sheeted dikes and is heated to ≈150°C, anhydrite begins to precipitate from seawater [e.g., *Bischoff and Seyfried*, 1978]. *Lowell and Yao* [2002] argue that if recharge zones were spatially limited, anhydrite precipitation would rapidly clog the permeability. Crustal stresses may reopen permeability, but this would tend to occur episodically and on time scales that may be greater than the rate of clogging by mineral precipitation.

The decline in hydrothermal temperatures and heat output that would result from such clogging is not observed. Hence, *Lowell and Yao* [2002] argue that the effective area of recharge must be at least 2 orders of magnitude larger than the area of discharge. The models used by *Lowell and Yao* [2002] are relatively simple and do not include other water-rock reactions or the possible effects of biological activity on sulfate removal.

Within the deep part of the circulation system, as fluid temperatures exceed 200°C, seawater Mg is exchanged for Ca in basalt [e.g., *Bischoff and Dickson*, 1975; *Mottl*, 1983; *Seyfried*, 1987] and essentially all the seawater sulfate is removed or inorganically reduced to H_2S [e.g., *Shanks et al.*, 1981, 1995]. As fluids near the top of the subaxial magma chamber, they are heated to ≈400°C [*Seewald and Seyfried*, 1990]. Numerous water-rock reactions take place, with the resulting hydrothermal fluid composition being largely buffered by solid phases [e.g., *Bowers and Taylor*, 1985; *Bowers et al.*, 1988; *Campbell et al.*, 1988a]. The hydrothermal fluid is thought to reach thermodynamic equilibrium [e.g., *Seyfried*, 1987] or at least steady state with the mineral assemblage [*Von Damm*, 1988]. *Alt* [1995], *Von Damm* [1995], and *German and Von Damm* [2003] provide excellent reviews of vent fluid chemistry. Relative to seawater, the resulting hydrothermal fluid is acidic with pH ~ 3.5–4.0, enriched in silica, metals such as iron, copper, and zinc, and depleted in magnesium and sulfate [e.g., *Von Damm*, 1995]. The fluids may also contain volatiles such as H_2S, H_2, CO_2, and CH_4. These volatiles, of which CO_2 and CH_4 are largely of magmatic origin, may be utilized by subsurface biota [e.g., *Von Damm and Lilley*, 2004; *Kelley et al.*, 2004]

Modeling of reactive transport processes in porous media has undergone dramatic advancement in the past two decades. The review by *Lichtner et al.* [1996] serves as a useful guide to reactive transport processes. Despite the broad application of reactive transport models to a variety of geologic processes, they have not been extensively applied to hydrothermal systems at ocean spreading centers. Much of the current understanding of high-temperature water-rock reactions in seafloor hydrothermal systems stems from laboratory experiments [e.g., *Bischoff and Dickson*, 1975; *Mottl and Seyfried*, 1980; *Seyfried and Ding*, 1995; *Seyfried et al.*, 2004]; and from reaction path modeling [e.g., *Bowers et al.*, 1985, 1988; *Janecky and Shanks*, 1988].

Recent work on reactive transport modeling in oceanic crust has focused on sedimented environments. This includes *Giambalvo et al.* [2002], who modeled sediments near the JdFR, and *Alt-Epping and Smith* [2001], who modeled the Middle Valley hydrothermal system on the JdFR. For a recent review, see *Alt-Epping and Diamond.* [this volume]. Modeling of reactive transport in ridge crest hydrothermal systems, in which models simultaneously reproduce vent temperatures, chemistry, and heat output, need to be developed.

3.3.1. Mixing in the shallow crust and the relationship between focused and diffuse flow. As high-temperature hydrothermal fluid ascends into the shallow crust, it may mix with seawater circulating in the extrusives. If mixing is extreme, the fluid may discharge as low temperature, relatively diffuse flow rather than as focused higher temperature black-smoker-like discharge. The hydrothermal discharge at the Galapagos Spreading Center (GSC) appears to represent this low temperature end-member case. Fluids discharging at the GSC emerge as warm springs with temperatures in the range of 3°–13°C [*Edmond et al.*, 1979]. Chemical analyses of the fluid, however, suggest that it had reached temperatures of ≈350 °C deeper in the system [*Edmond et al.*, 1979]. Sometimes high-temperature fluid that initially discharges over a broad area becomes focused into discrete vents [e.g., *Von Damm*, 2000], but usually diffuse flow occurs in patches adjacent to high-temperature focused flow [e.g., *Von Damm and Lilley*, 2004; *Ramondenc et al.*, 2006]. In this case, chemical data show that the diffuse flow is a mixture between seawater and high-temperature hydrothermal fluid [e.g., *Von Damm and Lilley*, 2004].

There are few mathematical or numerical models that address the relationship between diffuse and focused flow. *Pascoe and Cann* [1995] and *Lowell et al.* [2003] use single-pass models that involve two separate pipe-like circulation pathways to explore mixing between the deeply circulating high-temperature fluid and cooler seawater circulating through the extrusives (Figure 12). *Pascoe and Cann* [1995] showed that the resulting vent temperature depended in detail on the permeability structure of the two circulation limbs. For low-temperature diffuse flow to result, the upper limb must be relatively permeable compared to the deeper limb.

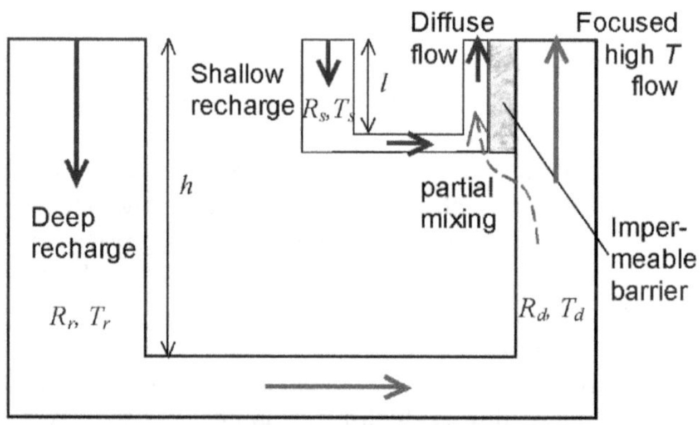

Figure 12. Cartoon of a single-pass hydrothermal circulation cell at a mid-ocean ridge consisting of deep and shallow fluid pathways. In this cartoon, it is assumed that mixing between the deep sulfate-free fluid with seawater in the shallow limb has resulted in precipitation of anhydrite, which tends to "unmix" the fluids. The symbol R corresponds to the integrated flow resistance in the various circulation branches, and T the average temperature in that branch, respectively [from *Lowell et al.*, 2003].

They further suggested that high-temperature, black smoker like venting may initiate as widespread diffuse flow that becomes focused as subsurface mineral precipitation clogs the permeability of the shallow crust. *Lowell et al.* [2003] used a similar model to that of Pascoe and Cann, but they explicitly investigated the role of anhydrite precipitation during mixing. They showed that focused black smoker flow could evolve in a matter of years provided the permeability of the deep part of the discharge zone was greater than or equal to that of the mixing zone and that the permeability of the mixing zone was initially greater than 10^{-12} m^2. In other cases, diffuse, low-temperature flow would result. *Lowell et al.* [2003] also showed that a relatively impermeable vertical layer of anhydrite ranging between 1 and 100 m thick could separate the focused black smoker discharge from the nearby low-temperature diffuse flow.

Lowell et al. [2007] have performed numerical simulations of single-pass circulation that included the effects of a high-permeability extrusive layer and temperature-dependent permeability. The goal of these simulations was to investigate mixing between high-temperature hydrothermal fluids and seawater circulating in the extrusives in order to understand the relationship between diffuse and focused flow. These models are incomplete because they do not include the effects of mineral precipitation upon mixing. The simulations show, however, that a low-permeability barrier, which may result from mineral precipitation, could separate high-temperature discharge from adjacent diffuse flow. The numerical simulations support the results of *Pascoe and Cann* [1995] and *Lowell et al.* [2003], which suggest that systems exhibiting only low-temperature diffuse discharge (e.g., GSC) require high permeability within the extrusives. Figure 13 from *Lowell et al.* [2007] plots expected vent temperature as a function of the permeability of the extrusive layer for different values of the extrusive layer thickness and permeability of the deep discharge zone. Figure 13 shows that the permeability of the extrusive layer exerts greater control on the vent temperature than its thickness. For the low vent temperatures exhibited by the GSC, the permeability ratio between the extrusives and the deep discharge zone exceeds 10^4.

To advance the understanding of the relationship between focused and diffuse flow, models that incorporate time-dependent

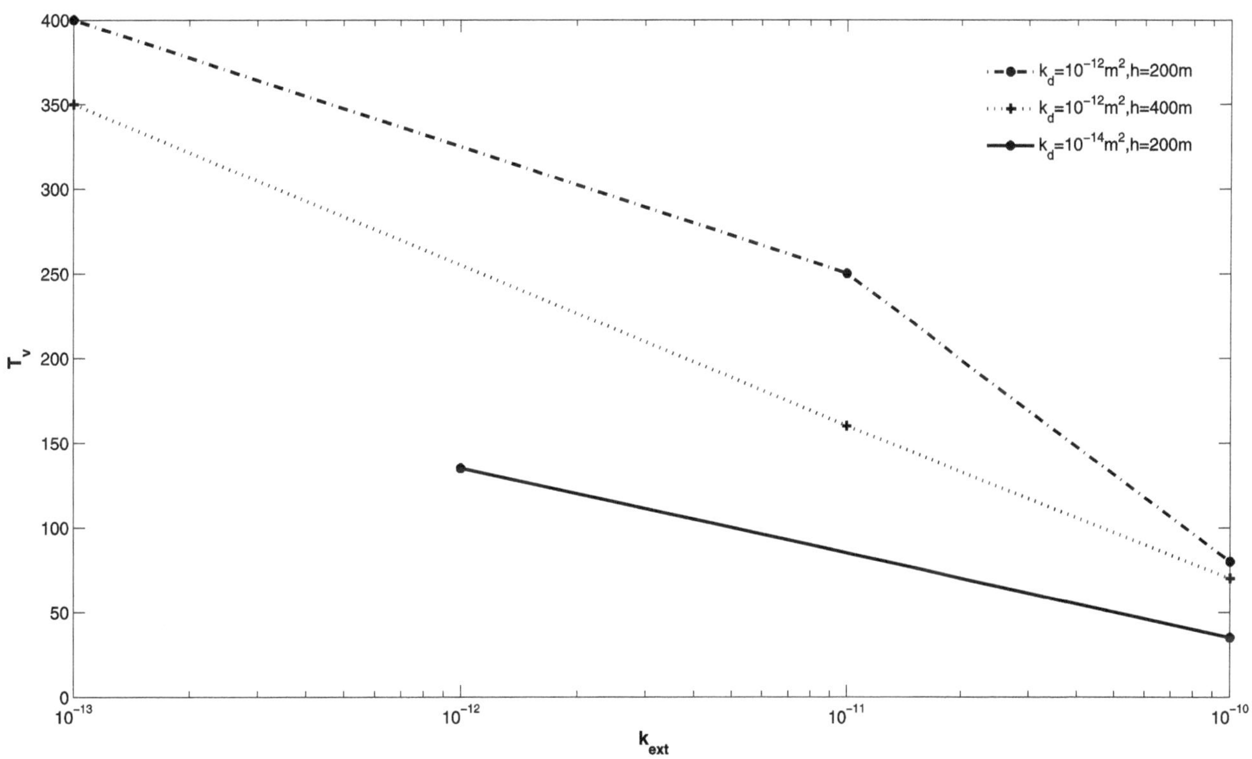

Figure 13. Plot of maximum vent temperature T_v versus k_{ext} for different values of k_d and h, where k_{ext}", k_d, and h are the permeability of the extrusive layer, permeability of the deep single-pass discharge limb, and the thickness of the extrusive layer, respectively. The results show that low-temperature venting requires that $k_{ext} > 10\ k_d$ for maximum vent temperatures to be below 150°C [from *Lowell et al.*, 2007].

permeability changes resulting from mineral precipitation need to be developed.

3.3.2. Biological processes in the shallow crust, reduction of porosity, and the evolution of "snowblower" vents. Since the discovery of biological ecosystems at oceanic spreading centers, there has been considerable interest in the nature of and the processes within the subsurface microbial biosphere [e.g., *Corliss et al.*, 1981; *Jannasch*, 1983, 1995; *Baross and Hoffman*, 1985; *Holden et al.,* 1998]. Reviews by *Kelley et al.* [2002], *Huber and Holden* [this volume], and *Schrenk et al.* [this volume] discuss a broad range of issues related to microbial activity. Of particular interest in this paper is the extent to which microbial growth might affect the porosity and permeability of the shallow crust. A related question concerns the observations of "snowblower" venting after volcanic eruptions on the seafloor. This term was coined by *Haymon et al.* [1993], who discovered the massive emission of "floc" of elemental sulfur [*Nelson et al.*, 1991] immediately after the 1991 eruption near 9°50′N on the EPR. Laboratory experiments showed that filamentous sulfur could be produced by H_2S-oxidizing microbes living near the oxic-anoxic interface [*Taylor and Wirsen*, 1997]. *Taylor et al.* [1999] report evidence of rapid in situ biogenic production of filamentous sulfur in a warm water hydrothermal vent at the 9°N vent field on the EPR.

Crowell [2007] and Crowell et al. (On the production of biological sulfur floc and "snowblower vents" at mid-ocean ridges, submitted to *Geochemistry, Geophysics, Geosystems*, 2008; hereinafter referred to as Crowell et al., submitted, 2008) estimated the rate of production of sulfur floc and the potential for porosity and permeability reduction in the oceanic crust by microbial activity by using geochemical data from the northern (Hole to Hell region) and southern (Tube Worm Pillar and Y vent) transects near 9°50′N on the EPR [*Von Damm and Lilley*, 2004]. They used the Si concentration in the vent fluid as a tracer to calculate the mixing ratio between high-temperature vent fluid and seawater in the nearby diffuse flow. Then, after correcting for pyrite precipitation, they estimated the amount of H_2S utilized by the subsurface biota by comparing the concentration of H_2S in diffuse flow emissions with the concentration of H_2S enter-

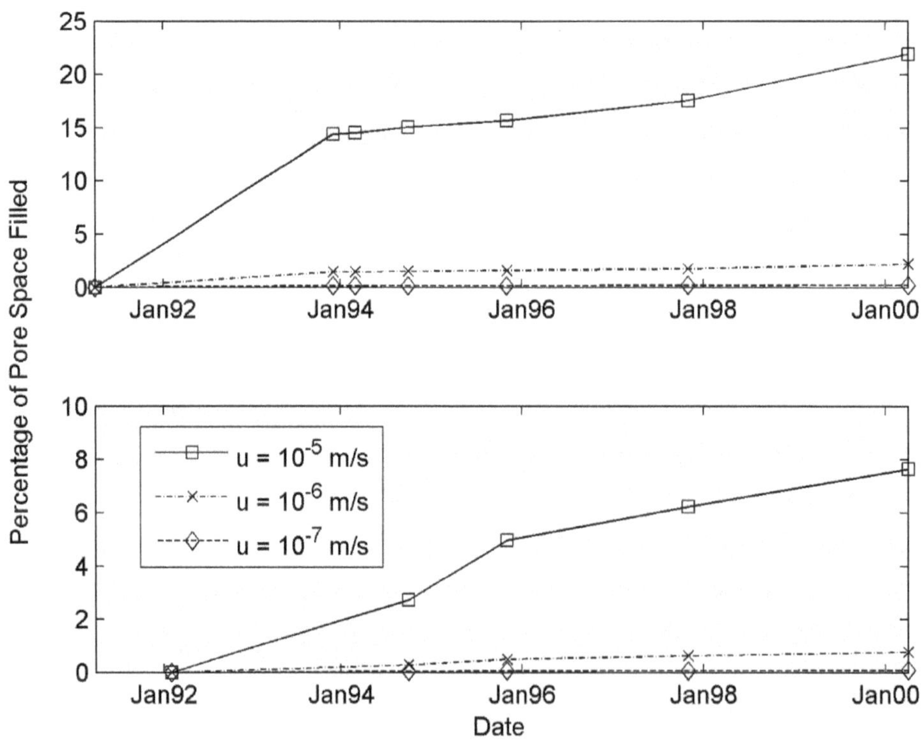

Figure 14. Percentage of pore occupied by sulfur floc since 1991 eruption. Panel (a) uses data from the Southern Transect and (b) uses data from the Northern Transect [*Von Damm and Lilley*, 2004]. The parameter u is the mean vertical Darcian velocity (m/s) in a boundary layer driven by side-wall heating. Different values of u arise as a result of different assumed values of permeability in the shallow crust, which here ranged between 10^{-10} and 10^{-12} m², respectively [*from Crowell*, 2007].

ing the base of the diffuse flow, using the concentration of H_2S in the high-temperature vent fluid and the mixing ratio as a constraint. From this analysis, the rate of sulfur floc production and the rate of porosity reduction resulting from the storage of sulfur floc were then calculated from the time of the 1991 eruption until 2000 as a function of area extent and mean upward velocity of diffuse flow. The latter parameters were estimated by assuming that the diffuse flow could be approximated as a convectively driven boundary layer flow [*Bejan*, 1995] along a 200-m-high vertical wall maintained at a temperature of 100°C. The results are shown in Figure 14.

The different values of the mean velocity u correspond to permeabilities in the shallow crust ranging between 10^{-10} and 10^{-12} m^2, respectively. The results have been extrapolated to the recent eruption in the same area in late 2005, early 2006 [e.g., *Rubin et al.*, 2006; *Shank et al.*, 2006]. To obtain the fraction of pore space filled as a result of biogenic sulfur production, an initial porosity of 10% was assumed; and deposition was assumed to occur in the upper 10 meters of the crust [*Crowell*, 2007; Crowell et al., submitted, 2008].

Figure 14 shows that the percentage of porosity reduction that results from biogenic sulfur production between the 1991 eruption and present appears to be relatively small (less than 10%) for the range of parameters used. The one exception is the estimate for the case $u = 10^{-5}$ m/s ($k_{ext} = 10^{-11}$ m^2) for the Southern Transect, where the estimated porosity loss is approximately 30%. The small changes in porosity estimated stemming from the generation of biogenic sulfur suggests that this process should have little impact on fluid flow patterns in the shallow crust. Comparison between the volume of sulfur floc deposited on the seafloor after the 1991 eruption and that which was generated in the 1991–2006 magmatic interval suggests that the volume of sulfur floc discharged during the 1991 eruption was likely a combination of biogenic sulfur generated and stored in the shallow crust during the preceding magmatic cycle and a microbial bloom, perhaps stemming from a fresh input of magma-derived nutrients [*Crowell*, 2007; Crowell et al., submitted, 2008].

4. CONCLUSIONS

This paper highlights a number of issues concerning the mathematical and numerical modeling of multiphase, multicomponent fluid flow in hydrothermal systems at oceanic spreading centers. These issues include: (a) the relationships between magma transport and heat transfer and high-temperature hydrothermal venting, (b) phase separation during high-temperature hydrothermal flow, and (c) geochemical and biogeochemical processes, particularly those that affect the porosity and permeability distribution in the crust, such as mineral dissolution and precipitation. The development of models to adequately describe these complex processes is in its infancy. For example, models coupling magmatic and hydrothermal heat transport are highly simplified. In this paper, simple zero-dimensional heat balance approaches have been used to determine magmatic and hydrothermal heat output from a convecting, crystallizing, replenished magma chamber. These models do not take into account details of crystal settling and chemical fractionation within the magma body, among other factors. Even though these simple models show the need for magma replenishment to maintain magmatic heat input to hydrothermal systems, models that incorporate the physics of magma replenishment and magma chamber inflation are not yet available. Coupling of magma-hydrothermal models to magmatic eruptions is also lacking. Numerical models that address phase separation in NaCl-H_2O fluids are also just becoming available. Initial results suggest that phase separation results in brine storage at the base of the hydrothermal system and vent fluids with salinity less than that of seawater. Brines may emerge later when the heat source is depleted. Finally, models that couple mineral dissolution and precipitation have been developed, but, at present, these models employ relatively simple geochemical systems, relatively simple hydrology, or both. Although some models of reactive transport in porous media are available, these models have not been used extensively in high-temperature hydrothermal systems at oceanic spreading centers. Part of the difficulty is that the products of high-temperature water-rock reactions in oceanic crust are strongly coupled to the salinity of the fluid. Because the necessary two-phase NaCl-H_2O hydrothermal codes have just recently become available, it has been difficult to link the reactive transport codes to the appropriate hydrology and heat transfer codes. This problem should be alleviated in coming years, provided better thermodynamic and kinetic data on the reaction between saline brines and basaltic rocks become available. Similarly, models of biogeochemical processes and models linking the hydrological and thermal aspects of hydrothermal flow to micro- and macrobiological ecosystems are yet to be developed.

We project that the next decade will produce more robust multicomponent and multiphase models that are directly applicable to hydrothermal processes at oceanic spreading centers. Such advances will play a critical role in understanding synergisms between magmatic, tectonic, biogeochemical, and hydrothermal processes at oceanic spreading centers.

Acknowledgments. The authors thank the two anonymous reviewers and the editor, Jeff Seewald, for their many helpful comments on the original version of this paper. This work has been supported by the National Science Foundation under grants OCE-0527208 and OCE-0351942.

REFERENCES

Alt, J. C. (1995), Subseafloor processes in mid-ocean ridge hydrothermal systems, in *Seafloor Hydrothermal Systems*, *Geophys. Monogr. Ser.*, vol. 91, edited by S. E. Humphris, et al., pp. 85–114, AGU, Washington, D. C.

Alt-Epping, P., and L. W. Diamond (2008), Reactive transport and numerical models of seafloor hydrothermal systems: a review, this volume.

Alt-Epping, P., and L. Smith (2001), Computing geochemical mass transfer and water/rock ratios in submarine hydrothermal systems: Implications for estimating the vigor of convection, *Geofluids, 1*, 163–181.

Anderko, A., and, K. S. Pitzer (1993), Equation-of-state representation of phase equilibria and volumetric properties of the system NaCl-H_2O above 573 K. *Geochim. Cosmochim. Acta*, *57*, 1657–1680.

Archer, D. G. (1992), Thermodynamic properties of the NaCl + H_2O system: II. Thermodynamic properties of NaCl(aq), NaCl·2H_2O(cr), and phase equilibria, *J. Phys. Chem. Ref. Data*, *21*, 793–820.

Baker, E. T. (1994), A six-year time series of hydrothermal plumes over the Cleft segment of the Juan de Fuca Ridge, *J. Geophys. Res.*, *99*, 4889–4904.

Baker, E. T., and G. J. Massoth (1986), Hydrothermal plume measurements: A regional perspective, *Science*, *234*, 980–982.

Baker, E. T., and G. J. Massoth (1987), Characteristics of hydrothermal plumes form two vent fields on the Juan de Fuca Ridge, northeast Pacific Ocean, *Earth Planet. Sci. Lett.*, *85*, 59–73.

Baker, E. T., G. J. Massoth, and R. A. Feely (1987), Cataclysmic hydrothermal venting on the Juan de Fuca Ridge, *Nature, 329*, 149–151.

Baker, E. T., R. E. McDuff, and G. J. Massoth (1990), Hydrothermal venting from the summit of a ridge axis seamount: Axial Volcano, Juan de Fuca Ridge, *J. Geophys. Res.*, *95*, 12,843–12,854.

Baker, E. T., G. J. Massoth, S. L. Walker, and R. W. Embley (1993), A method for quantitatively estimating diffuse and discrete hydrothermal discharge, *Earth Planet. Sci. Lett.*, *118*, 235–249.

Baker, E. T., G. J. Massoth, R. A. Feely, G. A. Cannon, and R. E. Thomson (1998), The rise and fall of the CoAxial hydrothermal site, 1993–1996, *J. Geophys. Res.*, *103*, 9791–9806.

Baker, E. T., G. J. Massoth, R. A. Feely, R. W. Embley, R. E. Thomson, and B. J. Burd (1995), Hydrothermal event plumes from the CoAxial eruption site, Juan de Fuca Ridge, *Geophys. Res. Lett.*, *22*, 147–150.

Baker, E. T., R. P. Lowell, J. A. Resing, R. A. Feely, R. W. Embley, G. J. Massoth, and S .L. Walker (2004), Decay of hydrothermal output following the 1998 seafloor eruption at Axial Volcano: Observations and models, *J. Geophys. Res.*, *109*, B01205 doi:10.1029/2003JB002618.

Baker, E. T., H. N. Edmonds, P. J. Michael, W. Bach, H. J. B. Dick, J. E. Snow, S. L. Walker, N. R. Banerjee, and C. H. Langmuir (2004), Hydrothermal venting in magma deserts: The ultraslow-spreading Gakkel and Southwest Indian ridges, *Geochem. Geophys. Geosyst.*, *5*, Q08002 doi:10.1029/2004GC000712.

Bai, W., W. Xu, and R. P. Lowell (2003), The dynamics of submarine geothermal heat pipes, *Geophys. Res. Lett.*, *30*(3), 1108, doi:10.1029/2002GL016176.

Baross, J. A., and S. E. Hoffman (1985), Submarine hydrothermal vents and associated gradient environments as sites for the origin and evolution of life, *Origins Life*, *15*, 327–345.

Bejan, A. (1995), *Convection Heat Transfer*, 2nd ed., John Wiley, New York, 623 pp.

Bemis, K. G., R. P. Von Herzen, and M. J. Mottl (1993), Geothermal heat flux from hydrothermal plumes on the Juan de Fuca Ridge, *J. Geophys. Res.*, *98*, 6351–6366.

Berndt, M. E., M. E. Person, W. E. Seyfried Jr. (2001), Phase separation and two phase flow in submarine hydrothermal systems: Geophysical modeling in the NaCl/H_2O system, presented at the *11th Goldschmidt Conference*, Hot Springs, Va, May 20–24.

Bischoff, J. L. (1991), Densities of liquids and vapors in boiling NaCl-H_2O solutions: A PVTX summary from 300° to 500°C, *Am. J. Sci.*, *291*, 309–338.

Bischoff, J. L., and F. W. Dickson (1975), Seawater-basalt interaction at 200°C and 500 bars: Implications for origin of seafloor heavy metal deposits and regulation of seawater chemistry, *Earth Planet. Sci. Lett.*, *25*, 385–397.

Bischoff, J. L., and K. Pitzer (1989), Liquid-vapor relations for the system NaCl-H_2O: Summary of the PTX surface from 300 to 500°C, *Am. J. Science, 289*, 217–248.

Bischoff, J. L., and J. R. Rosenbauer (1989), Salinity variations in submarine hydrothermal systems by layered double diffusive convection, *J. Geol.*, *97*, 613–623.

Bischoff, J. L., and W. E. Seyfried (1978), Hydrothermal chemistry of seawater from 25 to 350°C, *Am. J. Sci.*, *278*, 838–860.

Bodvarsson, G. (1982), Terrestrial energy currents and transfer in Iceland, in *Continental and Oceanic Rifts*, *Geodyn. Ser.*, vol. 8, edited by G. Palmason, pp. 271–282, AGU, Washington, D. C.

Bodvarsson, G., and R. P. Lowell (1972), Ocean-floor heat flow and the circulation of interstitial waters, *J. Geophys. Res.*, *77*, 4472–4475.

Bowers, T. S., and H. P. Taylor Jr. (1985), An J. integrated chemical and stable-isotope model of the origin of midocean ridge hot springs, *J. Geophys. Res.*, *90*, 12,583–12,606.

Bowers, T. S., K. L. Von Damm, and J. M. Edmond (1985), Chemical evolution of mid-ocean ridge hot springs, *Geochim. Cosmochim. Acta*, *49*, 2239–2252.

Bowers, T. S., A. C. Campbell, C. I. Measures, A. J. Spivack, M. Khadem, and J. M. Edmond (1988), Chemical controls on the composition of vent fluids at 13°–11°N and 21°N, East Pacific Rise, *J. Geophys. Res.*, *93*, 4522–4536.

Brandeis, G., and C. Jaupart (1986), On the interaction between convection and crystallization in cooling magma chambers, *Earth Planet. Sci. Lett.*, *77*, 345–361.

Bryan, W. B. (1983), Systematics of modal phenocryst assemblages in submarine basalts: petrologic implications, *Contrib. Mineral. Petrol.*, *83*, 62–74.

Butterfield, D. A., and G. J. Massoth (1994), Geochemistry of North Cleft segment vent fluids: temporal changes in chlorinity and their possible relation to recent volcanism, *J. Geophys. Res.*, *99*, 4951–4968.

Butterfield, D. A., R. E. McDuff, M. J. Mottl, M. D. Lilley, J. E. Lupton, and G. J. Massoth (1994), Gradients in the composition of hydrothermal fluids from Endeavour Ridge vent field: Phase separation and brine loss, *J. Geophys. Res.*, *99*, 9561–9583.

Butterfield, D. A., I. R. Jonasson, G. J. Massoth, R. A. Feely, K. K. Roe, R. E. Embley, J. F. Holden, R. E. McDuff, M. D. Lilly, and J. R. Delaney (1997), Seafloor eruptions and evolution of hydrothermal fluid chemistry, *Philos. Trans. R. Soc. London, Ser. A*, *355*, 369–386.

Campbell, A. C., et al. (1988), Chemistry of hot springs on the Mid-Atlantic Ridge, *Nature*, *335*, 514–519.

Campbell, A. C., T. S. Bowers, C. I. Measures, K. K. Falkner, M. Khadem, and J. M. Edmond (1988b), A time series of vent fluid compositions from 21°N, East Pacific Rise (1979, 1981, 1985), and the Guaymas Basin, Gulf of California (1982, 1985), *J. Geophys. Res.*, *93*, 4537–4549.

Canales, J. P., R. A. Sohn, and B. J. deMartin (2007), Crustal structure of the Trans-Atlantic Geotraverse (TAG) segment (Mid-Atlantic Ridge, 26°10′N): Implication for the nature of hydrothermal circulation and detachment faulting at slow spreading ridges, *Geochem. Geophys. Geosyst.*, *8*, Q08004, doi:10.1029/2007GC001629.

Cherkaoui, A. S. M., W. S. D. Wilcock, R. A. Dunn, and D. R. Toomey (2003), A numerical model of hydrothermal cooling and crustal accretion at a fast spreading mid-ocean ridge, *Geochem. Geophys. Geosyst.*, *4*(9), 8616, doi:10.1029/2001GC00215.

Christeson, G. L., K. D. McIntosh, and J. A. Karson (2007), Inconsistent correlation of seismic layer 2a and lava layer thickness in oceanic crust, *Nature*, *445*, 418–421.

Collier, J., and M. Sinha (1990), Seismic images of a magma chamber beneath the Lau Basin back-arc spreading centre, *Nature*, *346*, 646–648.

Converse, D. R., H. D. Holland, and J. M. Edmond (1984), Flow rates in the axial hot springs of the East Pacific Rise (21°N): Implications for the heat budget and the formation of massive sulfide deposits, *Earth Planet. Sci. Lett.*, *69*, 159–175.

Corliss, J. B., J. A. Baross, and S. E. Hoffman (1981), An hypothesis concerning the relationship between submarine hot springs and the origin of life on Earth, *Oceanol. Acta*, *4*, 59–69.

Crawford, W. C., S. C. Webb, and J. A. Hildebrand (1999), Constraints on melt in the lower crust and Moho at the East Pacific Rise, 9°48′N, using seafloor compliance measurements, *J. Geophys. Res.*, *104*, 2923–2939.

Crowell, B. W. (2007), Biological sulfur reactions and the influence on fluid flow at mid-ocean ridge hydrothermal systems, M.S. thesis, 53 pp., Georgia Institute of Technology, Atlanta.

Delaney, J. R., V. Robigou, R. E. McDuff, and M. K. Tivey (1992), Geology of a vigorous hydrothermal system on the Endeavor segment, Juan de Fuca Ridge, *J. Geophys. Res.*, *97*, 19,663–19,682.

deMartin, B. J., R. A. Sohn, J. P. Canales, and S. E. Humphris (2007), Kinetics and geometry of active detachment faulting beneath the Trans-Atlantic Geotraverse (TAG) hydrothermal field on the Mid-Atlantic Ridge, *Geology*, *35*, 711–714.

Detrick, R. S., P. Buhl, E. Vera, J. Mutter, J. Orcutt, J. Madsen, and T. Brocher (1987), Multi-channel seismic imaging of a crustal magma chamber along the East Pacific Rise, *Nature*, *326*, 35–41.

Driesner, T. (2001), A new model for thermodynamic and transport properties of the NaCl-water system from 0–700°C, 0.1 to 500 MPa, and X_{NaCl} from 0 to 1, presented at the *11th Goldschmidt Conference*, Hot Springs, Va, May 20–24.

Driesner T. (2007), The system NaCl-H_2O. Part II: Molar volume, enthalpy, and isobaric heat capacity from 0 to 1000°C, 0 to 5000 bar, and 0 to 1 X_{NaCl}, *Geochim. Cosmochim. Acta*, *71*, 4902–4919.

Driesner, T., and C. A. Heinrich (2007), The system NaCl-H_2O. Part I: Correlation formulae for phase relations in temperature-pressure-composition space from 0 to 1000°C, 0 to 5000 bar, and 0 to 1 X_{NaCl}, *Geochim. Cosmochim. Acta*, *71*, 4880–4901.

Dunn, R. A., D. R. Toomey, and S. C. Solomon (2000), Three-dimensional seismic structure and physical properties of the crust an shallow mantle beneath the East Pacific Rise at 9°30′N, *J. Geophys. Res.*, *105*, 23,537–23,555.

Dzurizin, D., R. Y. Koyanagi, and T. T. English (1984), Magma supply and storage at Kilauea volcano, Hawaii, 1956–1983, *J. Volcanol. Geotherm. Res.*, *21*, 177–206.

Edmond, J. M., A. C. Campbell, M. R. Palmer, G. P. Klinkhammer, C. R. German, H. N. Edmonds, H. Elderfield, G. Thompson, and P. A. Rona (1995), Time series studies of vent fluids from the TAG and MARK sites (1986, 1990) Mid-Atlantic Ridge: A new solution chemistry model and a mechanism for Cu/Zn zonation in massive sulphide orebodies, in *Hydrothermal Vents and Processes*, *Geol. Soc. London Spec. Publ.*, vol. 87, edited by L. M. Parson, C. L. Walker, and D. R. Dixon, pp. 77–86, Geological Society of London, London.

Edwards, K. J., W. Bach, T. McCollom, and D. Rogers (2004), Neutrophilic iron-oxidizing bacteria in the ocean: Their habitats, diversity, and roles in mineral deposition, rock alteration, and biomass production in the deep-sea, *Geomicrobiol. J.*, *21*, 393–404.

Elder, J. W. (1981), *Geothermal Systems*, 508 pp., Academic, Troy, Mass.

Emmanuel, S., and B. Berkowitz (2006), Suppression and stimulation of seafloor hydrothermal convection by exothermic mineral hydration, *Earth Planet. Sci. Lett.*, *243*, 657–668.

Faust, C., and J. Mercer (1979), Geothermal reservoir simulation, 1. Mathematical models for liquid- and vapor-dominated hydrothermal systems, *Water Resour. Res.*, *15*, 23–30.

Fialko, Y. (2001), On origin of near-axis volcanism and faulting at fast spreading mid-ocean ridges, *Earth Planet. Sci. Lett.*, *190*, 31–39.

Fisher, A. T., and K. Becker (1991), Heat flow, hydrothermal circulation and basalt intrusions in the Guaymas Basin, Gulf of California, *Earth Planet. Sci. Lett.*, *103*, 84–99.

Fontaine, F. J., and W. S. D. Wilcock (2006), Dynamics and storage of brine in mid-ocean ridge hydrothermal systems, *J. Geophys. Res.*, *111*, B06102, doi:10.1029/2005JB003866.

Fouquet, Y., et al. (1991), Hydrothermal activity and metallogenesis in the Lau back-arc basin, *Nature*, *349*, 778–781.

Geiger, S., T. Driesner, C. A. Heinrich, and S. K. Matthai (2006a), Multiphase thermohaline convection in the Earth's crust: I. A new finite element-finite volume solution technique combined with a new equation of state for NaCl-H_2O, *Transp. Porous Media, 63*, 399–434.

Geiger, S., T. Driesner, C. A. Heinrich, and S .K. Matthai (2006b), Multiphase thermohaline convection in the Earth's crust: II. Benchmarking and application of a finite element-finite volume solution technique combined with a new equation of state for NaCl-H_2O, *Transp. Porous Media., 63*, 435–461.

Gendron, J. F., J. F. Todd, R. A. Feely, E. T. Baker, and D. Kadko (1994), Excess ^{222}Rn over the Cleft segment, Juan de Fuca Ridge, *J. Geophys. Res., 99*, 5007–5015.

Gente, P., J. M. Auzende, V. Renard, Y. Fouquet, and D. Bideau (1986), detailed geological mapping by submersible of the East Pacific Rise axial graben near 13°N, *Earth Planet. Sci. Lett., 78*, 224–236.

German, C. R., and K. L. Von Damm (2003), Hydrothermal processes, in *Treatise on Geochemistry*, vol. 6, edited by H. Elderfield, pp. 181–222, Elsevier, Amsterdam.

Germanovich, L. N., and R. P. Lowell (1992), Percolation theory, thermoelasticity and discrete hydrothermal venting in the Earth's crust, *Science, 255,* 1564–1567.

Germanovich, L. N., R. P. Lowell, and D. K. Astakhov (2000), Stress dependent permeability and the formation of seafloor event plumes, *J. Geophys. Res., 105*, 8341–8354.

Germanovich, L. N., R. P. Lowell, and D. K. Astakhov (2001), Temperature-dependent permeability and bifurcations in seafloor hydrothermal flow, *J. Geophys. Res., 106*, 473–496.

Giambalvo, E. R., C. I. Steefel, A. T. Fisher, N. D. Rosenberg, and G. C. Wheat (2002), Effect of fluid-sediment reaction on hydrothermal fluxes of major elements, eastern flank of the Juan de Fuca Ridge, *Geochim. Cosmochim. Acta, 66*, 1739–1757.

Ginster, U., M. J. Mottl, and R. P. Von Herzen (1994), Heat flux from black smokers on the Endeavor and Cleft segments, Juan de Fuca Ridge, *J. Geophys. Res., 99,* 4937–4950.

Hashimoto, J., S. Ohta, T. Gamo, H. Chiba, T. Yamaguchi, S. Tsichida, T. Okudaira, H. Wanabe, T. Yamanaka, and M. Kitazawa (2001), Hydrothermal vents and associated biological communities in the Indian Ocean, *InterRidge News, 10*, 21–22.

Hayba, D. O., and S. E. Ingebritsen (1994), The computer model HYDROTHERM, A three dimensional, finite difference model to simulate ground-water flow and heat transport in the temperature range of 0 to 1200°C, *U. S. Geol. Surv. Water-Resour. Investig. Rep.* 94-4045.

Haymon, R. M., et al. (1993), Volcanic eruption of the mid-ocean ridge along East Pacific Rise crest at 9°45′–52′N: Direct submersible observations of seafloor phenomena associated with an eruption event in April 1991, *Earth Planet. Sci. Lett., 119,* 85–101.

Henstock, T. J., A. W. Woods, and R. S. White (1993), The accretion of oceanic crust by episodic sill intrusion, *J. Geophys. Res., 98*, 4143–4161.

Holden, J .F., and R. M. Daniel (2004), Upper temperature limit of life based on hyperthermophile culture experiments and field observations, in *The Subsurface Biosphere at Mid-Ocean Ridges, Geophys. Monogr. Ser.*, vol. 144, edited by W. S. D. Wilcock, et al., pp. 13–24, AGU, Washington, D. C.

Holden, J. F., M. Summit, and J. A. Baross (1998), Thermophilic and hyperthermophilic microorganisms in 3–30°C fluids following a deep-sea volcanic eruption, *FEMS Microbiol. Ecol., 25*, 33–41.

Houtz, R., and J. Ewing (1976), Upper crustal structure as a function of plate age, *J. Geophys. Res., 81*, 2490–2498.

Huber, J., and J. F. Holden (2008), Modeling the impact of diffuse vent microorganisms along mid-ocean ridges and flanks, this volume.

Humphris, S. E., and J. R. Cann (2000), Constraints on the energy and chemical balances of the modern TAG and ancient Cyprus seafloor sulfide deposits, *J. Geophys. Res., 105*, 28,477–28,488.

Huppert, H. E., and R. S. J. Sparks (1998), The generation of granitic magmas by intrusion of basalt into continental crust, *J. Petrol., 29*, 599–624.

Huppert, H. E., and A. W. Woods (2002), The role of volatiles in magma chamber dynamics, *Nature, 420*, 493–495.

Jacobsen, R. S. (1992), Impact of crustal evolution on changes of the seismic properties of the uppermost ocean crust, *Rev. Geophys., 30*, 23–42.

Jannasch, H. W. (1983), Microbial processes at deep-sea hydrothermal vents, in *Hydrothermal Processes at Sea Floor Spreading Centers*, edited by P. A. Rona, K. Bostrom, L. Laubier, and K. L. Smith, pp. 677–709, Plenum Press, New York.

Jannasch, H. W. (1995), Microbial interactions with hydrothermal fluids, in *Seafloor Hydrothermal Systems, Geophys. Monogr. Ser.*, vol. 91, edited by S. E. Humphris, et al., pp. 273–296, AGU, Washington, D. C.

Jellinek, A. M., and R. C. Kerr (1999), Mixing and compositional stratification produced by natural convection 2. Applications to the differentiation of basaltic and silicic magma chambers and komatiite lava flows, *J. Geophys. Res., 104*, 7203–7218.

Jellinek, A. M., and R. C. Kerr (2001), Magma dynamics, crystallization, and chemical differentiation of the 1959 Kilauea Iki lava lake, Hawaii, revisited, *J. Volcanol. Geotherm. Res., 110*, 235–263.

Johnson, H. P., M. Hutnak, R. P. Dziak, C. G., Fox, I. Urcuyo, J. P. Cowen, J. Nabelek, and C. Fisher (2000), Earthquake-induced changes in a hydrothermal system on the Juan de Fuca mid-ocean ridge, *Nature, 407*, 174–177.

Karson, J. A., and P. A. Rona (1990), Block-tilting, transfer faults, and structural control of magmatic and hydrothermal processes in the TAG area, Mid-Atlantic Ridge 26°N, *Geol. Soc. Am. Bull., 102,* 1635–1645.

Kelemen, P. B., and E. Aharonov (1998), Periodic formation of magma fractures and generation of layered gabbros in the lower crust beneath oceanic spreading ridges, in *Faulting and Magmatism at Mid-Ocean Ridges, Geophys. Monogr. Ser.*, vol. 106, edited by W. R. Buck, et al., pp. 267–289, AGU, Washington, D. C.

Kelemen, P. B., K. Koga, and N. Shimizu (1997), Geochemistry of gabbro sills in the crust-mantle transition zone of the Oman ophiolite: Implications for the origin of the oceanic lower crust, *Earth Planet. Sci. Lett., 146*, 475–488.

Kelley, D. S., K. M. Gillis, and G. Thompson (1993), Fluid evolution in submarine magma-hydrothermal systems at the Mid-Atlantic Ridge, *J. Geophys. Res.*, *98*, 19,579–19,596.

Kelley, D. S., et al. (2001), An off-axis hydrothermal vent field near the Mid-Atlantic Ridge at 30°N, *Nature*, *412*, 145–149.

Kelley, D. S., J. A. Baross, and J. R. Delaney (2002), Volcanoes, fluids and life at mid-ocean ridge spreading centers, *Annu. Rev. Earth Planet Sci.*, *30*, 385–491.

Kelley, D. S., M. D. Lilley, and G. L. Fruh-Green (2004), Volatiles in submarine environments: Food for life, in *The Subsurface Biosphere at Mid-Ocean Ridges, Geophys. Monogr. Ser.*, vol. 144, edited by W. S. D. Wilcock, et al., pp. 167–189, AGU, Washington, D. C.

Kissling, W. (2005a), Transport of three-phase hyper-saline brines in porous media: examples, *Transp. Porous Media*, *60*, 141–157.

Kissling, W. (2005b), Transport of three-phase hyper-saline brines in porous media: Theory and code implementation, *Transp. Porous Media*, *61*, 25–44.

Lewis, K. C. (2007), Numerical modeling of two-phase flow in the sodium chloride-water system with applications to seafloor hydrothermal systems, Ph.D. thesis, 143 pp., Georgia Institute of Technology, Atlanta.

Lewis, K. C. and R. P. Lowell (2004), Mathematical modeling of phase separation of seawater near an igneous dike, *Geofluids*, *4*, 197–209.

Lichtner, P. C., C. I. Steefel, and E. H. Oelkers, eds. (1996), *Reactive Transport in Porous Media, Rev. Mineral.*, vol. 34, Mineralogical Society of America, Washington, D. C., 438 pp.

Lilley, M. D., D. A. Butterfield, J. E. Lupton, and E. J. Olson (2003), Magmatic events produce rapid changes in hydrothermal vent chemistry, *Nature*, *422*, 878–881.

Lister, C. R. B. (1974), On the penetration of water into hot rock, *Geophys. J. R. Astron. Soc.*, *39*, 465–509.

Lister, C. R. B. (1982), "Active" and "passive" hydrothermal systems in the oceanic crust: Predicted physical conditions, in *The Dynamic Environment of the Ocean Floor*, edited by K. A. Fanning and F. T. Manheim, pp. 441–459, D. C. Heath, Lexington, MA.

Little, S. A., K. D. Stolzenbach, and R. P. Von Herzen (1987), Measurements of plume flow from a hydrothermal vent field, *J. Geophys. Res.*, *92*, 2587–2596.

Liu, L. (2007), The link between convection and crystallization in a sub-axial magma chamber and heat output in a seafloor hydrothermal system, M.S. thesis, 74 pp., Georgia Institute of Technology, Atlanta.

Lonsdale, P., and K. Becker (1985), Hydrothermal plumes, hot springs, and conductive heat flow in the Southern Trough of Guaymas Basin, *Earth Planet. Sci. Lett.*, *73*, 211–225.

Lowell, R. P. (1975), Circulation in fractures, hot springs, and convective heat transport on mid-ocean ridge crests, *Geophys. J. R. Astron. Soc.*, *40*, 351–365.

Lowell, R. P. (1991), Modeling continental and submarine hydrothermal systems, *Rev. Geophys.*, *29*, 457–476.

Lowell, R. P., and D. K. Burnell (1991), A numerical model for magma-hydrothermal boundary layer heat transfer in the oceanic crust, *Earth Planet. Sci. Lett.*, *104*, 59–69.

Lowell, R. P., and L. N. Germanovich (1994), On the temporal evolution of high-temperature hydrothermal systems at ocean ridge crests, *J. Geophys. Res.*, *99*, 565–575.

Lowell, R. P., and L. N. Germanovich (1995), Dike injection and the formation of megaplumes at ocean ridges, *Science*, *267*, 1804–1807.

Lowell, R. P., and L. N. Germanovich (1997), Evolution of a brine-saturated layer at the base of a ridge crest hydrothermal systems, *J. Geophys. Res.*, *102*, 10,245–10,255.

Lowell, R. P., and L. N. Germanovich (2004), Seafloor hydrothermal processes: Results from scale analysis and single-pass models, in *Mid-Ocean Ridges: Hydrothermal Interactions Between the Lithosphere and Oceans, Geophys. Monogr. Ser.*, vol. 148, edited by C. R. German, et al., pp. 219–244, AGU, Washington, D. C.

Lowell, R. P., and P. A. Rona (1985), Hydrothermal models for the generation of massive sulfide ore deposits, *J. Geophys. Res.*, *90*, 8769–8783.

Lowell, R. P., and P. A. Rona (2002), Seafloor hydrothermal systems driven by the serpentinization of peridotite, *Geophys. Res. Lett.*, *29*(11), 1531, doi:10.1029/2001GL014411.

Lowell, R. P., and Y. Yao (2002), Anhydrite precipitation and the extent of hydrothermal recharge zones at ocean ridge crests, *J. Geophys. Res.*, *107(B9)*, 2183, doi:10.1029/2001JB001289.

Lowell, R. P., Y. Yao, and L. N. Germanovich (2003), Anhydrite precipitation and the relationship between focused and diffuse flow in seafloor hydrothermal systems, *J. Geophys. Res.*, *108*(B9), 2424, doi:10.1029/2002JB002371.

Lowell, R. P., S. R. Gosnell, and Y. Yang (2007), Numerical simulations of single-pass hydrothermal convection at mid-ocean ridges: Effects of the extrusive layer and temperature-dependent permeability, *Geochem, Geophys. Geosyst.*, *8*, Q10011, doi:10.1029/2007GC001653.

Maclennan, J., T. Hulme, and S. C. Singh (2004), Thermal models of oceanic crustal accretion: Linking geophysical, geological and petrological observations, *Geochem. Geophys. Geosyst.*, *5*, Q02F25, doi:10.1029/2003GC000605.

Maclennan, J. (2008), The supply of heat to mid-ocean ridges by crystallization and cooling of mantle melts, this volume.

Macdonald, K. C., K. Becker, F. N. Spiess, and R. D. Ballard (1980), Hydrothermal heat flux of the "black smoker" vents on the East Pacific, *Earth Planet. Sci. Lett.*, *48*, 1–7.

Manning, C. E., and C. J. MacLeod (1996), Fracture-controlled metamorphism of Hess Deep gabbros, Site 894: Constraints on the roots of mid-ocean ridge hydrothermal systems at fast spreading centers, *Proc. Ocean Drill. Program, Sci. Res.*, vol. 47, pp. 189–212, Ocean Drilling Program, College Station, Tex.

Manning, C. E., C. J. MacLeod, and P. E. Weston (2000), Lower-crustal cracking front at fast-spreading ridges: Evidence from the East Pacific Rise and the Oman ophiolite, *Geol. Soc. Am., Spec. Pap.*, *349*, 261–272.

Marsh, B. D. (1981), On the crystallinity, probability of occurrence, and rheology of lava and magma, *Contrib. Mineral. Petrol.*, *78*, 85–98.

Martin, D., and R. Nokes (1989), A fluid dynamical study of crystal settling in convecting magmas, *J. Petrol.*, *30*, 1471–1500.

McBirney, A. R., and T. Murase (1984), Rheological properties of magmas, *Annu. Rev. Earth Planet. Sci.*, *12*, 337–357.

McConachy, T. F., R. D. Ballard, M. J. Mottl, and R. P. Von Herzen (1986), Geological form and setting of a hydrothermal vent field at latitude 10°56′N, East Pacific Rise: A detailed study using *Angus* and *Alvin*, *Geology, 14,* 295–298.

McKenzie, D. (1984), The generation and compaction of partially molten rock, *J. Petrol.*, *25*, 713–765.

Mével, C., and M. Cannat (1991), Lithospheric stretching and hydrothermal processes in oceanic gabbros from slow-spreading ridges, in *Ophiolite Genesis and Evolution of the Oceanic Lithosphere*, edited by T. Peters, A., Nicolas, and R. G. Coleman, pp. 293–312, Kluwer Acad. Publ., Dordrecht, Netherlands.

Michael, P. J., et al. (2003), Magmatic and amagmatic seafloor generation at the ultraslow-spreading Gakkel ridge, Arctic Ocean, *Nature*, *423*, 956–961.

Mottl, M. J. (1983), Metabasalts, axial hot springs, and the structure of hydrothermal systems at mid-ocean ridges, *Geol. Soc. Am. Bull.*, *94*, 161–180.

Mottl, M. J. (2003), Partitioning of energy and mass fluxes between mid-ocean ridge axes and flanks at high and low temperature, in *Energy and Mass Transfer in Marine Hydrothermal Systems*, edited by P. E. Hallbach, V. Tunnicliff, and J. R. Hein, pp. 271–286, Dahlem University Press, Berlin.

Mottl, M. J., and W. E. Seyfried (1980), Sub-seafloor hydrothermal systems rock-vs. seawater-dominated, in *Seafloor Spreading Centers*: *Hydrothermal Systems*, *Benchmark Pap. Geol.*, vol. 56, edited by P. A. Rona and R. P. Lowell, pp. 66–82, Dowden Hutchinson and Ross, Stroudsburg, Pa.

Murton, B. J., L. J. Redourn, C. R. German, and E. T. Baker (1999), Sources and fluxes of hydrothermal heat, chemicals and biology within a segment of the Mid-Atlantic Ridge, *Earth Planet. Sci. Lett.*, *171*, 301–317.

Navin, D. A., C. Pierce, and M. C. Sinha (1998), The RAMESSES Experiment II: Evidence for accumulated melt beneath a slow spreading ridge from wide-angle refraction and multichannel reflection seismic profiles, *Geophys. J. Int.*, *135*, 746–772.

Nelson, D. C., R. M. Haymon, and M. D. Lilley (1991), Rapid growth of unusual hydrothermal bacteria observed at new vents during AdVenture dive program to the EPR at 9°45′–52′N, *Eos Trans. AGU*, *72*, 481.

Palliser, C., and R. McKibbin (1998a), A model for deep geothermal brines, I: *T-P-X* state-space description, *Transp. Porous Media, 33*, 65–80.

Palliser, C., and R. McKibbin (1998b), A model for deep geothermal brines, II: Thermodynamic properties—density, *Transp. Porous Media, 33*, 129–154.

Palliser, C., and R. McKibbin (1998c), A model for deep geothermal brines, III: Thermodynamic properties—Enthalpy and viscosity, *Transp. Porous Media, 33*, 155–171

Pascoe, A. R., and J. R. Cann (1995), Modeling diffuse hydrothermal flow in black smoker vent fields, in *Hydrothermal Vents and Processes*, *Geol. Soc. Special Publ.* vol. 87, edited by L. M. Walker and C. L. Dixon, pp. 159–173, The Geological Society, London, UK.

Perfit, M. R., D. J. Fornari, M. C. Smith, J. F. Bender, C. H. Langmuir, and R. M. Haymon (1994), Small-scale spatial and temporal variations in mid-ocean ridge crest magmatic processes, *Geology*, *22*, 375–379.

Pruess, K. (1991). TOUGH2—A general-purpose numerical simulator for multiphase fluid and heat flow, *Report LBL-29400*, Lawrence Berkeley Laboratory.

Ramondenc, P. (2007), Effect of seismicity and diking on hydrothermal circulation at mid-ocean ridges, Ph.D. thesis, 226 pp., Georgia Institute of Technology, Atlanta.

Ramondenc, P., L. N. Germanovich, K. L. Von Damm, and R. P. Lowell (2006), The first measurements of hydrothermal heat output at 9°50′N, East Pacific Rise, *Earth Planet. Sci. Lett.*, *245*, 487–497, doi: 10.1016/j.epsl.2006.03.023.

Ramondenc, P., L. N. Germanovich, and R. P. Lowell (2008), Modeling hydrothermal response to earthquakes at oceanic spreading centers, this volume.

Rogers, D. R., C. M. Santelli, and K. J. Edwards (2003), Geomicrobiology of deep-sea deposits: estimating community diversity from low-temperature seafloor rocks and minerals, *Geobiology*, *1*(2), 109–117.

Rogers, P. S. Z., and K. S. Pitzer (1982), Volumetric properties of aqueous sodium chloride solutions, *J. Phys. Chem. Ref. Data*, *11*, 15–81.

Rona, P. A., and D. A. Trivett (1992), Discrete and diffuse heat transfer at ASHES vent field, Axial Volcano, Juan de Fuca Ridge, *Earth Planet. Sci. Lett., 109,* 57–71.

Rona, P. A., M. D. Hannington, C. V. Raman, G. Thompson, M. K. Tivey, S. E. Humphris, L. Lalou, and S. Petersen (1993), Major active and relict seafloor hydrothermal mineralization: TAG hydrothermal field, Mid-Atlantic Ridge 26°N, 45°N, *Econ. Geol., 88,* 1989–2017.

Rosenberg, N. D., J. E. Lupton, D. Kadko, R. Collier, M. D. Lilley, and H. Pak (1988), Estimation of heat and chemical fluxes from a seafloor hydrothermal vent field using radon measurements, *Nature, 344*, 604–607.

Rubin, K. H., M. R. Perfit, D. J. Fornari, S. A. Soule, M. Tolstoy, F. Waldhauser, et al. (2006), Geochronology and composition of the 2005–2006 volcanic eruptions of the East Pacific Rise, 9°46′–56′N, *Eos Trans. AGU*, *87(52)*, Fall Meet. Suppl., Abstract V23B-0602.

Rudnicki, M. D., and H. Elderfield (1992), Theory applied to the Mid-Atlantic Ridge hydrothermal plumes: The finite difference approach, *J. Volcanol. Geotherm. Res.*, *50*, 161–172.

Rudnicki, M. D., and C. R. German (2002), Temporal variability of the hydrothermal plume above the Kairei vent field, 25°S, Central Indian Ridge, *Geochem. Geophys. Geosyst.*, *3*(2), doi:10.1029/2001GC000240.

Schoofs, S., and U. Hanson (2000), Depletion of a brine layer at the base of the ridge-crest hydrothermal system, *Earth Planet. Sci. Lett.*, *180*, 341–353.

Schrenk, M., J. F. Holden, and J. A. Baross (2008), Magma-to-microbe networks in the context of sulfide hosted microbial ecosystems.

Schultz, A., J. M. Delaney, and R. E. McDuff (1992), On the partitioning of heat flux between diffuse and point source venting, *J. Geophys. Res., 97,* 12,229–12,314.

Seewald, J. S., and W. E. Seyfried Jr. (1990), The effect of temperature on metal mobility in subseafloor hydrothermal systems:

Constraints from basalt alteration experiments, *Earth Planet. Sci. Lett., 101*, 388–403.

Seyfried, W. E. (1987), Experimental and theoretical constraints on hydrothermal alteration processes at mid-ocean ridges, *Annu. Rev. Earth Planet. Sci., 15*, 317–355.

Seyfried, W. E., Jr., and K. Ding (1995), Phase equilibria in subseafloor hydrothermal systems: A review of the role of redox, temperature, pH and dissolved Cl on the chemistry of hot spring fluids at mid-ocean ridges, in *Seafloor Hydrothermal Systems, Geophys. Monogr. Ser.*, vol. 91, edited by S. E. Humphris, et al., pp. 248–272, AGU, Washington, D. C.

Seyfried, W. E., Jr., K. Ding, and M. E. Berndt (1991), Phase equilibria constraints on the chemistry of hot spring fluids at mid-ocean ridges, *Geochim. Cosmochim. Acta, 55*, 3559–3580.

Seyfried, W. E., Jr., D. I. Foustoukos, and D. E. Allen (2004), Ultramafic-hosted hydrothermal systems at mid-ocean ridges: Chemical and physical controls on pH, redox, and carbon reduction reactions, in *Mid-Ocean Ridges: Hydrothermal Interactions Between the Lithosphere and Oceans, Geophys. Monogr. Ser.*, vol. 148, edited by C. R. German, et al., pp. 267–284, AGU, Washington, D. C.

Seyfried, W. E., Jr., D. R. Janecky, and M. J. Mottl (1984), Alteration of the oceanic crust: Implications for the geochemical cycles of lithium and boron, *Geochim. Cosmochim. Acta, 48*, 557–569.

Seyfried, W. E., Jr., J. S. Seewald, M. E. Berndt, K. Ding, and D. I. Foustoukos (2003), Chemistry of hydrothermal vent fluids from the Main Endeavour Field, northern Juan de Fuca Ridge: Geochemical controls in the aftermath of June, 1999 seismic events, *J. Geophys. Res., 108*(B9), 2429, doi:10.1029/2002JB001957.

Shank, T. M., et al. (2006), Initial biological, chemical and geological observations after the 2005–6 volcanic eruptions on the East Pacific Rise, *Eos Trans. AGU, 87(52)*, Fall Meet. Suppl., Abstract V13C-04.

Shanks, W. C., III, J. L. Bischoff, and R. J. Rosenbauer (1981), Seawater sulfate reduction and sulfur isotope fractionation in basaltic systems: Interaction of seawater with fayalite and magnetite at 200–350°C, *Geochim. Cosmochm. Acta, 45*, 1977–1995.

Shanks, W. C., III, J. K. Bolke, and R. R. Seal II (1995), Stable isotopes in mid-ocean ridge hydrothermal systems: Interactions between fluids, minerals, and organisms, in *Seafloor Hydrothermal Systems, Geophys. Monogr. Ser.*, vol. 91, edited by S. E. Humphris, et al., pp. 222–247, AGU, Washington, D. C.

Shea, K., A. Metaxas, C. R. Young, and C. R. Fisher (2008), Process and interactions in macrofaunal assemblages at hydrothermal vents: a modeling perspective, this volume.

Sim, Y. (2004), Mechanics of complex hydraulic fractures in the Earth's crust, Ph.D. thesis, 324 pp., Georgia Institute of Technology, Atlanta.

Singh, S. C., J. S. Collier, A. J. Harding, G. M. Kent, and J. A. Orcutt (1999), Seismic evidence for a hydrothermal layer above the solid roof of the axial magma chamber at the southern East Pacific Rise, *Geology, 27*, 219–222.

Singh, S. C., W. C. Crawford, H. Carton, T. Seher, V. Combier, M. Cannat, J. P. Canales, D. Dusunur, J. Escartin, and J. M. Miranda (2006), Discovery of a magma chamber and faults beneath a Mid-Atlantic Ridge hydrothermal field, *Nature, 442*, 1029–1032.

Sinha, M. C., S. C. Constable, C. Pierce, A. White, G. Heinson, L. M. MacGregor, and D. A. Navin (1998), Magmatic processes at slow spreading ridges: implications of the RAMESSES experiment at 57°45′N on the Mid-Atlantic Ridge, *Geophys. J. Int., 135*, 731–745.

Sinton, J. M., and R. S. Detrick (1992), Mid-ocean ridge magma chambers, *J. Geophys. Res., 97*, 197–216.

Sohn, R. A., D. J. Fornari, K. L. Von Damm, J. A. Hildebrand, and S. C. Webb (1998), Seismic and hydrothermal evidence for a cracking event on the East Pacific Rise crest at 9°50′N, *Nature, 396*, 159–161.

Sohn, R. A., J. A. Hildebrand, and S. C. Webb (1999), A microearthquake survey of the high-temperature vent fields on the volcanically active East Pacific Rise (9°50'N), *J. Geophys. Res., 104*, 25,367–25,377.

Sourirajan S. and G. C. Kennedy (1962), The system H_2O-NaCl at elevated temperatures and pressures. *Am. J. Sci., 260*, 115–141.

Spera, F. J. (2000), Physical properties of magma, in *Encyclopedia of Volcanoes*, edited by H. Sigurdsson, B. Houghton, H. Rymer, J. Stix, and S. McNutt, pp. 171–190, Elsevier, New York, NY.

Spiegelman, M. (1993a), Flow in deformable porous media. Part 1. Simple analysis, *J. Fluid Mech., 247*, 17–38.

Spiegelman, M. (1993b), Flow in deformable porous media. Part 2. Numerical analysis-the relationship between shock waves and solitary waves, *J. Fluid Mech., 247*, 39–63.

Stein, J. S., and A. T. Fisher (2001), Multiple scales of hydrothermal circulation in Middle Valley, northern Juan de Fuca Ridge: Physical constraints and geologic models, *J. Geophys. Res., 106*, 8563–8580.

Strens, M. R., and J. R. Cann (1982), A model of hydrothermal circulation in fault zones at mid-ocean ridge crests, *Geophys. J. R. Astron. Soc., 71*, 225–240.

Tanger, J. C., and K. S. Pitzer (1989), Thermodynamics of NaCl-H_2O: a new equation of state for the near-critical region and comparisons with other equations for adjoining regions, *Geochim. Cosmochim. Acta, 53*, 973–987.

Taylor, C. D., and C. O. Wirsen (1997), Microbiology and ecology of filamentous sulfur formation, *Science, 277*, 1483–1485.

Taylor, C. D., C. O. Wirsen, and F. Gaill (1999), Rapid microbial production of filamentous sulfur mats at hydrothermal vents, *Appl. Environ. Microbiol., 65*, 2253–2255.

Thompson, R. E., J. R. Delaney, R. E. McDuff, D. R. Janecky, and J. S. McClain (1992), Physical characteristics of the Endeavor Ridge hydrothermal plume during July 1988, *Earth Planet. Sci. Lett., 111*, 141–154.

Tolstoy, M. (2008), Seismological constraints on magmatic and hydrothermal processes at mid-ocean ridges, this volume.

Van Ark, E. M., R. S. Detrick, J. P. Canales, S. M. Carbotte, A. J. Harding, G. M. Kent, M. R. Nedimovic, W. S. D. Wilcock, J. B. Diebold, and J. M. Babcock (2007), Seismic structure of the Endeavour Segment, Juan de Fuca Ridge: Correlations with seismicity and hydrothermal activity, *J. Geophys. Res., 112*, B02401, doi: 10.1029/2005JB004210.

Veirs, S. R., R. E. McDuff, F. R. Stahr (2006), Magnitude and variance of near-bottom horizontal heat flux at the Main Endeavour hydrothermal vent field, *Geochem. Geophys. Geosyst.*, *7*, Q02004, doi:10.1029/2005GC000952.

Von Damm, K. L. (1988), Systematics and postulated controls on submarine hydrothermal solution chemistry, *J. Geophys. Res.*, *93*, 4551–4561.

Von Damm, K. L. (1995), Controls on the chemistry and temporal variability of seafloor hydrothermal fluids, in *Seafloor Hydrothermal Systems, Geophys. Monogr. Ser.*, vol. 91, edited by S. E. Humphris, et al., pp. 222–247, AGU, Washington, D. C.

Von Damm, K. L. (2000), Chemistry of hydrothermal vent fluids from 9°–10°N, East Pacific Rise: "Time Zero," the immediate posteruptive period, *J. Geophys. Res.*, *105*, 11,203–11,222.

Von Damm K. L. (2004) Evolution of the hydrothermal system at East Pacific Rise 9°50′N: Geochemical evidence for changes in the upper oceanic crust, in *Mid-Ocean Ridges: Hydrothermal Interactions Between the Lithosphere and Oceans, Geophys. Monogr. Ser.*, vol. 148, edited by C. R. German, et al., pp. 285–304, AGU, Washington, D. C.

Von Damm, K. L., L. G. Buttermore, S. E. Oosting, A. M. Bray, D. J. Fornari, M. D. Lilley, and W. C. Shanks III (1997), Direct observation of the evolution of a seafloor 'black smoker' from vapor to brine, *Earth Planet. Sci. Lett.*, *149*, 101–111.

Von Damm, K. L, J. M. Edmond, B. Grant, C. I. Measures, B. Waldon, and R. F. Weiss (1985), Chemistry of submarine hydrothermal solutions at 21°N, East Pacific Rise, *Geochim. Cosmochim. Acta*, *49*, 2221–2237.

Von Damm, K. L. and M. D. Lilley (2004), Diffuse flow hydrothermal fluids from 9°50′N East Pacific Rise: Origin, evolution and biogeochemical controls, in *The Subsurface Biosphere at Mid-Ocean Ridges, Geophys. Monogr. Ser.*, vol. 144, edited by W. S. D. Wilcock, et al., pp. 245–268, AGU, Washington, D. C.

Von Damm, K. L., S. E. Oosting, R. Kozlowski, L. G. Buttermore, D. C. Colodner, H. N. Edmonds, J. M. Edmond, and J. M. Grebmeier (1995), Evolution of East Pacific Rise hydrothermal vent fluids following a volcanic eruption, *Nature*, *375*, 47–50.

Von Damm, K. L., C. M. Parker, R. M. Gallant, and J. P. Loveless (2002), Chemical evolution of hydrothermal fluids from EPR 21°N: 23 years later in a phase separating world, *Eos Trans. AGU*, *83*(47), Fall Meet. Suppl., Abstract V61B-1365.

Wilcock, W. S. D., and J. R. Delaney (1996), Mid-ocean ridge sulfide deposits: Evidence for heat extraction from magma chambers or cracking fronts?, *Earth Planet. Sci. Lett.*, *145*, 49–64.

Wilcock, W. S. D., and A. T. Fisher (2004), Geophysical constraints on the sub-seafloor environment near mid-ocean ridges, in *The Subsurface Biosphere at Mid-Ocean Ridges, Geophys. Monogr. Ser.*, vol. 144, edited by W. S. D. Wilcock, et al., pp. 51–74, AGU, Washington, D. C.

Wilkens, R. H., G. J. Fryer, and J. L. Karsten (1991), Evolution of porosity and seismic structure of upper oceanic crust: Importance of aspect ratios, *J. Geophys. Res.*, *96*, 17,981–17,995.

Worster, M. G., H. E. Huppert, and R. S. J. Sparks (1990), Convection and crystallization in magma cooled from above, *Earth Planet. Sci. Lett.*, *101*, 78–89.

Xu, W. (2004), Modeling dynamic marine gas hydrate systems, *Am. Mineral.*, *89*, 1271–1279.

Xu, W., and R. P. Lowell (1998), Oscillatory instability of one-dimensional two-phase hydrothermal flow in heterogeneous porous media, *J. Geophys. Res.*, *103*, 20,859–20,868.

Xu, W., R. P. Lowell, and W. Bai (2001), GTHSW—A numerical simulator of two-phase NaCl-Water hydrothermal systems: Applications to continental and seafloor environment, *11th Annual Goldschmidt Conference*, Hot Springs, Va, May 20–24.

Robert P. Lowell, Department of Geosciences, Virginia Polytechnic Institute and State University, 4044 Derring Hall, Blacksburg, VA, 24060. (email: rlowell@vt.edu)

The Supply of Heat to Mid-Ocean Ridges by Crystallization and Cooling of Mantle Melts

John Maclennan

Department of Earth Sciences, University of Cambridge, UK

Supply of mantle melt to ridges provides material for accretion of the solid crust and heat to drive hydrothermal flow. Cooling of mantle melt to ambient seawater temperatures provides 2.0 ± 0.2 MJ kg^{-1} and a potential global heat supply of 3.2 ± 0.3 TW. Half of this heat is released from latent heat of solidification and specific heat on cooling from ~1300° to ~1000°C. The remainder is released during cooling from ~1000° to 2°C. Sub-solidus hydration reactions release less than 0.1 MJ kg^{-1}. More than 75% of the heat release from cooling and crystallisation of mantle melts occurs within ~10 km of the ridge axis. Injection of dykes and eruption of lava flows provide episodic heat supply to the upper crust that drives transient high-temperature venting. Episodic heat supply constitutes 1/3 of the total magmatic supply. The remainder of the magmatic heat release is associated with crystallisation and cooling of rocks, principally gabbros, that form the lower oceanic crust. This crystallisation occurs in axial magma chambers and melt sills. The uppermost of these magma chambers have been imaged at 1- to 4-km depth and are associated with persistent high-temperature venting. Partitioning of latent heat release between these shallow axial magma chambers and sills deeper in the crust is poorly constrained. If this lower crustal crystallisation is important at fast-spreading ridges, then deep, near-axis hydrothermal circulation is required to cool the lower crust. There is a pressing need for more observations and models that can be used to improve understanding of the geometry of hydrothermal flow and distribution of hydrothermal cooling.

1. INTRODUCTION

The total rate of heat loss from the solid earth through rocks of the ocean floor is 31.2 ± 0.7 TW [*Pollack et al.*, 1993]. Although incident solar radiation can provide power of 1020 W m^{-2} of earth surface near the equator, the average heat loss across the ocean floor is a modest 0.1 W m^{-2}. While most of this heat loss occurs by conductive transfer, 11 ± 4 TW is accommodated by the flow of hydrothermal fluids [*Sclater et al., 1980*; *Stein and Stein*, 1994]. The best-known manifestations of such flow are black smokers at active mid-ocean ridges, which provide highly focussed heat supply from vents with areas of as little as ~0.1 m^{-2} at powers of up to ~10^8 W [*Lowell et al.*, 1995]. Much of the heat supply for these black smoker systems is likely to be provided by cooling and crystallisation of basaltic mantle-derived melts at mid-ocean ridges. The aim of this contribution is to review the supply of heat from cooling mantle melts close to active mid-ocean ridges. The interested reader should also refer to

Magma to Microbe: Modeling Hydrothermal Processes at Ocean Spreading Centers
Geophysical Monograph Series 178

10.1029/178GM04

recent review papers by *Lowell et al.* [1995], *Elderfield and Schulz* [1996], *Mottl* [2003], *Cannat et al.* [2004] and *Sinha and Evans* [2004], which each emphasise and detail different parts of the system.

Heat is supplied to mid-ocean ridges principally by the advection of hot molten basalt into the crust. Contributions by conduction from mantle into crust or possibly by transfer of deep hydrothermal fluids are likely to be of secondary importance within a few tens of kilometres of ridge axes. Ultra-slow spreading ridges, with a full spreading rate of less than 20 mm yr^{-1} and long amagmatic stretches, provide an exception to this rule [*Edmonds et al.*, 2003; *Michael et al.*, 2003]. This review focusses on the release of heat during cooling of mantle-derived melts. This cooling is associated with three types of heat release: specific heat release as the temperature drops, latent heat of solidification of the melt, and latent of heat of sub-solidus reactions. The quantification of these three sources of heat is one of the principal goals of this paper (Sections 2 and 4).

The spatial and temporal distribution of the heat release is controlled by the distribution of crystallisation of melt within the crust. A wide range of geological, geophysical, geochemical and petrological observations both from active ridges and from crust exposed in ophiolites has been used to constrain this distribution of crystallisation (reviewed by *Maclennan et al.* [2004]). Thermal models are required to integrate the observations and show that the heat release by crystallisation is coupled to the ability of hydrothermal systems to penetrate and cool the crust [*Cherkaoui et al.*, 2003; *Maclennan et al.*, 2005]. Unfortunately, our understanding of the geometry of the hydrothermal systems remains rudimentary and improvement of this understanding is a key goal for future research. Schematic representations of the distribution of latent heat release and hydrothermal cooling at fast- and slow-spreading mid-ocean ridges are given in Figures 1 and 2.

The cool, solid rocks that constitute almost all of the oceanic crust had their origins as hot basaltic melt in the mantle. A selection of important events in the cooling of a batch of

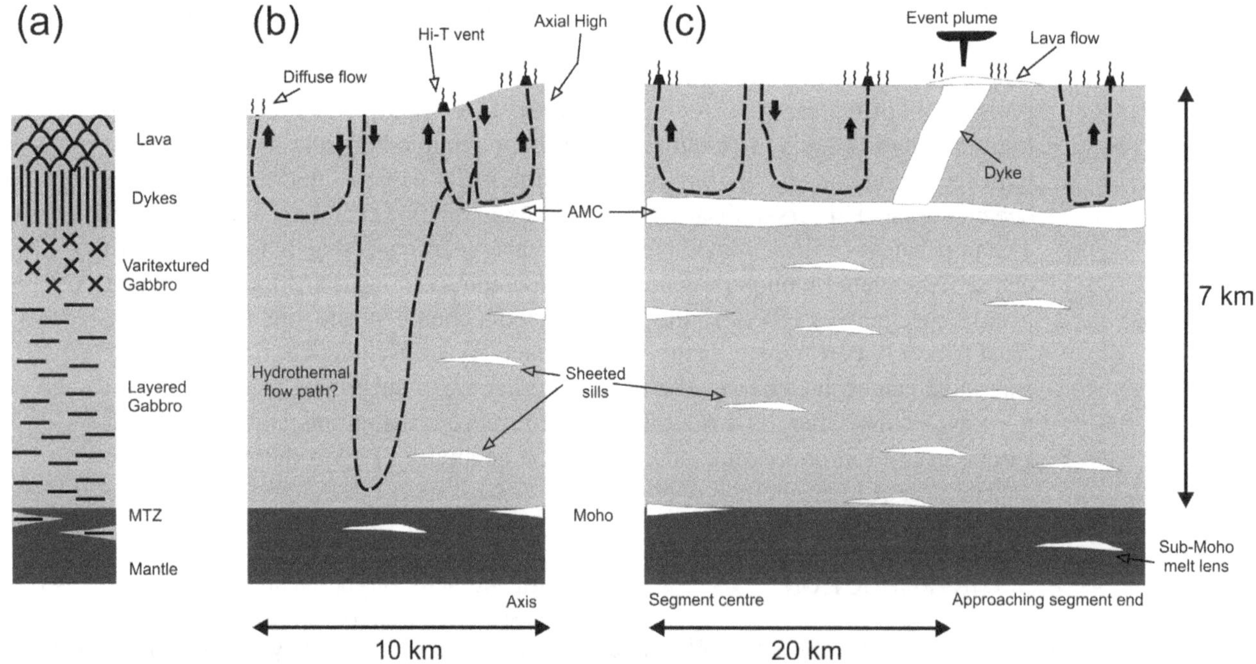

Figure 1. A schematic model of the distribution of heat release during magmatic solidification and hydrothermal cooling at a fast-spreading ridge. See main text for details. (a) Lithological column resulting from crustal accretion at fast-spreading ridges. (b) Across-axis section with approximate scale given at the base. The white lenses show the possible location of magma bodies where solidification occurs. The AMC is the stable upper axial magma chamber. Sheeted sills may be present anywhere within the axial lower crust and uppermost mantle. Substantial latent heat of crystallisation released in these sills must be removed by deep hydrothermal circulation. The pathways of hydrothermal flow are highly speculative and poorly constrained. Little is known about their 3-D structure, especially in the lower crust. Axis-parallel faults, which are not shown here, may play a key role in the focussing of fluid flow. Pervasive background recharge flow may also be important [*Teagle et al.*, 2003]. (c) Along-axis section. A dyke event with lava flow and hydrothermal megaplume are shown. These events are likely to be episodic and superimposed on longer-term heat supply from crystallisation of melt in the AMC and sheeted sills.

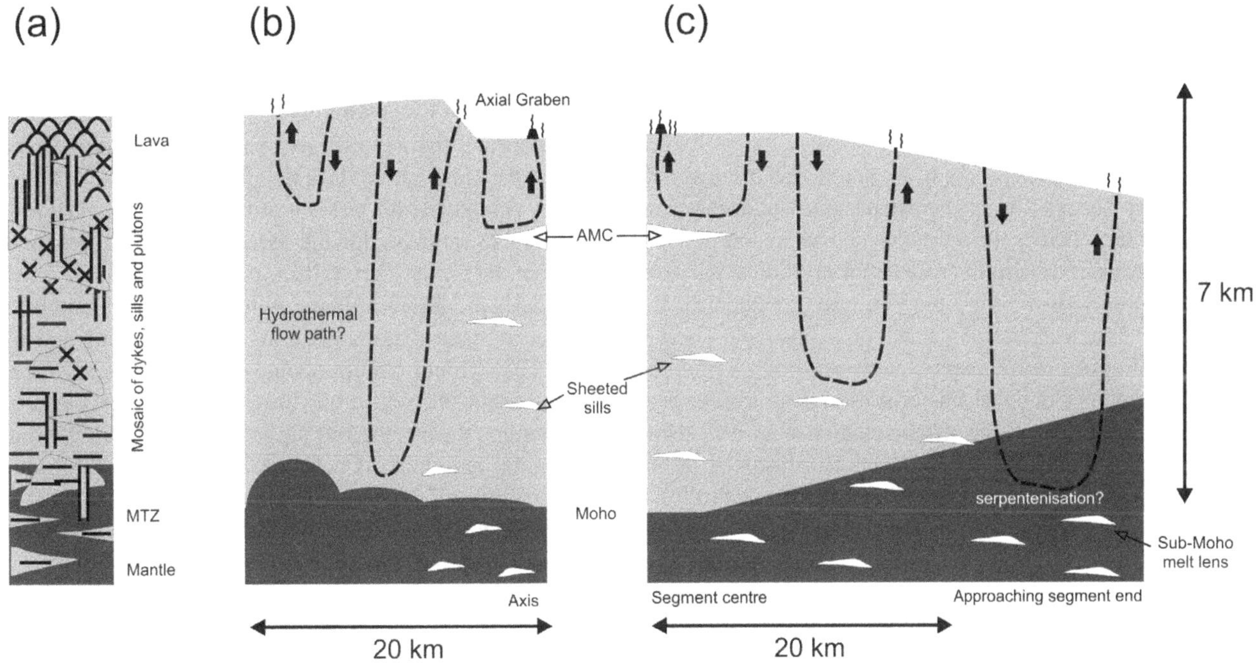

Figure 2. A schematic model of the distribution of heat release during magmatic solidification and hydrothermal cooling at a melt-dominated slow-spreading ridge. See main text for details. (a) Lithological column for slow-spreading ridge. Note the absence of well-developed lithological layering. (b) Across-axis section with approximate scale given at the base. The topography on the Moho reflects variation in the melt supply to the slow-spreading ridge over timescales of 10^5–10^6 years. (c) Along-axis section. The decrease in crustal thickness towards the segment end is likely to be accompanied by deeper hydrothermal penetration and crystallisation of melts in the lithospheric mantle.

mantle melt to seafloor temperatures is summarised in Figure 3. The thermal history can be split up into magmatic cooling and sub-solidus cooling. The magmatic cooling interval stretches from the temperature of unmodified mantle melts, 1350°–1250°C, to that of solidification, 1050°–820°C. This cooling is discussed in Sections 2 and 3. The sub-solidus interval spans from the solidification temperature down to the final temperature of cooling which, in stable oceanic lithosphere, is controlled by the depth of the material and its age. Sub-solidus cooling is the subject of Section 4.

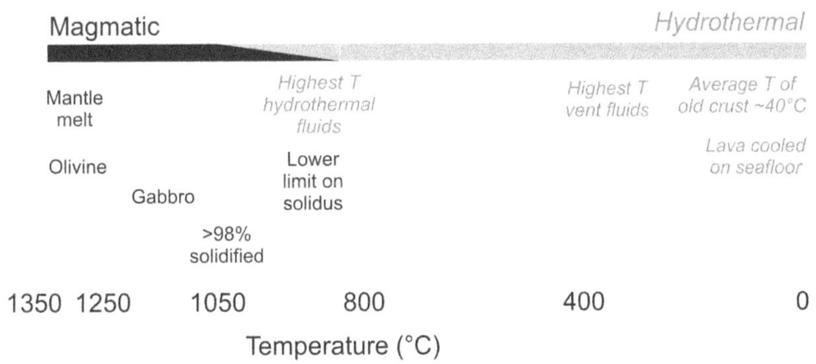

Figure 3. Figure showing major events in the cooling history of a batch of magma. Magmatic events are shown in bold black text, while hydrothermal events are shown in grey italicised text. Details are given in the main text. 'Olivine' and 'gabbro' refer to the temperature intervals where these are the principal solid materials added to the crust during solidification of the mantle melt.

2. HEAT SUPPLY DURING SOLIDIFICATION OF MANTLE MELTS

2.1. Crystallisation Models

Both latent and specific heat are released during cooling of magma. The relationship between latent heat release and cooling is controlled by the crystallisation behaviour of the mantle melt. A brief description of a set of crystallisation models is therefore merited before they are used to calculate the heat supply from a crystallising batch of mantle melt. There is substantial agreement amongst many petrologists upon aspects of the compositional characteristics of mantle melts produced under mid-ocean ridges [*Langmuir et al.*, 1992]. However, the composition of these products is influenced by melting conditions in the mantle, which are in turn controlled by the mantle temperature and the plate spreading rate. Fortunately, as summarised in Section 2.3, the relatively small compositional variations that arise as a result of changes in the melting conditions are not associated with large shifts in the latent or specific heat release during crystallisation.

A set of fractional crystallisation models for a putative mantle melt composition, KG1 [*Kinzler and Grove*, 1992], are shown in Figure 4a and the melt composition is provided in Table 1. The models were produced by different sets of authors and involve different parameterisations of experimental data relevant to the crystallisation of basalt. The pressure of crystallisation was set at 0.1 GPa, suitable for studying crystallisation within the oceanic crust. All three models find that the onset of crystallisation occurs at ~1270°C with olivine as the liquidus phase. After about 7% crystallisation has taken place, plagioclase joins the crystallising assemblage at ~1235°C and clinopyroxene follows at ~1200°C. The plutonic rocks produced by these crystallisation paths would be dunites during the olivine crystallisation interval and troctolites and gabbros between 7 and 90% crystallisation. The first 30% of this interval in the 0.1 GPa crystallisation models is dominated by the formation of troctolites. These rocks are commonly predicted as early products of low-pressure crystallisation of primitive mid-ocean ridge basalt (MORB) compositions because clinopyroxene appears on the liquidus after plagioclase at pressures of less than about 0.8 GPa

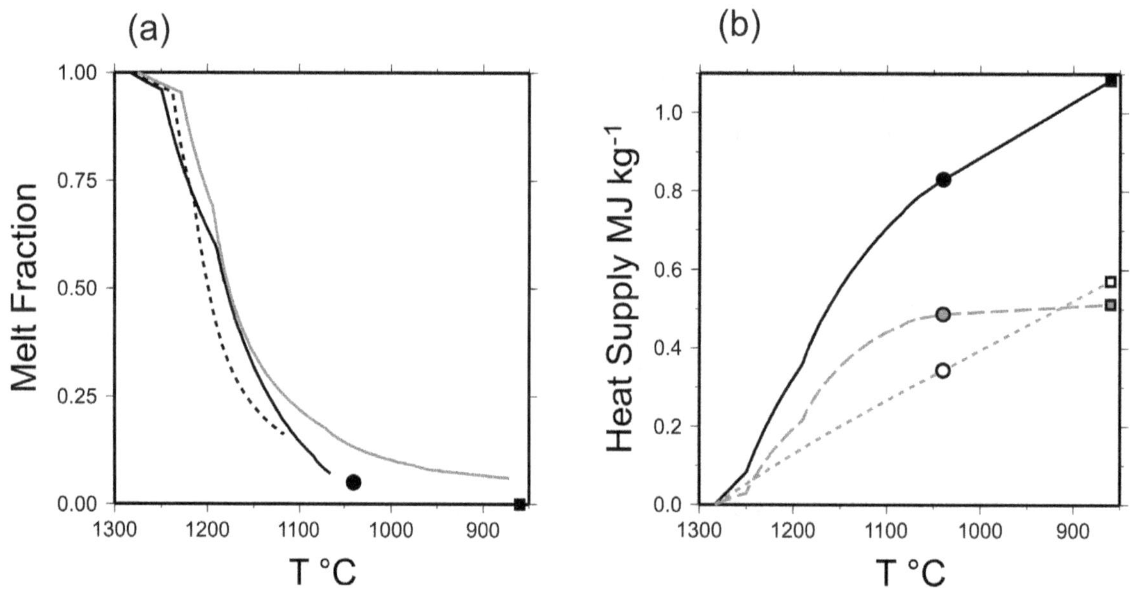

Figure 4. (a) Results of fractional crystallisation models. Note that temperature decreases from left to right, reflecting the progression of heat loss during cooling. The initial composition used as an estimate for a mantle melt at spreading ridges is KG1 of *Kinzler and Grove* [1992]. Black line, COMAGMAT version 3.57 model [*Ariskin et al.*, 1993]; black dashed line, MELTS Adiabat1_ph version 1.7 [*Smith and Asimow*, 2005; *Ghiorso and Sack*, 1995]; grey line, PETROLOG version 2.1.3a [*Danyushevsky*, 2001]. All models run at 0.1 GPa and oxygen fugacity one log unit below the fayalite-magnetite-quartz buffer. Filled circle, experimental constraint, described in main text [*Juster et al.*, 1989]; filled square, observational constraint from amphibole-plagioclase thermometry [*Coogan et al.*, 2001]. (b) Heat release from a batch of magma of composition KG1, calculated using COMAGMAT model in (a) for temperatures above 1060°C and the reference points of the *Juster et al.* [1989] experiment (circles) and the solidus estimate (squares) of *Coogan et al.* [2001]. The heat release is shown with reference to the liquidus temperature, with no super-liquidus cooling included. The black line shows the total heat release, the grey long-dashed line shows the latent heat release and the grey short-dashed line shows the specific heat release.

Table 1. Compositions for Thermodynamic or Crystallisation Modelling.

	MORB[a]	Gabbro[b]	Harzburgite[a]	KG1[c]
SiO_2	52.4	49.9	44.5	49.0
TiO_2	1.6	0.3	0.0	0.88
Al_2O_3	16.9	16.9	1.7	16.8
FeO	10.3	6.0	9.6	7.85
MgO	7.1	10.5	42.6	11.6
CaO	10.1	14.6	1.4	11.9
Na_2O	3.2	1.5	0.1	1.92
K_2O	–	–	–	0.05

Data sources: [a]*Schmidt and Poli* [1998]; [b]*Pallister* [1984]; [c]*Kinzler and Grove* [1992].

[*Grove et al.*, 1992]. Fractional crystallisation at a pressure of 0.2 GPa, which corresponds to the pressure encountered near the base of typical oceanic crust, involves a narrower interval of troctolite crystallisation (10–20%). Gabbro and troctolite crystallisation provide similar quantities of heat for each increment of solidification, so the distinction between them is not important for the overall heat budget at ridges. However, as shall be described in Section 3.3, the distribution of these rock types in the lower crust provides constraints upon the spatial variations in release of heat.

The crystallisation models are in broad agreement at temperatures above ~1100°C, when 15–25% of the original melt mass remains. However, at lower temperatures, the parameterisations run into problems and either fail or produce results which differ widely between models. One reason for this behaviour is the difficulty in acquiring near-solidus experimental data for natural basaltic compositions. One of the lowest temperature experiments on a MORB composition was that of *Juster et al.* [1989]. They held a natural MORB sample at a temperature of ~1040°C and, on inspection of the experimental charge, found that the sample was 13% molten at this temperature. The natural MORB sample that they used in their experiments was not a pristine mantle melt and is estimated to have lost about 50% of its original liquid mass through fractional crystallisation in the crust. Therefore, it is likely that mantle melt compositions are ~95% crystalline at ~1040°C. While experiments in hydrous basaltic systems have been carried out in the temperature range 960–1040°C [*Feig et al.*, 2006], controls on the crystallisation of natural mid-ocean ridge basalts without introduction of extra water are not present. Amphibole-plagioclase thermometry of interstitial phases in oceanic gabbros indicates that the solidus temperature of mid-ocean ridge basalt is 860 ± 30°C [*Coogan et al.*, 2001].

The crystallisation models shown in Figure 3a all involve fractional crystallisation, where crystals are chemically isolated from the melt as soon as they are formed. Evidence that fractional crystallisation plays an important role in the generation of the oceanic crust can be found in the large-scale compositional stratification into cumulate gabbro in the lower crust and trapped liquid compositions in the upper crust (see summary of *Maclennan et al.* [2004]). The upper crust has the composition of a basaltic liquid that is in equilibrium with solids that have a composition similar to that of the average lower crust. In broad terms, this liquid has separated from the solids of the lower crust and risen to form the upper crust. Furthermore, the preservation of correlated trace element variations on length scales of centimetres to metres in the lower layered gabbros of the Oman ophiolite indicates that chemical interaction between melt and crystals after crystal growth was limited [*Korenaga and Kelemen,* 1998]. If, however, equilibrium crystallisation, where crystals are not removed and remain in equilibrium with the melt, is important in any part of the crust, it will narrow the solidification interval whilst preserving the same liquidus temperature. For example, a COMAGMAT [*Ariskin et al.*, 1993] equilibrium crystallisation model that is in all other respects identical to the model shown in Figure 3a reaches 90% crystallisation at 1125°C. Such narrowing of the crystallisation interval is also expected in processes where melts re-equilibrate with cumulates formed from similar magma, such as the assimilation and infiltration proposed by, for example, *Bédard* [1993], *Coogan et al.* [2000] and *Ridley et al.* [2006].

The pressure of crystallisation also shifts the relationship between temperature and the extent of crystallisation. COMAGMAT fractional crystallisation models at 0.5 and 0.9 GPa have liquidus temperatures 15° and 29°C higher, respectively, than the model shown in Figure 4a. Likewise, the 90% crystallisation temperature is increased by 19° and 47°C for the two models. Therefore, increasing pressure appears to shift both the liquidus and solidus temperatures to higher temperatures and has little effect on the width of the crystallisation interval at fixed pressure. If, however, crystallisation is polybaric, the temperature range of crystallisation will be expanded, and more specific heat will be released during solidification.

2.2. *Heat Release on Cooling and Crystallization*

If the relationship between temperature and extent of crystallisation is known, then it is possible to calculate the heat release across the solidification interval. An example of such a calculation for the COMAGMAT model of Figure 4a is shown in Figure 4b. The heat release, ΔH, on cooling from the liquidus temperature, T_l, to a temperature, T, in the solidification interval can be expressed as an integration of the specific and latent heat contributions,

$$\Delta H = \int_T^{T_l} \left[C_p + \frac{dF}{dT} L_h \right] dT \tag{1}$$

where C_p is the specific heat capacity, L_h is the latent heat and F is the melt fraction which is obtained from crystallisation models. In order to account for differences between the C_p of melt and solid phases, the following expression is used.

$$C_p = \begin{cases} FC_p^{liq} + (1-F)C_p^{ol} & \text{if } (1-F) \leq F_{ol} \\ FC_p^{liq} + F_{ol}C_p^{ol} + (1-F-F_{ol})C_p^{gab} & \text{if } (1-F) > F_{ol} \end{cases} \tag{2}$$

where the specific heat capacities of the melt, olivine and gabbro are given as C_p^{liq}, C_p^{ol} and C_p^{gab}, respectively, and F_{ol} is the fraction crystallised as olivine before the arrival of plagioclase in the crystallising assemblage. Differences in the latent heat of crystallisation of olivine and gabbroic rocks can be accounted for using

$$L_h = \begin{cases} L_h^{ol} & \text{if } (1-F) \leq F_{ol} \\ L_h^{gab} & \text{if } (1-F) > F_{ol} \end{cases} \tag{3}$$

Experimental estimates C_p and L_h over the crystallisation interval of basalts have recently been compiled by *Cannat et al.* [2004] and the following values from their compilation are used here: C_p^{liq} of 1600 J kg^{-1} K^{-1} [*Lange and Navrotsky*, 1992; *Kojitani and Akaogi*, 1995]; C_p^{ol} of 1350 J kg^{-1} K^{-1} [*Richet and Fiquet*, 1991]; C_p^{gab} of 1250 J kg^{-1} K^{-1} [*Cannat et al.*, 2004; *Richet and Fiquet*, 1991; *Kojitani and Akaogi*, 1995]; L_h^{ol} of 7.48 × 10^5 J kg^{-1} [*Richet et al.*, 1993] and L_h^{gab} of 5 × 10^5 J kg^{-1} [*Kojitani and Akaogi*, 1995]. In equation (1) it is assumed that the latent heat release is uniform across the crystallisation interval. While experimental evidence indicates that this is a reasonable assumption for tholeiitic basalts similar to those found at mid-ocean ridges, crystallisation models run using the MELTS algorithm feature enhanced latent heat release associated with the arrival of new solid phases on the liquidus [*Lange et al.*, 1994; *Ghiorso*, 1997]. While the model results in Figure 4a are based on fractional crystallisation, the evaluations of equation (1) in Figure 4b include the assumption that melt and solid remain in thermal equilibrium. This assumption is unlikely to be correct because significant differential motions between crystals and the melt are expected. Nevertheless, the heat release models in Figure 4b provide a useful guide for evaluating heat release in different stages of the crystallisation history of melts at mid-ocean ridges and also the total heat release available from each batch of mantle melt as it cools within the crust.

The results shown in Figure 4b show that, as expected, 90% of the latent heat of crystallisation is released between 1250° and 1050°C. Over this interval, the contribution of the latent heat release is ~50% more important than the specific heat. However, outside this interval, and most notably in the final stages of crystallisation at 1050°–860°C, specific heat release dominates the heat output. Over the entire crystallisation history of the melt, from 1280° to 860°C, the contribution of latent and specific heat is almost equal.

2.3. Uncertainties Associated With Magmatic Heat Release Estimates

2.3.1. Errors in C_p and L_h estimates. The results of the simple heat release model proposed above can be compared with the results of the MELTS algorithm which can also be used to predict the heat release across the crystallisation interval of basalts [*Ghiorso and Carmichael*, 1985]. The MELTS algorithm predicts an integrated release across the crystallisation interval of KG1 at 0.1 GPa pressure, which is about 15% lower than that obtained from equation (1). This difference in heat release provides a good guide to the uncertainties in the calculations. The discrepancy is dominated by uncertainty in the value of the latent heat of crystallisation of basalt. Recent experimental studies have provided estimates of the latent heat of basalt crystallisation of 4–7 × 10^5 J kg^{-1} [*Lange et al.*, 1994; *Kojitani and Akaogi*, 1995; *Bouhifd et al.*, 2007]. A latent heat of 5 × 10^5 J kg^{-1} was used to produce the results shown in Figure 4 because this value closely matches the experimental results of *Kojitani and Akaogi* [1995] and is close to the value proposed of *Bouhifd et al.* [2007] for crystallisation at 1257°C. The equivalent latent heat release is expected to decrease at lower temperatures, perhaps dropping by as much as 20% over the crystallisation interval of basalt composition KG1 [*Bouhifd et al.*, 2007]. Further experimental constraints upon the latent heat release during MORB crystallisation are required to improve the accuracy of the heat budget calculations.

The use of a constant specific heat capacity in equation (1) is also likely to be a cause of error. Experimental measurements of the heat capacities of the principal solid phases involved in basalt crystallisation and the paramaterisations used to match them agree to within 1% [*Holland and Powell*, 1998; *Berman*, 1988; *Richet and Fiquet*, 1991]. The predicted heat capacity of the solid phases drops by about

4% across the crystallisation interval of basalt composition KG1. The specific heat capacity of basalt melts of known composition can be predicted with an uncertainty of ~3% [*Lange and Navrotsky*, 1992]. The *Lange and Navrotsky* [1992] model was used to calculate the variation in the specific heat capacity of the liquid during crystallisation of KG1 and showed a drop of 12% from the liquidus temperature to 90% solidification. The use of constant specific heats in equation (1) probably results in an error of ~5%.

2.3.2. The upper temperature limit of heat release and magmatic superheat. The model results shown in Figure 4 only include heat that is released during the crystallisation interval of basalts at mid-ocean ridges. However, if melts are supplied to the ridge axis at temperatures above their liquidus, then specific heat can be released by super-liquidus cooling. Geological, geochemical, petrological, experimental and theoretical reasoning has been used to support arguments for the importance of fractional melting and rapid melt extraction under mid-ocean ridges [*Kelemen et al.*, 1997]. *Asimow* [2002] has shown that if melt is both chemically and thermally isolated from its mantle source before it rises towards the surface, then melt temperatures at the top of the melting region are not only higher than the liquidus of the melt composition but also higher than that of the mantle. For a typical mid-ocean ridge melting geometry, with a mantle potential temperature of 1288°C, the melt is about 30°C above its liquidus temperature and 45°C hotter than the surrounding mantle [*Asimow*, 2002]. Under these extreme conditions of total thermal isolation from the surrounding mantle during upwards melt transport, ~0.05 MJ kg^{-1} of specific heat can be released during super-liquidus cooling. This cooling therefore increases the total heat release on solidification of mantle melt by 5% or less. Any thermal interaction between the rising melt and the surrounding mantle will further reduce the thermal input from super-liquidus cooling. Evidence that such interaction is likely to take place is provided by the presence of dunite channels in ophiolite complexes [*Kelemen et al.*, 1997].

2.3.3. Global variations in mantle temperature. Variations in mantle temperature under mid-ocean ridges result in variations in the composition, temperature and volume of melt generated during mantle melting. The melt composition used for the calculation in Figure 4 is KG1, an estimate of mantle melt composition generated under a mid-ocean ridge of with a typical oceanic crustal thickness (6–7 km) and overall melt production. However, the range of thicknesses of crust generated by solidification of melt at spreading ridges is large, from <2 km on portions of ultra-slow spreading ridges such as Gakkel [*Jokat et al.*, 2003; *Michael et al.*, 2003] or the South West Indian Ridge [*Cannat et al.*, 2006] up to 40 km under central Iceland [*Darbyshire et al.*, 2000]. This full range in crustal thicknesses reflects variation in melt production resulting from changes in mantle temperature and the plate separation rate at the ridge. The large crustal thickness under Iceland is thought to result from a combination of mantle temperatures 100°–200°C higher than standard mid-ocean ridges and plume-driven upwelling [*White et al.*, 1992; *Langmuir et al.*, 1992; *Maclennan et al.*, 2001b; *Putirka*, 2005, 2007; *Bourdon et al.*, 2006; *Herzberg et al.*, 2007].

If an Icelandic basalt sample with a composition similar to that expected for unmodified mantle melts is used as the input composition for the COMAGMAT low-pressure fractional crystallisation models described in Section 2.1, then the liquidus temperature is raised to 1327°C. This temperature is in good agreement with the highest olivine-melt equilibrium temperatures in primitive Icelandic basalts found by *Maclennan et al.* [2001b]. After an interval of 10% crystallisation with olivine alone as the liquidus phase, plagioclase and clinopyroxene join the crystallising assemblage at similar temperatures as they did in the models of crystallisation for composition KG1 and the temperature of 75% crystallisation is close to 1130°C for the two different initial compositions. Therefore, it appears that melts generated in areas with higher degrees of mantle melting crystallise over a wider temperature interval and will therefore release more specific heat during their crystallisation. In the case of Iceland, the total heat release across the solidification interval may be 10% higher than for melt composition KG1 due to this effect. The difference in basalt composition that results from melting of mantle of differing potential temperature has a small influence on the specific and latent heat capacity of the melt: the method of *Lange and Navrotsky* [1992] predicts a specific heat that is 1% higher for primitive Icelandic basalts than KG1. The principal influence of mantle potential temperature on the heat supply is through its control of the volume of basalt that is supplied to the ridge axis.

3. THE LOCATION OF MAGMATIC HEAT RELEASE

The past 10 years have seen a great deal of investigation of the depth of crystallisation at mid-ocean ridge spreading centres. Through a combination of geological and petrological observations from ophiolite complexes and other tectonic windows into the deep oceanic crust, geophysical and petrological observations at active mid-ocean ridges and thermal and mechanical models of crustal accretion, it has been possible to develop and refine conceptual models of the distribution of crystallisation under spreading ridges. This distribution of crystallisation and the associated heat release is closely coupled to hydrothermal systems and further cooling of the oceanic crust. Therefore, information about the distribution of hydrothermal cooling of the oceanic crust can provide constraints on the distribution of crystallisation. However, this part of the paper will focus on petrological, geophysical and geological observations that pertain directly to the magmatic activity at spreading ridges. This review builds on the recent global reviews of crystallisation depths provided by *Michael and Cornell* [1998], *Cannat et al.* [2004] and *Herzberg* [2004]. The constraints on heat release are discussed in detail below for slow-spreading ridges (full spreading rate of <55 mm yr^{-1}) and for fast-spreading ridges (full spreading rate of >80 mm yr^{-1}). Ridges that spread at an intermediate rate, such as the Juan de Fuca Ridge, are not considered in detail herein, but what is known about their structure from geophysical and geochemical studies indicates that the heat release mechanisms are transitional between the fast and slow regimes.

3.1. Crystallisation Depths at Slow-Spreading Ridges

3.1.1. Iceland. Iceland is sometimes viewed with suspicion by the mid-ocean ridge community because it dares to stick its head out of the waters of the North Atlantic and bare its surface to the chilly climes of the maritime sub-arctic. Iceland is sub-aerial because it has unusually thick oceanic crust (20–40 km), which results from large amounts of melting of the mantle. However, the basic processes of plate spreading, adiabatic decompression melting of mantle to produce basalts, their rise towards the surface and cooling to form gabbros, dykes and lavas are common to Iceland and almost every other part of the mid-ocean ridge system. Therefore, the study of magmatic processes under Iceland is the study of processes that occur under mid-ocean ridges. The great attraction of Iceland is that very detailed and well-controlled sampling programmes can be conducted in the sub-aerial environment. For the purposes of this review, the most important difference between Iceland and submerged ridges is the lack of a significant supply of seawater into the hydrothermal system.

The northern part of the active Northern Volcanic Zone (NVZ) of Iceland has now been extensively studied and the samples provide a unique record of the distribution of crystallisation in oceanic crust [*Nicholson et al.*, 1991; *Slater et al.*, 2001; *Maclennan et al.*, 2001b]. The collection of lava samples spanning a wide range of compositions and containing phenocrysts and cognate xenoliths allows a number of different types of barometers to be applied [*Maclennan et al.*, 2003]. The broad agreement in the results of these barometers indicates that crystallisation takes place throughout the axial crust and uppermost mantle under Iceland, up to depths of 30 km.

The observation that the composition of clinopyroxene crystals in equilibrium with basaltic melt is sensitive to the pressure of their crystallisation has prompted the development of a number of barometers. With increasing pressure, clinopyroxene is able to accommodate more Na into its crystal structure and the partitioning of Na between melt and clinopyroxene is the basis of the barometer of *Putirka et al.* [1996]. The standard error of estimation of this barometer is quoted at 0.14 GPa and, for the range of melt compositions found in the NVZ, no significant systematic errors in crystallisation pressure estimates have been found. In particular, the barometer of *Putirka et al.* [1996] does not systematically overestimate the equilibration pressure for the products of experiments with bulk compositions similar to those of Icelandic basalts. The barometric results indicate that clinopyroxene crystallisation starts at pressures of ~1 GPa [*Maclennan et al.*, 2001b], and the lowest crystallisation pressures recovered from the basalts from the NVZ is ~0.4 GPa. Basalts that return the lowest crystallisation pressures are estimated to have lost 75% of their original liquid mass through fractional crystallisation (Figure 5b).

One drawback of the *Putirka et al.* [1996] method is that a clinopyroxene composition is required. Clinopyroxene phenocrysts are rare in MORB and only a handful of compositions are available for the global ridge system. An alternative barometric method relies on the observation that the composition of melt in equilibrium with olivine, plagioclase and augitic clinopyroxene varies significantly with pressure [*Grove et al.*, 1992; *Yang et al.*, 1996; *Herzberg*, 2004; *Villiger et al.*, 2007]. The attraction of this method is that only a liquid composition is required in order to make an estimate of the pressure of crystallisation. For this pressure estimate to be meaningful, however, it is necessary to confirm that the melt compositions are in equilibrium with olivine, plagioclase and augite. In Iceland, many of the samples fall along compositional trends that are controlled by fractional crystallisation of these three phases. Furthermore, Icelandic samples often contain phenocrysts of all three phases, with the phenocryst compositions being close to equilibrium with the melt in which they are carried. When such samples are used to estimate crystallisation pressures under Iceland, the results are in good agreement with those from the clinopyroxene-melt barometry. The parameterisations of *Yang et al.* [1996] and *Villiger et al.* [2007] return maximum crystallisation pressures of ~1 GPa from basalts that have lost little of their original liquid mass through crystallisation, whereas more evolved basalts give crystallisation pressures of ~0.5 GPa [*Maclennan et al.*, 2001b]. In contrast, the method of *Herzberg* [2004] provides maximum crystallisation pressures of ~0.3 GPa for the primitive Icelandic samples. These results are in keeping with Herzberg's suggestion that the

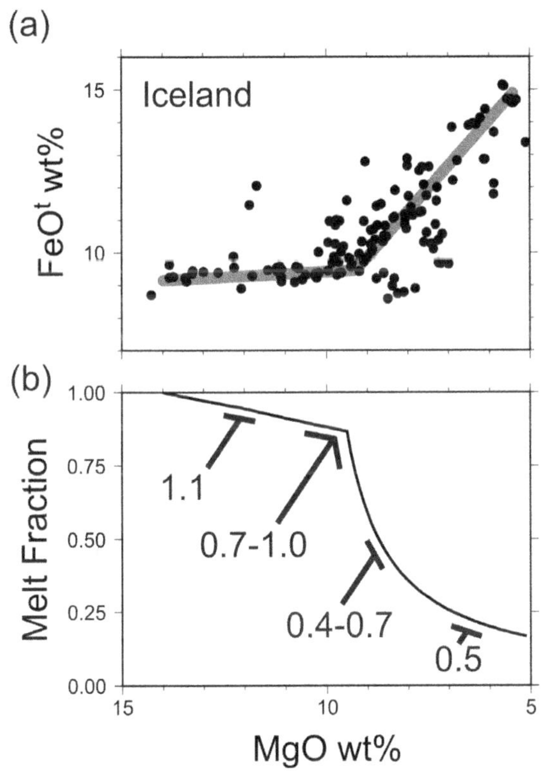

Figure 5. (a) Black filled circles show the composition of whole-rock samples from the Krafla and Theistareykir volcanic systems of Iceland. Following the convention of Figure 4, where high temperatures were shown decreasing from left to right, MgO is plotted decreasing to the right. The MgO content of basaltic melt drops during fractional crystallisation and is therefore correlated with temperature. The thick grey lines show fits to the major element data determined using principal component analysis [*Maclennan et al.*, 2001a]. The flat-lying trend at high MgO is caused by crystallisation of olivine-rich cumulates such as dunites and wehrlites from the melts, while the steeper trend at lower MgO corresponds to removal of a gabbroic rock, very similar in composition to the average cumulate gabbro observed in Oman [*Pallister*, 1984; *Maclennan et al.*, 2001a]. Samples that plot to the left of the grey lines often contain accumulated olivine crystals, and those to the right contain accumulated plagioclase. (b) The black continuous line shows the variation in remaining melt fraction during fractional crystallisation of a mantle melt composition with 14 wt % MgO. The fractional crystallisation generates the cumulate materials described in (a). As expected from the crystallisation modelling in Section 2, about 10% olivine-dominated crystallisation occurs before the onset of gabbro formation. The numbers and their attached lines show the range of pressures estimated for different segments of the crystallisation path, derived from the three barometric methods described in the text.

submerged ridges adjacent to Iceland only host crystallisation in the shallow crust at pressures of under ~0.3 GPa. However, recent experimental data compiled by *Villiger et al.* [2007] indicates that the *Herzberg* [2004] parameterisation systematically underestimates the pressure of crystallisation of experimental liquids produced at 0.8–1.0 GPa. Furthermore, the *Herzberg* [2004] model underestimates the pressure of equilibration of liquids produced in experiments on Icelandic basalt compositions, with an underestimate of ~0.3 GPa for the ~1.0 GPa experiments of *Thy* [1991] and an ~0.2 GPa underestimate for the atmospheric pressure experiments of *Yang et al.* [1996].

The order of appearance of crystal phases during cooling of a basalt is dependent on pressure. For typical MORB compositions, it has long been known that at pressures of <0.8 GPa the order of appearance is olivine, followed by plagioclase, followed by clinopyroxene. At pressures >0.8 GPa , olivine remains the first phase on the liquidus, but clinopyroxene appears before plagioclase. *Maclennan et al.* [2003] used crystal compositional variation within a set of xenoliths in a primitive basalt from the NVZ to demonstrate that clinopyroxene preceded plagioclase in the crystallising assemblage, indicating crystallisation pressures in excess of 0.8 GPa. These pressure estimates also concur with the two other barometric methods for the lava flow that hosts the xenoliths. One source of error on this barometric estimate is caused by the influence of mantle melt composition on the switchover pressure from olivine followed by clinopyroxene to olivine followed by plagioclase. Large extents of melting under Iceland generate melts with Ca/Al ratios higher than those of typical MORB, which tends to favour clinopyroxene crystallisation. Use of the COMAGMAT, MELTS and PETROLOG [*Danyushevsky*, 2001] crystallisation models shows that even for the highest Ca/Al samples, which are all picrites, the switchover pressure is close to >0.6 GPa. Further experimental studies of Icelandic basalt composition are required in order to refine the estimates of crystallisation depth. However, the available experimental data and parameterisations indicate that it is likely that crystallisation commences at pressures close to 1.0 GPa under Iceland.

While the above results indicate that crystallisation occurs at a range of depths throughout the crust and uppermost mantle, these pressure estimates cannot be tied to estimates of heat release unless the extent of crystallisation at depth is known. Figure 5 shows the estimates of crystallisation depth referenced to a crystallisation model for the NVZ. Samples that crystallise at 1 GPa have lost very little of their liquid mass. The crustal thickness in this part of the NVZ is 20–25 km [*Darbyshire et al.*, 2000], so pressures at the Moho will be approximately 0.7–0.8 GPa. It is of interest to note that the first 10% of crystallisation generates olivine-rich cumulates at sub-Moho depths. The physical properties of such cumulates are similar to those of the harzburgite residue of mantle melting, so the seismic Moho is unlikely to be the lower limit of crystallisation and heat release under Iceland. Extensive crystallisation occurs between the Moho and mid-crustal depths, with some basalts crystallising at 12–15 km having lost 60–80% of their original liquid mass through crystallisation. It is clear that large quantities of latent heat are being released in the lower crust under Iceland. By applying simple mass balance methods, such as those described by *Cannat et al.* [2004], to the range of basalt compositions found in northern Iceland, it is possible to estimate that for each unit mass of melt supplied from the mantle, 13% is left as ultramafic cumulates, 40% as gabbroic cumulates and the remaining 47% solidify into rock with compositions similar to those of basaltic melts. The mass balance is based on the solid cumulate compositions determined for northern Iceland by *Maclennan et al.* [2001a] and are therefore consistent with the observed compositional range of Icelandic basalts. A simple synthesis of such a mass balance with the barometric results is that the fractional crystallisation of basalts to form the cumulate gabbros takes place in the lower crust, and that the resulting evolved melt compositions are fed upwards to freeze in lava flows, dykes or magma chambers in the upper crust. Significant differential vertical motions of solidified cumulate gabbroic material are not required in order to construct the Icelandic crust.

3.1.2. Submerged slow-spreading ridges: petrological constraints. Unfortunately, the detailed mass balance and barometric analysis presented in the previous section is not possible for submerged ridges due to the challenging conditions of sampling. However, there is copious evidence that crystallisation at sub-Moho depths occurs under slow-spreading ridges [*Michael and Cornell*, 1998; *Cannat et al.*, 2004; *Herzberg*, 2004]. Studies based on both clinopyroxene-melt equilibrium [*Cannat et al.*, 2004; *Hellevang and Pedersen*, 2005] and melt composition alone [*Michael and Cornell*, 1998; *Cannat et al.*, 2004; *Herzberg et al.*, 2004; *le Roux et al.*, 2002] find crystallisation pressures ranging from 0.1 to almost 1 GPa. It is important to note that it is often hard to verify the co-saturation of a melt with olivine, plagioclase and augite in samples suites from submerged ridges due to poorly defined evolutionary trends in the melt compositions and the lack of phenocryst compositions available from the samples. The inadequacy of the simple one-dimensional (1-D) models that have been considered so far herein is highlighted by a clear difference in the depth range of crystallisation between fracture zones and segment centres. *Herzberg* [2004] found maximum crystallisation pressures of close to 1 GPa near fracture zones but only 0.5 GPa in the centres of segments. Possible causes of this variation in the depth of crystallisation and its importance for heat release at segmented slow-spreading ridges will be discussed in a later section.

Evidence for high-pressure crystallisation in the composition of basalts from slow-spreading ridges is supported by the record from intrusive rocks that have been sampled at the seafloor on from drilling campaigns. *Cannat et al.* [2004] summarised the evidence for crystallisation of gabbroic material within abyssal peridotites, and used the width of the reaction zone around gabbroic bodies as indirect evidence for high pressure crystallisation. Thermochronological studies of gabbros from the Atlantis Bank, which formed at the slow-spreading South West Indian Ridge, have shown that the magnetic age and zircon U-Pb ages differ by up to 2.5 Ma [*Schwartz et al.*, 2005]. Under the assumption that mantle upwelling rates are similar to plate spreading rates, these age differences can be converted into crystallisation depths of up to 15km beneath the base of the crust. Further studies of the Atlantis bank by *Coogan et al.* [2001] and *Dick et al.* [2000] have found that the composition of the gabbroic lower crust has too low a Mg/Fe ratio to have been constructed from unmodified mantle melts. As noted by *Herzberg* [2004], this low Mg/Fe in a lower crustal section may result if significant crystallisation had occurred beneath the Moho.

3.1.3. Submerged slow-spreading ridges: geophysical and modelling constraints. Crystallisation is likely to occur in association with magma bodies in the crust, and geophysical observations of the positions of magma chambers can therefore be used to better understand the distribution of crystallisation in the crust. Magma chambers at slow-spreading ridges are much more restricted in time and space than those at fast-spreading ridges and have accordingly been difficult to image. However, *Singh et al.* [2006a] have recently used reflection seismic data to image the roof of an axial magma chamber (AMC) at a depth of 3km beneath the seafloor at the Lucky Strike area of the Mid-Atlantic Ridge. These results are consistent with the findings of wide-angle seismic surveys that found evidence for the presence of low seismic velocity anomalies in the central parts of slow-spreading segments [*Magde et al.*, 2000; *Dunn et al.*, 2005]. These along-axis variations in geophysical properties are mirrored in bathymetric and lithologic variations in slow-spreading segments [*Cannat et al.*, 1992]. This 3-D accretion geometry, along with the possibility of important temporal variation in melt supply, has hampered the development of adequate thermal models of slow-spreading ridges. Nevertheless, a simple 2-D thermal model by *Sleep and Barth* [1997] showed that limited sub-Moho crystallisation is expected at slow-spreading ridges. This crystallisation leads to the development of a thick Moho-transition zone, which appears to be consistent with the results of seismic surveys at slow-spreading ridges. However, the details of the distribution of crystallisation cannot be well constrained by these models because they do not include 3-D cooling and, as is explained in detail below for fast-spreading ridges, the pattern of hydrothermal cooling is poorly constrained.

3.2. Crystallisation Depths at Fast-Spreading Ridges

3.2.1. MORB compositions. Direct petrological evidence from the composition of MORB for deep crystallisation under fast-spreading ridges is scarce. Global studies of crystallisation depths have found that the overwhelming majority of MORB glasses from fast-spreading ridges record low-pressure crystallisation [*Grove et al.*, 1992; *Michael and Cornell*, 1998; *Herzberg*, 2004]. While 65% of glass compositions from the East Pacific Rise (EPR) reported by *Herzberg* [2004] have crystallisation pressures equivalent to crustal depths (0–0.2 GPa), the errors involved in such barometry are substantial ($2\sigma = 0.28$ GPa) such that more than 95% of EPR glass compositions have pressures of crystallisation within error of crustal pressures. Those few remaining glasses with estimated crystallisation pressures of >0.5 GPa come from close to fracture zones where the magmatic plumbing system is clearly different to that of the remainder of the segment. In these cases, *Herzberg* [2004] has noted that it is not straightforward to demonstrate that the melt compositions are in equilibrium with olivine, plagioclase and augite, because the composition of these melts may be generated either by mixing of melts on a low-pressure differentiation path or by high-pressure crystallisation. Therefore, the constraints on the distribution of crystallisation at fast-spreading ridges from MORB compositions are weak: crystallisation occurs within the crust. Fortunately, by turning to different types of observations from active ridges and ophiolite complexes, it is possible to better understand the distribution of crystallisation within the crust and mantle.

3.2.2. Geophysical observations. A number of geophysical techniques have been used to map the distribution of melt within the crust at fast-spreading ridges. It is likely that regions where melt accumulates and can be imaged are also important sites of cooling and crystallisation and heat release. Bright seismic reflectors have been observed at 1–2 km depth beneath the seafloor at the East Pacific Rise [e.g., *Detrick et al.*, 1987], with an across-axis width of 1–2 km and impressive continuity along individual segments. Such reflectors are thought to correspond to the roof of a shallow AMC, and further investigation has found that this magma chamber has a vertical thickness of 30–100 m [*Singh et al.*, 1998], and often consists of a lens of almost pure melt sandwiched between a roof and floor that are close to solid. However, over distances of kilometres, the layer of pure melt grades into melt-solid mush. The association of melt-rich regions of the lens with hydrothermal activity [*Singh et al.*, 1998] highlights the close link between the release of heat by cooling melt and the driving of hydrothermal systems. The shallow AMC reflector is the most striking geophysical manifestation of the presence of melt within the crust and its discovery has played an important part in the development of models of the development of the crust. In extreme models, solidification only takes place in lava flows, in dykes above the AMC, and in the AMC itself. In such cases, known as gabbro glacier models (Figures 6a and 6b), the full thickness of the crust is attained by extensive downwards flow of solidified material from the AMC towards the Moho.

The region between the AMC and the Moho is predominantly solid, with porosity of 1–20% [*Dunn and Toomey*, 2000; *Crawford and Webb*, 2002]. This mushy region has an across-axis extent of 5–7 km. It is not yet clear whether the melt in this region is distributed evenly in the intergranular pore-space of a uniform crystal mush or if it is contained in sills of similar form to the shallow AMC. The lack of continuous lower crustal reflectors in recent high-resolution studies indicates that if the melt is held in magma chambers, then these bodies must be thinner than 50 m or less than 500 m in horizontal dimension [*Singh et al.*, 2006b].

Evidence for melt ponding and crystallisation at the Moho and in a thin crust-mantle transition zone (MTZ) has been provided by seismic refraction [*Garmany*, 1989], seismic reflection [*Nedimovic et al.*, 2005] and seafloor compliance [*Crawford and Webb*, 2002] studies of intermediate- and fast-spreading ridges. Of these studies, only *Crawford and Webb* [2002] identified a melt body beneath the AMC. The other studies found evidence of melt or frozen melt that may indicate the presence of near-Moho melt bodies up to 10 km from the ridge axis. There is no geophysical evidence for magma chambers forming >1 km deeper than the Moho at fast-spreading ridges.

3.2.3. Ophiolite studies. Slivers of oceanic crust and mantle that have been exposed by tectonic processes have provided important constraints on the lithological structure of oceanic crust formed at fast-spreading ridges. The Oman ophiolite is one of the most intensely studied due to the preservation of magmatic features and the lack of dismemberment of the complex by faulting. Extraordinary exposure of young fast-spreading oceanic crust in rifting environments (e.g., Hess Deep; *Francheteau et al.* [1990]; *Karson et al.* [2002]) or by ocean drilling [*Wilson et al.*, 2006] have verified the relative lithological stratigraphy based on ophiolites and reinforces the utility of ophiolite sections as an analogue for oceanic crust.

The upper parts of the ophiolite igneous stratigraphy, the lava flows, dykes, and uppermost plutonic section are likely to have solidified in situ. After solidification of these bodies, they are buried by subsequent lava flows, with a maximum burial of several hundred metres [*Karson et al.*, 2002]. The effect of the lateral spreading of lava flows is that ~25% of the latent and specific heat release from the upper oceanic crust takes place as lava flows cool on the seafloor, with the remaining ~75% occurring evenly throughout the rest of the upper crust. The heat release associated with the rocks now found in the upper third of the crust occurred within the upper third of the crust. Some heat release also occurs close to the Moho; *Boudier et al.* [1996] reported the existence of gabbroic sills encased in ultramafic rocks in the Oman MTZ, which are likely to have crystallised close to the base of the crust.

An important and unresolved question in the understanding of crustal accretion at fast-spreading ridges is the distribution of crystallisation in the lower axial crust, that is to say, between the AMC and the Moho. In the gabbro glacier model (Figures 6a and 6b), the solids that eventually form the lower oceanic crust are generated by crystallisation in the AMC. Focussed heat release at AMC depths is therefore associated with this mode of accretion, which will be available to drive vigorous hydrothermal circulation in the crust above the AMC. The other extreme model can be referred to as the sheeted sills model (Figures 6c and 6d), where the lower crust is built by crystallisation in a series of stacked sills at lower crustal depths. In this case, the release of heat of solidification will be fairly even with depth, and little differential vertical motion of the solids in the lower crust is required. Authors such as *Kelemen et al.* [1997] have argued for a sheeted sills mode of accretion in the layered gabbros of the lowermost 1/2 to 1/3 of the Oman crustal section. These arguments are based on the similarity of the igneous fabrics and geochemistry of the lower crustal layered gabbros with those of the MTZ gabbros, which are likely to have formed in situ. However, even the models proposed by *Kelemen et al.* [1997] involve a layer of material of several hundred metres in thickness sitting beneath the AMC which contains solid material that has been fed downwards after crystallising in the AMC. This material corresponds to the 100- to 2000-m-thick layer of gabbro found beneath the putative AMC depth in Oman which has the composition of the solid products of fractional crystallisation of basalt and contains a strong vertical fabric based on alignment of plagioclase crystals [*Nicolas et al.*, 1996; *MacLeod and Yaouancq*, 2000]. These vertically layered fabrics are different to the horizontal modal layering found in gabbro nearer the Moho [*Nicolas et al.*, 1996; *MacLeod and Yaouancq*, 2000].

The observations from ophiolites and active spreading ridges therefore indicate that a hybrid model of crustal accretion is suitable for fast-spreading ridges, where some focussing of crystallisation occurs in the AMC, and some crystallisation occurs in stacked sills in the MTZ and lowermost crust (see *Maclennan et al.* [2004] for further discussion). In order to determine the relative importance of these two processes in crustal accretion, it is necessary to collect detailed petrological and geochemical data from vertical traverses of oceanic crust formed at high spreading rates. If the gabbro glacier mode of accretion is dominant, then the crust would not be expected to show systematic variations in the compositions of the cumulates as a function of depth. Care must be taken to isolate the compositional effects of variable extents of trapped interstitial melt in the gabbros from those of the progress of fractional crystallisation. If the stacked sills mode of accretion operates, then mass balance and crystallisation model constraints indicate that a gradual systematic variation in cumulate composition should be observed from the Moho, where the most primitive cumulates will be present, up to the AMC level, where the gabbro should be in equilibrium with a basalt that has lost ~2/3 of its original liquid mass through fractional crystallisation [*Maclennan et al.*, 2004].

The results of crystallisation models indicate that the Mg# (i.e., the molar Mg/Mg + Fe ratio) of the solids generated is a robust indicator of the extent of crystallisation. Some cumulate rocks from the MTZ of Oman have Mg# ~91, in equilibrium with primary mantle melts, and are likely to represent the very first increment of crystallisation of such melts [*Koga et al.*, 2001]. A compilation of data from the gabbroic section of the Oman crust shows that the lower quarter of this plutonic crust has an average Mg# of 81 and is significantly different in composition to the underlying MTZ cumulates [*Garrido et al.*, 2001; *MacLeod and Yaouancq*, 2000; *Coogan*, 2007]. The middle portion of the gabbroic crust has an average Mg# of 77 and while this unit is different in composition to the underlying and overlying gabbros, it shows no systematic internal variation in composition. The uppermost quarter of gabbros is dominated by foliated gabbros and its low average Mg# of 66 is likely to reflect the presence of trapped basaltic melt. The observed drop in Mg# with decreasing depth in the gabbroic section is broadly consistent with the predictions of a stacked-sills mode of accretion. However, two important observations indicate that processes in addition to the stacked sills accretion must be taking place. The first is that while the stacked sills model predicts systematic variation in the Mg# of the gabbros in the middle portion of the crust, none is observed. The second is that the Mg# and dominance of gabbros over troctolites in the cumulates in the lowest quarter of the plutonic crust indicate that they were generated from melts that had lost an average of about 50% of their original mass through fractional crystallisation. The stacked sills models predict that the gabbros in the lowest quarter should be generated from melts that have lost no more than 20% on average of their original liquid mass. The inability of the stacked sills model to match observations of the compositional variation with height in gabbros exposed in the Pito Deep has been recently described by *Perk et al.* [2007]. Further collection of compositional data from ophiolites [e.g., *Lissenberg et al.*, 2004] and oceanic exposures of gabbros [*Perk et al.*, 2007] and their comparison with the petrological predictions of simple accretion models will be required to better determine the relative importance of stacked sills and gabbro glaciers. It is furthermore hoped that extended drilling of IODP hole 1256D will be able to provide a suitable section of young oceanic crust preserved in an ocean basin [*Wilson et al.*, 2006]. As a final note, ophiolite studies provide no evidence for crystallisation at depths more than 1 km beneath the Moho at fast-spreading ridges.

3.3. Summary of Vertical Distribution of Heat Release

The thermal budget calculations from Section 2 can be combined with the distribution of crystallisation estimates in this section to obtain an estimate of the variation in heat release associated with solidification with depth in different tectonic settings. The results of such calculations are shown in Figures 6, 7 and 8.

Each of these figures shows the temperature profile required to generate a solid crust. These temperatures do not necessarily correspond to any geotherm that exists in the crust close to the spreading axis, but are used to calculate the total heat released during crystallisation of magma in the crust and uppermost mantle. The fractional crystallisation model of Figure 4a was used to generate these temperatures. The upper part of the crust shows a crystallisation temperature of 860°C because this material has the same solidus as the fractionating basalt from Figure 4a. It should be reiterated, however, that ~95% of the latent heat has been released during crystallisation at temperatures of >1050°C. The lower part of the crust in all the models has a higher solidus temperature because this material is the solid residue left behind by fractional crystallisation. In the gabbro glacier model shown in Figure 6a, the lower crust has a uniform solidification temperature of >1152°C, which corresponds to the temperature of 67% crystallisation of the melt in Figure 4a. This figure was chosen because the crust shown in Figure 6 has an upper third that forms by solidification in dykes and lava flows, and a lower two-thirds that is generated by fractional crystallisation. In the gabbro glacier model all of this crystallisation takes place in the AMC, and the solid gabbro fed downwards to generate the crust is the accumulated range of solid fractional crystallisation products over the interval 1280–1152°C. In the stacked sills models of Figure 6, and the models of Figure 7 and 8, fractional crystallisation to produce the lower crust occurs within the lower crust, so the temperature of solidification drops gradually as depth decreases.

By following the mass of melt supplied to a given depth in the crust, and the temperature change and extent of crystallisation required to generate solid material at that depth, it is possible to estimate the heat released by a batch of mantle melt as it solidifies to form the crust. The total heat release from each model is slightly lower than that shown in Figure 4b because the models in Figures 6–8 contain solid material at range of temperatures, while the calculations in Figure 4b were made under the assumption of thermal equilibration of all of the melt and solid. The integrated fraction of the total heat release associated with solidification of the primary melt in the crust and uppermost mantle is shown as F_h in Figures 6 and 7. This quantity increases from 0 at the level of the deepest crystallisation of primary mantle melts to 1 at the surface.

As expected, large variations in F_h are directly related to variation in the distribution of crystallisation. In the gabbro glacier model of Figures 6a and 6b there is large peak in heat release at the AMC depth caused by the focussed crystallisation of the lower crust at this level. It is important to recall that in such models, the AMC sits at the boundary between the upper and lower crust, and that the solid products of crystallisation flow downwards to construct the full thickness of the oceanic crust. The peak in heat supply at the base of the upper crust may drive hydrothermal circulation above the AMC. In the models displayed in Figures 6a and 6b, no crystallisation occurs in the lower crust. For simplicity, in the upper crust of dykes and lavas it has been assumed that solidification takes place uniformly with depth. However, if off-axis thickening of the lava pile occurs [*Hooft et al.*, 1996; *Fornari et al.*, 2004] then heat release will be focussed at the seafloor and this focussing is responsible for the step of F_h from ~0.83 to 1 at the seafloor in Figures 6–8. This 17% step in the heat release is caused by the solidification of 17% of the crust on the seafloor as lava flows. The bulk of the heat release from lava flows will heat seawater in the water column above the flow, and will therefore not be available to hydrothermal systems in the crust.

The stacked sills model of Figures 6c and 6d shows even release of heat of crystallisation with depth in the crust, which follows directly from the uniform distribution of crystallisation assumed in the model. The slight increase in heat release close to the Moho corresponds to the depth range where olivine-rich cumulates are formed. The temperature decrease required for each unit of crystallisation is greater

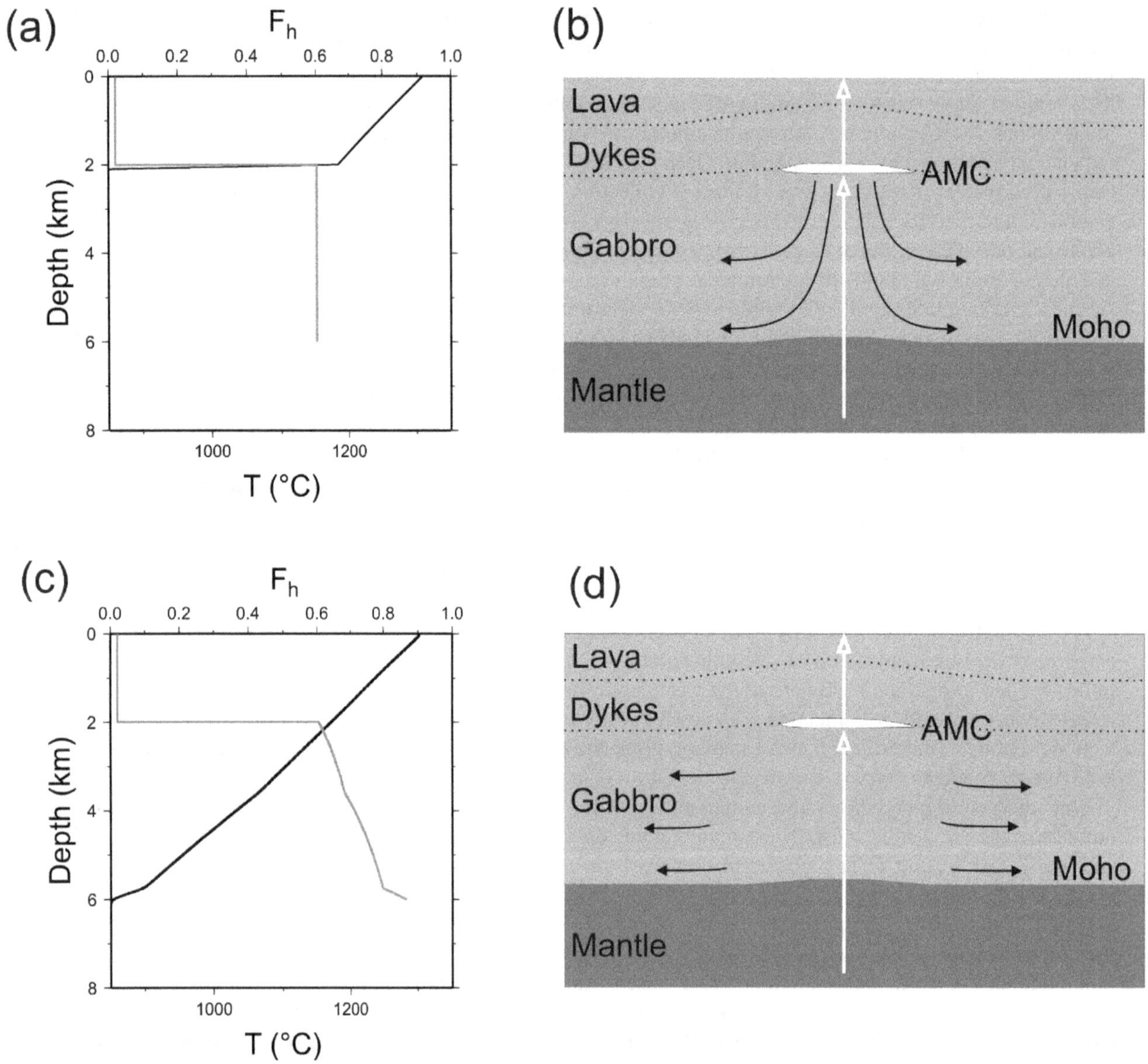

Figure 6. Heat release and crustal accretion models for fast-spreading ridges. See text for detailed discussion. (a) Gabbro glacier accretion model. Black line: F_h, integrated fraction of heat release associated with solidification of primary mantle melt, increasing from 0 at the deepest level of solidification to 1 at the surface. F_h steps from ~0.83 to 1 on the seafloor due to the focussed release of heat associated with lava flows. Grey line: temperature of solidification. (b) Gabbro glacier model. White lines with white arrow heads show melt movement. The white lens marked AMC is the axial magma chamber. Black solid arrows show downwards flow of solidified gabbro from AMC. Dotted lines mark lithological boundaries between lavas, dykes and plutonic rocks. (c) Same as (a) for sheeted-sills model. (d) Same as (b) for sheeted-sills model. The dark band at the Moho shows the layer of olivine-rich cumulates that form at the base of the crust in such models. The seismic Moho lies above this layer. The layer marked gabbro should be subdivided into a lower portion of troctolite and upper portion of gabbro if the increments of crystallisation in each sill are small, as described in the text.

during olivine crystallisation than gabbro crystallisation and the spike in heat release is therefore a result of enhanced specific heat release at near-Moho depths. The increase in heat release required for solidification during crystallisation of olivine-rich material may be partly responsible for the presence of the near-Moho melt bodies that have been imaged by seafloor compliance methods at the East Pacific Rise [*Crawford and Webb*, 2002].

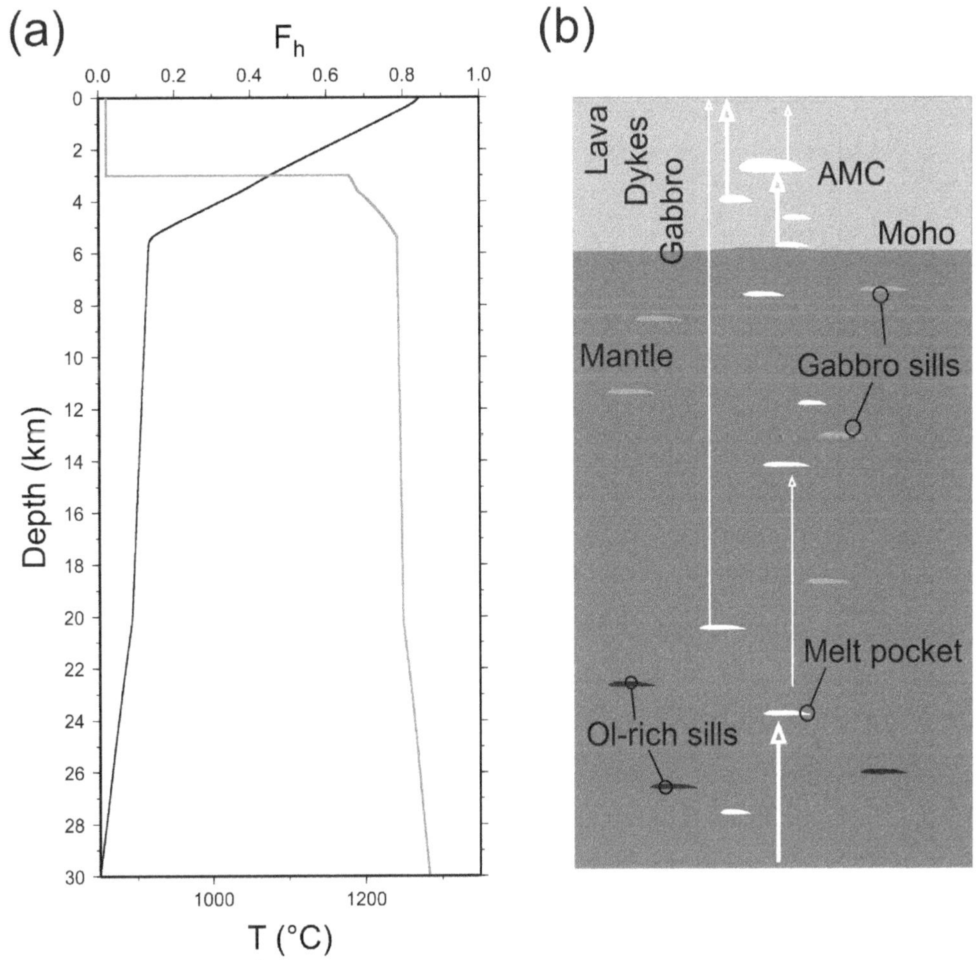

Figure 7. Heat release and crustal accretion model for magmatic portions of slow-spreading ridges. See text for detailed discussion. (a) Same as Figure 6a. (b) Same as Figure 6b. Note the presence of both melt pockets in the mantle (white lenses) and gabbros (grey lenses) and olivine-rock cumulates (dark lenses), the products of crystallisation in such sills.

The schematic crustal accretion models for a magma-rich segment of a slow-spreading ridge and the Krafla volcanic system of Iceland are shown in Figures 7 and 8, respectively. The lithologic structure of slow-spreading ridges does not conform to the well-defined stratigraphic sequence preferred for fast-spreading crust, and the transitions between lavas, dykes and plutonic rocks are interleaved and out of sequence in many instances. The relative importance of magmatic and tectonic processes in the generation of this poorly stratified crust is not yet clear [*Cannat et al.*, 2004]. Little is known about the details of the vertical distribution of crystallisation within the crust, and levels where focussing may occur are not well defined. Furthermore, slow-spreading ridges show marked variations in melt supply in space and time. The heat release plot in Figure 7a must therefore be treated with caution, as this involves the assumptions of 1-D, steady-state crustal construction, with the lower half of the crust built by fractional crystallisation (as in the stacked sills model of Figure 6d) and the upper half of the crust by solidification of melts in quick-cooling plutons, dykes or lava flows. The switchover from stacked-sills accretion behaviour to freezing of melt in the upper crust is chosen to take place at a depth of 3 km in order to match the AMC depth recently determined by *Singh et al.* [2006a] for the Lucky Strike area of the Mid-Atlantic ridge.

The depth distribution of crystallisation under Iceland is presently better constrained than that of other slow-spreading ridges and the mass balance arguments detailed in Section 3.1 give a useful estimate of the large-scale variation in heat release with depth: Half of the heat release associated with crystallisation occurs in the upper half of the crust, and about 40% of the heat release occurs in the lower half of the crust. Further focussing at specific levels in the crust have not been resolved by petrological or geophysical methods,

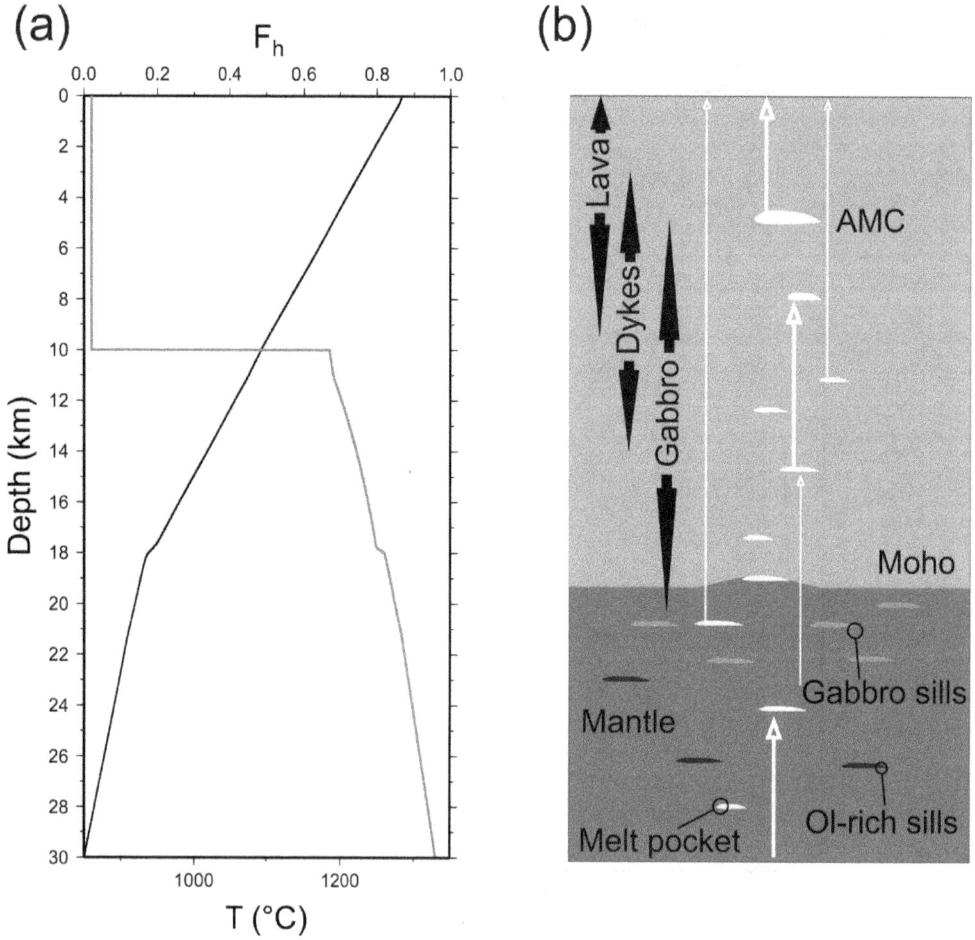

Figure 8. Heat release and crustal accretion model for the Krafla volcanic system of northern Iceland. See text for detailed discussion. (a) Same as Figure 6a. (b) Same as Figure 6b. Dark wedges show possible extent of lava, dykes and gabbro in the Icelandic crust.

so an even supply of heat from solidification is assumed throughout the crust. It should be noted that this distribution of heat release is quite different to that used in the thermal model proposed by *Menke and Sparks* [1995], who sought to match geophysical observations from Iceland with a gabbro glacier style emplacement model. However, *Maclennan et al.* [2001a] showed that a stacked sills model based on their thermobarometric results could fit both the geophysical and the petrological observations from the Icelandic rift zones. In Iceland, ~10% of the total heat release on solidification is associated with sub-Moho crystallisation. The heat release associated with this deep crystallisation is much less focussed than that in the crust, with the gradient in F_h about 5 times lower in the sub-Moho crystallisation zone than in the crust. If the sub-Moho crystallisation at submerged slow-spreading ridges is of a similar extent to Iceland, then the heat release beneath the seismic Moho will be weak. In all four of the heat release models shown in Figures 6–8, heat release by solidification is modest beneath the seismic Moho.

The models of heat release shown in Figures 6–8 are one-dimensional and steady-state. In the following three subsections, the importance of spatial and temporal variations in the heat release is considered.

3.4. Across-Axis Variations

The schematic crustal accretion models shown in Figure 6–8 only show melt supply and crystallisation in crust that lies close to the spreading ridge axis. Three separate sets of observations can be used to constrain the across-axis variation in heat release. The first is simply the distribution of young volcanic rocks at the ridge axis, which reflects the width of the magmatic plumbing system. The second set of observations consists of geophysical images of the melt distribution at ridge axes. The third entails geophysical observations of systematic variations in crustal thickness as a function of distance from the ridge axis. If significant quantities of melt are being supplied to the ridge away from the ridge axis, then off-axis crustal thickening will occur.

At the fast-spreading East Pacific Rise, the volcanic output appears to be dominated by flows which are sourced from fissures that define the ridge axis [*Sims et al.*, 2003; *Soule et al.*, 2005]. However, U-series and geochemical studies have shown that some eruptions take place at distance of >4 km from the ridge axis [*Perfit and Chadwick*, 1998; *Sims et al.*, 2003]. Geophysical studies indicate that the bulk of the melt imaged in the crust lies within a kilometre of the ridge, with a possible low porosity region in the lower crust extending up to 5 km from the axis [*Barth and Mutter*, 1996; *Dunn and Toomey*, 2000; *Crawford and Webb*, 2002]. Restricted accumulations of melt may also be present near the Moho at greater distances from the axis [*Garmany*, 1989; *Crawford and Webb*, 2002]. Geophysical studies have now shown that the Moho is well formed at the ridge axis under the East Pacific Rise [*Singh et al.*, 2006b], with muted variations in crustal thickness (<300 m) along a crustal flow line out to off-axis distances of 20km and more [*Vera*, 1990; *Barth and Mutter*, 1996; *Canales et al.*, 2003]. Taken together, these sets of observations indicate that 90% of the heat supply from solidifying magma occurs within 2 km of the axis, with the remainder being supplied at greater distances.

Evidence from ridges spreading at intermediate rates, such as the Cleft segment of the Juan de Fuca Ridge [*Canales et al.*, 2005; *Stakes et al.*, 2006] or the Valu Fa Ridge [*Turner et al.*, 1999], also indicates that heat release from solidification is focussed into a region within 2 km of the axis. At slower spreading rates, the scarcity of magma chamber images and dating of young lava flows hinders assessment of the across-axis distribution of heat release. Detailed geological maps of Iceland show that young eruptions may emanate from fissures in axial zone with an across-axis width of ~20 km. The crust thickens by ~15% from the centre to the edge of this axial zone [*Staples et al.*, 1997], and the shallow magma chamber has an across-axis width of 8–10 km [*Brandsdóttir et al.*, 1997]. The zone of magmatic heat release appears to be wider in Iceland than in fast- and intermediate-spreading ridges. Images of high-porosity magma chambers at submerged slow-spreading ridges show that their across-axis width is ~4 km [*Sinha et al.*, 1998; *Singh et al.*, 2006] and that low porosities may exist in regions of >10 km in width. The observations indicate that the zone of crustal accretion and magmatic heat release is broader at slow than fast-spreading ridges, but the scarcity of the available data precludes useful quantification of these differences.

3.5. Along-Axis Variations

Variations in the supply of magma along the length of individual segments of ridge have been thoroughly reviewed by *Cannat et al.* [2004]. Fast-spreading ridges show muted variations in crustal thickness and host axial-magma chambers that are relatively continuous along axis. These features indicate that heat release is likely to be uniform along the ridge axis. In contrast, slow-spreading ridges exhibit large variations in crustal thickness, from 8 km or more in segment centres to <3 km towards the ends of segments. Part of the seismic crust at segment ends is likely to be composed of serpentenised peridotite rather than solidified magma. Furthermore, imaged magma chambers at slow-spreading ridges are found only close to segment centres, with along-axis extents of <10 km. Thus, magmatic heat release must vary by a factor of 2 or more along the length of a spreading centre.

It has been noted in an earlier section that basalts from segment ends return greater depths of crystallisation under slow- and possibly fast-spreading ridges. It is likely that this effect is related to enhanced cooling of the axial lithosphere at the transform offset [*Phipps Morgan and Forsyth*, 1988]. One possibility is that deep crystallisation only happens at segment ends because this is the only place where the lithosphere is sufficiently thick. Another option is that deep crystallisation happens everywhere, but that the shallow magma chambers which are more likely to be stable near segment centres are able to filter any magma supplied from deep and overprint the petrological record of deep crystallisation. In this case, the thermal effects of the transform need only influence the upper few kilometres of the system.

3.6. Temporal Variation in Magmatic Heat Supply

The supply rate of melt from the mantle to mid-ocean ridges varies substantially on a range of timescales. It is known, for example, that the development of ocean basins by rifting and reorganisations of plate motions over 10^6–10^7 years must influence the supply of magmatic heat. Variations in oceanic crustal thickness in the north Atlantic have also been used to pick out variation in the magmatic heat supply over timescales of 10^6 years [*Jones et al.*, 2002; *Escartin et al.*, 2001]. These variations have been linked to periodic variations in mantle temperature under ridges that may be related to mantle plume activity. Propagation of individual ridge segments is also likely to influence the spatial distribution of magmatic heat supply over 10^5–10^6 years.

Studies on recent eruptive events have also shown that episodic variation in melt supply and heat release occur over timescales of 10^0 years and less. The 9°50′N area of the EPR has hosted two eruptions in the last 20 years, in 1991 [*Rubin et al*., 1994] and in 2005 [*Tolstoy et al*., 2006]. A number of recent eruptions have also been reported from the Juan de Fuca Ridge, along the CoAxial segment in 1993 [*Embley et al*., 2000], at the Axial volcano in 1998 [*Embley et al*., 1999] and in 1996 at the Gorda Ridge [*Rubin et al*., 1998]. These eruption events are associated with the focussed supply of large amounts of heat to the water column, which produce buoyant upwellings known as event plumes. However, the detection and dating of such submerged eruptions requires intensive surveillance that has only been available for the last few decades. In addition, research vessels must be available in order to provide a rapid response to these events. Therefore, the record of temporal variations over timescales of 10^2–10^5 years is extremely difficult to reconstruct. Two sets of observations can be used to inform our understanding of these temporal variations: the post-glacial eruptive record from Iceland and the geochemical speedometry of magmatic and cooling processes at ridges.

The historical record of volcanic activity in Iceland stretches back over 1000 years and has been recently summarised by *Thordarson and Larsen* [2007]. In segments such as Krafla in the north of Iceland, the historical record contains two major eruptive episodes: Mývatnseldar (1724–1729) and Kröflueldar (1975–1984). The Kröflueldar episode involved a series of nine eruptions over a 9-year period. The pre-historical record of activity can be reconstructed using tephrochronology and detailed studies such as those of *Sæmundsson* [1991] and *Sinton et al*. [2005] provide an excellent record of eruptive activity over the last 12 ka. Once more, these records show temporal variation on a number of scales. Segments may remain volcanically inactive for thousands of years before undergoing episodes of rejuvenation. For example, the period from 7 to 3 ka saw only one small eruption in the Krafla system, while the last 3 ka have seen five eruptive episodes. Therefore, the Icelandic record provides evidence for variation in the volcanic output over timescales of 10^0–10^3 years. Episodicity is likely to be present in all ridge volcanic systems. However, it is known that volcanism in Iceland is modulated by glacial activity [*Maclennan et al*., 2002] and the Icelandic crust is unusually thick, so the Icelandic volcanic record can only be used as a rough guide to the expected behaviour on other parts of the ridge system.

A number of geochemical techniques have been developed over the last 20 years in order to examine timescales of magmatic processes at mid-ocean ridges. In particular, U-series disequilibrium measurements on young basalts and measurement of diffusion profiles in MORB phenocrysts have been useful in defining timescales of processes. U-series evidence and observations of the relationship between deglaciation and volcanism on Iceland now indicate that melt is supplied from its mantle source regions to the surface in less than 1 ka [*Kelemen et al*., 1997; *Maclennan et al*., 2002]. Recently, *Rubin et al*. [2005] have presented ^{210}Pb-^{226}Ra disequilibrium data in young MORB which indicates that the timescale between generation of disequilibrium and eruption is on the order of decades. If this ^{210}Pb-^{226}Ra disequilibrium arises during mantle melting, then melt must travel upwards through the mantle at rates of several km yr^{-1}. U-series measurements have also been used by *Blake and Rogers* [2004] to estimate the power output of mid-ocean ridge magma chambers. While their method shows promise for estimating the heat supply rate from crustal magma chambers, their models are not able to concurrently match the ^{226}Ra-^{230}Th and ^{230}Th-^{238}U disequilibrium data and appear to be missing a key magmatic process. This process is likely to be magma mixing. A study of diffusion profiles in olivine phenocrysts from the EPR has found that magma residence times in axial magma chambers are on the order of months [*Pan and Batiza*, 2002]. Combination of these timescales with information about the relative rates of mixing and crystallisation from melt inclusion studies may provide further important constraints on the power output of mid-ocean ridge magma chambers [*Maclennan et al*., 2003]. Unfortunately, while the present array of geochemical observations indicates that a range of magmatic process operate rapidly at ridges, we lack the necessary volumetric information in order to estimate variation in the rates of magmatic heat supply over the same timescales.

4. SUB-SOLIDUS HEAT SUPPLY

More than 90% of the heat associated with crystallisation of mantle-derived melt is released by cooling melt at temperatures of greater than 1000°C (Figure 4). However, sub-solidus cooling of lava, dykes and gabbros can provide a similar quantity of heat to that released during crystallisation. This heat is supplied both from specific heat on cooling of rocks and from heat generated during hydration reactions that occur when hot rocks come into contact with hydrothermal fluids. In the case of serpentenisation of ultramafic rocks in the Moho Transition Zone or uppermost mantle, it has been suggested that up to 0.25 MJ kg^{-1} is released on completion of the hydration reaction [*Fyfe and Lonsdale*, 1981; *Cannat et al*., 2004]. This latent heat release is similar to that expected for crystallisation of basalt, and has therefore been proposed as the heat source that drives hydrothermal circulation in peridotite [*Kelley et al*., 2001]. More recently, *Lowell and Rona* [2002] and *Allen and Seyfreid* [2004] have examined the circumstances under which serpentenisation can provide a suitable heat source to drive the hydrothermal circulation associated with high-temperature serpentinite-hosted vent sites such as Lost City on the northern Mid-Atlantic Ridge.

The following section contains a description of some model results designed to constrain the heat release on cooling of rocks from the oceanic crust and uppermost mantle. Such rocks release 1.0–1.2 MJ kg^{-1} of specific heat on cooling from 1000°C to seawater temperatures. The total heat release from hydration reactions is less easy to quantify because this heat release is controlled by the extent of reaction. Nevertheless, in terms of the total supply of heat from mantle-derived material to seawater, it is likely that the average contribution of hydration reactions is small, at less than 0.1 MJ kg^{-1}.

4.1. Thermodynamic Modelling

Table 1 shows three compositions that were chosen as representative of upper ocean crust (MORB), lower oceanic crust (cumulate gabbro) and uppermost oceanic mantle (harzburgite). The mineralogy of these compositions was calculated for a range of pressures and temperatures using VERTEX, a Gibbs free energy gridded minimisation technique which has been described in detail by *Connolly* [2005]. The equilibrium solid mineralogy was calculated both for the dry case and for saturation with a hydrous fluid. The mineral end-member properties were taken from *Holland and Powell* [1998] and the thermodynamic properties of minerals with solid solutions were calculated using the VERTEX implementation of the solution models of *Holland and Powell* [2003] and *Dale et al.* [2005]. These models operate under an assumption of equilibrium which is unlikely to be satisfied in rocks of the oceanic crust. The model results shown below should therefore only be used as a rough guide to the heat release on cooling, particularly the heat that can be released by hydration reactions.

4.2. Heat Release From Specific Heat on Cooling

If no water is introduced to the system, the heat release on cooling of all three rock types is controlled by the specific heat capacity. The black lines in Figures 9, 10 and 11a show that, on cooling from 1000° to 0°C, 1.07 MJ kg^{-1} is released from specific heat by MORB and 1.18 MJ kg^{-1} by cumulate gabbro and harzburgite. The calculations for MORB were carried out at a pressure of 0.1 GPa and those for gabbro and harzburgite at 0.2 GPa. The specific heat capacity of the rocks drops with temperature leading to a convex-up shape in the heat release curves. For example, the specific heat capacity of the gabbro composition drops from 1.16 kJ K^{-1} kg^{-1} at 1000°C to 0.70 kJ K^{-1} kg^{-1} at 0°C. The errors in the esti-

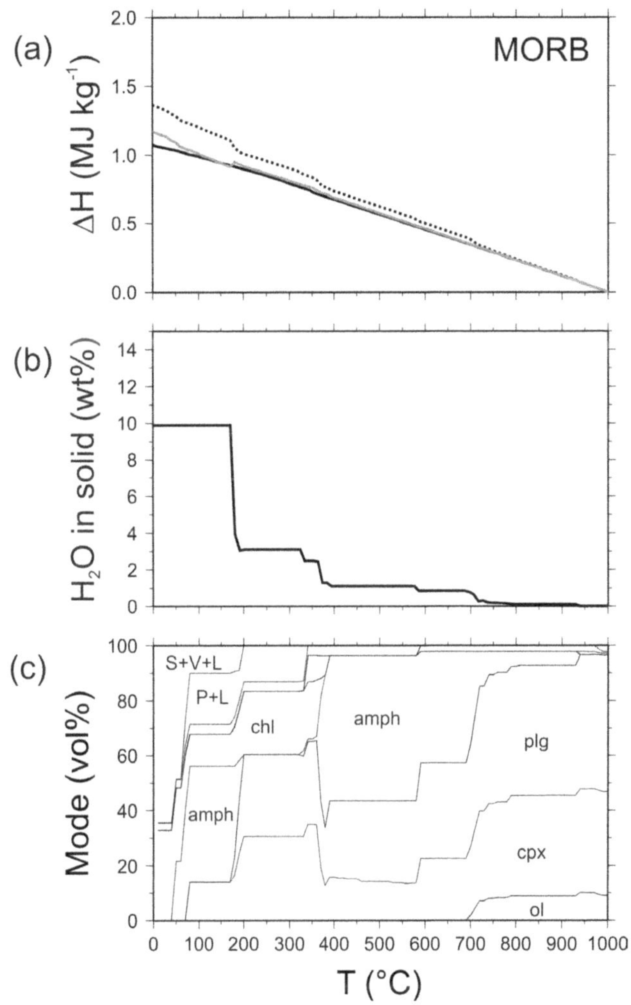

Figure 9. Heat release and reactions on cooling of a typical MORB composition at 0.1 GPa pressure. (a) Heat release on cooling a unit mass of solid from 1000°C. When hydration reactions are included the total mass of solid increases, but the result shown here is corrected to show the heat release for each unit mass of material supplied from the mantle. The solid black line shows heat release in the absence of hydration reactions. The upper dotted black line shows the heat release when hydration reactions occur to completion at each temperature and the hydrous fluid is added at the reaction temperature. The grey solid line shows heat release when hydration reactions go to completion but the water for the reaction is added at 2°C. See discussion in the main text. (b) Weight percentage of H_2O present in the solid when hydration reactions reach completion. (c) Modal mineralogy in volume percent of the rocks produced by hydration reactions during cooling: ol, olivine; cpx, clinopyroxene; plg, plagioclase feldspar; amph, amphibole; chl, chlorite; P + L, prehnite and laumontite; S + V + L, stilbite, vesuvianite and lawsonite.

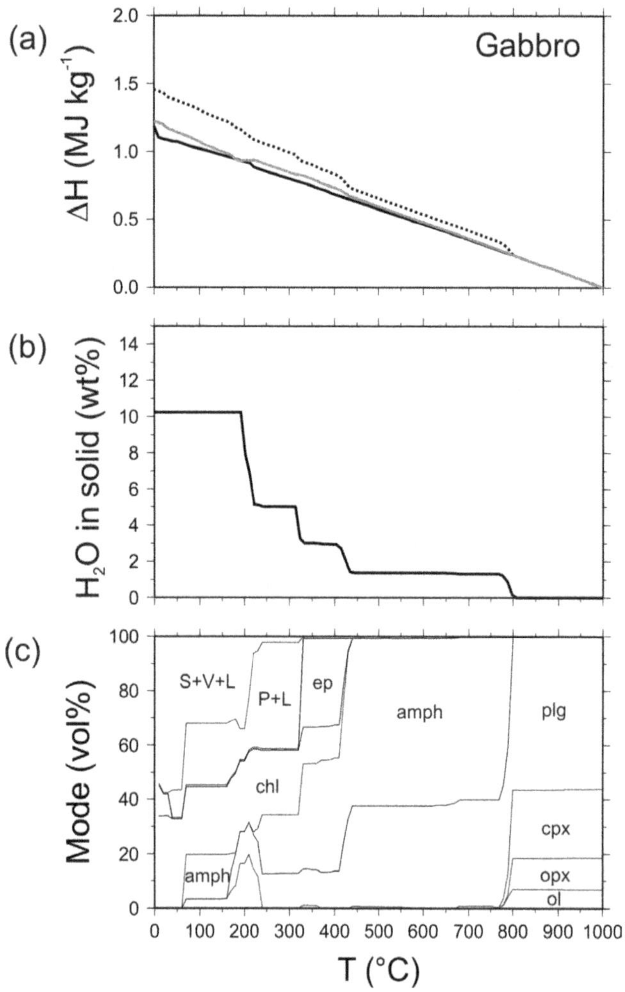

Figure 10. Heat release and reactions on cooling of a average Oman lower crustal cumulate gabbro composition at 0.2 GPa pressure. Same legend as for Figure 8. Additional mineral abbreviations: opx, orthopyroxene; ep, epidote.

mates of heat capacities for rocks of known composition is likely to be <1%. However, uncertainty in the bulk composition of the cooling crust introduces an uncertainty of ~10% into the heat release estimates.

4.3. Heat Release From Hydration Reactions

If the rocks are saturated with respect to a hydrous fluid throughout their cooling, then hydration reactions can occur. The incorporation of water into the solids is shown in Figures 9, 10 and 11b. No hydrous minerals are stable in the chosen compositions at temperatures exceeding 940°C. The highest temperature hydrous mineral that is predicted from the models is a pargasitic amphibole, which is in agreement with observations of near-solidus phase relationships in experiments involving hydrated MORB compositions [*Feig et al.*, 2006]. Silicate melts are not included as phases in the thermodynamic modelling presented herein. In the MORB and gabbro compositions the modal mineralogy is dominated by amphibole and plagioclase between 400° and 800°C and the bulk solid contains ~1.5 wt % H_2O. If pervasive equilibrium could be reached in rocks of the oceanic crust, the growth of amphibole below 800°C would correspond to a retrograde transition from the granulite to the amphibolite metabasite facies. The heat release curves indicate that this hydration reaction is exothermic because the dotted black lines step away from the solid black lines in Figures 8–10a over the down-temperate interval where the reaction occurs. The significance of the dotted lines and grey lines is discussed in more detail below. On cooling from 800° to 400°C the

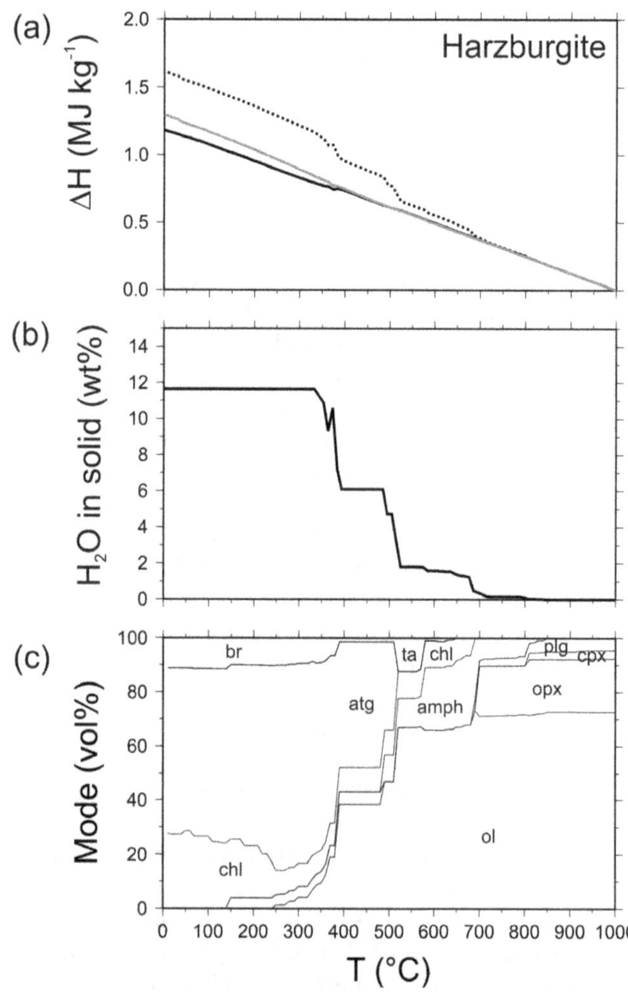

Figure 11. Heat release and reactions on cooling of harzburgite 0.2 GPa pressure. Same legend as for Figure 8. Additional mineral abbreviations: ta, talc; atg, antigorite; br, brucite.

composition of the amphibole shifts from pargasite towards hornblende sensu stricto.

The next hydrous phase to stabilise on cooling is chlorite, which appears at about 650°C in the harzburgite composition and at about 400°C along with epidote for MORB and gabbro. This transition corresponds to the amphibolite-greenschist boundary for metabasites, and 2–3 wt % H_2O is held in the solid. Once again, this hydration reaction is exothermic. With a continuing decrease in temperature, the predicted MORB and gabbro mineralogy shifts through the hydration reactions associated with the transition from the greenschist to prehnite and zeolite metabasite facies. All of these reactions are mildly exothermic. In Figures 9a and 10a, these reactions are associated with steps in the dotted line of approximately 0.1 MJ kg^{-1}. However, the approach to equilibrium is likely to be extremely limited for such low-temperature reactions.

A variety of serpentine called antigorite is predicted to be stable at temperatures of 500°C and below for the harzburgite bulk composition. Formation of antigorite occurs via two reactions on cooling, with the reaction at 500°C releasing 0.2 MJ kg^{-1}, and another at 400°C releasing 0.1 MJ kg^{-1}.

4.4. Heat Balance With Seawater

The heat release associated with cooling under water-saturated conditions is shown in Figures 9–11a. This heat release is calculated with respect to each unit mass of rock at the beginning of sub-solidus cooling at approximately 1000°C. The final mass of each batch of rock will increase during cooling due to the incorporation of water into the hydrated solids. The heat release during hydration is often estimated from individual reactions such as those shown below for serpentenisation [*Macdonald and Fyfe*, 1985].

$$\text{olivine} + \text{pyroxene} + \text{water} \rightarrow \text{serpentine} \quad (4)$$

$$\text{olivine} + \text{water} \rightarrow \text{serpentine} + \text{brucite} \quad (5)$$

Both of these reactions release about 0.25 MJ kg^{-1}, and these values have been widely used as estimates of the latent heat release upon serpentenisation [e.g., *Lowell and Rona*, 2002]. These reactions roughly correspond to those that occur at 500° and 400°C, respectively, in Figure 11. It is important to note that the results of the modelling presented in Figures 9–11 indicate that the heat release from hydration may occur in steps over a range of temperatures rather than at a single temperature from a single reaction. The 0.25 MJ kg^{-1} heat release estimated for the reactions above by *Macdonald and Fyfe* [1985] is for the case when the water is supplied as a reactant at the reaction temperature. The heat release from the hydration reactions where water is supplied at the reaction temperature is shown as a dotted black line in Figures 9–11a. This water must have been taken to the reaction temperature by heat derived from elsewhere in the crust or uppermost mantle. Therefore, the heat release on hydration shown by the dotted black lines corresponds to a maximum that can only be achieved locally. If the global role of hydration reactions in the supply of heat to hydrothermal systems is to be considered, it is necessary to examine the heat balance between cold seawater and hot rock. When the energy required to heat the cold seawater to the reaction temperature is subtracted from the heat release on hydration, the heat release differs little from that expected for cooling without hydration. Such heat release curves are shown as grey solid lines in Figures 9–11a. The results show that while hydration reactions may be an important localised source of heat under the right conditions [*Lowell and Rona*, 2002], the contribution of such heat release over significant spatial or temporal scales is unimportant, with a maximum theoretical heat release of less than 0.1 MJ kg^{-1}. These results are in accord with the study of the Lost City vent field by *Allen and Seyfried* [2004], who combined geochemical and local thermal constraints to infer a limited temperature increase in hydrothermal fluids as a result of serpentenisation.

4.5. Extent of Hydration Reactions

A further factor that limits the importance of hydration reactions in any global heat balance is that the heat release estimates above are based on complete equilibrium reaction. However, the oceanic lithosphere is well known to show heterogeneous and incomplete alteration patterns, often focussed around veins, and containing evidence for disequilbrium between minerals in individual rock samples. For example, *Laverne et al.* [2001] conducted a study of alteration in the upper crustal section of ODP hole 504B and estimated that the extent of alteration in the background rock lies in the range 5–40%, with more altered haloes around veins comprising under 20% of the rock volume. The total extent of alteration is therefore certainly <50%, and likely to be closer to 25%. The extent of alteration of the lower oceanic crust is likely to be even more limited. For example, *Bach et al.* [2001] studied the gabbros sampled in ODP hole 735B and found the alteration to be heterogeneous, with an average of roughly 15% and range of 0–30%. It is important to note that this lack of alteration products does not preclude the removal of heat from the lower oceanic crust, due to the different diffusion rates of heat and chemical signatures in solids. Samples of serpentinite from seafloor dredges and ocean drilling programme cores show much higher extents of alteration than basalts and gabbros. For example, *Bach et al.* [2004] reported ~80% alteration in harzburgite cored during ODP Leg 209 and *Cannat et al.* [1992] found extensive alteration of dredged samples from the Mid-Atlantic Ridge. However, the overall structure of serpentenisation and its relationship to faulting and the thermal structure at slow-spreading ridges remains poorly constrained. The modelling results presented herein and the geochemical work of *Allen and Seyfried* [2004] indicate that heat release by serpentenisation does not play an important role in either the generation of localised high-temperature vent sites or the global heat balance.

4.6. Geochemical Constraints on Rates of Sub-Solidus Heat Supply

A geochemical diffusion-based method for estimating the rate of sub-solidus heat supply from cooling oceanic crust has been developed by *Coogan et al.* [2002, 2005] who use down-temperature exchange of Ca and Li between crystals to estimate cooling rates. Their recovered cooling rates in the upper part of the sheeted dykes (more than 450°C per hour) are sufficient to power hydrothermal megaplume formation, and their estimated cooling rates in lower crustal gabbros have been matched by ridge thermal models that involve extensive cooling of the off-axis lower crust by deep hydrothermal circulation [*Maclennan et al.*, 2005].

5. LOCAL, REGIONAL AND GLOBAL HEAT SUPPLY

The arguments presented above and the results shown in Figures 4, 9 and 10 indicate that if a mantle derived batch of basaltic melt is cooled to seawater temperatures then roughly 2.0 ± 0.2 MJ kg^{-1} is released. About 45% of this heat is released during solidification of the melt between 1300° and 1000°C, followed by 15% of the heat release between 1000° and 800°C, 23% between 800° and 400°C, and 17% from 400°C down to seawater temperatures.

If the magmatic supply rates to a mid-ocean ridge can be estimated, then it is possible to use these figures to estimate the heat available from cooling primitive mantle melts. In Table 2, the melt supply rate to three selected parts of the mid-ocean ridge system is constrained using crustal thickness, T_c, and full spreading rate, u, observations. The cross-sectional flux is the mass of melt supplied per metre of ridge axis and is given by $J_x = T_c u \rho_c Y$, where ρ_c is the estimated average crustal density of 2.8 kg m^{-3} and Y is a constant that accounts for the conversion from years to seconds. The magmatic heat flux per metre of ridge is then calculated from $Q_x = J_x \Delta H_0$, where ΔH_0 is the heat released on cooling to seawater temperatures.

The crustal thickness and spreading rate estimates were taken from *Canales et al.* [2003] for the East Pacific Rise, *Sinha et al.* [1998] for the Mid-Atlantic Ridge and *Staples et al.* [1997] for Iceland. If the crust is cooled to 400°C, then the values in the final column of Table 2 should be multiplied by 0.83.

It is of interest to note that the power plant at the Krafla area of Iceland is producing at a capacity of 60 MW, equivalent to the heat released by cooling the steady-state mantle melt supply to 1 km of ridge to 0°C. Recent estimates of hydrothermal heat flux from a 2-km stretch of the EPR at 9°50′N are 21 ± 10 kW m^{-1} for the high-temperature vents and 163 ± 80 kW m^{-1} total [*Ramondec et al.*, 2006]. These values are similar to those expected if the mantle melt flux is cooled to seawater temperatures at the EPR axis. However, time dependence is likely to be important both at the 9°50′N region of the EPR and Krafla because both of these areas have hosted eruptions within the past 25 years. Such time dependence will only be better understood when longer-term and more accurate hydrothermal heat-flux estimates are available.

A global compilation of ocean crustal thicknesses and spreading rates was recently used by *Sinha and Evans* [2004] to estimate a global magmatic volumetric flux of 18 km^3 yr^{-1}, which corresponds to a mass flux of 1.6 × 10^6 kg s^{-1} and a heat flux of 3.2 ± 0.3 TW. Two key questions follow from this calculation. Firstly, how much of this potential global heat flux is actually supplied to hydrothermal systems close to the ridge axis? Secondly, where is this heat transferred from the basalt to the hydrothermal fluids?

The answers to these questions are intimately linked to the nature of oceanic crustal accretion, and the spatial distribution of crystallisation and hydrothermal cooling within the crust. The oceanic crust is thought to cool until it has been moved far from the ridge axis by plate spreading and reached an age of 80 Ma. At this age, the thickness of the lithospheric plate is close to 100 km and the average temperature of the crust

Table 2. Supply at Mass and Heat at Three Different Ridge Segments.

Area	T_c, km	u, mm yr^{-1}	J_x, kg s^{-1} m^{-1}	Q_x, kW m^{-1}
East Pacific Rise, 9°N	6.7	110	6.7×10^{-2}	134
Mid-Atlantic Ridge, 57°N	6.0	20	1.1×10^{-2}	22
Iceland, Krafla area	20.0	20	3.6×10^{-2}	72

is ~40°C. The maximum heat release of 2 MJ kg^{-1} quoted above is realised on cooling to ~0°C, and cooling crust to an average of ~40°C will release ~1.95 MJ kg^{-1}. The absence of widespread crustal low-velocity zones at distances of more than 10 km from the ridge axis indicates that solidification of magma to form the crust takes place close to the axis. Therefore, the heat associated with solidification of basalt, at least 0.9 MJ kg^{-1}, or 45% of the available heat, must be supplied close to the axis. Further constraints on the thermal structure of the near-axis oceanic crust must come from a variety of geophysical, geological and petrological observations. Thermal models of fast-spreading ridges that are based on the gabbro-glacier style of crustal accretion (Figures 6a and 6b) involve removal of the latent heat of crystallisation at shallow levels [*Henstock et al.*, 1993; *Phipps-Morgan and Chen*, 1993]. These models predict an average crustal temperature of 250°–500°C at a distance of 10 km from the axis which, in the absence of significant conductive heating from the mantle, indicates that 1.5–1.7 MJ kg^{-1}, or 75–85% of the available heat, has been released. A recent thermal model of sheeted-sills crustal accretion [*Maclennan et al.*, 2004, 2005] predicts average crustal temperatures of 100°C at a distance of 10 km, implying that 95% of the available heat has been released.

A number of observations indicate that the axial crust at slow-spreading ridges is cooler than that at fast-spreading ridges. The base of the seismogenic zone at slow-spreading ridge axes varies from 4 km depth at segment centres to 8 km depth at segment ends at slow-spreading ridges [*Barclay et al.*, 2001; *Tilmann et al.*, 2004; *Tryggvason et al.*, 2002], in contrast to the maximum earthquake depth of 1.1 km recorded above the shallow axial melt lens at the fast-spreading East Pacific Rise [*Sohn et al.*, 1999]. The base of the seismogenic layer is thought to correspond to the 600°C isotherm in the oceanic crust [*McKenzie et al.*, 2005]. Further evidence for cooler axial crust at slow-spreading ridges comes from the increasing magma chamber depth [*Purdy et al.*, 1992], the development of axial valleys [*Phipps-Morgan and Chen*, 1993], petrological observations of high-pressure crystallisation [*Grove et al.*, 1992; *Michael and Cornell*, 1998; *Herzberg*, 2004; *Villiger et al.*, 2007] and the apparent drop in extent of mantle melting at slower spreading ridges [*Langmuir et al.*, 1992; *White et al.*, 2001]. It is therefore likely that the proportion of the available heat from mantle melt released at slow spreading rates is greater than that at fast-spreading rates. However, observational evidence and thermal models of the near off-axis crust from slow-spreading ridges are sparse. In summary, it is likely that more than 85% of the available 2 MJ kg^{-1} is released from the magmatic crust within 10–15 km of the axes of both fast- and slow-spreading ridges.

This magmatic heat drives various different types of hydrothermal activity. Seafloor eruptions, dyking or rifting events may be associated with event plumes which are transient features that decay over a time scale of weeks to decades [e.g., *Baker et al.*, 2004; *Lowell and Germanovich*, 1994; *Cann and Strens*, 1989]. Much of the cooling of the upper crust may be linked to the development of such plumes: lavas and dykes typically constitute about 1/3 of the thickness of oceanic crust, and therefore can release about 1/3 of the available magmatic heat. High-temperature axial venting which may be stable for decades or more is also likely to be driven by cooling of the shallow melt lens [*Lister*, 1974; *Cann and Strens*, 1982] and possibly by heat supplied by cooling the flanks of the axial hot zone [*Maclennan et al.*, 2005]. There is a growing body of evidence which indicates that high-temperature vent sites may be active at distances of 5–15km from the ridge axis, with *Haymon et al.* [2005] suggesting that off-axis sites in the abyssal hills flanking the EPR may have hosted vents with maximum temperatures exceeding 250°C. The Lost City vent site exhales fluids at 80°C and sits about 15km from the axis of the Mid-Atlantic Ridge [*Kelley et al.*, 2001]. Diffuse flow of lower temperature fluids has been observed to remove heat from the axial crust and is also likely to be important in the off-axis regions [*Haymon et al.*, 2005; *Stakes et al.*, 2006].

The above estimates of heat supply only include that which might be provided by cooling of mantle melts near the ridge axis. If extensive hydrothermal cooling of the crust occurs near the ridge axis, as proposed by *Maclennan et al.* [2005], then conduction of heat from the lithospheric mantle into the crust may also provide heat to crustal hydrothermal systems. The extreme models of *Maclennan et al.* [2005] indicate that up to 10 kW m^{-1} may be provided by conductive cooling of the mantle within 10 km of the ridge axis. This conductive supply of heat from the mantle becomes increasingly important with respect to magmatic heat with increasing distance from the axis and must be the source of heat which drives much of the hydrothermal flow that is required to match the misfit between the observed heat flux in the oceanic basins and that predicted from lithospheric cooling models.

6. CONCLUSIONS

The cooling of a batch of primitive mantle-derived melt to ambient seawater temperatures of 2°C releases 2 ± 0.2 MJ kg^{-1}. About half of this heat is provided by latent heat of crystallisation and specific heat over the solidification interval from ~1300° to 1000°C. The remainder is supplied by specific heat and latent heat of sub-solidus reactions from 1000° to 0°C. The global heat supply from cooling of mantle melts at mid-ocean ridges is ~3.2 ± 0.3 TW.

Sub-solidus hydration reactions are unlikely to be important in the overall global supply of heat, with extensive reaction between cool seawater and hot crustal rocks providing less than 0.1 MJ kg^{-1} in latent heat.

Geophysical estimates of the distribution of melt at ridges indicate that more than half of the total heat release associated with solidification of magma and cooling of magmatic rocks takes place within the zone of crustal accretion, which is less than 10 km from the ridge axis at fast-spreading ridges, and perhaps 15 km at slow-spreading ridges. The location of sub-solidus cooling and heat release is less well constrained, but the results of recent thermal models of fast-spreading ridges indicate that much of this heat is also supplied close to the ridge axis.

Episodes of heat supply to the shallow crust in lava flows and dykes is associated with megaplumes and localised high-temperature venting. Approximately 1/3 of the total magmatic heat release occurs in such events. Crystallisation and cooling in the shallow axial melt lens is important in driving hydrothermal circulation and there is a clear association between locations where a shallow melt lens has been imaged and high temperature vent sites, both at fast-spreading ridges [*Singh et al.*, 1998] and at slow-spreading ridges [*Singh et al.*, 2006a; *Brandsdóttir et al.*, 1997]. The relative importance of heat release from the shallow melt lens and heat release from crystallisation in deeper sheeted sills is not yet clear, but recent models indicate that 1/4 to 1/3 of the lower crust crystallises in sheeted sills [*Maclennan et al.*, 2004].

The sheeted-sills crustal accretion models for fast-spreading ridges involve crystallisation within the lower axial crust. In order to satisfy geophysical estimates of the melt fraction in these regions, the heat release must be supplied to hydrothermal fluids circulating deep in the crust on the flanks of the ridge axis [*Maclennan et al.*, 2004, 2005]. A number of geochemical studies of exposed lower-crustal rocks indicate that seawater-derived hydrothermal fluids may indeed penetrate the deep crust [*Nicolas and Mainprice*, 2005]. However, it is not known if this deep circulating hydrothermal fluid is entrained in the high-temperature axial system on its return to shallow levels, or if it emanates in separate off-axis sites such as those recently reported by *Haymon et al.* [2005]. The presence of a geochemical signature of deep, high-temperature reaction in axial vent fluids would provide important support for crustal accretion models that require deep circulation.

Deep crystallisation, at depths of up to 30 km is known to occur at slow-spreading ridges. It is not yet clear if this deep crystallisation drives hydrothermal flow. Thermal models of slow-spreading ridges will allow quantification of the requirement for hydrothermal cooling of the lower crust and uppermost mantle at slow-spreading ridges. Acquisition of data that can be used to improve our understanding of the geometry of the hydrothermal cooling systems must be a key objective.

Acknowledgments. Mike Perfit encouraged me to pursue an interest in this topic, attend the Mammoth meeting, and make this paper clearer. Joe Cann was full of questions and syntheses from Heathrow to Reno and back. Many of the workshop participants provided fascinating points of discussion, not least amongst them being Jason Phipps Morgan and Bob Lowell. Mike Bickle, Tim Holland and Jamie Connolly were happy to answer my queries during the preparation of the manuscript and Steve Jones most kindly performed an awk scripting masterclass to turn VERTEX into modes. Laurence Coogan and Claude Herzberg delivered thoughtful and encouraging reviews that led to important improvements of this work.

REFERENCES

Allen, D. E., and W. E Seyfreid (2004), Serpentinization and heat generation: Constraints from Lost City and rainbow hydrothermal systems, *Geochim. Cosmochim. Acta, 68*, 1347–1354.

Ariskin, A. A., G. S. Barmina, M. Ya. Frenkel, and R. L. Nielsen (1993), COMAGMAT: A Fortran program to model magma differentiation processes, *Comput. Geosci., 19*, 1155–1170.

Asimow, P. D. (2002), Steady-state mantle-melt interactions in one dimension: II. Thermal interactions and irreversible terms, *J. Petrol., 43*, 1707–1724.

Bach, W., J. C. Alt, Y. Niu, S. E. Humphris, J. Erzinger, and H. J. B. Dick (2001), The geochemical consequences of late-stage low-grade alteration of lower ocean crust at the SW Indian Ridge: Results from ODP Hole 735B (Leg 176), *Geochim. Cosmochim. Acta, 65*, 3267–3287.

Bach, W., C. J. Garrido, H. Paulick, J. Harvey, and M. Rosner (2004), Seawater-peridotite interactions: First insights from ODP Leg 209, MAR 15°N, *Geochem. Geophys. Geosyst. 5*, Q09F26, doi:10.1029/2004GC000744.

Baker, E. T., R. P. Lowell, J. A. Resing, R. A. Feely, R. W. Embley, G. J. Massoth, and S. L. Walker (2004), Decay of hydrothermal output following the 1998 seafloor eruption at Axial Volcano: Observations and models, *J. Geophys. Res., 109*, B01205, doi:10.1029/2003JB002618.

Barclay, A. H., D. R. Toomey, and S. C. Solomon (2001), Microearthquake characteristics and crustal V_P/V_S structure at the Mid-Atlantic Ridge, 35°N, *J. Geophys. Res., 106*, 2017–2034.

Barth, G. A., and J. C. Mutter (1996), Variability in oceanic crustal thickness and structure: Multichannel seismic reflection results from the northern East Pacific Rise, *J. Geophys. Res., 101*, 17,951–17,975.

Bédard, J. H. (1993), Oceanic crust as a reactive filter—Synkinematic intrusion, hybridization, and assimilation in an ophiolitic magma chamber, Western Newfoundland, *Geology, 21*, 77–80.

Berman, R. G. (1988), Internally-consistent thermodynamic data for minerals in the system Na_2O-K_2O-CaO-MgO-FeO-Fe_2-O_3-Al_2O_3-SiO_2-TiO_2-H_2O-CO_2, *J. Petrol., 29*, 445–522.

Boudier, F., A. Nicolas, and B. Ildefonse (1996), Magma chambers in the Oman ophiolite: Fed from the top and the bottom, *Earth Planet. Sci. Lett., 144*, 239–250.

Bouhifd, M. A., P. Besson, P. Courtail, C. Gérardin, A. Navrotsky, and P. Richet (2007), Thermochemistry and melting properties of basalt, *Contrib. Mineral. Petrol., 153*, 689–698.

Bourdon, B., N. M. Ribe, A. Stracke, A. E. Saal, and S. Turner (2006), Insights into the dynamics of mantle plumes from uranium-series geochemistry, *Nature, 444*, 713–717.

Brandsdóttir, B., W. Menke, P. Einarsson, R. S. White, and R. Staples (1997), Färoe-Iceland Ridge Experiment. 2. Crustal structure of the Krafla central volcano, *J. Geophys. Res., 102*, 7867–7886.

Canales, J. P., R. S. Detrick, D. R. Toomey, and W. S. D. Wilcock (2003), Segment-scale variations in the crustal structure of 150–300 kyr old fast spreading oceanic crust (East Pacific Rise, 8°15′N–10°5′N) from wide-angle seismic refraction profiles, *Geophys. J. Int., 152*, 766–794.

Canales, J. P., R. S. Detrick, S. M. Carbotte, G. M. Kent, J. B. Diebold, A. Harding, J. Babcock, M. R. Nedimović, and E. van Ark (2005), Upper crustal structure and axial topography at intermediate spreading ridges: Seismic constraints from the southern Juan de Fuca Ridge, *J. Geophys. Res., 110,* B12104, doi:10.1029/2005JB003630.

Cann, J. R., and M. R. Strens (1982), Black smokers fuelled by freezing magma, *Nature, 298*, 147–149.

Cann, J. R., and M. R. Strens (1989), Modeling periodic megaplume emission by black smoker systems, *J. Geophys. Res., 94*, 12,227–12,237.

Cannat, M., D. Bideau, and H. Bougault (1992), Serpentenized peridotites and gabbros in the Mid-Atlantic Ridge axial valley at 15°37′N and 16°52′N, *Earth Planet. Sci. Lett., 109*, 87–106.

Cannat, M., J. Cann, and J. Maclennan (2004), Some hard rock constraints on the supply of heat to mid-ocean ridges, in *Mid-Ocean Ridges: Hydrothermal Interactions Between the Lithosphere and Oceans, Geophys. Monogr. Ser.*, vol. 148, edited by C. R. German, J. Lin, and L. M. Parson, pp. 111–149, AGU, Washington, D.C.

Cannat, M., D. Sauter, V. Mendel, E. Ruellan, K. Okino, J. Escartin, V. Combier, and M. Baala (2006), Modes of seafloor generation at a melt-poor ultraslow-spreading ridge, *Geology, 34*, 605–608.

Cherkaoui, A. S. M., W. S. D. Wilcock, R. A. Dunn, and D. R. Toomey (2003), A numerical model of hydrothermal cooling and crustal accretion at a fast spreading mid-ocean ridge, *Geochem. Geophys. Geosyst., 4*(9), 8616, doi:10.1029/2001GC000215.

Connolly, J. A. D. (2005), Computation of phase equilibria by linear programming: A tool for geodynamic modeling and its application to subduction zone decarbonation, *Earth Planet. Sci. Lett., 236*, 524–541.

Coogan, L. A. (2007), The lower oceanic crust, in *Treatise on Geochemistry*, edited by K. Turekian and H. D. Holland, Elsevier, Oxford, UK, in press.

Coogan, L. A., P. D. Kempton, A. D. Saunders, and M. J. Norry (2000), Melt aggregation within the crust beneath the Mid-Atlantic Ridge: Evidence from plagioclase and clinopyroxene major and trace element compositions, *Earth Planet. Sci. Lett., 176*, 245–257.

Coogan, L. A., R. N. Wilson, K. M. Gillis, and C. J. MacLeod (2001), Near-solidus evolution of oceanic gabbros: Insights from amphibole geochemistry, *Geochim. Cosmochim. Acta, 65*, 4339–4357.

Coogan, L. A, C. J. MacLeod, H. J. B. Dick, S. J. Edwards, A. Kvassnes, J. H. Natland, P. T. Robinson, G. Thompson, and M. J. O'Hara (2001), Whole-rock geochemistry of gabbros from the Southwest Indian Ridge: Constraints on geochemical fractionations between the upper and lower oceanic crust and magma chamber processes at (very) slow-spreading ridges, *Chem. Geol., 178*, 1–22.

Coogan, L. A., G. R. T. Jenkin, and R. N. Wilson (2002), Constraining the cooling rate of the lower oceanic crust: A new approach applied to the Oman ophiolite, *Earth Planet. Sci. Lett., 199*, 127–146.

Coogan, L. A., S. A. Kasemann, and S. Chakraborty (2005), Rates of hydrothermal cooling of new oceanic upper crust derived from lithium-geospeedometry, *Earth Planet. Sci. Lett., 240*, 415–424.

Crawford, W. C., and S. C. Webb (2002), Variations in the distribution of magma in the lower crust and at the Moho beneath the East Pacific Rise at 9°–10°N, *Earth Planet. Sci. Lett., 203*, 117–130.

Dale, J., R. Powell, R. W. White, F. L. Elmer, and T. J. B. Holland (2005), A thermodynamic model for Ca-Na clinoamphiboles in Na_2O-CaO-FeO-MgO-Al_2O_3-SiO_2-H_2O-O for petrological calculations, *J. Metamorph. Geol., 23*, 771–791.

Danyushevsky, L. V. (2001), The effect of small amounts of H_2O crystallisation of mid-ocean ridge and backarc basin magmas, *J. Volcanol. Geotherm. Res., 110*, 265–280.

Darbyshire, F. A., R. S. White, and K. F. Priestley (2000), Structure of the crust and uppermost mantle of Iceland from a combined seismic and gravity study, *Earth Planet. Sci. Lett., 181*, 409–428.

Detrick, R. S., P. Buhl, E. Vera, J. Mutter, J. Orcutt, J. Madsen, and T. Brocher (1987), Multi-channel seismic imaging of a crustal magma chamber along the East Pacific Rise, *Nature, 326*, 35–41.

Dick, H. J. B., et al. (2000), A long *in situ* section of the lower ocean crust: results of ODP Leg 176 drilling at the Southwest Indian Ridge, *Earth Planet. Sci. Lett., 179*, 31–51.

Dunn, R. A., D. R. Toomey, and S. C. Solomon (2000), Three-dimensional seismic structure and physical properties of the crust and shallow mantle beneath the East Pacific Rise at 9°30′N, *J. Geophys. Res., 105*, 23,537–23,555.

Dunn, R. A., V. Lekić, R. S. Detrick, and D. R. Toomey (2005), Three-dimensional seismic structure of the Mid-Atlantic Ridge (35°N): Evidence for focused melt supply and lower crustal dike injection, *J. Geophys. Res,, 110*, B09101, doi:10.1029/2004JB003473.

Edmonds, H. N., P. J. Michael, E. T. Baker, D. P. Connelly, J. E. Snow, C. H. Langmuir, H. J. B. Dick, R. Mühe, C. R. German, and D. W. Graham (2003), Discovery of abundant hydrothermal vernting on the ultraslow-spreading Gakkel ridge in the Arctic Ocean, *Nature, 421*, 252–256.

Elderfield, H., and A. Schultz (1996), Mid-ocean ridge hydrothermal fluxes and the chemical composition of the ocean, *Annu. Rev. Earth Planet. Sci., 24*, 191–224.

Embley, R. W., W. W. Chadwick Jr., D. Clague, and D, Stakes (1998), 1998 Eruption of Axial Volcano: Multibeam anomalies and sea-floor observations, *Geophys. Res. Lett., 26*, 3425–3428.

Embley, R. W., W. W. Chadwick, M. R. Perfit, M. C. Smith, and J. R. Delaney (2000), Recent eruptions on the CoAxial segment of the Juan de Fuca Ridge: Implications for mid-ocean ridge accretion processes, *J. Geophys. Res., 105*, 16,501–16,525.

Escartín, J., M. Cannat, G. Pouliquen, A. Rabain, and J. Lin (2001), Crustal thickness of V-shaped ridges south of the Azores: Interaction of the Mid-Atlantic Ridge (36°–39°N) and the Azores hot spot, *J. Geophys. Res., 106*, 21,719–21,735.

Feig, S. T., J. Koepke, and J. E. Snow (2006), Effect of water on tholeiitic basalt phase equilibria: An experimental study under oxidizing conditions, *Contrib. Mineral. Petrol., 152*, 611–638.

Fornari, D. J., et al. (2004), Submarine lava flow emplacement at the East Pacific Rise 9°50′N: Implications for uppermost ocean crust stratigraphy and hydrothermal fluid circulation, in *Mid-Ocean Ridges: Hydrothermal Interactions Between the Lithosphere and Oceans, Geophys. Monogr. Ser.*, vol. 148, edited by C. R. German, J. Lin, and L. M. Parson, pp. 187–218, AGU, Washington, D.C.

Francheteau, J., R. Armijo, J.-L. Cheminée, R. Hekinian, P. Lonsdale, and N. Blum (1990), 1 Ma East Pacific Rise Oceanic crust and uppermost mantle exposed by rifting in Hess Deep (Equatorial Pacific Ocean), *Earth Planet. Sci. Lett., 101*, 281–295.

Fyfe, W. S., and P. Lonsdale (1981), Ocean floor hydrothermal activity, in *The Sea, vol. 7: The Oceanic Lithosphere*, edited by C. Emiliani, pp. 589–638, Wiley, New York.

Garmany, J. (1989), Accumulations of melt at the base of young oceanic crust, *Nature, 340*, 628–632.

Garrido, C. J., P. B. Kelemen, and G. Hirth (2001), Variation of cooling rate with depth in lower crust formed at an oceanic spreading ridge: Plagioclase crystal size distributions in gabbros from the Oman ophiolite, *Geochem. Geophys. Geosyst., 2*(10), doi:10.1029/2000GC000136.

Ghiorso, M. S. (1997), Thermodynamic models of igneous processes, *Annu. Rev. Earth Planet. Sci., 25*, 221–241.

Ghiorso, M. S., and I. S. E Carmichael (1985), Chemical mass transfer in magmatic processes. II. Applications in equilibrium crystallization, fractionation and assimilation, *Contrib. Mineral. Petrol. 90*, 121–141.

Ghiorso, M. S., and R. O. Sack (1995), Chemical mass transfer in magmatic processes. 4. A revised and internally consistent thermodynamic model for the interpolation and extrapolation of liquid-solid equilibria in magmatic systems at elevated temperatures and pressures, *Contrib. Mineral. Petrol., 119*, 197–212.

Grove, T. L., R. J. Kinzler, and W. B. Bryan (1992), Fractionation of mid-ocean ridge basalt (MORB), in *Mantle Flow and Melt Generation at Mid-Ocean Ridges, Geophys. Monogr. Ser.*, vol. 71, edited by J. Phipps Morgan, D. K. Blackman, and J. M. Sinton, pp. 281–310, AGU, Washington, D.C.

Haymon, R. M., K. C. Macdonald, S. B. Benjamin, and C. J. Ehrhardt (2005), Manifestations of hydrothermal discharge from young abyssal hills on the fast-spreading East Pacific Rise flank, *Geology, 33*, 153–156.

Hellevang, B., and R. B. Pedersen (2005), Magmatic segmentation of the northern Knipovich Ridge: Evidence for high-pressure fractionation at an ultraslow spreading ridge, *Geochem. Geophys. Geosyst., 6*, Q09007, doi:10.1029/2004GC000898.

Henstock, T. J., A. W. Woods, and R. S. White (1993), The accretion of oceanic crust by episodic sill intrusion, *J. Geophys. Res., 98*, 4143–4161.

Herzberg, C. (2004), Partial crystallization of mid-ocean ridge basalts in the crust and mantle, *J. Petrol., 45*, 2389–2405.

Herzberg, C., P. D. Asimow, N. Arndt, Y. Niu, C. M. Lesher, J. G. Fitton, M. J. Cheadle, and A. D. Saunders (2007), Temperatures in ambient mantle and plumes: Constraints from basalts, picrites, and komatiites, *Geochem., Geophys. Geosyst., 8*, Q02006, 2006GC001390.

Holland, T. J. B., and R. Powell (1998), An internally consistent thermodynamic data set for phases of petrological interest, *J. Metamorph. Geol., 16*, 309–343.

Holland, T. J. B., and R. Powell (2003), Activity-composition relations for phases in petrological calculations: an asymmetric multicomponent formulation, *Contrib. Mineral. Petrol., 145*, 492–501.

Hooft, E. E. E., H. Schouten, and R. S. Detrick (1996), Constraining crustal emplacement processes from the variation in seismic layer 2A thickness at the East Pacific Rise, *Earth Planet. Sci. Lett., 142*, 289–309.

Jokat, W., O. Ritzmann, M. C. Schmidt-Aursch, S. Drachev, S. Gauger, and J. Snow (2003), Geophysical evidence for reduced melt production on the Arctic ultraslow Gakkel mid-ocean ridge, *Nature, 423*, 962–965.

Jones, S. M., N. White, and J. Maclennan (2002), V-shaped ridges around Iceland: Implications for spatial and temporal patterns of mantle convection, *Geochem. Geophys. Geosyst., 3*(10), 1059, doi:10.1029/2002GC000361.

Juster, T. C., T. L. Grove, and M. R. Perfit (1989), Experimental constraints on the generation of FeTi basalts, andesites, and rhyodacites at the Galapagos Spreading Center, 85°W and 95°W, *J. Geophys. Res., 94*, 9251–9274.

Karson, J. A., et al. (2002), Structure of uppermost fast-spread oceanic crust exposed at the Hess Deep Rift: Implications for subaxial processes at the East Pacific Rise, *Geochem. Geophys. Geosyst., 3*(1), 1002, doi:10.1029/2001GC000155.

Kelemen, P. B., K. Koga, and N. Shimizu (1997a), Geochemistry of gabbro sills in the crust-mantle transition zone of the Oman ophiolite: Implications for the origin of the oceanic lower crust, *Earth Planet. Sci. Lett., 146*, 475–488.

Kelemen, P. B., G. Hirth, N. Shimizu, M. Spiegelman, and H. J. B. Dick (1997b), A review of melt migration processes in the adiabatically upwelling mantle beneath oceanic spreading ridges, *Philos. Trans. R. Soc. Lond., A355*, 283–318.

Kelley, D. S., et al. (2001), An off-axis hydrothermal vent field near the Mid-Atlantic Ridge at 30°N, *Nature, 412*, 145–149.

Kinzler, R. J., and T. L. Grove (1992), Primary magmas of mid-ocean ridge basalts 2. Applications, *J. Geophys. Res., 97*, 6907–6926.

Kojitani, H., and M. Akaogi (1995), Measurement of heat of fusion of model basalt in the system diopside - forsterite - anorthite, *Geophys. Res. Lett., 22*, 2329–2332.

Korenaga, J., and P. B. Kelemen (1998), Melt migration through the oceanic lower crust: A constraint from melt percolation modeling with finite solid diffusion, *Earth Planet. Sci. Lett., 156*, 1–11.

Lange, R. A., and A. Navrotsky (1992), Heat capacities of Fe_2O_3 bearing silicate liquids, *Contrib. Mineral. Petrol., 110*, 311–320.

Lange, R. A., K. V. Cashman, and A. Navrotsky (1994), Direct measurements of latent heat during crystallisation and melting of a ugandite and an olivine basalt, *Contrib. Mineral. Petrol., 118*, 169–181.

Langmuir, C. H., E. M. Klein and T. Plank (1992), Petrological systematics of mid-ocean ridge basalts: Constraints on melt generation beneath ocean ridges, in *Mantle Flow and Melt Generation at Mid-Ocean Ridges, Geophys. Monogr. Ser.*, vol. 71, edited by J. Phipps Morgan, D. K. Blackman, and J. M. Sinton, pp. 183–280, AGU, Washington, D.C.

Laverne, C., P. Agrinier, D. Hermitte and M. Bohn, Chemical fluxes during hydrothermal alteration of a 1200-m long section of dikes in the oceanic crust, DSDP/ODP Hole 504B, *Chem. Geol., 181*, 73–98, 2001.

le Roux, P. J., A. P. le Roex, and J. G. Schilling (2002), Crystallization processes beneath the southern Mid-Atlantic Ridge (40–55 degrees S), evidence for high-pressure initiation of crystallization, *Contrib. Mineral. Petrol., 142*, 582–602.

Lissenberg, C. J., J. H. Bédard, and C. R. van Staal (2004), The structure and geochemistry of the gabbro zone of the Annieopsquotch ophiolite, Newfoundland: Implications for lower crustal accretion at spreading ridges, *Earth Planet. Sci. Lett., 229*, 105–123.

Lowell, R. P., and L. N. Germanovich (1994), On the temporal evolution of high-temperature hydrothermal systems at ocean ridge crests, *J. Geophys. Res., 99*, 565–575.

Lowell, R. P., and P. A. Rona (2002), Seafloor hydrothermal systems driven by the serpentinization of peridotite, *Geophys. Res. Lett., 29*(11), 1531, doi:10.1029/2001GL014411.

Lowell, R. P., P. A. Rona, and R. P. Von Herzen (1995), Seafloor hydrothermal systems, *J. Geophys. Res., 100*, 327–352.

Macdonald, A. H., and W. S. Fyfe (1985), Rate of serpentenization in seafloor environments, *Tectonophysics, 116*, 123–135.

Maclennan, J., D. McKenzie, K. Grönvold, and L. Slater (2001a), Crustal accretion under northern Iceland, *Earth Planet. Sci. Lett., 191*, 295–310.

Maclennan, J., D. McKenzie, and K. Grönvold (2001b), Plume-driven upwelling under central Iceland, *Earth Planet. Sci. Lett., 194*, 67–82.

Maclennan, J., M. Jull, D. McKenzie, L. Slater, and K. Grönvold (2002), The link between volcanism and deglaciation in Iceland, *Geochem. Geophys. Geosyst., 3*(11), 1062, doi:10.1029/2001GC000282.

Maclennan, J., D. McKenzie, K. Grönvold, N. Shimizu, J. M. Eiler, and N. Kitchen (2003), Melt mixing and crystallization under Theistareykir, northeast Iceland, *Geochem. Geophys. Geosyst., 4*(11), 8624, doi:10.1029/2003GC000558.

Maclennan, J., T. Hulme, and S. C. Singh (2004), Thermal models of oceanic crustal accretion: Linking geophysical, geological and petrological observations, *Geochem. Geophys. Geosyst., 5*, Q02F25, doi:10.1029/2003GC000605.

Maclennan, J., T. Hulme, and S. C. Singh (2005), Cooling of the lower oceanic crust, *Geology, 33*, 357–360.

MacLeod, C. J., and G. Yaouancq (2000), A fossil melt lens in the Oman ophiolite: Implications for magma chamber processes at fast spreading ridges, *Earth Planet. Sci. Lett., 176*, 357–373.

Magde, L. S., A. H. Barclay, D. R. Toomey, R. S. Detrick, and J. A. Collins (2000), Crustal magma plumbing within a segment of the Mid-Atlantic Ridge, 35°N, *Earth Planet. Sci. Lett., 175*, 55–67.

McKenzie, D., J. A. Jackson, and K. Priestley (2005), Thermal structure of oceanic and continental lithosphere, *Earth Planet. Sci. Lett., 233*, 337–349.

Menke, W., and D. Sparks (1995), Crustal accretion model for Iceland predicts 'cold' crust, *Geophys. Res. Lett., 22*, 1673–1676.

Michael, P. J., and W. C. Cornell (1998), Influence of spreading rate and magma supply on crystallization and assimilation beneath mid-ocean ridges: Evidence from chlorine and major element chemistry of mid-ocean ridge basalts, *J. Geophys. Res., 103*, 18,325–18,356.

Michael, P. J., et al. (2003), Magmatic and amagmatic seafloor generation at the ultraslow-spreading Gakkel ridge, Arctic Ocean, *Nature, 423*, 956-U1.

Mottl, M. J. (2003), Partitioning of energy and mass fluxes between mid-ocean ridge axes and flanks at high and low temperature, in *Energy and Mass Transfer in Marine Hydrothermal Systems*, edited by P. E. Halbach, V. Tunnicliffe, and J. R. Hein, pp. 271–286, Dahlem University Press, Berlin.

Nedimovic, M. R., S. M. Carbotte, A. J. Harding, R. S. Detrick, J. P. Canales, J. B. Diebold, G. M. Kent, M. Tischer, and J. M. Babcock (2005), Frozen magma lenses below the oceanic crust, *Nature, 436*, 1149–1152.

Nicolas, A. and D. Mainprice (2005), Burst of high-temperature seawater injection throughout accreting oceanic crust: A case study in Oman ophiolite, *Terra Nova, 17*, 326–330.

Nicolas, A., F. Boudier, and B. Ildefonse (1996), Variable crustal thickness in the Oman ophiolite: Implication for oceanic crust, *J. Geophys. Res., 101*, 17,941–17,950.

Nicholson, H., M. Condomines, J. G. Fitton, A. E. Fallick, K. Grönvold, and N. Rogers (1991), Geochemical and isotopic evidence for crustal assimilation beneath Krafla, Iceland, *J. Petrol., 32*, 1005–1020.

Pallister, J. S. (1984), Parent magma of the Semail ophiolite, Oman, *Geol. Soc. London Spec. Publ. 13*, 63–70.

Pan, Y. C., and R. Batiza (2002), Mid-ocean ridge magma chamber processes: Constraints from olivine zonation in lavas from the East Pacific Rise at 9°30′N and 10°30′N, *J. Geophys. Res., 107*(B1), 2022, doi:10.1029/2001JB000435.

Morgan, J. P., and Y. J. Chen (1993a), The genesis of oceanic crust: Magma injection, hydrothermal circulation, and crustal flow, *J. Geophys. Res., 98*, 6283–6297.

Morgan, J. P., and Y. J. Chen (1993b), Dependence of ridge-axis morphology on magma supply and spreading rate, *Nature, 364*, 706–708.

Morgan, J. P., and D. W. Forsyth (1988), Three-dimensional flow and temperature perturbations due to a transform offset: Effects on oceanic crustal and upper mantle structure, *J. Geophys. Res.*, *93*, 2955–2966.

Perfit, M. R., and W. W. Chadwick (1998), Magmatism at mid-ocean ridges: Constraints from volcanological and geochemical investigation, in *Faulting and Magmatism at Mid-Ocean Ridges, Geophys. Monogr. Ser.*, vol. 106, edited by W. R. Buck, et al., pp. 59–115, AGU, Washington, D.C.

Perk, N. W., L. A. Coogan, J. A. Karson, E. M. Klein, and H. D. Hanna (2007), Petrology and geochemistry of primitive lower oceanic crust from Pito Deep: Implications for the accretion of the lower crust at the Southern East Pacific Rise, *Contrib. Mineral. Petrol.*, DOI 10.1007/s00410-007-0210-z.

Pollack, H. N., S. J. Hurter, and J. R. Johnson (1993), Heat flow from the Earth's interior: Analysis of the global data set, *Rev. Geophys.*, *31*, 267–280.

Purdy, G. M., L. S. L. Kong, G. L. Christeson, and S. C. Solomon (1992), Relationship between spreading rate and the seismic structure of midocean ridges, *Nature, 355*, 815–817.

Putirka, K., M. Johnson, R. Kinzler, J. Longhi, and D. Walker (1996), Thermobarometry of mafic igneous rocks based on clinopyroxene-liquid equilibria, 0–30 kbar, *Contrib. Mineral. Petrol., 123*, 92–108.

Putirka, K. D. (2005), Mantle potential temperatures at Hawaii, Iceland, and the mid-ocean ridge system, as inferred from olivine phenocrysts: Evidence for thermally driven mantle plumes, *Geochem. Geophys. Geosyst., 6*, Q05L08, doi:10.1029/2005GC000915.

Putirka, K. D., M. Perfit, F. J. Ryerson, and M. G. Jackson (2007), Ambient and excess mantle temperatures, olivine thermometry, and active vs. passive upwelling, *Chem. Geol., 241*, 177–206.

Ramondenc, P., N. G. A. Leonid, K. L. Von Damm, and R. P. Lowell (2006), The first measurements of hydrothermal heat output at 90°50′N, East Pacific Rise, *Earth Planet. Sci. Lett., 245*, 487–497.

Richet, P., and G. Fiquet (1991), High-temperature heat capacity and premelting of minerals in the system $MgO-CaO-Al_2O_3-SiO_2$, *J. Geophys. Res., 96*, 445–456.

Richet, P., F. Leclerc, and L. Benoist (1993), Melting of forsterite and spinel, with implications for the glass transition of Mg_2SiO_4 liquid, *Geophys. Res. Lett., 20*, 1675–1678.

Ridley, W. I., M. R. Perfit, M. C. Smith, and D. J. Fornari (2006), Magmatic processes in developing oceanic crust revealed in a cumulate xenolith collected at the East Pacific Rise, 9°50′N, *Geochem. Geophys. Geosyst., 7*, Q12O04, doi:10.1029/2006GC001316.

Rubin, K. H., J. D. Macdougall, and M. R. Perfit (1994), ^{210}Po-^{210}Pb dating of recent volcanic eruptions on the sea floor, *Nature, 368*, 841–844.

Rubin, K. H., M. C. Smith, M. R. Perfit, D. M. Christie, and L. F. Sacks (1998), Geochronology and geochemistry of lavas from the 1996 North Gorda Ridge eruption, *Deep-Sea Res. Part II, 45*, 2571–2597.

Rubin, K. H., I. van der Zander, M. C. Smith, and E. C. Bergmanis (2005), Minimum speed limit for ocean ridge magmatism from ^{210}Pb-^{226}Ra-^{230}Th disequilibria, *Nature, 437*, 534–538.

Sæmundsson, K. (1991), Geology of the Krafla system, in *Náttúra Myvatns*, edited by A. Gardasson and A. Einarsson A, pp. 25–95, Hid Islenka Náttúrfraedifélag, Reykjavik.

Schmidt, M. W., and S. Poli (1998), Experimentally based water budgets for dehydrating slabs and consequences for arc magma generation, *Earth Planet. Sci. Lett., 163*, 361–379.

Schwartz, J. J., B. E. John, M. J. Cheadle, E. A. Miranda, C. B. Grimes, J. L. Wooden, and H. J. B. Dick (2005), Dating the growth of oceanic crust at a slow-spreading ridge *Science, 310*, 654–657.

Sclater, J. G., C. Jaupart, and D. Galson (1980), The heat flow through oceanic and continental crust and the heat loss of the Earth, *Rev. Geophys., 18*, 2269–2311.

Sims, K. W. W., et al. (2003), Aberrant youth: Chemical and isotopic constraints on the origin of off-axis lavas from the East Pacific Rise, 9°–10°N, *Geochem. Geophys. Geosyst., 4*(10), 8621, doi:10.1029/2002GC000443.

Sinha, M. C., and R. L. Evans (2004), Geophysical constraints upon the thermal regime of the ocean crust *AGU Monogr., 148*, 19–62.

Singh, S. C., G. M. Kent, J. S. Collier, A. J. Harding, and J. A. Orcutt (1998), Melt to mush variations in crustal magma properties along the ridge crest at the southern East Pacific Rise, *Nature, 394*, 874–878.

Singh, S. C., W. C. Crawford, H. Carton, T. Seher, V. Combier, M. Cannat, J. P. Canales, D. Dusunur, J. Escartin, and J. M. Miranda (2006a), Discovery of a magma chamber and faults beneath a Mid-Atlantic Ridge hydrothermal field, *Nature, 442*, 1029–1032.

Singh, S. C., et al. (2006b), Seismic reflection images of the Moho underlying melt sills at the East Pacific Rise, *Nature, 442*, 287–290.

Sinha, M. C., and R. L. Evans (2004), Geophysical constraints upon the thermal regime of the ocean crust AGU Monogr., *148*, 19–62.

Sinha, M. C., S. C. Constable, C. Peirce, A. White, G. Heinson, L. M. MacGregor, and D. A. Navin (1998), Magmatic processes at slow spreading ridges: implications of the RAMESSES experiment at 57°45′N on the Mid-Atlantic Ridge, *Geophys. J. Int., 135*, 731–745.

Sinton, J., K. Grönvold, and K. Sæmundsson (2005), Postglacial eruptive history of the Western Volcanic Zone, Iceland, *Geochem. Geophys. Geosyst., 6*, Q12009, doi:10.1029/2005GC001021.

Slater, L., D. McKenzie, K. Grönvold, and N. Shimizu (2001), Melt generation and movement beneath Theistareykir, NE Iceland, *J. Petrol., 42*, 321–354.

Sleep, N. H., and G. A. Barth (1997), The nature of oceanic lower crust and shallow mantle emplaced at low spreading rates, *Tectonophysics, 279*, 181–191.

Smith, P. M., and P. D. Asimow (2005), Adiabat_1ph: A new public front-end to the MELTS, pMELTS, and pHMELTS models, *Geochem, Geophys. Geosyst., 6*, Q02004, doi:10.1029/2004GC000816.

Sohn, R. A., J. A. Hildebrand, and S. C. Webb (1999), A microearthquake survey of the high-temperature vent fields on the volcanically active East Pacific Rise (9°50′N), *J. Geophys. Res., 104*, 25,367–25,377.

Soule, S. A., D. J. Fornari, M. R. Perfit, M. A. Tivey, W. I. Ridley, and H. Schouten (2005), Channelized lava flows at the East Pacific Rise crest 9°–10°N: The importance of off-axis lava transport in developing the architecture of young oceanic crust, *Geochem. Geophys. Geosyst., 6*, Q08005, doi:10.1029/2005GC000912.

Stakes, D. S., M. R. Perfit, M. A. Tivey, D. W. Caress, T. M. Ramirez, and N. Maher (2006), The Cleft revealed: Geologic, magnetic, and morphologic evidence for construction of upper oceanic crust along the southern Juan de Fuca Ridge, *Geochem. Geophys. Geosyst., 7*, Q04003, doi:10.1029/2005GC001038.

Staples, R. K., R. S. White, B. Brandsdóttir, W. Menke, P. K. H. Maguire, and J. H. McBride (1997), Färoe-Iceland Ridge Experiment 1. Crustal structure of northeastern Iceland, *J. Geophys. Res., 102*, 7849–7866.

Stein, C. A., and S. Stein (1994), Constraints on hydrothermal heat flux through the oceanic lithosphere from global heat flow, *J. Geophys. Res., 99*, 3081–3095.

Teagle, D. A. H., M. J. Bickle, and J. C. Alt (2003), Recharge flux to ocean-ridge black smoker systems: A geochemical estimate from ODP Hole 504B, *Earth Planet. Sci. Lett., 210*, 81–89.

Tilmann, F., E. Flueh, L. Planert, T. Reston, and W. Weinrebe (2004), Microearthquake seismicity of the Mid-Atlantic Ridge at 5°S: A view of tectonic extension, *J. Geophys. Res., 109*, B06102, doi:10.1029/2003JB002827.

Thordarson, T., and G. Larsen (2007), Volcanism in Iceland in historical time: Volcano types, eruption styles and eruptive history, *J. Geodyn., 43*, 118–152.

Thy, P. (1991), High and low pressure phase equilibria of a mildly alkalic lava from the 1965 Surtsey eruption: Experimental results, *Lithos, 26*, 223–243.

Tolstoy, M., et al. (2006), A sea-floor spreading event captured by seismometers, *Science, 314*, 1920–1922.

Tryggvason, A., S. T. Rognvaldsson, and O. G. Flovenz (2002), Three-dimensional imaging of the P- and S-wave velocity structure and earthquake locations beneath Southwest Iceland, *Geophys. J. Int., 151*, 848–866.

Turner, I. M., C. Peirce, and M. C. Sinha (1999), Seismic imaging of the axial region of the Valu Fa Ridge, Lau Basin—The accretionary processes of an intermediate back-arc spreading ridge, *Geophys. J. Int., 138*, 495–519.

Vera, E. E., J. C. Mutter, P. Buhl, J. A. Orcutt, A. J. Harding, M. E. Kappus, R. S. Detrick, and T. M. Brocher (1990), The structure of 0- to 0.2-m.y.-old oceanic crust at 9°N on the East Pacific Rise from expanded spread profiles, *J. Geophys. Res., 95*, 15,529–15,556.

Villiger, S., O. Müntener, and P. Ulmer (2007), Crystallization pressures of mid-ocean ridge basalts derived from major element variations of glasses from equilibrium and fractional crystallization experiments, *J. Geophys. Res., 112*, B01202, doi:10.1029/2006JB004342.

White, R. S., D. McKenzie, and R. K. O'Nions (1992), Oceanic crustal thickness from seismic measurements and rare earth element inversions, *J. Geophys. Res., 97*, 19,683–19,715.

White, R. S., T. A. Minshull, M. J. Bickle, and C. J. Robinson (2001), Melt generation at very slow-spreading oceanic ridges: Constraints from geochemical and geophysical data, *J. Petrol., 42*, 1171–1196.

Wilson, D. S., D. A. H. Teagle, J. C. Alt, N. R. Banerjee, S. Umino, S. Miyashita, G. D. Acton, R. Anma, S. R. Barr, A. Belghoul, J. Carlut, D. M. Christie, R. M. Coggon, K. M. Cooper, C. Cordier, L. Crispini, S. R. Durand, F. Einaudi, L. Galli, Y. Gao, J. Geldmacher, L. A. Gilbert, N. W. Hayman, E. Herrero-Bervera, N. Hirano, S. Holter, S. Ingle, S. Jiang, U. Kalberkamp, M. Kerneklian, J. Koepke, C. Laverne, H. L. Lledo Vasquez, J. Maclennan, S. Morgan, N. Neo, H. J. Nichols, S.-H. Park, M. K. Reichow, T. Sakuyama, T. Sano, R. Sandwell, B. Scheibner, C. E. Smith-Duque, S. A. Swift, P. Tartarotti, A. A. Tikku, M. Tominaga, E. A. Veloso, T. Yamasaki, S. Yamazaki, and C. Ziegler (2006), Drilling to gabbro in intact ocean crust, *Science, 312*, 1016–1020.

Yang, H. J., R. J. Kinzler, and T. L. Grove (1996), Experiments and models of anhydrous, basaltic olivine-plagioclase-augite saturated melts from 0.001 to 10 kbar, *Contrib. Mineral. Petrol., 124*, 1–18.

J. Maclennan, Department of Earth Sciences, University of Cambridge, CB2 3EQ, UK (jmac@esc.cam.ac.uk)

Seismological Constraints on Magmatic and Hydrothermal Processes at Mid-Ocean Ridges

Maya Tolstoy

Lamont-Doherty Earth Observatory of Columbia University

Seismic methods have been used for decades to study the basic structure of oceanic crust and, in particular, the structure on and near the mid-ocean ridge axis where new crust is actively forming. At the ridge axis, significant heat and chemical exchange occur with the overlying ocean, fueling numerous active hydrothermal venting systems and associated biological communities. Important structural features controlling hydrothermal venting include the presence/absence, depth, and size of a magma body, as well as the fault and porosity structure of the upper crust, all of which can be mapped to some degree using active source seismic techniques with ever-increasing precision. In addition, passive seismic monitoring can allow some dynamic hydrothermal processes to be illuminated, as well as highlight areas of ongoing faulting or increased tectonic stresses that might provide conduits for hydrothermal flow. Recent long-term data have shown that microearthquakes at seafloor hydrothermal systems are strongly tidally influenced, implying a seismic pulsing of the seafloor that could have implications for hydrothermal flow mechanisms as well as biological cycles for the communities relying on hydrothermal nutrients. Passive seismic monitoring, particularly hydroacoustic monitoring, has revolutionized our understanding of the most fundamental temporal event at the seafloor, a dike injection. Comparisons between hydrothermal venting sites show variability in the volume of the hydrothermal cracking zone with spreading rate as well as variability in eruptive character and may help explain some of the fundamental differences in the scale of venting at different spreading rates.

1. INTRODUCTION

Fundamental questions in the study of magmatic and hydrothermal processes at mid-ocean ridges include the presence or absence of a magma body or hydrothermal heat source, and the fault, porosity, and permeability structure surrounding this heat source. Seismic methods provide powerful tools for monitoring basic processes and imaging structures fueling hydrothermal circulation. Mid-ocean ridge hydrothermal systems are dominantly fueled by heat from an axial magma chamber (AMC) and episodic diking and eruptive events. The magma chamber may be inflating where it is building toward an eruption, it may be in a relatively steady state, or it may be cooling where heat is being mined more rapidly than it is being replaced and the crust is cooling over a prolonged period. Seismic methods have been effectively used to image an AMC reflector associated with actively

Magma to Microbe: Modeling Hydrothermal Processes at Ocean Spreading Centers
Geophysical Monograph Series 178

10.1029/178GM05

venting hydrothermal fields at slow, intermediate, and fast spreading ridges [e.g., *Detrick et al.*, 1987; *West et al.*, 2001; *Singh et al.*, 2006a; *Van Ark et al.*, 2007].

Active source methods are most useful for imaging structures that might provide heat sources for hydrothermal circulation or provide conduits allowing fluid flow within the solid crust. Of particular interest for mid-ocean ridge hydrothermal systems are the presence, depth, width, and potentially the % melt of an axial magma body. It is not thought that hydrothermal fluid penetrates the AMC, which is believed to have a thin, impervious, nonbrittle layer directly overlying it. AMCs have been imaged beneath the majority of major active hydrothermal fields on axis, for example, 9°50′N—East Pacific Rise (EPR) [*Detrick et al., 1987*; *Kent et al.*, 1993], Endeavour Segment—Juan de Fuca Ridge (JdFR) [*Carbotte et al.*, 2006], and Lucky Strike—Mid-Atlantic Ridge (MAR) [*Singh et al.*, 2006a], although not all, for example, trans-Atlantic geotraverse (TAG)—MAR [*Canales et al.*, 2007]. For the purposes of this paper, off-axis hydrothermal systems (e.g., Lost City—MAR; *Kelley et al.* [2005]) will not be specifically considered, although many of the techniques discussed are also applicable to such systems. As active source imaging methods improve, so too will the level of detail that they provide on AMC structure and properties. Although the presence or absence of an AMC is now relatively straightforward to establish, finer scale mapping can reveal subtler features, such as estimates of the % melt and presence of small discontinuities, that may link to hydrothermal features at the surface.

Hydrothermal systems are highly dynamic and as such lend themselves to seismic monitoring techniques. Hydrothermal circulation depends in part on subtle structural variations, such as faults, lithology, and porosity, that are amenable to seismic imaging, and in particular high-resolution imaging. Earthquakes are an important source of porosity and permeability at mid-ocean ridges, vital to hydrothermal circulation. Earthquakes may help maintain circulation by potentially limiting clogging of pathways from mineral precipitation and microbial growth [e.g., *Lowell et al.*, 2008, this volume]. Faults and cracks created through tectonically driven earthquakes can provide the initial conduits for seawater to enter the crust and allow it to penetrate to depths where the thermal contrast permits hydrothermal cracking to begin. Earthquakes and their associated changes in permeability may have fundamental impacts on vent chemistry and thus the biological ecosystems associated with them.

As geographic coverage and duration of earthquake monitoring increases, a more refined understanding of magmatic and hydrothermal processes should emerge. Capturing a range of seafloor spreading events on seismic instruments is vital to understanding the character and variability of this fundamental process throughout the world's oceans. Knowledge of precise timing of such events leads to better understanding of the response of biological and hydrothermal systems. Present knowledge of microearthquake activity associated with hydrothermal venting is also limited in its geographical and temporal coverage. Continuous monitoring of microearthquake activity in these environments should illuminate the variability within and between vent sites, providing information about the driving forces and structures of hydrothermal circulation.

Important limitations exist with seismic monitoring. First, permeability cannot be determined directly although it might be inferred from geological structures and associated activity. Second, some deformation happens either aseismically (as predicted by some models particularly in the upper crust [e.g., *Cowie et al.*, 1993]) or below the detection capabilities of a given array. Therefore, caution should be used in determining that an area is not seismically, hydrothermally, magmatically, or tectonically active, just because earthquakes are not observed.

There is a growing understanding of the interactions at mid-ocean ridges between the geophysical structures and processes, and the chemical, biological, and thermal character of hydrothermal systems. These different characteristics, generally monitored by different scientists with different disciplinary backgrounds, are inherently interdependent. To understand the system as a whole, scientists with different specializations need to understand each other's results, and in particular the limitations of given techniques. This paper outlines how seismic methods are applied to understanding magmatic and hydrothermal processes at mid-ocean ridges. Some general conclusions are drawn on the links between the observed crustal structure, seismicity, and the character of hydrothermal venting.

2. ACTIVE SOURCE SEISMOLOGY: BASIC PRINCIPLES AND CONSTRAINTS ON CRUSTAL STRUCTURE

Creating sound sources with precisely known and controlled origin times and locations provides an opportunity to map out subtle crustal structures in great detail [e.g., *Sheriff and Geldart*, 1995]. Early studies dominantly relied on explosive sources to map crustal structure including mid-ocean ridge on-axis structure [e.g., *Ewing and Ewing*, 1959; *Spudich and Orcutt*, 1980; *Purdy and Detrick*, 1986; *Tolstoy et al.*, 1993]. The method, although very successful at imaging crustal structure, was relatively crude with large shot spacing and considerable hazards associated with handling explosives, and so in the last decade and a half, airguns have become the source of choice [e.g., *Detrick et al.*, 1993; *Kent et al.*,

2000; *Carbotte et al.*, 2006; *Singh et al.*, 2006a, 2006b]. Airguns provide source repeatability, more rapid repeat rates, and precise timing, and allow handlers to avoid many of the dangers associated with handling explosives. Concern that airguns would not be able to image the crust as deeply as explosives were mitigated by the ability to accurately define,

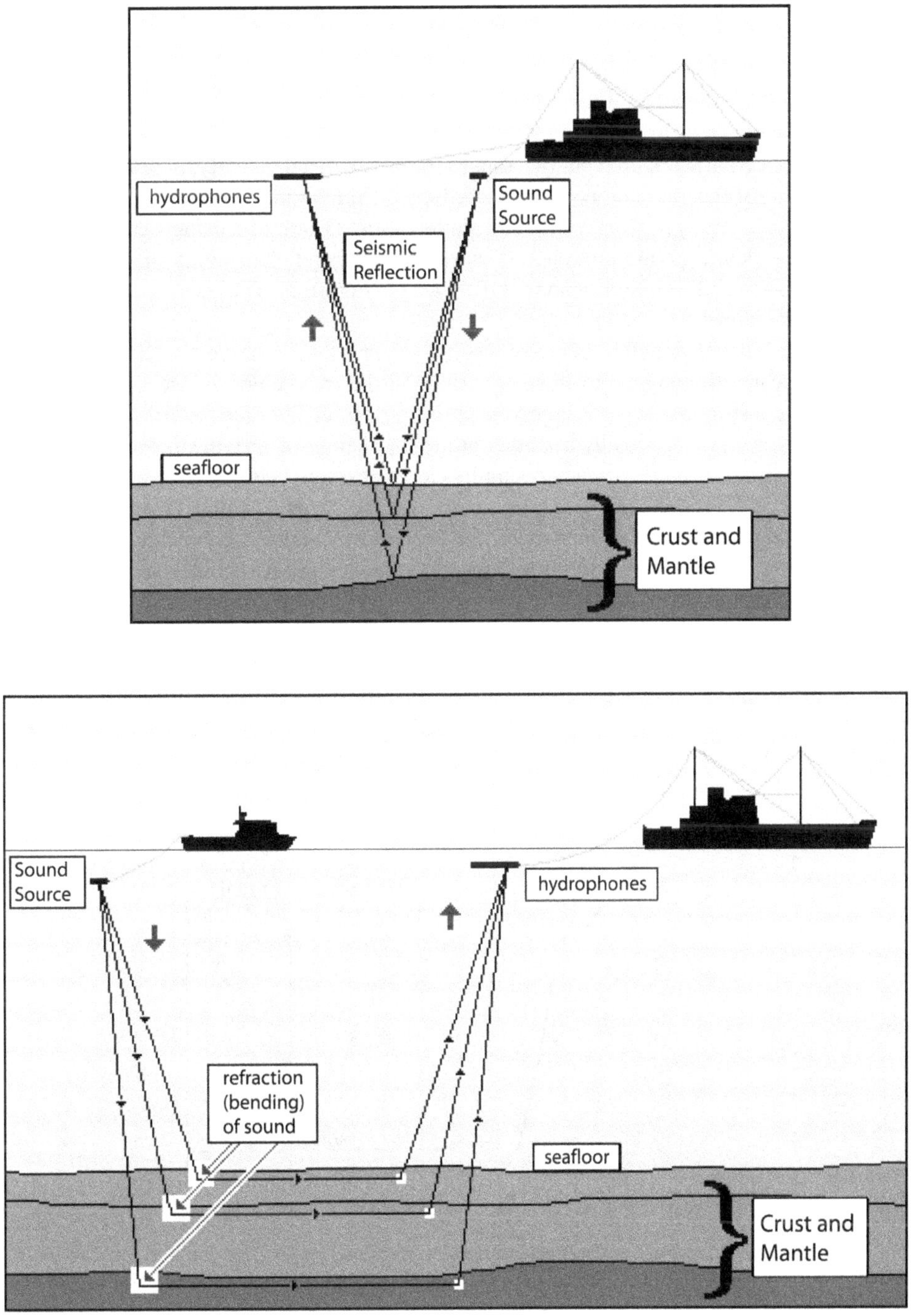

Figure 1. Illustration of the difference between reflection surveys and refraction surveys, adapted from http://www.dosits.org. Note that the larger distances required by refraction might be achieved through using two ships (as illustrated), a longer hydrophone streamer, or ocean bottom seismometers/hydrophones.

tune, and repeat source signatures, and in some cases stack arrivals for better resolution at great depth [e.g., *Gaherty et al.*, 2004]. Seismic reflection and refraction methods (Figure 1) have characterized solid crustal structure as well as magma chamber and Moho properties at many sites around the globe. To understand the resolution and limitations of these observations some of the basic principles involved in active source seismology are discussed below.

Active source seismology signals are recorded on either an ocean bottom seismometer/hydrophone (OBS/H) for refraction work, or for reflection work on a multichannel seismic (MCS) streamer. OBS/Hs are autonomous recording packages that are free-fall deployed from a ship and sink to the seafloor where they record ground motion and/or water pressure on a seismometer and/or a hydrophone. OBSs are also used for passive monitoring work, and can be outfitted to record data for more than a year before being acoustically released from their anchor, floating up to the sea surface, and being recovered by a ship. MCS work involves towing a long line of closely spaced (~12.5 m) hydrophone groups behind the ship while firing airguns. Streamer lengths can be adjusted to suit specific needs and are now commonly in the 6- to 8-km length range, but can be as long as 12 km. For three-dimensional (3-D) seismic work, multiple streamers (commonly four or more) are pulled along side each other, at a fixed distance (~100 m) apart.

When using active source data the analysis is primarily based on arrivals of two seismic phases: *P*-waves and *S*-waves. Primary waves, or *P*-waves, travel as compressional waves through the crust and can be recorded on seismometers (particularly on the vertical channel) and hydrophones. Secondary waves, or *S*-waves, travel through the crust as shear waves but originate as *P*-to-*S* conversions at the water-crust interface, and are only apparent on hydrophones as subsequent *S*-to-*P* converted arrivals because shear waves do not travel through water. *S*-waves are best recorded on horizontal channels of ocean bottom seismometers (OBSs) although coupling to the seafloor for OBSs is often poor relative to land stations. *P*-waves travel more quickly than *S*-waves, and because they tend to be better recorded, lithologies are most commonly interpreted in terms of *P*-wave velocities.

Crustal velocity, measured using the travel time of a sound from source to receiver, is one of the most essential properties mapped using active source seismology and can be interpreted in terms of lithology [e.g., *Christensen*, 1979] (Figure 2). The accuracy of this method is very dependent on the coverage—that is, the number of source-receiver pairs and their relative ranges and azimuths; the heterogeneity of the crust being imaged; as well as the methods used to interpret the data. Refraction/tomography work looks at the change of arrival times with distance which can constrain absolute velocities, whereas reflection does not constrain velocity per se, rather two-way travel time to a reflective interface (Figure 1), which can be interpreted as depth based on assumptions about velocity structure or a prior knowledge from refracted arrivals. Velocities in the upper crust can be determined from reflection data using what are essentially short refraction profiles (gathers), but the depth is limited by the length of the streamer. Uncertainties in determining velocity structure can arise from nonuniqueness of the results both from subtle variations in velocity inferred and in terms of interpreting the velocity as lithology.

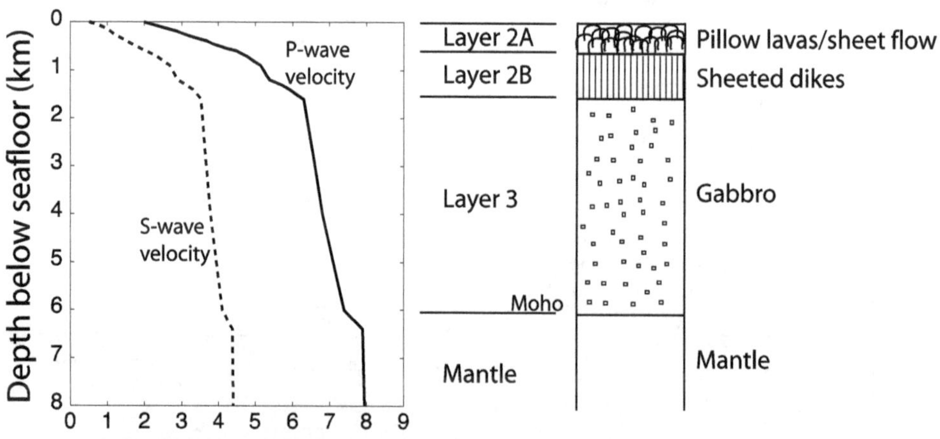

Figure 2. Illustration of the seismic velocity and lithological interpretations of layer 2A, layer 2B, layer 3, and Seismic Moho. Velocity profiles are based on EPR 9°50′N (*Tolstoy et al.*, 2008), and the thickness of the different layers and the overall crust will vary with spreading rate and geographical location.

Some recent studies have suggested that the seismic boundaries do not always correspond with lithological boundaries [e.g., *Detrick et al.*, 1994, 1998; *Christeson et al.*, 2007]; however, other researchers agree with interpreted refraction results [*Wilson et al.*, 2006], although they suggest that porosity and alteration may be more important than lithology. Interpretation of the seismic position of the layer 2/3 boundary at hole 504B [*Detrick et al.*, 1994, 1998] could be questioned in that the seismic data did not extend to a large enough distance to confidently image the layer 2/3 boundary triplication (e.g., Figure 4). Instead, a velocity of 6.5 km/s was used to interpret the boundary depth, whereas other studies have used 6.7 km/s [e.g., *Tolstoy et al.*, 1993]. Interpretation of the layer 2A/2B boundary by *Christeson et al.* [2007] could also be debated in that the ground truth of the boundary position was fracture zone outcrops, and the seismic lines necessarily measured thickness a few kilometers back from the transform. Rapid changes in crustal thickness are known to occur at transforms, and so here again there is uncertainty as to how well the seismic lines might match the outcrop observations.

Normal oceanic crustal thickness at slow- to fast-spreading ridges is between 5 and 8.5 km [e.g., *White et al.*, 1992] and is composed at the ridge axis of a thin (100–800 m) layer 2A [e.g., *Baran et al.*, 2005] made of extrusive basalt (dominantly pillow lavas and sheet flows); a thicker layer 2B of intrusive basalt (sheeted dike complex), which is anywhere from 1 to 3 km thick; and a gabbroic layer 3, which is generally 0.5–5 km thick, with the thinnest areas of layer 3 being near segment ends at slow-spreading ridges [*Tolstoy et al.*, 1994]. Layer 2A has the highest porosity because of void space from drain back, space between pillows, flow contacts, and brecciation [*Carbotte and Scheirer*, 2004], and is thus a critical zone where cold seawater might enter the crust, particularly in the absence of active faulting. Less is known about crustal structure at ultraslow spreading ridges, but it is generally found to be thinner than faster spreading ridges [e.g., *Muller et al.*, 1999; *Jokat and Schmidt-Aursch*, 2007] and in some places spreading amagmatically [*Michael et al.*, 2003].

Snell's law describes how waves traveling through a medium with varying velocity will bend (refract) at velocity interfaces (Figure 3). It is this process that leads to seismic phases traveling from the surface to deep in the crust and back up to the surface again as velocities increase with depth. This is known as refraction and it is the seismic technique used to determine the true velocity of the crust and upper mantle. Where velocity inversions exist such as with an AMC, seismic ray paths are bent downward rather than upward and so refracted arrivals disappear. This process makes tomographic studies of deep crustal structure problematic at mid-ocean ridges sites with magma chambers. Nevertheless, there have been a number of successful tomographic studies on mid-ocean ridge magma chambers [e.g., *Toomey et al.*, 1990; *West et al.*, 2001] or zones of slightly decreased velocities interpreted as warmer areas [e.g., *Dunn et al.*, 2005]. Although arrival times are the most utilized data from refraction experiments, waveform and, in particular, amplitude studies can also be extremely valuable in interpreting velocity gradients. For instance, layer 2B, which has a steep velocity gradient (e.g., changing by 2–2.5 km/s over 1–3 km), induces strong ray bending of rays and arrivals from that layer have a relatively strong amplitude. However, layer 3 has substantially less of a velocity gradient (e.g., changing only by about 0.2–0.4 km/s over 1–3 km) and thus very little energy is turned, producing relatively low amplitude arrivals (Figure 4). Without considering the amplitude of arrivals as well as the timing, analysis techniques are prone to smearing out what should be fairly abrupt boundaries (e.g., Moho or AMC). In particular, many tomographic results on ridges to date have considered primarily travel time, which may result in considerably less abrupt horizontal boundaries than observed through forward modeling that include S-waveform techniques such as Wentzel-Kramers-Brillouin-Jeffreys (WKBJ) [e.g., *Chapman*, 1985]. However, application of finite-frequency and full-waveform inversions may lead to substantial improvements in tomographic image quality in the future.

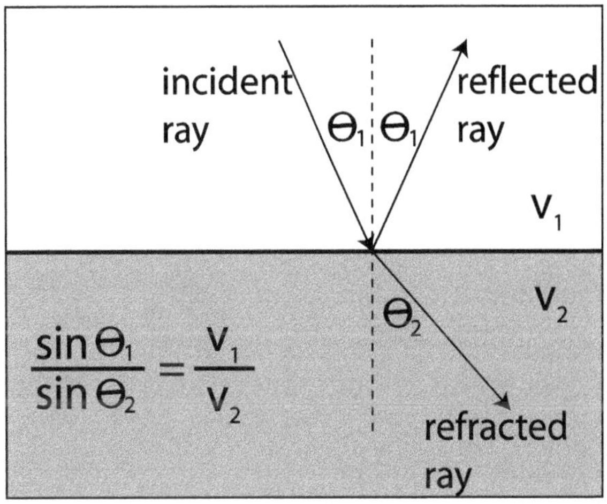

Figure 3. Illustration of Snell's law showing that an arriving (incident) ray will have some energy reflected, and some refracted when it comes to a boundary where there is a velocity contrast. The angle of the refracted ray is dependent on the velocity of both mediums and the angle of the incident ray.

Reflection mapping of crustal structure relies on source returns reflected back from interfaces with a velocity or

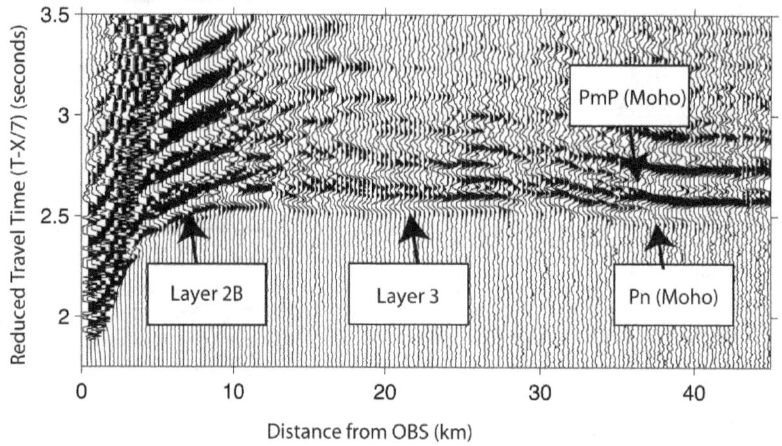

Figure 4. Refraction record section from an off-axis line on the South East Indian Ridge [*Holmes et al.*, 2005]. Vertical axis is travel time with a reduction velocity of 7 km/s (so that phases turning at a velocity of 7 km/s will appear horizontal). The clipped arrivals in the first few kilometers are from the direct water wave arrival of the shot at the OBS. At distances past ~4 km, crustal refracted arrivals are coming in before the water waves. Strong arrivals from layer 2B are evident in the 4- to 14-km range, followed by the lower amplitude layer 3 arrivals. Pn, the refracted phase from the top of the mantle or head wave at the Moho, is faintly evident at about 35–40 km. A strong reflection from the Moho (PmP) is also apparent at these ranges coming in later than the refracted arrivals.

impedance contrast, and is interpreted in terms of two-way travel time of arrivals, which is a function of depth and velocity structure (Figure 5). Impedance contrasts, dependent on the seismic velocity and density of materials, can arise from changes in lithology, porosity, fault structures, or transitions between solid crust and melt. The strength of the reflection is determined by the reflection and transmission coefficients, which are dependent on the impedance contrast,

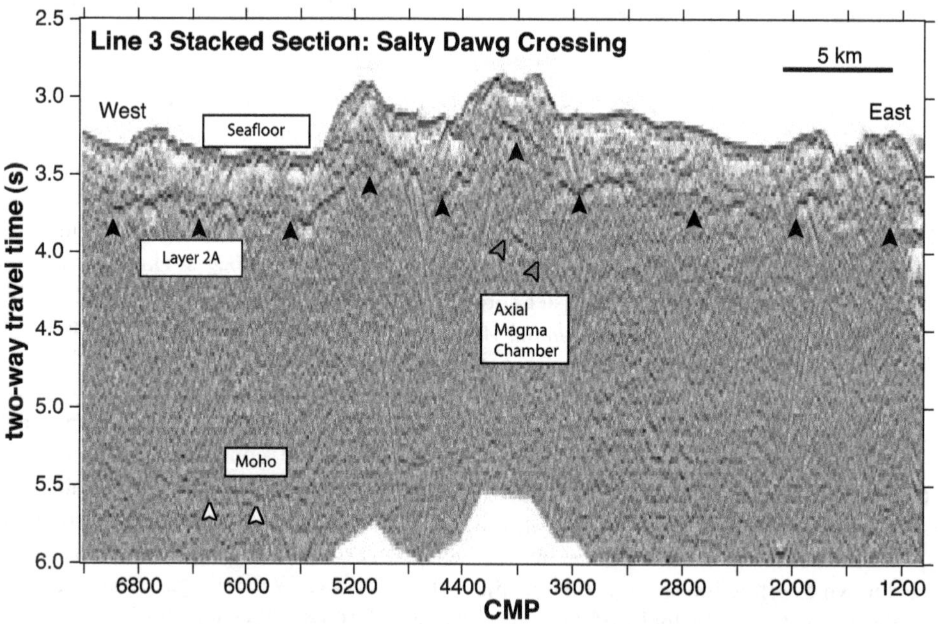

Figure 5. Across-axis reflection profile from the Endeavour segment of the Juan de Fuca ridge, adapted from *Von Ark et al.* [2007]. Reflections from the seafloor, the base of the pillow basalts (black arrowheads), an axial magma chamber (gray arrowheads), and the moho (white arrowheads) are apparent. Note that the vertical scale is two-way travel time rather than depth, and horizontal axis is line number.

and significant geological information can be obtained by looking at the amplitude of the reflected arrival.

The horizontal resolution of seismic techniques is determined by the Fresnel zone that defines a circular region, the size of which is dependent on the signal wavelength. In general, higher frequencies (or shorter wavelengths) give better resolution, but attenuate more rapidly and are thus not able to sample the crust as deeply as lower frequencies. Trade-offs between resolution and depth to which information is desired must therefore be considered for these types of data. One of the big advantages of 3-D multichannel seismic techniques is that the Fresnel zone is collapsed through migration, and thus 3-D imaging can provide much finer resolution images of structures within the crust than the theoretical 1-D Fresnel Zone [e.g., *Biondi*, 2006]. Collection of 3-D data, a capability to be provided by the R/V *Marcus G. Langseth*, can essentially provide high-resolution bathymetric maps of interfaces within the crust in depth slices, and promises to greatly enhance our understanding of ridge-axis structures associated with hydrothermal circulation.

Vertical resolution is controlled by the Rayleigh criterion and is also a function of wavelength. The rule of thumb is that for two nearby reflective interfaces to be distinguished well they must be at least a 1/4 wavelength apart in distance. So, for example, a 30-Hz wave traveling in 6 km/s crust has a 200-m wavelength, and thus would be able to image interfaces that are >50 m apart [e.g., *Hussenoeder et al.*, 1996; *Singh et al.*, 1998].

Reflection techniques have been successfully used at mid-ocean ridges to not only map out the basic layering of the crust and the depth of magma bodies, but also to examine subtler properties of the crust and AMC reflector. In the context of hydrothermal circulation, reflection data has been particularly valuable in identifying the base of layer 2A [e.g., *Carbotte and Scheirer*, 2004], and the presence/absence, width, thickness, and internal properties of the AMC [e.g., *Kent et al.*, 1993; *Hussenoeder et al.*, 1996; *Singh et al.*, 1998].

Where a melt lens is present, it has been observed to deepen with decreasing spreading rate (Figure 6) [e.g., *Purdy et al.*, 1992; *Baran et al.*, 2005], and is not commonly observed on slow-spreading ridges except near hotspot influenced areas [e.g., *Bunch and Kennet*, 1980; *Singh et al.*, 2006a]. The width of magma bodies is highly variable even within adjacent areas of the same spreading rate (e.g., 0.25–4.15 km between 9° and 10°N on the EPR; *Kent et al.* [1993]), and has been found to be particularly large at a fast-spreading overlapping spreading center [*Kent et al.*, 2000]. The thickness of the AMC lens at fast-spreading ridges is generally less than 100 m [*Hussenoeder et al.*, 1996].

Along-axis multichannel seismic measurements have revealed that the AMC reflector is nearly continuous along fast-spreading ridges [*Detrick et al.*, 1987, 1993], and relatively stable in depth over long distances, with slight deepening toward fracture zones [e.g., *Tolstoy et al.*, 1997] and the tips of overlapping spreading centers [*Kent et al.*, 2000]. However, significant variability in the character of the AMC reflector that is interpreted largely as variability in the % melt has been observed [e.g., *Hussenoeder et al.*, 1996; *Singh et al.*, 1998], although care must be taken to ensure that variability is not due to navigation uncertainties, or the streamer drifting slightly off-axis. These melt-to-mush transitions have been proposed to correlate with sites of hydrothermal venting [*Singh et al.*, 1998] based on the presence of water column plumes [*Baker and Urabe*, 1996], with venting focused over sights of greater melt. The interpretation is that areas of greater % melt are hotter and perhaps inflating and thus generate hydrothermal venting. An alternative

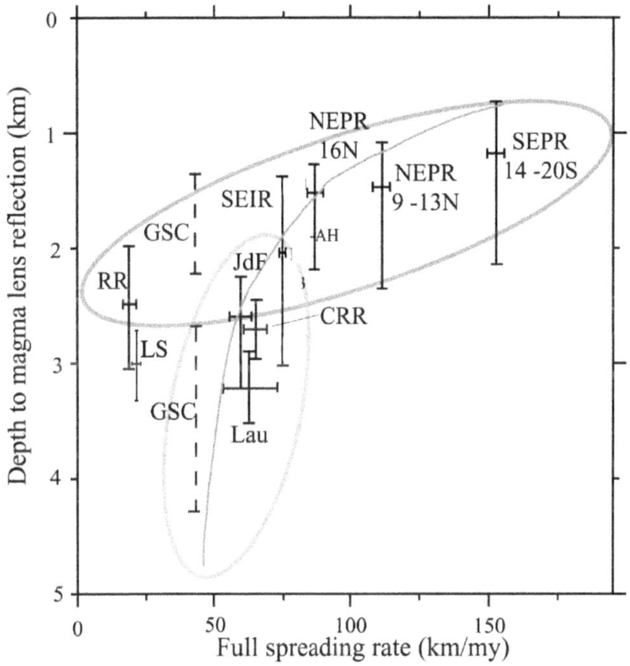

Figure 6. Spreading rate versus average depth to AMC reflector, adapted from *Baran et al.* [2005]. Curved line is the 1200°C isotherm calculated from *Phipps Morgan and Chen* [1993]. The ellipses show areas of axial highs (top) and rifted axial highs (bottom). Results from *Singh et al.* [2006a] have been added and would fall into the rifted axial high category. Other ridges included are Reykjanes Ridge (RR) [*Sinha et al.*, 1997], JdFR [*Morton et al.*, 1987], Costa Rica Rift (CRR) [*Mutter et al.*, 1995], Lau Basin (Lau) [*Collier and Sinha*, 1990], Northern EPR (NEPR) and Southern EPR(SEPR) [*Detrick et al.*, 1987; *Purdy et al.*, 1992; *Carbotte et al.*, 1998], Galapagos Spreading Center (GSC) [*Detrick et al.*, 2002], and Southeast Indian Ridge (SEIR) [*Baran et al.*, 2005].

explanation might be that there is more melt in the AMC at this site because it is a site of up-flow rather than down-flow (Tolstoy et al., Seismic anatomy of a fast-spreading ridge hydrothermal system, submitted to *Nature*, 2007, hereinafter referred to as *Tolstoy et al.*, 2008) and thus is exposed primarily to already warmed fluids. Along intermediate spreading ridges, AMC reflectors are more variable in depth [e.g., *Baran et al.*, 2005; *Carbotte et al.*, 2006] but are also frequently observed except in areas with deep rift valleys [*Baran et al.*, 2005].

Full-waveform inversion of wide aperture multichannel seismic results from the fast-spreading Southern EPR show a 150- to 200-m-thick layer of slightly lowered velocities lying 50–60 m above the roof of a 50-m-thick AMC [*Singh et al.*, 1999], and similar features have been imaged in refraction studies [e.g., *Vera et al.*, 1990; *Toomey et al.*, 1994]. This zone of lowered velocities is interpreted as a zone of hydrothermal cracking overlying a 50- to 60-m-thick impermeable AMC lid.

A major limitation of reflection data is that it does not constrain velocities to as great a depth as OBS (wide aperture) refraction, although this limitation is getting less significant in the mid-ocean ridge setting as multichannel streamers get longer. The length of the streamer determines the distance to which refraction data allows constraint of velocities. The depth to which refracted rays are sampling decreases with decreasing offset distance of refraction data. Reflection surveys are often supplemented with OBSs that can provide the advantage of wider source-receiver aperture than allowed by the streamer, and thus better velocity control. OBSs also have the advantage of having a receiver at the seafloor that provides earlier emergence of the arrivals of interest from behind the water wave, and generally better signal to noise ratio.

Tomographic studies on the EPR have suggested a steep-sided magma body continuing from the AMC to near Moho depths [*Dunn et al.*, 2000]. To thermally maintain such a shape is interpreted as requiring hydrothermal circulation along the sides of magma chamber [e.g., *Cherkaoui et al.*, 2003; *Maclennan et al.*, 2004]. Multichannel and compliance measurements both show magma bodies in the shallow crust and near the Moho, but are not interpreted as showing a large lower crustal magma body [e.g., *Singh et al.*, 2006b; *Crawford et al.*, 1999]. Limitations of tomography exist in terms of how effectively rays can sample such a velocity structure, and how well ray paths are truly known. Nevertheless, the width and depth (although not the asymmetry) of the lower crustal low velocity zone are fairly consistent between the different methods. Ophiolite studies do show evidence for hydrothermal circulation in the lower crust [e.g., *Gregory and Taylor*, 1981], although this may have occurred after the lithosphere moved off-axis. However, evidence from ophiolites for penetration of fluids at very high temperatures is an active area of ongoing research [e.g., *Garrido et al.*, 2001; *Bosch et al.*, 2004]. Recent analysis of rocks collected at Hole 1256D in the eastern Pacific do not show evidence for the large volumes of water that would need to pass through the upper crust to reach and cool the lower crust [*Wilson et al.*, 2006].

Our understanding of the crustal structure and its spatial relationship to hydrothermal venting is most mature at fast-spreading ridges. Until recently, active source data had failed to clearly image a melt lens beneath a slow-spreading ridge vent field. However, recent reflection data from the Lucky Strike segment of the MAR has imaged a large (3 km wide by at least 7 km long) AMC at 3 km depth underlying the Lucky Strike vent field. Axial valley bounding faults appear to penetrate down to AMC depths beneath the vent field and may provide fluid conduits. Although this site is influenced by its proximity to the Azores hotspot [*Cannat et al.*, 1999; *Scheirer et al.*, 2000], it nevertheless provides valuable insight into the potential plumbing of slow-spreading (21 mm/yr; *Cannat et al.* [1999]) ridge vent system. Tomographic results in the MARK area of the Mid-Atlantic Ridge at 23°N indicate an anomalously warm lower crust beneath the Snake Pit hydrothermal field [*Canales et al.*, 2000] that may contain partial melt fueling the venting, but refraction results beneath the TAG hydrothermal field at 26°N show no evidence of warmer than normal upper or lower crust [*Canales et al.*, 2007].

When the crust or the mantle includes features that are dominantly in one direction (e.g., ridge parallel faulting in upper crust or olivine alignment in the upper mantle), waves traveling at the same depth through the same material may travel at different speeds depending on the direction they are traveling with respect to the dominant fabric of the rock. This is known as anisotropy, and although it is generally quite a small signal (0–5% of the velocity) it can be significant in interpreting arrivals and can yield valuable geological information. Anisotropy can be apparent in *P*-waves or *S*-waves. *P*-wave anisotropy is assessed by examining the difference in arrival times of *P*-waves from different directions [e.g., *White and Whitmarsh*, 1984; *Shearer and Orcutt*, 1985]. *S*-wave anisotropy is studied with three-component OBSs, by looking at the difference in travel time arrivals of *S*-waves on the two orthogonal horizontal channels [e.g., *Shearer and Orcutt*, 1986; *Barclay et al.*, 1998; *Almendros et al.*, 2000]. However, because of inherent uncertainties in overlying crustal structures that may be of variable thickness, great care must be taken in interpreting these relatively small signals.

Active source seismology with its capability to image interfaces, their velocity contrast, and layer velocities, can be valuable in terms of mapping major features controlling a hydrothermal system including faulting, crustal layering, porosity, and the characteristics of a magma body if it exists. Careful analysis and some assumptions about velocity interpretation can allow properties such as lithology, porosity, % melt, temperature, and direction of cracking to be inferred. Nevertheless, it should be recognized that the local scale of hydrothermal processes (millimeters or less) is well below the best resolution of seismic methods (tens to hundreds of meters). In short, active source seismology is able to map the basic geological framework in which hydrothermal circulation takes place, but likely not the fine-scale details. To understand the dynamics of a hydrothermal system, earthquake seismology is a more effective tool than active source seismology unless repeat surveys can be done.

3. EARTHQUAKE SEISMOLOGY—BASIC PRINCIPLES

Earthquakes provide information on a number of critical factors controlling hydrothermal circulation. The presence of earthquakes is telling us that the environment being examined is dynamic and changing. Earthquakes may define newly created zones of porosity, and thus possible zones of permeability vital for hydrothermal circulation to be maintained. Earthquakes also provide information on the location and timing of some perturbations to the hydrothermal system, thus providing critical data to interpret thermal and chemical perturbations observed at the seafloor output. In addition, earthquakes can tell us the location of a hydrothermal cracking zone. Ultimately, long-term monitoring of microearthquakes has the potential to illuminate some primary structures of a hydrothermal circulation cell.

The ability to accurately locate an earthquake is critical to being able to usefully interpret it. The absolute accuracy of an earthquake location depends on the accuracy of the velocity model, the reliability of phase arrival picks (Figure 7), and the location techniques applied. Recent advances in relative relocation techniques (earthquakes relocated relative to each other) now applied widely to terrestrial earthquakes [*Shearer*, 1997; *Waldhauser and Ellsworth*, 2000] have re-

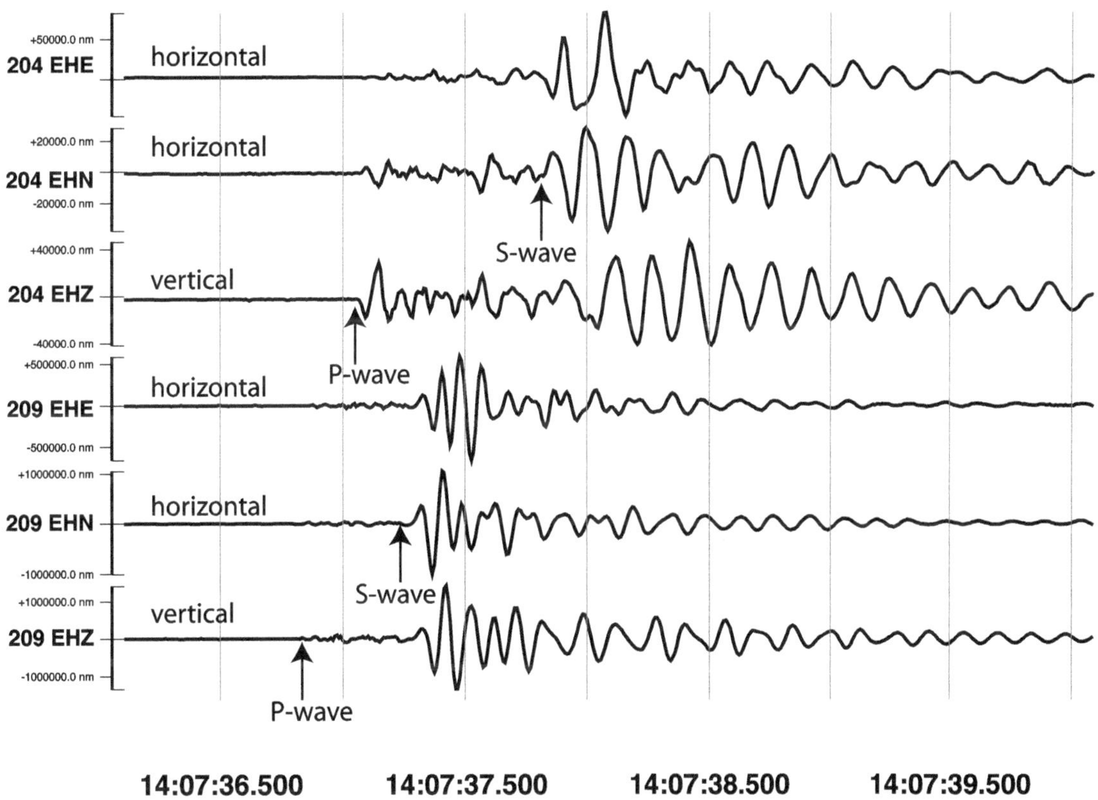

Figure 7. Example of an earthquake on two OBSs (nos. 204 and 209) adapted from *Tolstoy et al.* [2006]. The earthquake is well recorded on both instruments, and is closer to instrument no. 209 as evidenced by the earlier arrivals and the shorter *P*-wave to *S*-wave interval. The *P*-wave is best recorded on the vertical channel, and the *S*-wave is best recorded on the horizontal channels.

cently been applied to the seafloor [*Sohn et al.*, 1998; 1999; *Wilcock et al.*, 2002; *Tolstoy et al.*, 2008]. Good relative locations rely more on consistency of picks or uniformity of waveforms than the velocity model because the techniques interpret earthquake locations relative to other nearby earthquakes and thus far-field velocity uncertainties are removed. This technique can result in diffuse clouds of seismicity with relatively large absolute errors coming into focus as tightly clustered features (e.g., Figure 8), often outlining subtle structures that would otherwise go unresolved [e.g., *Waldhauser et al.*, 1999]. Although reasonable absolute locations are still required, precise relative locations are often good enough to put the seismic observations in the context of the finer scale geological structure. Recent work has also shown that relative location techniques can correct for large absolute location errors [*Waldhauser and Ellsworth*, 2001; *Menke and Schaff*, 2004]. Waveform relative relocation involves the cross-correlation of waveforms to more accurately determine the relative timing of arrivals that are critical to relative relocation. However, in the noisy ocean environment with poor station coupling to the seafloor, waveforms can be inconsistent, and waveform cross-correlation can reduce the number of events that can be located to 25 to 50% of the original catalog [e.g., *Sohn et al.*, 1998; *Wilcock et al.*, 2002]. Travel time relocation based purely on pick data is able to precisely relocate most of the events (~80%), but relies on consistent excellent quality picks [e.g., *Tolstoy et al.*, 2008]. Combining travel time data and waveform cross-correlation for events with consistent waveforms can further improve locations [*Waldhauser and Ellsworth*, 2000].

It is important to distinguish between location accuracy (the true location) that depends primarily on the velocity model and precision (the degree of mutual agreement between measurements) that depends more on network geometry. Events that have been relatively relocated can be very precise with respect to each other, thus clearly outlining structural features in the seismicity, whereas their absolute accuracy is more dependent on the quality of the velocity model. Depth is generally the most poorly constrained dimension, and the rule of thumb is that earthquake depths are best constrained below about half the station spacing. An earthquake can be within an array in terms of latitude and longitude, but in the vertical direction an event is essentially always outside of the array, barring extreme topography, in the sense that an instrument cannot be on both sides of the earthquake in the vertical dimension. An earthquake epicenter is the latitude and longitude surface position of an event as commonly shown on location maps. An earthquake hypocenter includes the depth information and might be

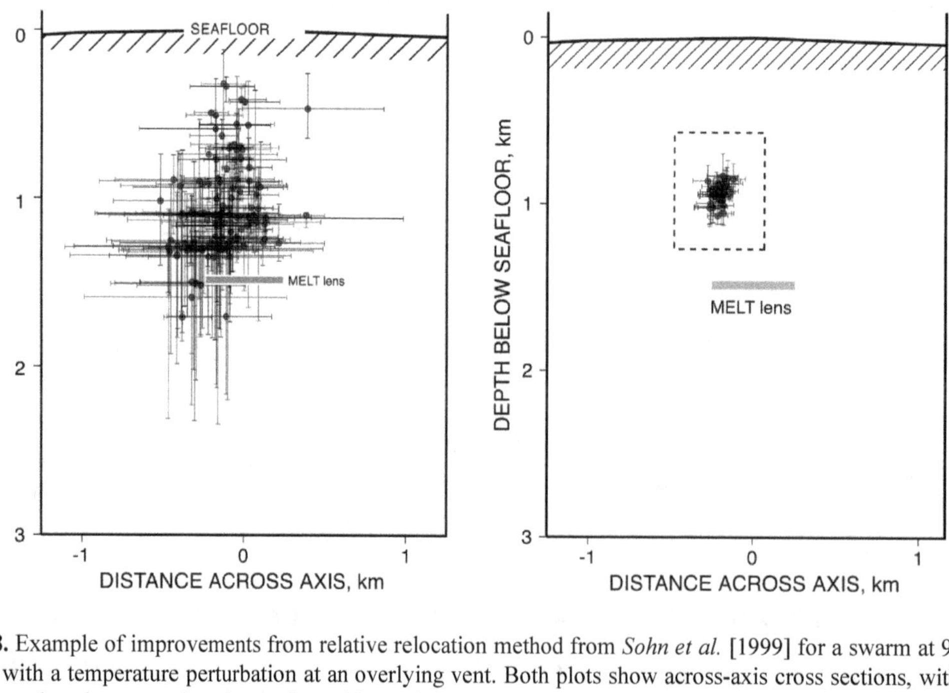

Figure 8. Example of improvements from relative relocation method from *Sohn et al.* [1999] for a swarm at 9°50′N associated with a temperature perturbation at an overlying vent. Both plots show across-axis cross sections, with left plot showing earthquake swarm locations using grid search localization, and the right plot showing earthquake locations following relative relocation. The dashed box indicates 1σ confidence limits for the relocated group, and 1σ error bars are shown for both groups.

shown as a depth cross section. Probably the most important controls for nonseismologists to understand when relating earthquake activity to their data is that absolute depth is often poorly constrained, and locations get worse very rapidly outside of an array. Moreover, the absence of earthquakes, even within an array, may reflect gaps in spatial coverage where magnitudes are just too small to be picked up by enough instruments for a location to be determined.

P-waves are generally considered well recorded on OBSs, although for small hydrothermal cracking events they can be fairly emergent in character and thus hard to pick the first arrival. *S*-waves are historically more difficult, particularly when recorded only on a vertical channel. Some studies have expressed concern over picking arrivals that are too early [e.g., *Trehu and Solomon*, 1983; *Christeson et al.*, 1997]. However, examples provided by *Trehu and Solomon* [1983] seem to show arrivals from an earthquake distant enough that a surface bounce of the *P*-wave arrives before the *S*-wave. Such arrivals (*P*-wave multiples) can be distinguished from *S*-waves by their higher frequency content, and are unlikely to be a problem for hydrothermal earthquakes because they are small enough magnitude that they can only be recorded at very nearby instruments and thus the reflected arrival does not arrive before the *S*-wave. Other studies [e.g., *Christenson et al.*, 1997] have expressed concern about picking arrivals too early because the *S*-wave may convert to a *P*-wave at the layer 2A base and thus, what would appear to be the start of the *S*-wave would actually be a wave that traveled as a *P*-wave for the final 100–300 m and thereby arrives slightly earlier than a true *S*-wave. However, Christenson et al. are looking at *S* arrivals for active source (airgun data). This data is a purely compressional source, and all *S*-wave energy arriving in the crust will be converted energy from one interface or another. An earthquake, on the other hand, is an excellent source of shear energy and therefore the true *S*-wave arrival is likely to be more distinctive than an *S-P* phase converted at the base of layer 2A, although *Wilcock et al.* [2002] have also discussed this limitation for earthquake data. Accuracy of picking is a critical concern, and all these possible sources of errors should be considered. *S*-waves are generally given higher uncertainties (or lower weights) in location work. It is particularly valuable when doing relative travel-time location work that arrivals are picked consistently, and preferably by the same analyst, so that potential errors for position of phase picks are minimized. Waveform cross-correlation techniques are also very valuable for measuring relative travel times but can limit the number of events that are useable. Resolution tests to check for consistency in picks can be done by locating events using only *P*-waves as well as only *S*-waves, and then comparing these single-phase locations with joint *P*-wave and *S*-wave locations.

Careful analysis of waveforms and distribution of polarity of the *P*-wave on different instruments for the same event can provide information on the orientation of a fault and the direction of slip, defined as the focal mechanism [e.g., *Stein and Wysession*, 2003]. Focal mechanisms are typically shown as "beach ball" plots that illustrate the areas of compressional and tensional stress and the strike and dip of the fault planes. This information can be critical in interpreting the geological cause of an earthquake. However, there is also some ambiguity in focal mechanisms in that there are two possible fault planes. For hydrothermal earthquakes, because magnitudes are so small, the polarity of the first arrival is often hard to determine [e.g., *Sohn et al.*, 1999]. Therefore, hydrothermal earthquake focal mechanisms are commonly looked at in "composite" form, where multiple earthquakes with similar waveforms, generally from the same swarm or same area, are stacked to make the polarity of first arrivals more apparent. Even then, polarities are often hard to determine and the limited number of stations usually available for an OBS survey means that good focal mechanisms are very hard to obtain.

In theory, a hydrothermal cracking event (caused by thermal stresses from cold water encountering hot rock) may be distinguished from a simple tectonic event because it would contain a significant nondouble couple component [e.g., *Foulger and Long*, 1984; *Miller et al.*, 1998]; however, in practice this is generally hard to resolve given the limits of data. Because hydrothermal cracking occurs preferentially at times of peak extension stress [*Wilcock*, 2001; *Stroup et al.*, 2007], it is likely that these earthquakes contain both a tectonic and thermal component. Hydrothermal events are therefore primarily distinguished by their geological setting with respect a heat source, vents, and local faults, as well as their swarm characteristics and low magnitude. For instance, on the EPR at 9°50′N the events in the cracking zone generally have a magnitude of less than 1 M_L, frequently occur in small swarms, lie directly over the AMC, do not fall along a clear fault line, and lie below the main vents [*Sohn et al.*, 1999; *Tolstoy et al.*, 2008].

Earthquake magnitude, a logarithmic scale, is a property that has been somewhat underutilized in the mid-ocean ridge environment, perhaps in part because of the lack of standardized instrumentation with well-known response functions. However, magnitude of events as well as the size, space, and temporal distribution of earthquakes can yield important information about the geological environment. For events relevant to the hydrothermal process, either local magnitude, M_L (a measure of the amplitude of the maximum trace deflection versus distance from the source) [*Bakun and Joyner*, 1984], or moment, M_O (energy based on rupture area, offset, and shear modulus), is the most appropriate way to describe the size of these types of earthquakes. Hydrothermal seismicity is commonly less than a magnitude 1 M_L, and magnitudes as low as −1.8 M_L have been reported (*Tolstoy et al.*, 2008), corresponding to a slip area on the scale of less than a meter.

The Gutenberg-Richter relationship looks at the distribution of earthquake magnitudes in a given area over a given period [*Gutenberg and Richter*, 1944]. The slope of the magnitude frequency distribution, that is, the cumulative number of events above a given magnitude on a logarithmic scale versus the magnitude threshold, defines the *b*-value. A *b*-value of ~1 is characteristic of tectonic areas, whereas a *b*-value of >1 is characteristic of volcanic and hydrothermal areas. Higher *b*-values may result from lowering the effective stress through high pore pressure [e.g., *Wyss*, 1973] or from increasing heterogeneity of the medium [e.g., *Mogi*, 1963].

Tertiary waves, or *T*-waves, are the waterborne arrival of an earthquake [*Tolstoy and Ewing*, 1952], which in the mid-ocean ridge setting are generated through scattering of earthquake arrivals into the water column from rough seafloor topography. Some of this energy becomes trapped in a low-velocity wave guide known as the SOund Fixing And Ranging (SOFAR) channel, which allows the waves to travel great distances with relatively little loss in energy compared to earthquake waves (which travel in the solid earth). *T*-waves are generally used for monitoring events of a magnitude 2–3 and higher ([e.g., *Fox et al.*, 1993, 2001; *Smith et al.*, 2002; *Bohnenstiehl et al.*, 2002]), and are thus not used for monitoring hydrothermal processes. They are, however, extremely valuable for studying magmatic and tectonic processes, and have been used extensively for this purpose using regional-scale arrays of hydrophones [e.g., *Fox*, 1995; *Fox et al.*, 1995, 2001; *Fox and Dziak*, 1998; *Dziak et al.*, 1995, 1999, 2004; *Smith et al.*, 2002, 2003]. The accuracy of *T*-wave locations is not as good as locations obtained using OBS arrays, and there is little depth information, although shoaling trends have been inferred based on rise times [*Schreiner et al.*, 1995]. However, hydrophone arrays have the ability to monitor very large areas (on the scale of hundreds to thousands of kilometers) with relatively few instruments, and with 1–2 orders-of-magnitude more sensitive coverage, as well as about 1 order-of-magnitude better location accuracy, than teleseismic arrays [e.g., *Bohnenstiehl and Tolstoy*, 2003]. For instance, in the equatorial Pacific, six hydrophones monitor the 10°N–10°S area of the EPR, over more than 2000 km worth of ridge axis [*Fox et al.*, 2001], whereas 7–12 OBSs were used to monitor ~10 km of the EPR at 9°50′N [*Tolstoy et al.*, 2006; 2008]. Hydroacoustic and seismic methods are both valuable monitoring tools and are complementary rather than alternatives to one another.

4. EARTHQUAKE CONSTRAINTS ON HYDROTHERMAL PROCESSES

The association of venting with small microearthquakes has been known for some time [e.g., *Riedesel et al.*, 1982; *McClain et al.*, 1993]. Earthquakes are a vital component of hydrothermal circulation. Faulting and cracking provide pathways for fluids to penetrate the crust, and hydrothermal cracking itself helps maintain permeability. In effect, earthquakes can be used to map at least some aspects of the hydrothermal circulation cell. Earthquakes can also signal disruptions or changes in the hydrothermal flow regime. Both nearby microearthquakes [e.g., *Sohn et al.*, 1998, 1999; *Johnson et al.*, 2000] and regional or distant larger earthquakes [e.g., *Davis et al.*, 2001; *Dziak et al.*, 2003] can influence hydrothermal systems through regional stress changes, passage of surface waves, pore pressure changes, changes in fluid pathways, magma movement, or cracking into the AMC lid. The disruptions caused by swarms or individual earthquakes, when well documented and recorded contiguously with other data types, can provide vital information on the plumbing of hydrothermal systems.

4.1. Disruptions of the Hydrothermal System

There are two well-documented non-eruptive on-axis seismic swarms that have had measurable impacts on the overlying hydrothermal output. On the fast-spreading (120 mm/yr; *Carbotte and MacDonald* [1994]) EPR at 9°50′N an ~3-h swarm of microearthquakes with local magnitudes between −1 and 0 resulted in a temperature excursion at the overlying high-temperature Bio 9 vent (Figure 8). The temperature increase started 4 days after the swarm, and temperatures rose by about 1° per day for 7 days before gradually decaying back to temperatures just above the pre-swarm temperature [*Sohn et al.*, 1998; *Fornari et al.*, 1998]. The swarm, forming an ~300-m-tall and ~100-m-wide pipe (Figure 8), occurred at a depth of ~300 m above the western edge of the AMC reflector [*Kent et al.*, 1993], and fits within the broader hydrothermal cracking zone observed in later data (*Tolstoy et al.*, 2008). The swarm was interpreted as a cracking event that allowed a previously isolated area of the crust to be tapped [*Sohn et al.*, 1998, 1999; *Fornari et al.*, 1998]. It was also interpreted as having shown an up-flow residence time of only 4 days, which was considerably smaller than previous estimates. However, *Wilcock* [2004] shows that up-flow residence time is even smaller because a lot of heated fluid must pass through the crack and heat the wall rock before a temperature perturbation is observed at the seafloor. *Ramondenc et al.* [2008, this volume] show that modest changes in permeability within the deep discharge zone can lead to measurable thermal perturbations at the vents. They make a case that the 1995 EPR temperature excursion may have been associated with a diking event that did not reach the surface, but enhanced permeability.

A very different type of swarm was observed on the Endeavor segment of the JdFR in 1999, where >2600 earthquakes were detected over a 5-day period using the SOund SUrveillance System (SOSUS) hydrophone network [*Johnson et al.*, 2000] that resulted in significant increases in temperature at four diffuse vents along the segment beginning 4–11 days following the start of the swarm. The swarm initiated with a magnitude 4.5 earthquake leading it to be initially characterized as tectonic in nature, supported by the absence of fresh lavas in the area in subsequent field work. However, later measurements from vent fluid chemistry changes [*Lilley et al.*, 2003; *Seyfried et al.*, 2003] and reanalysis of the swarm characteristics of the seismic data [*Bohnenstiehl et al.*, 2004] have shown that the event was most likely magmatic in origin even though it appears not to have resulted in a seafloor eruption. The similarity in response time of the vent systems to the EPR swarm is quite striking and suggests a common hydrothermal mechanism transporting the temperature change from the site of the swarm to the seafloor venting.

Both the EPR and Endeavour physical responses to local earthquakes were modeled by *Wilcock* [2004] using a temperature perturbation model and a pressure perturbation model. The EPR event was well fit by the temperature perturbation model, with a new path opening to a heat source and the temperature of fluid thus increasing. The Endeavor event was less well constrained and could be fit by both models.

With increasing numbers of in situ seismic and vent monitoring data sets, the number of "disruption" events will increase, allowing better insight into the feedbacks. Indeed, a few such events have already been identified over the 2003–2007 EPR OBS deployment but more work remains to be done in characterizing them. Numerous individual vent temperature perturbations were recorded at the TAG hydrothermal field [*Sohn*, 2007] over a 1-year deployment of temperature probes, but these also have yet to be linked to specific seismic events.

4.2. Continuous Earthquake Monitoring

As the capabilities, and in particular the deployment duration of OBSs has improved, our understanding of the hydrothermal cracking associated with vent fields has similarly improved. Many early OBS studies required recording to be event triggered, with buffers imposed on the number of events that could be recorded in a given time interval [e.g., *Wolfe et al.*, 1995]. This may have resulted in an incomplete picture from these short deployments because events were undoubtedly missed from imperfect triggering algorithms, or too high a triggering threshold, and buffers often filled up—meaning that no matter how large an event was, it would not be recorded unless it occurred near the beginning of the buffering interval. It is by recording and picking the smallest events that the most complete picture of hydrothermal cracking can be developed, particularly because hydrothermal earthquakes are commonly less than a magnitude 1. However, given that these cracking events may be occurring on the spatial scale of 1 m or less (local magnitude 0 to −1 and below), the challenge is to space instruments closely enough and pick the smallest events to be able to locate these microearthquakes. The ability to record the smallest events will similarly drop off with depth (distance from the array).

Lowell and Yao [2002] have hypothesized that down-flow zones would rapidly clog with anhydrite precipitation, and thus that down-flow must take place over a large area. An alternative interpretation is that focused down-flow could only be maintained through near constant tectonically induced cracking (*Tolstoy et al.*, 2008), and it has been hypothesized that vigorous hydrothermal circulation might require continuous earthquake activity [*McClain et al.*, 1993]. The presence of ongoing seismic activity away from the hydrothermal cracking zone directly overlying the AMC may thus indicate areas of potential up- or down-flow with permeability maintained by constantly opening new fluid pathways.

Long-term OBS deployments have been undertaken at hydrothermal venting sites on slow-, intermediate-, and fast-spreading ridges [e.g., *McClain et al.*, 1993; *Sohn et al.*, 1999; *Wilcock et al.*, 2002; *Golden et al.*, 2003; *Tolstoy et al.*, 2006; *deMartin et al.*, 2007; *Tolstoy et al.*, 2008]. The spacing of these different OBS arrays, number of instruments used, and deployment duration varied from site to site, but general comparisons can nevertheless be made.

At the slow-spreading MAR, the TAG hydrothermal vent field was monitored for a period of 8 months in 2003–2004 using 13 short-period OBSs spaced between ~1 and 3 km apart, in a 7 × 7 km array [*deMartin et al.*, 2007]. More than 19,000 earthquakes were recorded with local magnitudes between 1 and 4 along a 30-km stretch of ridge axis. The earthquakes highlight a number of zones of activity, but in particular an ~3-km-thick zone of seismicity in the lower part of the crust inferred to fall along a detachment fault. The results are interpreted as suggesting that hydrothermal fluids at TAG are extracting heat from a region near the crust/mantle interface by utilizing the underlying detachment fault. A crustal melt reservoir is not imaged [*Canales et al.*, 2007], and based on calculations of the heat source necessary to sustain TAG over prolonged periods [*Humphris and Cann*, 2000] the penetration of fluids to the crust mantle interface is an alternative source of heat.

At the intermediate spreading Endeavour segment of the JdFR (56–57 mm/yr; *Wilson* [1993]; *DeMets et al.* [1994]), an array of 15 OBSs were deployed for 55 days in 1995 along an ~5-km section of the ridge axis, and extending 15 km off-axis to the west [*Wilcock et al.*, 2002]. Seven on-axis instruments were centered around Main vent field with 1- to 2-km instrument spacing. Beneath the ridge axis, earthquakes were seen to concentrate in an intense band between 1.5 and 3.5 km depth. Focal mechanisms were variable but dominantly extensional often with a strike-slip component, and subhorizontal tension axis, illustrating a hydrothermal and spreading component to the seismicity. Comparisons with AMC depths imaged from multichannel seismic data show that this intense band directly overlies the AMC [*Van Ark et al.*, 2007]. A similarly intense band of earthquakes was observed in the 1- to 3-km depth range beneath Dead Dog vent field in Middle Valley [*Golden et al.*, 2003]. A network of eight OBSs with 0.5-km instrument spacing recorded more than 3000 earthquakes during a 6-month period from August 1996 to January 1997. Seismicity at both Middle Valley and Endeavour segment was dominated by swarm activity.

At the fast-spreading EPR, two long-term OBS deployments have taken place: one for 3 months in 1995 using 9 OBS [*Sohn et al.*, 1999] and one for over 3.3 years (October 2003–January 2007) using 7–12 OBSs [*Tolstoy et al.*, 2006, 2008]. The spatial distribution of the OBSs were similar for both experiments, covering an ~4 × 4 km area centered around the vents at 9°50′N, with ~1-km instrument spacing. The 1995 deployment detected ~283 earthquakes, whereas the later deployments recorded well in excess of 300,000 earthquakes [*Tolstoy et al.*, 2006]. More than 16,000 earthquakes have been located to date from the October 2003–April 2004 deployment (*Tolstoy et al.*, 2008). An intense band of seismicity ~500 m thick in depth was observed covering a similar area to the events located by *Sohn et al.* [1999]. This concentrated band of activity was centered directly over the AMC imaged in this area in 1985 [*Kent et al.*, 1993], and is interpreted as the hydrothermal cracking zone.

The presence of a concentrated band of hydrothermal cracking directly overlying the interpreted heat source appears to be consistent regardless of the spreading rate. Because the heat source varies in depth with spreading rate, so does the depth of the hydrothermal cracking zone. The thickness of the cracking zone also appears to vary with depth, and this suggests that the area of crust hot enough to generate hydrothermal cracking increases with depth, consistent with a stretching apart of the thermal contours as the distance between seafloor temperatures and melting temperatures increases (Figure 9). This is consistent with the conclusion of *German and Lin* [2004] that the heat fluxes required to sustain the long-lived vent systems at slower spreading ridges appear to require cooling of the upper and lower crust as well as possible serpentinization.

In addition to a thickening of the hydrothermal cracking zone with decreasing spreading rate, there also appears to be a horizontal widening with decreasing spreading rate. At TAG, the cracking zone appears ~1 to 2 km wide, at Endeavour, ~0.5 to 1 km wide, and at EPR it is ~0.3 to 0.5 km wide. This widening is perhaps consistent with lateral cooling occurring less rapidly at depth, or with a broadening of the heat source.

The resulting increase in the volume of the cracking zone with decreasing spreading rate may help explain both the relative longevity and increase in size of hydrothermal vent structures with decreasing spreading rate (Figure 9). A larger volume from which to tap heat and minerals may result in a longer-lived and volumetrically more substantive hydrothermal edifice [e.g., *Wilcock and Delaney*, 2006].

4.3. Tidal Triggering and Temporal Variability

Microearthquakes at hydrothermal vents systems have been observed to be strongly correlated with tidal periodicity [*Wilcock*, 2001; *Tolstoy et al.*, 2002; *Stroup et al.*, 2007]. Along the JdFR earthquakes are observed to occur preferentially during low ocean tides, when extensional stresses are at a maximum due to unloading of the crust associated with the temporary removal of several meters of water [*Wilcock*, 2001; *Tolstoy et al.*, 2002]. On the EPR, however, earthquakes are observed to occur preferentially at high ocean tides [*Stroup et al.*, 2007].

Although the EPR results initially seem to run counter to the findings on the JdFR, they can be explained by the fact the EPR site studied (9°50′N) sits near a tidal node, with exceptionally small ocean tides (a few cm). The stresses associated with Earth tide deformation (tidal deformation of the solid crust rather than the ocean) oppose those of ocean tide deformation. That is, with respect to ocean tides, crustal extensional stresses peak at low tide due to unloading, whereas with respect to earth tides, crustal extensional stresses peak at high tide due to pulling upward of the crust by tidal forces. At this EPR location with very small ocean tides, the stresses associated with earth tides are more dominant, and hence, when combined with effects of ocean loading, the earthquakes occur preferentially near high tide. Therefore, although the EPR and JdFR are triggering on different tidal phases, they are both triggering at the time of maximum extensional stress, and so it is important to consider total tidal stress rather than individual components.

Hydrothermal circulation is known to be tidally modulated [e.g., *Schultz and Elderfield*, 1997]. Tidal loading of

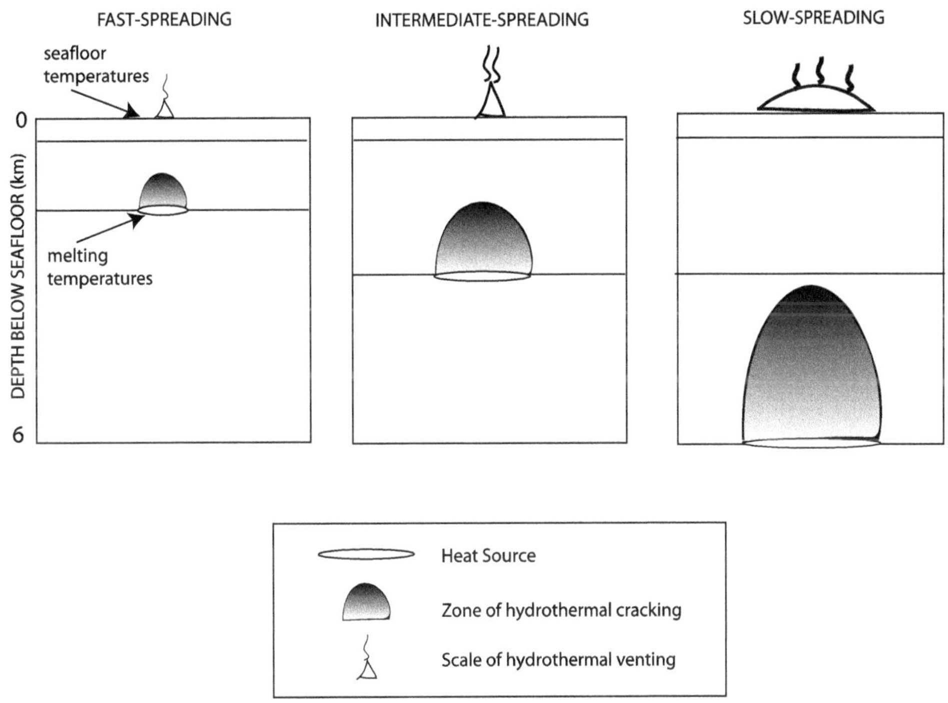

Figure 9. Cartoon illustrating how the thickening of the hydrothermal cracking zone with decreasing spreading rate results in a larger volume for the cracking zone and possibly larger, longer-lived hydrothermal venting sites. With a deepening heat source, the thermal contours would stretch as the distance between the cold seafloor and melting temperatures increased. The layers are broadly illustrating layer 2A, layer 2B, and layer 3, and the position of the heat source and cracking zone within them. Note that for the examples given in this paper (9°50′N EPR, Endeavour, and TAG) there is an apparently widening of the heat source and/or zone of cracking with decreasing spreading rate, but this does not necessarily apply to other ridges. However, a wider heat source is not required to increasing the cracking volume, because the thickening alone will do this.

a poroelastic media [*Jupp and Schultz*, 2004; *Crone and Wilcock*, 2005] can induce pressure changes that could lead to tidal modulation of vent output. The impact of seismicity versus poroelastic loading effects on vent outflow is not well quantified. However, it seems likely that the poroelastic effects will be most important for diffuse vents, which are thought to have relatively shallow circulation [e.g., *Scheirer et al.*, 2006], and seismicity may be more significant at high-temperature vents, which appear to have a deep plumbing source [e.g., *Sohn et al.*, 1998, 1999; *Fornari et al.*, 1998]. Recent analysis of sound data from high-temperature vents suggests that outflow at these sites may vary significantly with tides [*Crone et al.*, 2006].

Hydrothermal seismicity reported to date appears to be dominated by swarm activity, with some indication of long-term tidal trends, such as fortnightly peaks [*Stroup et al.*, 2007]. Event rates appear to fluctuate on weekly and monthly time scales, but to date only one study has reported a significant long-term trend. At 9°50′N EPR, seismicity is seen to continuously increase by more than 10-fold over a >2-year period building up to an eruption in January 2006 [*Tolstoy et al.*, 2006]. Locations for the first 7 months of that deployment (October 2003–April 2004) show that although the rates continue to climb, the basic structure of the seismicity, and in particular the dense cloud overlying the AMC, does not change in depth or thickness (*Tolstoy et al.*, 2008). This is consistent with an increase in hydrothermal cracking resulting from a buildup in tectonic stresses with increasing time following the last eruption (Buck, manuscript in preparation).

5. EARTHQUAKE CONSTRAINTS ON ERUPTION PROCESSES

Before the recent EPR eruption [*Tolstoy et al.*, 2006; *Cowen et al.*, 2007], most of what we know from seismicity about mid-ocean ridge eruption processes comes from *T*-wave studies, with some contributions from terrestrial seismic stations. Access to the U.S. Navy's SOSUS arrays in the area of the JdFR has revolutionized our understanding of seafloor eruptions [e.g., *Fox*, 1995; *Fox and Dziak*, 1998; *Dziak and Fox*, 1999]. This work, pioneered by Dr. C. Fox of the National Oceanic and Atmospheric Administration-Pacific Marine Environmental Laboratory, was so successful that it was extended to monitoring in the equatorial Pacific [*Fox et al.*, 2001] and the northern Atlantic [e.g., *Smith et al.*, 2002, 2003]. From these data, we have learned enormous amounts about magmatic processes at intermediate-, fast-, and slow-spreading ridges.

Early work on *T*-waves met with some skepticism about the accuracy of *T*-wave locations, despite response cruises being successful in finding fresh lava at the sites they were sent to [e.g., *Embley et al.*, 1995, 1999]. Dike propagation may or may not trigger large magnitude seismicity in and of itself. However, the stress field at the tip of a propagating dike can trigger earthquakes on nearby faults [*Rubin*, 1992]. Therefore, a propagating dike at a mid-ocean ridge may trigger earthquakes on an axial valley wall or other nearby faults that may result in events that look too far off axis to be along the path of dike propagation (Figure 10).

Confirmed eruptions at the intermediate spreading Co-Axial segment [*Fox*, 1995], Axial Volcano [*Dziak and Fox*, 1999], and Gorda [*Fox and Dziak*, 1998] have illustrated

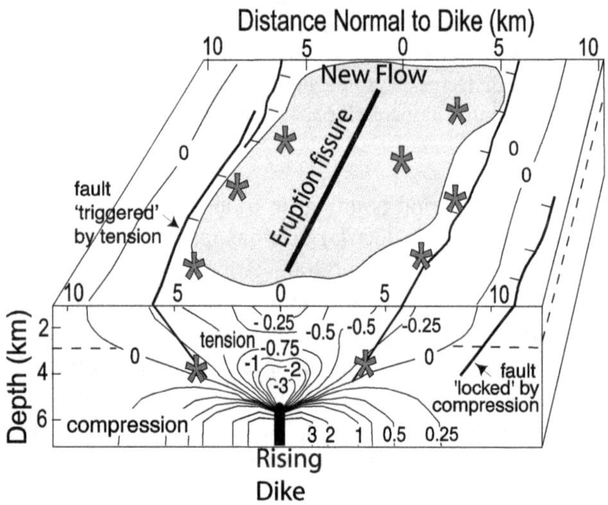

Figure 10. Schematic diagram showing the influence of dike emplacement on the axial stress field, adapted from *Tolstoy et al.* [2001]. A dike-like magma body is modeled with internal magma pressure of 5 MPa above lithostatic pressure extending from a depth of 5–8 km within the brittle mantle. Contours show changes in dike-normal stress and are labeled in MPa. Asterisks show epicenters of earthquakes that may be triggered on bounding wall faults due to the stress changes induced by the propagating dike, with hypocenters shown on the depth cross section. Shallow sources would inhibit slip by increasing normal stresses to the sides.

how volcanic activity at these sites is frequently accompanied by dike propagation at rates of 1–2 km/h in the early part of the eruptive cycle, with activity including thousands of earthquakes, and generally lasting on the order of 10–20 days. The vast majority of the earthquakes are less than a magnitude 4, because they are not detected by nearby land seismic stations. A few events on the order of magnitude 4 or greater are usually observed, however. For the 1998 Axial Volcano eruption, three events of magnitude 4.8, 4.9, and 4.8 were observed on nearby land arrays [*Dziak and Fox*, 1999]. The Gorda eruption had 18 events observed on nearby land arrays, with a maximum magnitude of 4.5, whereas the Co-Axial eruption had none.

For the fast-spreading EPR, the most convincing cases for eruptive activity recorded on the EPR Autonomous hydrophone array [*Fox et al.*, 2001] have been at 8°40′N [*Bohnenstiehl et al.*, 2003], 3°20′N, and possibly 1°45′N [*Tolstoy et al.*, 1999]. The 3°20′N case is the most convincing of these events with dike propagation rates of ~2 km/h in the early part of the swarm, consistent with what has been observed at the Juan de Fuca during confirmed eruptions [*Dziak et al.*, 2007a]. Although not definitively confirmed to be eruptive, largely because of the difficulty in dating very recent activity at such an active spreading center, these events are generally consistent with what was observed using OBSs at 9°50′N in 2006 [*Tolstoy et al.*, 2006]. All four of these events took place on the scale of a few days with most of the activity taking place within the first 24 h, or in one case all the activity taking place in 10 h [*Bohnenstiehl et al.*, 2003]. Between 0 and 4 teleseismic earthquakes were listed in the most sensitive teleseismic catalog (the PIDC) for each of these swarm events, with magnitudes from 3.4 to 4.3. The 2006 eruption at the EPR did not produce any teleseismically recorded events, and the most intense activity appeared to be concentrated in an ~6-h period on both the OBS records [*Tolstoy et al.*, 2006], and recently recovered hydrophone records [*Dziak et al.*, 2007b].

For the slow-spreading MAR, only one confirmed magmatic event has been captured with hydrophone data, and that occurred at the Lucky Strike segment at 36°N near the Azores hotspot in March 2001 [*Dziak et al.*, 2003]. Activity lasted only 1.2 days although 33 events were recorded teleseismically with magnitudes up to a magnitude 5.0 *mb*. The event is interpreted as a dike intrusion rather than a seafloor eruption, based on both the earthquake data and seafloor observations, but nevertheless it provides valuable comparisons with other magmatic activity elsewhere on the global ridge system.

The ultraslow spreading Gakkel Ridge near the North Pole was once thought to be spreading so slowly that it might be amagmatic. However, an enormous seismic swarm in 1999

[*Tolstoy et al.*, 2001], confirmed by sidescan observations to be a likely eruption [*Edwards et al.*, 2001], put this idea to rest. The magmatic activity at this site was recorded entirely using teleseismic stations because no hydrophone or OBS data were available in the Arctic ocean. Swarm activity lasted for ~7 months and 252 teleseismic events were recorded over an ~20-km-long section of ridge, with a maximum magnitude of 5.4 *mb* and 14 events of magnitude 5.0 or greater. The size of earthquakes and distance to the activated faults can also yield information on the depth from which the dike propagated. In the case of the Gakkel Ridge, the dike is interpreted to have been tapped from mantle depths (>5 km), consistent with very cold crust and low magma supply in this area [*Tolstoy et al.*, 2001]. Deep tapping of magma at ultraslow spreading rates is consistent with hydrothermal venting tapping a much deeper heat source at slower-spreading mid-ocean ridges.

A scaling of eruptions with spreading rate is apparent in looking at both the duration of eruptions and maximum magnitude of seismicity associated with them (Figure 11). This suggests that for the colder and slower spreading regimes, a larger scale of eruption is required to break through the thicker brittle layer overlying it. Presumably, magma bodies, although deeper, must be larger to support these larger scale eruptions. This may mean that if eruptions are larger at a

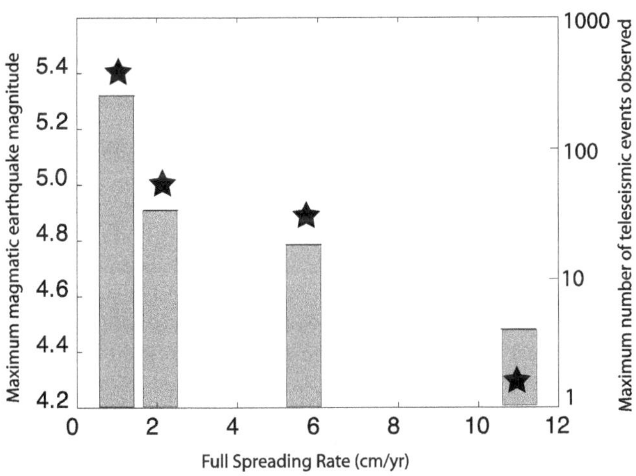

Figure 11. Spreading rate versus the maximum magnitude observed (black star, left scale) as well as maximum number of teleseismic events (gray bars, right scale) for an eruptive event/magmatic swarm. Note that both *y*-axis scales are logarithmic because magnitude is a logarithmic scale. The number of maximum teleseismic events observed is a rather imprecise scale because it depends on proximity to land, and the catalogue used. Nevertheless, a general pattern in number of events observed is apparent, and is consistent with the maximum magnitude observed, as would be predicted from the Guttenberg-Richter power law.

slower spreading rate, then they might be even less frequent than would be anticipated from the simple calculations of expected recurrence rate based on an ~1-m dike intrusion [e.g., *Perfit and Chadwick*, 1998].

6. CONCLUSIONS AND FUTURE DIRECTIONS

Our knowledge of crustal structure and continually improving understanding of dynamic seismic processes provide a picture of a global ridge system where spreading rate, magmatic activity, and hydrothermal processes are inextricably linked [e.g., *Wilcock and Delaney*, 1996]. The less stable, but perhaps larger, heat sources at slower spreading ridges provide larger longer-lived but sparser hydrothermal venting sites. The steady, near-continuous, shallow magma supply at the fast spreading ridges leads to frequent but small-scale vent fields.

Both active and passive source seismology bring a great deal of valuable information to hydrothermal and magmatic processes. 3-D seismic methods bring the promise of fine-scale structural maps of the crust at different depths. 3-D imaging will enable the resolution of much smaller features than previously possible with 2-D methods. Time lapse 3-D imaging can create a 4-D picture of the variability of fine-scale structures and magma bodies, as is now being routinely done in the oil industry to monitor reservoirs. However, the limitation that permeability cannot be imaged through seismic methods remains, and other techniques must be applied to address this critical issue.

Passive monitoring has the ability to capture dynamic processes, and earthquakes are one of the most active and defining characteristics of dynamic processes at mid-ocean ridges. From the meter scale cracking associated with hydrothermal circulation to segment scale shifts that can happen during a magmatic event, earthquake monitoring can capture the movement of fluids as well as tectonic plates. Maintenance of existing hydroacoustic arrays and expansion to other areas, or access to other existing arrays will allow continued monitoring of seafloor spreading events in areas we understand well, as well as potentially covering less studied parts of the globe. As OBS technology becomes more cost-effective and readily available, the ability to deploy more instruments in tightly spaced arrays can provide continuous monitoring of the small earthquakes, potentially illuminating important features of hydrothermal circulation cells through full volcanic cycles.

The advent of the Ocean Observatory Initiative has promise to provide real-time interactive seafloor laboratories that will allow sophisticated experiments to be conducted. For sites without real-time data access, continuing improvements in seafloor technology means that the capability now exists for us to collect multidisciplinary data streams, including seismic data, for extended periods, that when combined will provide new insights into ridge system linkages at scales from the mantle to the microbe.

Acknowledgments. This paper has benefited greatly from discussions with S. M. Carbotte, D. R. Bohnenstiehl, and F. Waldhauser. The paper was significantly improved by constructive comments from W. Wilcock, R. Lowell, and an anonymous reviewer. This work was supported by the National Science Foundation (OCE03-27283). LDEO contribution number 7081.

REFERENCES

Almendros, J., A. H. Barclay, W. S. D. Wilcock, and G. M. Purdy (2000), Seismic anisotropy of the shallow crust at the Juan de Fuca Ridge, *Geophys. Res. Lett.*, *27*, 3109–3112.

Baker, E. T., and T. Urabe (1996), Extensive distribution of hydrothermal plumes along the superfast spreading East Pacific Rise, 13°30′–18°40′S. *J. Geophys. Res.*, *101*, 8685–8695.

Bakun, W. H., and W. B. Joyner (1984), The M_L scale in central California, *Bull. Seismol. Soc. Am.*, *74*, 1827–1843.

Baran, J. M., J. R. Cochran, S. M. Carbotte, and M. R. Nedimović (2005), Variations in upper crustal structure due to variable mantle temperature along the Southeast Indian Ridge, *Geochem., Geophys. Geosyst.*, *6*, Q11002, doi: 10.1029/2005GC000943.

Barclay, A. H., D. R. Toomey, and S. C. Solomon (1998), Seismic structure and crustal magmatism at the Mid-Atlantic Ridge, 35° N, *J. Geophys. Res.*, *103*, 17,827–17,844.

Biondi, B. (2006), 3-D seismic imaging, in *Society of Exploration Geophysicists, Series: Investigations in Geophysics*, edited by G. Gardner, Covington Group, Kansas City, Missouri, 240.

Bohnenstiehl, D. R., and M. Tolstoy (2003), Comparison of teleseismically and hydroacoustically derived earthquake locations along the north-central Mid-Atlantic Ridge and Equatorial East Pacific Rise, *Seismol. Res. Lett.*, *74*, 790–801.

Bohnenstiehl, D. R., M. Tolstoy, D. K. Smith, C. G. Fox, and R. Dziak (2002), Aftershock sequences in the mid-ocean ridge environment: An analysis using hydroacoustic data, *Tectonophysics*, *354*, 49–70.

Bohnenstiehl, D. R., et al. (2003), Anomalous seismic activity at 8°37–42′N on the East Pacific Rise: Hydroacoustic detection and site investigation, *Ridge2000 Events Newsl.*, *1*(1), 18–20.

Bohnenstiehl, D. R., R. P. Dziak, M. Tolstoy, C. G. Fox, and M. Fowler (2004), Temporal and spatial history of the 1999–2000 Endeavour Segment seismic series, Juan de Fuca Ridge, *Geochem., Geophys. Geosyst.*, *5*, Q09003, doi:10.1029/2004GC000735.

Bosch, D., M. Jamais, F. Boudier, A. Nicolas, J. M. Dautria, and P. Agrinier (2004), Deep and high-temperature hydrothermal circulation in the Oman ophiolite—Petrological and isotopic evidence, *J. Petrol.*, *6*, 1181–1208.

Bunch, A. W. H., and B. L. N. Kennett (1980), The crustal structure of the Reykjanes Ridge at 59°30′N, *Geophys. J. R. Astron. Soc.*, *61*, 141–166.

Canales, J. P., J. A. Collins, J. Escartín, and R. S. Detrick (2000), Seismic structure across the rift valley of the Mid-Atlantic Rise at 23°20′ (MARK area): Implications for crustal accretion processes at slow spreading ridges, *J. Geophys. Res.*, *105*, 28,411–28,425.

Canales, J. P., R. A. Sohn, and B. J. deMartin (2007), Crustal structure of the Trans-Atlantic Geotraverse (TAG) segment (Mid-Atlantic Ride, 26°10′N): Implications for the nature of hydrothermal circulation and detachment faulting at slow spreading ridges, *Geochem., Geophys. Geosyst.*, *8*, Q08004, doi:10.1029/2007GC001629.

Cannat, M., et al. (1999), Mid-Atlantic Ridge-Azores hotspot interactions: Along-axis migration of a hotspot-derived event of enhanced magmatism 10 to 3 Ma ago, *Earth Planet. Sci. Lett.*, *173*, 257–269.

Carbotte, S. M., and K. C. Macdonald (1994), Comparison of seafloor tectonic fabric at intermediate, fast, and super fast spreading ridges: Influence of spreading rate, plate motions, and ridge segmentation on fault patterns, *J. Geophys. Res.*, *99*, 13,609–13,631.

Carbotte, S. M., and D. S. Scheirer (2004), Variability of ocean crustal structure created along the global mid-ocean ridge, in *Hydrology of the Oceanic Lithosphere*, edited by E. E. Davis and H. Elderfield, pp. 59–107, Cambridge Univ. Press, New York.

Carbotte, S. M., C. Z. Mutter, J. Mutter, and L. Wu (1998), Insights from a fast-intermediate spreading ridge on the influence of magma supply and spreading rate on crustal magma bodies and emplacement of the extrusive layer, *Geology*, *26*, 455–458.

Carbotte, S. M., R. S. Detrick, A. Harding, J. P. Canales, J. Babcock, G. Kent, E. Van Ark, M. Nedimovic, and J. Diebold (2006), Rift topography linked to magmatism at the intermediate spreading Juan de Fuca Ridge, *Geology*, *34*, 209–212.

Chapman, C. H. (1985), Ray theory and its extensions—WKBJ and Maslov seismograms, *J. Geophys.*, *58*, 27–43.

Cherkaoui, A. S. M., W. S. D. Wilcock, R. A. Dunn, and D. R. Toomey (2003), A numerical model of hydrothermal cooling and crustal accretion at a fast spreading mid-ocean ridge, *Geochem. Geophys. Geosyst.*, *4*(9), 8616, doi: 10.1029/2001GC000215.

Christensen, N. I. (1979), Compressional wave velocities in rocks at high temperatures and pressures, critical thermal gradients, and low-velocity zones, *J. Geophys. Res.*, *84*, 6849–6857.

Christeson, G. L., P. R. Shaw, and J. D. Garmany (1997), Shear and compressional wave structure of the East Pacific Rise, 9°–10°N, *J. Geophys. Res.*, *102*, 7821–7835.

Christeson, G. L., K. D. McIntosh, and J. A. Karson (2007), Inconsistent correlation of seismic layer 2a and lava layer thickness in oceanic crust, *Nature*, *445*, doi:10.1038/nature05517.

Collier, J., and M. C. Sinha (1990), Seismic images of a magma chamber beneath the Lau Basin back-arc spreading center, *Nature*, *346*(6285), 646–648.

Cowen, J. P., et al. (2007), Volcanic eruptions at East Pacific Rise near 9°50′N, *Eos Trans. AGU*, *88*(7), 81.

Cowie, P. A., C. H. Scholz, M. Edwards, and A. Malinverno (1993), Fault strain and seismic coupling on mid-ocean ridges, *J. Geophys. Res.*, *98*, 17,911–17,920.

Crawford, W. C., S. C. Webb, and J. A. Hildebrand (1999), Constraints on melt in the lower crust and Moho at the East Pacific Rise, 9°48′N, using seafloor compliance measurements, *J. Geophys. Res.*, *104*, 2923–2939.

Crone, T. J., and W. S. D. Wilcock (2005), Modeling the effects of tidal loading on mid-ocean ridge hydrothermal systems, *Geochem. Geophys. Geosyst.*, *6*, Q07001, doi:10.1029/2004GC000905.

Crone, T. J., W. S. D. Wilcock, A. H. Barclay, and J. D. Parsons (2006), The sound generated by Mid-Ocean Ridge black smoker hydrothermal vents, *PLoS ONE*, *1*(1), e133, doi:10.1371/journal pone.0000133.

Davis, E. E., K. Wang, R. E. Thomson, K. Becker, and J. F. Cassidy (2001), An episode of seafloor spreading and associated plate deformation inferred from crustal fluid pressure transients, *J. Geophys. Res.*, *106*, 21,953–21,963.

deMartin, B. J., R. A. Sohn, J. P. Canales, and S. E. Humphris (2007), Kinematics and geometry of active detachment faulting beneath the TAG hydrothermal field on the Mid-Atlantic Ridge, *Geology*, *35*(8), 711.

DeMets, C., R. G. Gordon, D. F. Argus, and S. Stein (1994), Effect of recent revision to the geomagnetic reversal time scale on estimates of current plate motions, *Geophys. Res. Lett.*, *21*, 2192–2194.

Detrick, R. S., P. Buhl, E. Vera, J. Mutter, J. Orcutt, J. Madsen, and T. Brocher (1987), Multichannel seismic imaging of a crustal magma chamber along the East Pacific Rise, *Nature*, *326*, 35–41.

Detrick, R. S., A. J. Harding, G. M. Kent, J. A. Orcutt, J. C. Mutter, and P. Buhl (1993), Seismic structure of the southern East Pacific Rise, *Science*, *259*, 499–503.

Detrick, R. S., J. Collins, R. Stephen, and S. Swift (1994), In-situ evidence for the nature of the seismic layer 2/3 boundary in oceanic-crust, *Nature*, 370, 288–290.

Detrick, R. S., D. R. Toomey, and J. A. Collins (1998), Three-dimensional upper crustal heterogeneity and anisotropy around Hole 504B from seismic tomography, *J. Geophys. Res.*, *103*, 30,485–30,504.

Detrick, R. S., J. M. Sinton, G. Ito, J. P. Canales, M. Behn, T. Blacic, B. Cushman, J. E. Dixon, D. W. Graham, and J. J. Mahoney (2002), Correlated geophysical, geochemical, and volcanological manifestations of plume-ridge interaction along the Galápagos Spreading Center, *Geochem. Geophys. Geosyst.*, *3*(10), 8501, doi:10.1029/2002GC000350.

Dunn, R. A., D. R. Toomey, and S. C. Solomon (2000), Three-dimensional seismic structure and physical properties of the crust and shallow mantle beneath the East Pacific Rise at 9°30′N, *J. Geophys. Res.*, *105*, 23,537–23,555.

Dunn, R. A., V. Lekić, R. S. Detrick, and D. R. Toomey (2005), Three-dimensional seismic structure of the Mid-Atlantic Ridge (35°N): Evidence for focused melt supply and lower crustal dike injection, *J. Geophys. Res.*, *110*, B09101, doi: 10.1029/2004JB003473.

Dziak, R. P., and Time Critical Studies Committee Members (2007), Time critical studies update, *Ridge2000 Events Newsl.*, in press.

Dziak, R. P., C. G. Fox, and A. E. Schreiner (1995), The June-July 1993 Seismo-acoustic event at CoAxial segment, Juan de Fuca Ridge: Evidence for a lateral dike injection, *Geophys. Res. Lett.*, *22*, 135–138.

Dziak, R. P., and C. G. Fox (1999), The January 1998 earthquake swarm at Axial Volcano, Juan de Fuca Ridge: Hydroacoustic evidence of seafloor volcanic activity, *Geophys. Res. Lett.*, *26*, 3429–3432.

Dziak, R. P., W. W. Chadwick, C. G. Fox, and R. W. Embley (2003), Hydrothermal temperature changes at the southern Juan de Fuca Ridge associated with Mw 6.2 Blanco Transform earthquake, *Geology*, *31*, 119–122.

Dziak, R. P., D. K. Smith, D. R. Bohnenstiehl, C. G. Fox, D. Desbruyeres, H. Matsumoto, M. Tolstoy, and D. J. Fornari (2004), Evidence of a recent magma dike intrusion at the slow spreading Lucky Strike segment, Mid-Atlantic Ridge, *J. Geophys. Res.*, *109*, B12102, doi:10.1029/2004JB003141.

Dziak, R., B. Chadwick, J. Cowen, E. Baker, R. Embley, D. Bohnenstiehl, and J. Resing (2006), Detecting volcanic events in the Northeast Pacific, *Eos Trans. Am. Geophys. Union*, *87*(4), 37.

Edwards, M. H., G. J. Kurras, M. Tolstoy, D. R. Bohnenstiehl, B. J. Coakley, and J. R. Cochran (2001), Evidence of recent volcanic activity on the ultra-slow spreading Gakkel Ridge, *Nature*, *409*, 808–812.

Embley, R. W., W. W. Chadwick Jr., I. R. Jonasson, D. A. Butterfield, and E. T. Baker (1995), Initial results of the rapid response to the 1993 CoAxial event: Relationships between hydrothermal and volcanic processes, *Geophys. Res. Lett.*, *22*, 143–146.

Embley, R. W., W. W. Chadwick Jr., D. Clague, and D. Stakes (1999), 1998 eruption of Axial Volcano: Multibeam anomalies and sea-floor observations, *Geophys. Res. Lett.*, *26*, 3425–3428.

Ewing, J., and M. Ewing (1959), Seismic refraction measurements in the Atlantic Ocean Basins, in the Mediterranean Sea, on the Mid-Atlantic Ridge, and in the Norwegian Sea, *Geol. Soc. Am. Bull.*, *70*, 291.

Fornari, D. J., T. Shank, K. L. Von Damm, T. K. P. Gregg, M. Lilley, G. Levai, A. Bray, R. M. Haymon, M. R. Perfit, and R. Lutz (1998), Time-series temperature measurements at high-temperature hydrothermal vents, East Pacific Rise 9 degrees 49′–51′N: Evidence for monitoring a crustal cracking event, *Earth Planet. Sci. Lett.*, *160*, 419–431.

Foulger, G. R., and R. E. Long (1984), Anomalous focal mechanism solutions: Evidence for tensile crack formation on an accreting plate boundary, *Nature*, *310*, 43–45.

Fox, C. G. (1995), Special collection on the June 1993 volcanic eruption on the CoAxial segment, Juan de Fuca Ridge, *Geophys. Res. Lett.*, *22*, 129–130.

Fox, C. G., and R. P. Dziak (1998), Hydroacoustic detection of volcanic activity on the Gorda Ridge, February–March 1996, *Deep-Sea Res. Part II*, *45*, 2513–2530.

Fox, C. G., R. P. Dziak, H. Matsumoto, and A. E. Schreiner (1993), Potential for monitoring low-level seismicity on the Juan de Fuca Ridge using military hydrophone arrays, *Mar. Technol. Soc. J.*, *27*, 22–30.

Fox, C. G., W. E. Radford, R. P. Dziak, T.-K. Lau, H. Matsumoto, and A. E. Schreiner (1995), Acoustic detection of a seafloor spreading episode on the Juan de Fuca Rise using military hydrophone arrays, *Geophys. Res. Lett.*, *22*, 131–134.

Fox, C. G., H. Matsumoto, and T.-K. A. Lau (2001), Monitoring Pacific Ocean seismicity from an autonomous hydrophone array, *J. Geophys. Res.*, *106*, 4183–4206.

Gaherty, J. B., D. Lizarralde, J. A. Collins, G. Hirth, and S. Kim (2004), Mantle deformation during slow seafloor spreading constrained by observations of seismic anisotropy in the western Atlantic, *Earth Planet. Sci. Lett.*, *228*, 255–265.

Garrido, C. J., P. B. Kelemen, and G. Hirth (2001), Variation of cool rate with depth in lower crust formed at an oceanic spreading ridge: Plagioclase crystal size distributions in gabbros from the Oman ophiolite, *Geochem., Geophys. Geosyst.*, *2*(10), doi:10.1029/2000GC000136.

German, C. R., and J. Lin (2004), The thermal structure of the oceanic crust, ridge-spreading and hydrothermal circulation: How well do we understand their inter-connections? in *Mid-Ocean Ridges: Hydrothermal Interactions Between the Lithosphere and Oceans*, *Geophys. Monogr. Ser.*, vol. 148, edited by C. R. German, J. Lin, and L. M. Parson, pp. 1–18, AGU, Washington, D. C.

Golden, C. E., S. C. Webb, and R. A. Sohn (2003), Hydrothermal microearthquake swarms beneath active vents at Middle Valley, northern Juan de Fuca Ridge, *J. Geophys. Res.*, *108*(B1), 2027, doi: 10.1029/2001JB000226.

Gregory, R. T., and H. P. Taylor Jr. (1981), An oxygen isotope profile in a section of Cretaceous oceanic crust, Samail ophiolite, Oman: Evidence for $\sigma^{18}O$ buffering of oceans by deep (>5 km) seawater-hydrothermal circulation at mid-ocean ridges, *J. Geophys. Res.*, *86*, 2737–2755.

Gutenberg, B., and C. F. Richter (1944), Frequency of earthquakes in California, *Bull. Seismol. Soc. Am.*, *34*, 185–188.

Holmes, R., M. Tolstoy, J. R. Cochran, and J. S. Floyd (2005), Crustal structure along the Southeast Indian Ridge from seismic refraction data, *Eos Trans. AGU*, *86*(52), Fall Meet Suppl., Abstract T33A-0513.

Humphris, S. E., and J. R. Cann (2000), Constraints on the energy and chemical balances of the modern TAG and ancient Cyprus seafloor sulfide deposits, *J. Geophys. Res.*, *105*, 28,477–28,488.

Hussenoeder, S. A., J. A. Collins, G. M. Kent, R. S. Detrick, and the TERA Group (1996), Seismic analysis of the axial magma chamber reflector along the southern East Pacific Rise from conventional reflection profiling, *J. Geophys. Res.*, *101*, 22,087–22,105.

Johnson, H. P., M. Hutnak, R. P. Dziak, C. G. Fox, I. Urcuyo, J. P. Cowen, J. Nabelek, and C. Fisher (2000), Earthquake-induced changes in the hydrothermal system on the Juan de Fuca mid-ocean ridge, *Nature*, *407*, 174–177.

Johnson, H. P., R. P. Dziak, C. R. Fisher, C. G. Fox, and M. J. Pruis (2001), Earthquakes' impact on hydrothermal systems may be far-reaching, *Eos Trans. AGU*, *82*(21), 233.

Jokat, W., and M. C. Schmidt-Aursch (2007), Geophysical characteristics of the ultraslow spreading Gakkel Ridge, Arctic Ocean, *Geophys. J. Int.*, *168*, 983–998.

Jupp, T. E., and A. Schultz (2004), A poroelastic model for the tidal modulation of seafloor hydrothermal systems, *J. Geophys. Res.*, *109*, B03105, doi:10.1029/2003JB002583.

Kelley, D. S., et al. (2005), A serpentinite-hosted ecosystem: The Lost City hydrothermal field, *Science*, *307*, 1428, doi:10.1126/science.1102556.

Kent, G. M., A. J. Harding, and J. A. Orcutt (1993), Distribution of magma beneath the East Pacific Rise between the Clipperton Transform and the 9°17′N Deval from forward modeling of common depth point data, *J. Geophys. Res.*, *98*, 13,945–13,969.

Kent, G. M., et al. (2000), Evidence from three-dimensional seismic reflectivity images for enhanced melt supply beneath mid-ocean ridge discontinuities, *Nature*, *406*, 614–618.

Lilley, M. D., D. A. Butterfield, J. E. Lupton, and E. J. Olson (2003), Magmatic events can produce rapid changes in hydrothermal vent chemistry, *Nature*, *442*, 878–881.

Lowell, R. P., and Y. Yao (2002), Anhydrite precipitation and the extent of hydrothermal recharge zones at ocean ridge crests, *J. Geophys. Res.*, *107*(B9), 2183, doi: 10.1029/2001JB001289.

Lowell, R. P., B. W. Crowell, S. R. Gosnell, K. C. Lewis, L. Liu, and Y. Yang (2007), Modeling multiphase, multi-component processes at oceanic spreading centers: Magma to microbe, this volume.

Maclennan, J., T. Hulme, and S. C. Singh (2004), Thermal models of oceanic crustal accretion: Linking geophysical, geological and petrological observations, *Geochem. Geophys. Geosys.*, *5*, Q02F25, doi:10.1029/2003GC000605.

McClain, J. S., M. L. Begnaud, M. A. Wright, J. Fondrk, and G. K. Von Damm (1993), Seismicity and tremor in a submarine hydrothermal field: The northern Juan de Fuca Ridge, *Geophys. Res. Lett.*, *20*, 1883–1886.

Menke, W., and D. Schaff (2004), Absolute earthquake locations with differential data, *Bull. Seismol. Soc. Am.*, *94*, 2254–2264.

Michael, P. J., et al. (2003), Magmatic and amagmatic seafloor generation at the ultraslow-spreading Gakkel Ridge, Arctic Ocean, *Nature*, *423*, 956–961.

Miller, A. D., B. R. Julian, and G. R. Foulger (1998), Three-dimensional seismic structure and moment tensors of non-double-couple earthquakes at the Hengill-Grensdalur volcanic complex, Iceland, *Geophys. J. Int.*, *133*, 309–325.

Mogi, K. (1963), Some discussions on aftershocks, foreshocks and earthquake swarms—The fracture of a semi-infinite body caused by an inner stress origin and its relation to the earthquake phenomena, *Bull. Earthquake Res. Inst. Univ. Tokyo*, *41*, 615–658.

Morton, J. L., N. H. Sleep, W. R. Normark, and D. H. Tompkins (1987), Structure of the southern Juan de Fuca Ridge from seismic reflection records, *J. Geophys. Res.*, *92*, 11,315–11,326.

Muller, M. R., T. A. Minshull, and R. S. White (1999), Segmentation and melt supply at the Southwest Indian Ridge, *Geology*, *10*, 867–870.

Mutter, J. C., S. M. Carbotte, W. Su, L. Xu, P. Buhl, R. S. Detrick, G. A. Kent, J. A. Orcutt, and A. J. Harding (1995), Seismic images of active magma systems beneath the East Pacific Rise between 17 050 and 17 350S, *Science*, *268*, 391–395.

Perfit, M. R., and W. W. Chadwick Jr. (1998), Magmatism at mid-ocean ridges: Constraints from volcanological and geochemical investigations, in *Faulting and Magmatism at Mid-Ocean Ridges*, *Geophys. Monogr. Ser.*, vol. 106, edited by W. R. Buck et al., pp. 59–115, AGU, Washington, D. C.

Phipps Morgan, J., and Y. J. Chen (1993), Dependence of ridge-axis morphology on magma supply and spreading rate, *Nature*, *364*, 706–708.

Purdy, G. M., and R. S. Detrick (1986), Crustal structure of the Mid-Atlantic Ridge at 23°N from seismic refraction studies, *J. Geophys. Res.*, *91*, 3739–3762.

Purdy, G. M., L. S. L. Kong, G. L. Christeson, and S. C. Solomon (1992), Relation between spreading rate and the seismic structure of mid-ocean ridges, *Nature*, *355*, 815–817.

Ramondenc, P., L. N. Germanovich, and R. P. Lowell (2007), Modeling the hydrothermal response to earthquakes with application to the 1995 event at 9°50′N, East Pacific Rise, this volume.

Riedesel, M., J. A. Orcutt, K. C. MacDonald, and J. S. McClain (1982), Microearthquakes in the black smoker hydrothermal field, East Pacific Rise at 21°N, *J. Geophys. Res.*, *87*, 10,613–10,623.

Rubin, A. M. (1992), Dike-induced faulting and graben subsidence in volcanic rift zones, *J. Geophys. Res.*, *97*, 1839–1858.

Scheirer, D. S., D. J. Fornari, S. E. Humphris, and S. Lerner (2000), High-resolution seafloor mapping using the DSL-120 sonar system: Quantitative assessment of sidescan and phase-bathymetry data from the Lucky Strike Segment of the Mid-Atlantic Ridge, *Mar. Geophys. Res.*, *21*, 121–142.

Scheirer, D. S., Shank, T. M., and D. J. Fornari (2006), Temperature variations at diffuse and focused flow hydrothermal vent sites along the northern East Pacific Rise. *Geochem. Geophys. Geosyst.*, *7*, Q03002, doi:10.1029/2005GC001094.

Schreiner, A. E., C. G. Fox, and R. P. Dziak (1995), Spectra and magnitudes of T-waves from the 1993 earthquake swarm on the Juan de Fuca Ridge, *Geophys. Res. Lett.*, *22*, 139–142.

Schultz, A., and H. Elderfield (1997), Controls on the physics and chemistry of seafloor hydrothermal circulation, *Philos. Trans. R. Soc. Lond.*, *355*, 387–425.

Seyfried, W. E., Jr., J. S. Seewald, M. E. Berndt, K. Ding, and D. I. Foustoukos (2003), Chemistry of hydrothermal vent fluids from the Main Endeavour Field, northern Juan de Fuca Ridge: Geochemical controls in the aftermath of June 1999 seismic events, *J. Geophys. Res.*, *108*(B9), 2429, doi:10.1029/2002JB001957.

Shearer, P., and J. A. Orcutt (1985), Anisotropy in the oceanic lithosphere—Theory and observations from the Ngendei seismic refraction experiment in the southwest Pacific, *Geophys. J. R. Astron. Soc.*, 80, 493–526.

Shearer, P., and J. A. Orcutt (1986), Compressional and shear-wave anisotropy in the oceanic lithosphere—The Ngendei seismic refraction experiment, *Geophys. J. R. Astron. Soc.*, *87*, 967–1003.

Shearer, P. M. (1997), Improving local earthquake locations using the L1 norm and waveform cross correlation: Application to the Whittier Narrows, California, aftershock sequence, *J. Geophys. Res.*, *102*, 8269–8283.

Sheriff, R. E., and L. P. Geldart (1995), *Exploration Seismology*, Cambridge University Press, Cambridge, U.K.

Singh, S. C., G. M. Kent, J. S. Collier, A. J. Harding, and J. A. Orcutt (1998), Melt to mush variations in crustal magma properties along the ridge crest at the southern East Pacific Rise, *Nature*, *394*, 874–878.

Singh, S. C., J. S. Collier, A. J. Harding, G. M. Kent, and J. A. Orcutt (1999), Seismic Evidence for a hydrothermal layer above the solid roof of the axial magma chamber at the southern East Pacific Rise, *Geology*, *27*, 219–222.

Singh, S. C., W. C. Crawford, H. Carton, T. Seher, V. Combier, M. Canat, J. P. Canales, D. Dusunur, J. Escartin, and J. M. Miranda (2006a), Discovery of a magma chamber and faults beneath a Mid-Atlantic Ridge hydrothermal field, *Nature*, *442*, 1029–1032.

Singh, S. C., et al. (2006b), Seismic reflection images of the Moho underlying sills at the East Pacific Rise, *Nature*, *442*, 287–290.

Sinha, M. C., D. A. Navin, L. M. MacGregor, S. Constable, C. Pierce, A. White, G. Heinson, and M. A. Inglis (1997), Evidence for accumulated melt beneath the slow spreading Mid-Atlantic Ridge, in *Mid-Ocean Ridges: Dynamics of Processes Associated With Creation of New Ocean Crust, Philos. Trans. R. Soc. London,* Ser. A, vol. 355, edited by J. R. Cann, H. Elderfield, and A. Laughton, pp. 233–253, Cambridge University Press, Cambridge, U.K.

Smith, D. K., M. Tolstoy, C. G. Fox, D. R. Bohnenstiehl, H. Matsumoto, and M. J. Fowler (2002), Hydroacoustic monitoring of seismicity at the slow-spreading Mid-Atlantic Ridge, *Geophys. Res. Lett.*, *29*(11), 1518, doi:10.1029/2001GL013912.

Smith, D. K., J. Escartin, M. Cannat, M. Tolstoy, C. G. Fox, D. R. Bohnenstiehl, and S. Bazin (2003), Spatial and temporal distribution of seismicity along the northern Mid-Atlantic Ridge (15°–35°N), *J. Geophys. Res.*, *108*(B3), 2167, doi: 10.1029/2002JB001964.

Spudich, P., and J. Orcutt (1980), A new look at the seismic velocity structure of the oceanic crust, *Rev. Geophys.*, *18*, 627–645.

Sohn, R., D. J. Fornari, K. Von Damm, S. Webb, and J. Hildebrand (1998), Seismic and hydrothermal evidence for a propagating cracking event on the East Pacific Rise crest at 9°50′N, *Nature*, *396*, 159–161.

Sohn, R. A. (2007), Stochastic analysis of exit fluid temperature records from the active TAG hydrothermal mound (Mid-Atlantic Ridge, 26°N): 2. Hidden Markov Models of flow episodes, *J. Geophys. Res.*, *112*, B09102, doi:10.1029/2007JB004961.

Sohn, R. A., J. A. Hildebrand, and S. C. Webb (1999), A microearthquake survey of the high-temperature vent fields on the volcanically active East Pacific Rise (9°50′N), *J. Geophs. Res.*, *104*, 25,367–25,377.

Sohn, R. A., A. H. Barclay, and S. C. Webb (2004), Microearthquake patterns following the 1998 eruption of Axial Volcano, Juan de Fuca Ridge: Mechanical relaxation and thermal strain, *J. Geophys. Res.*, *109*, B01101, doi:10.1029/2003JB002499.

Stein, S., and M. Wysession (2003), *An Introduction to Seismology, Earthquakes, and Earth Structure*, Blackwell Publishing, Oxford, 498 pp., ISBN 0-86542-087-5.

Stroup, D. F., D. R. Bohnenstiehl, M. Tolstoy, F. Waldhauser, and R. T. Weekly (2007), Pulse of the seafloor: Tidal triggering of microearthquakes at 9°50′N East Pacific Rise, *Geophys. Res. Lett.*, *34*, L15301, doi:10.1029/2007GL030088.

Tolstoy, I., and W. M. Ewing (1950), The T phase of shallow-focus earthquakes, *Bull. Seismol. Soc. Am.*, *40*, 25–51.

Tolstoy, M., A. J. Harding, and J. A. Orcutt (1993), Crustal thickness on the Mid-Atlantic Ridge—Bull's eye gravity anomalies and focused accretion, *Science*, *262*, 726–729.

Tolstoy, M., A. J. Harding, J. A. Orcutt, and TERA Group (1997), Deepening of the axial magma chamber on the southern East Pacific Rise toward the Garret Fracture Zone, *J. Geophys. Res.*, *102*, 3097–3108.

Tolstoy, M., D. J. Fornari, and C. G. Fox (1999), Detailed Investigation of *T*-phase swarms on the East Pacific Rise, *Eos Trans. AGU.*

Tolstoy, M., D. Bohnenstiehl, M. Edwards, and G. Kurras (2001), The seismic character of volcanic activity at the ultra-slow spreading Gakkel Ridge, *Geology*, *29*, 1139–1142.

Tolstoy, M., F. L. Vernon, J. A. Orcutt, and F. K. Wyatt (2002), The breathing of the seafloor: Tidal correlations of seismicity on Axial volcano, *Geology*, *30*, 503–506.

Tolstoy, M., et al. (2006), A seafloor spreading event captured by seismometers, *Science*, DOI: 10.1126/science.1133950.

Tolstoy, M., F. Waldhauser, D. R. Bohnenstiehl, R. T. Weekly, and W.–Y. Kim (2008), Seismic identification of along-axis hydrothermal flow on the East Pacific Rise, *Nature*, doi:10.1038/nature06424.

Toomey, D. R., G. M. Purdy, S. C. Solomon, and W. S. D. Wilcock (1990), The 3-dimensional seismic velocity structure of the East Pacific Rise near latitude 9-degrees–30′N, *Nature*, *347*, 639–645.

Toomey, D. R., S. C. Solomon, and G. M. Purdy, (1994), Tomographic imaging of the shallow crustal structure of the East Pacific Rise at 9°30′N, *J. Geophys. Res.*, *99*, 24,135–24,157.

Tréhu, A. M., and S. C. Solomon (1983), Earthquakes in the Orozco Transform Zone: Seismicity, source mechanisms and tectonics, *J. Geophys. Res.*, *88*, 8203–8225.

Van Ark, E. M., R. S. Detrick, J. P. Canales, S. M. Carbotte, A. J. Harding, G. M. Kent, M. R. Nedimovic, W. S. D. Wilcock, J. B. Diebold, and J. M. Babcock (2007), Seismic structure of the Endeavour Segment, Juan de Fuca Ridge: Correlations with seismicity and hydrothermal activity, *J. Geophys. Res.*, *112*, B02401, doi:10.1029/2005JB004210.

Vera, E. E., J. C. Mutter, P. Buhl, J. A. Orcutt, A. J. Harding, M. E. Kappus, R. S. Detrick, and T. M. Brocher (1990), The structure of 0- to 0.2-m.y.-old oceanic crust at 9°N on the East Pacific Rise from expanded spread profiles, *J. Geophys. Res.*, *95*, 15,529–15,556.

Waldhauser, F., and W. L. Ellsworth (2000), A double-difference earthquake location algorithm: Method and application to the northern Hayward fault, California, *Bull. Seismol. Soc. Am.*, *6*, 1,353–1,368.

Waldhauser, F., W. L. Ellsworth, and A. Cole (1999), Slip-parallel seismic lineations on the northern Hayward fault, California, *Geophys. Res. Lett.*, *26*, 3525–3528.

West, M. E., W. Menke, M. Tolstoy, S. C. Webb, and R. Sohn (2001), Magma storage beneath Axial volcano on the Juan de Fuca mid-ocean ridge, *Nature*, *413*, 833–836.

White, R. S., and R. B. Whitmarsh (1984), An investigation of seismic anisotropy due to cracks in the upper oceanic-crust at 45-degrees N, Mid-Atlantic Ridge, *Geophys. J. Royal Astron. Soc.*, *79*, 439–467.

White, R. S., D. McKenzie, and R. K. O'Nions (1992), Oceanic crustal thickness from seismic measurements and rare earth element inversions, *J. Geophys. Res.*, *97*, 19,683–19,715.

Wilcock, W. S. D. (2001), Tidal triggering of microearthquakes on the Juan de Fuca Ridge, *Geophys. Res. Lett.*, *28*, 3999–4002.

Wilcock, W. S. D. (2004), Physical response of mid-ocean ridge hydrothermal systems to local earthquakes, *Geochem. Geophys. Geosyst.*, *5*, Q11009, doi:10.1029/2004GC000701.

Wilcock, W. S. D., and J. R. Delaney (1996), Mid-ocean ridge sulfide deposits: Evidence for heat extraction from magma chambers or cracking fronts? *Earth Planet. Sci. Lett.*, *145*, 49–64.

Wilcock, W. S. D., S. D. Archer, and G. M. Purdy (2002), Microearthquakes on the Endeavour segment of the Juan de Fuca Ridge. *J. Geophys. Res.*, *107*(B12), 2336, doi:10.1029/2001JB000505.

Wilson, D. S. (1993), Confidence intervals for motion and deformation of the Juan de Fuca plate, *J. Geophys. Res.*, *98*, 16,053–16,071.

Wilson, D. S., et al. (2006), Drilling to gabbro in intact ocean crust, *Science*, *312*, 1016–1020.

Wolfe, C. J., G. M. Purdy, D. R. Toomey, and S. C. Solomon (1995), Microearthquake characteristics and crustal velocity structure at 29°N on the Mid-Atlantic Ridge: The architecture of a slow spreading segment, *J. Geophys. Res.*, *100*, 24,449–24,472.

Wyss, M. (1973), Towards a physical understanding of earthquake frequency distribution, *Geophys. J. R. Astron. Soc.*, *31*, 341–359.

M. Tolstoy, Lamont-Doherty Earth Observatory of Columbia University, 61 Route 9W, Palisades, NY 10964-8000, USA.

Modeling Hydrothermal Response to Earthquakes at Oceanic Spreading Centers

Pierre Ramondenc[1] and Leonid N. Germanovich

Georgia Institute of Technology, School of Civil and Environmental Engineering, Atlanta, Georgia, USA

Robert P. Lowell

Virginia Polytechnic Institute & State University, Department of Geosciences, Blacksburg, Virginia, USA

We develop a two-branch single-pass modeling approach to investigate the response of hydrothermal systems to earthquakes at oceanic spreading centers. In this approach, deep circulation gives rise to high-temperature fluids that are assumed to arrive at the seafloor as focused, black smoker-like vents, whereas shallow circulation within the extrusives results in low-temperature diffuse flow. The diffuse flow is a mixture of black smoker fluid and seawater; consequently, seismic events that alter the permeability of the crust may affect both high-temperature vents and the low-temperature diffuse flow. In this model, ascending fluid in both the high- and low-temperature branches is nearly isothermal except within thin boundary layers at the seafloor and at the junction where fluid mixing occurs. As a result, thermal perturbations at depth do not need to traverse the entire upflow zone, but rather just need to affect the boundary layers. We show that cracking events, which generate new permeability and thereby increase the rate of heat transfer in a narrow zone at the base of the system, do not give rise to rapid temperature changes at the seafloor; however, only modest increases in permeability occurring over a region of a few hundred meters within the deep discharge zone are sufficient to generate observable thermal perturbations. Moreover, the response of the coupled circulation systems simultaneously results in an initial positive temperature pulse as well as the observed longer-lived temperature decrease in both the high- and low-temperature fluid.

1. INTRODUCTION

High-temperature hydrothermal systems at oceanic spreading centers are characterized by a complex interplay among magmatic, tectonic, and biogeochemical reactive transport processes. Magmatic processes provide the principal heat source, whereas all three processes affect the permeability structures through which the fluid circulates. A complete understanding of hydrothermal systems at oceanic spreading centers thus requires the development of integrated mathematical models that describe the combined effects of these processes. Although steady-state models provide important insight into the behavior of hydrothermal systems, much information can be obtained by investigating the response of the system as a result of temporal changes in various forcing

[1]Now at Schlumberger, Integrated Productivity and Conveyance Centre, Sugar Land, Texas 77478-3154, USA.

Magma to Microbe: Modeling Hydrothermal Processes at Ocean Spreading Centers
Geophysical Monograph Series 178

10.1029/178GM06

functions. *Lowell et al.* [2008] discuss some temporal aspects of magmatic and biogeochemical processes in the context of multiphase, multicomponent fluid flow. In this work we focus on the volcanic-tectonic aspects by investigating the response of hydrothermal systems to earthquakes. Because tectonic processes, including earthquakes, may be associated with magmatic processes such as diking, the separation between magmatic and tectonic processes is not pure. We introduce this material by briefly reviewing issues related to the detection and monitoring of the earthquakes themselves and then by reviewing the limited data on the hydrothermal response. A more thorough analysis of earthquakes at oceanic spreading centers is given by *Tolstoy* [2008].

1.1. Magmatic and Tectonic Events at Mid-Ocean Ridges

Despite considerable improvements over the past two decades, the characterization of magmatic and tectonic events at mid-ocean ridges remains difficult. The first discoveries of diking and eruptive events at the seafloor were serendipitous [*Baker et al.*, 1987; *Haymon et al.*, 1993], because earthquakes associated with volcanic events at fast and intermediate spreading ridges generally have magnitudes below $M = 4$, which is the threshold for global detection by land-based seismometers [*Fox et al.*, 2001]. *Fox et al.* [1993/1994] showed that hydrophone arrays recording tertiary waves (or T waves) could detect seismicity (as low as $M = 1.8$). Their work and the use of the U.S. Navy's Sound Surveillance System (SOSUS) [e.g., *Cowen et al.*, 2004], allowed a better monitoring of the activity of the intermediate spreading Juan de Fuca Ridge in the northeast Pacific Ocean, thought to be "aseismic" up to that point [*Dziak et al.*, 2006]. Today, Iceland and the northeast Pacific are the only parts of the mid-ocean ridge system covered by a continuous real-time hydroacoustic monitoring system [*Embley and Lupton*, 2004]. This remains, at present, the best available method to remotely detect seafloor volcanic and tectonic activity. Other places, such as the northern East Pacific Rise [*Fox et al.*, 2001] and the northern Mid-Atlantic Ridge [*Smith et al.*, 2002], have benefited from autonomous hydrophone arrays, whose performances are comparable to SOSUS. Deployed in the late 1990s, these systems are permanent but preclude real-time monitoring of data being collected. They need to be maintained periodically, and earthquakes are only available after a certain time lag, which can reach ~2.5 years [*Dziak et al.*, 2004]. Both hydrophone arrays allowed the detection of several earthquake swarms that have been associated with volcanic activity [*Fox et al.*, 2001; *Smith et al.*, 2002]. In particular, the March 2001 Lucky Strike event [*Dziak et al.*, 2004] on the Mid-Atlantic Ridge was interpreted as a dike intrusion in the shallow crust, the first dike emplacement episode that was detected along the slow-spreading Mid-Atlantic Ridge.

The detection of seismic activity by hydroacoustic arrays needs to be followed by response cruises for a detailed assessment of the events and actual in situ measurements [*Cowen et al.*, 2004]. In this respect, SOSUS real-time monitoring is particularly interesting because it considerably reduces the response time. Since its implementation in June 1993, six response cruises have been organized to look for magmatic and/or hydrothermal activity based on the seismic characteristics of the detected swarms (see Figure 1 of *Dziak et al.* [2006]). Three of these events were interpreted as dike injection with eruption at the seafloor, as response cruises found fresh lava flows, microbial-mineral flocs and enhanced chronic venting: in June 1993 on the CoAxial segment (Juan de Fuca Ridge) [*Baker et al.*, 1995, 1998; *Dziak et al.*, 1995; *Embley et al.*, 1995; *Fox*, 1995; *Fox et al.*, 1995; *Juniper et al.*, 1995], in February 1996 on the North Gorda Ridge [*Baker*, 1998; *Fox and Dziak*, 1998], and in January 1998 at Axial Volcano (Juan de Fuca Ridge) [*Baker et al.*, 1999; *Dziak and Fox*, 1999; *Embley and Baker*, 1999]. The other three events provided valuable information through their unexpected responses. For example, the 2001 Gorda Ridge swarm showed that the lack of lateral migration of earthquake epicenters may be an indication that magma failed to reach the seafloor. The 2001 Middle Valley (Juan de Fuca Ridge) event was the first to occur at a sedimented ridge and the absence of water column signal may have been due to the sediment cap (~100 m thick) that prevented the dike from reaching the seafloor [*Cowen et al.*, 2004]. Also, in contrast to the usual increase in hydrothermal discharge, the fluid pressure transients were found to be negative [*Davis et al.*, 2004], perhaps as the result of dilatation of the upper permeable crust. The most recent earthquake swarm on the Juan de Fuca Ridge occurred on January 27, 2005. Despite a lower rate and distance of migration, a response cruise was sent out, but it did not find any sign of a seafloor eruption or the release of an event plume [*Dziak et al.*, 2006]. Yet, this sixth response effort, along with the analysis of all the others, allowed *Dziak et al.* [2006] to identify two factors that may indicate eruption and hydrothermal discharge during an event: the earthquake migration rate and timing.

1.2. The Hydrothermal Response

Although the detection and analysis of seafloor seismic events has progressed over the past two decades, it provides only one piece of the information needed to understand the response of seafloor hydrothermal systems to tectonic and/or magmatic perturbations in the crust. Vent fluid temperature and flux data are also needed. Yet, instrumentation is seldom deployed and recording during the course of a perturbation. Most examples concern short-term experiments that happened to be deployed at the right time. Axial Volcano, on the Juan de Fuca Ridge, represents perhaps the best example of a long-term, multidisciplinary seafloor observatory that has recorded data during a seafloor spreading event [*Baker et al.*, 1999; *Embley and Baker*, 1999]. The temperature data were of particular interest, featuring a nearly instantaneous increase in the hydrothermal output just after the eruption, followed by a gradual decrease for 2 weeks, and again a slight increase to stabilize at a value approximately 0.1°C higher than the pre-event level [*Baker et al.*, 1999]. The 9°50′N area of the East Pacific Rise has also benefited from continuous temperature monitoring for nearly two decades [*Scheirer et al.*, 2006]. The most recent spreading event was recorded in this area in 2006 [*Tolstoy et al.*, 2006].

The best documented cases of hydrothermal response to crustal perturbation are associated with two noneruptive events. On the Endeavour segment of the Juan de Fuca Ridge, the pattern of the June 1999 swarm contrasted with the usual submarine volcanic activity pattern. Instead of steady low-level seismicity, it featured a main shock followed by 5 days of aftershocks that decayed quasi-exponentially and seemed devoid of volcanic tremor [*Johnson et al.*, 2000]. The ridge community decided not to send a response cruise after the detection by SOSUS [*Cowen et al.*, 2004]. Submersible dives in the area, which occurred 3 months later, did not find evidence of volcanic activity, but observed an increase of the same particulate matter ejected by "snowblower" vents associated with eruptive events [*Johnson et al.*, 2000]. Vent fluid chemistry changes [*Lilley et al.*, 2003; *Seewald et al.*, 2003; *Seyfried et al.*, 2003] showed strong evidence of subseafloor magmatic activity. Moreover, a multichannel seismic reflection experiment conducted across and along the full length of the Juan de Fuca Ridge in 2002 clearly detected an axial magma chamber 2.3–2.6 km below the zone where the swarm occurred [*Detrick et al.*, 2002]. Based on these data and reanalysis of the hydroacoustic data, *Bohnenstiehl et al.* [2004] suggested that the swarm resulted from lateral propagation of a dike initiated from the shallow magma chamber.

Continuous recording of temperature evolution on the Endeavour segment at three diffuse vents situated tens of kilometers from the swarm epicenters showed that the previous 10 months were stable. Fluid temperatures first started increasing 4 (±1) days after the onset of seismicity at the Easter Island site, situated within 10 m of a high-temperature smoker. The other two sites, situated several hundreds of meters away from the adjacent high-temperature fields and 2 km apart from each other, showed a temperature increase approximately 11 days after the initiation of seismicity. Once the maximum rise was reached, large-amplitude oscillations (up to ±5°C) took place at the three sites, with periods ranging between 8 and 12 days, and lasted for approximately a month. Although the oscillations were correlated within each site, they showed poor correlation between them [*Johnson et al.*, 2000]. *Johnson et al.* [2000] suggested that the oscillations resulted from the nonuniform advance of a cracking front at depth to mine heat, plugging and unblocking due to biological activity in the subsurface, or the unsteady evolution of convection toward a new equilibrium after the disturbance. Based on the temperature records and assuming the reservoir of crustal fluid was a porous medium, they estimated that the vent flow rate had increased by 1 order of magnitude over the course of at least 80 days along the whole ridge segment. Vent fluid monitoring at Axial Seamount, approximately 220 km from the epicenter, also showed a temperature increase after the onset of seismicity. This occurred no more than 8 hours after the main shock, sooner than at vent fields located much closer to the epicenters [*Johnson et al.*, 2001]. *Davis et al.* [2001] also reported crustal fluid pressure variations in a series of Ocean Drilling Program boreholes, which they interpreted as the result of partially aseismic fault-related slip.

Another significant event occurred in March 1995 in the 9°50′N area of the East Pacific Rise, where an array of nine ocean-bottom seismometers had been deployed for a microearthquake survey [*Sohn et al.*, 1998, 1999]. This site has been the focus of continuous investigation since the fortuitous discovery of a fresh volcanic eruption in 1991 [*Haymon et al.*, 1993]. The three-month deployment of an ocean bottom seismometer array allowed the detection of 283 microearthquakes (with local magnitudes of $-1 \leq M_L \leq 0$) whose epicenters formed two separate clusters along the ridge axis. The larger was situated north of the 9°50′N area and close to Bio9 and P vents, and the smaller was located further south near the Tube Worm Pillar and Y vent. The north earthquake cluster occurred in two distinct swarms. The first occurred on March 22, with 162 events in less than 3 hours, and the second occurred 22 days later, with 25 events. The south cluster showed a nearly continuous low-level seismicity over the span of the survey. *Sohn et al.* [1999] estimated 147 hypocenters from the recovered seismic data (115 for the north group, 32 for the south one), all in the upper 1.2 km of the crust. Relocation of 65 of these hypocenters suggested a particularly high density between 0.7 and 1.1 km depth, just above the axial magma chamber whose depth has been estimated at 1.4 km [*Kent et al.*, 1993]. Based on this hypocentral pattern, *Sohn et al.* [1999] considered a diking event unlikely. They also discounted a tectonic origin. Rather, the microearthquake activity was attributed to the release of thermal stresses at the base of the hydrothermal system. The analysis of vent fluid chemistry also supported the hypothesis of crack-deepening of the hydrothermal reaction zone [*Fornari et al.*, 1998; *Sohn et al.*, 1998].

Monitoring of vent fluid temperature and chemistry of the 9°50′N area after the 1991 magmatic eruption has allowed correlation between the March 1995 seismic activity and changes in the hydrothermal systems situated above the epicenters [*Fornari et al.*, 1998; *Sohn et al.*, 1998, 1999]. The temperature data (Figure 1) collected by *Fornari et al.* [1998] for the Bio9 vent showed that vent temperature was stable for approximately 15 months prior to the March 1995 event, at 365 ± 1°C. On March 26, a 7°C increase was observed over 8 days. Temperature stabilized at 372 ± 1°C until April 16, then gradually decreased to 366 ± 1°C over approximately 3.5 months before steadily increasing again for several months. In November 1995, the temperature probe was found embedded into the chimney and replaced. The replacement probe showed the same steady temperature increase, with temperature reaching 373 ± 1°C in November 1997 (Figure 1). Two surveys of animal communities around the Bio9 area (October 1994 and November 1995) showed signs of an overall increase in venting temperature. In particular, they featured the appearance of alvinellids, usually considered a distinctive class of high-temperature vent fauna [*Fornari et al.*, 1998]. Bio9 is the only vent in the 9°50′N area that had continuous monitoring between October 1994 and November 1995. Point measurements and other probe records showed that temperatures steadily increased by more than 40°C at M vent between March 1992 and November 1995, while they remained nearly constant at Biovent, however.

In this paper, we focus on the March 1995 East Pacific Rise events due to the availability of well-documented data during, before, and after to the earthquake swarms [*Fornari et al.*, 1998; *Sohn et al.*, 1998, 1999]. Records of diffuse venting do not exist [*Scheirer et al.*, 2006]; consequently, we focused on the high-temperature variations near the epicenters. Our model, however, can be used to study the combined evolution of low-temperature diffuse and high-temperature focused venting systems.

2. MECHANISMS OF HYDROTHERMAL RESPONSE

For both the March 1995 East Pacific Rise and June 1999 Juan de Fuca events, the interval between the onset of seismicity and temperature variations at the seafloor was interpreted as the fluid residence time [*Fornari et al.*, 1998; *Sohn et al.*, 1998; *Johnson et al.*, 2000]. This residence time was significantly smaller than the previous constraint of ~3 years that *Kadko and Moore* [1988] suggested based on a study of ^{210}Pb/Pb and ^{228}Ra/^{226}Ra ratios on the Endeavour segment of the Juan de Fuca Ridge. *Johnson et al.* [2000] suggested that the time delay may be a characteristic time constant associated with the response of the hydrothermal system to the disturbance. Comparison of temperature time series obtained on both the Main Endeavour field and Axial Seamount (Juan de Fuca Ridge) during the June 1999 swarm showed that this time constant may vary over a wide range, from 8 hours to 36 days [*Johnson et al.*, 2001]. In comparison, land-based hydrothermal systems tend to respond nearly instantaneously to seismic activity [e.g., *Hill et al.*, 1993; *Roeloffs*, 1996]. *Dziak and Johnson* [2002] explain this dissimilarity by a difference of crustal architecture, noting that on-land systems present diverse fluid reservoir geometries, and that the observed hydrological response is usually linked to seismic waves compressing and dilating the pore space of these reservoirs. At mid-ocean ridges, the aquifer is thought to be more uniform and continuous; and the hydrothermal circulation is thermally driven.

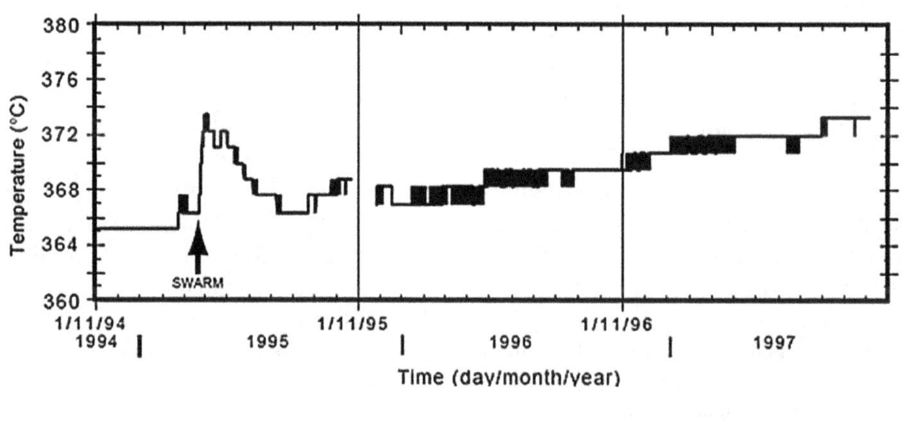

Figure 1. Thermal response of the Bio9 vent at East Pacific Rise 9°50′N [from *Sohn et al.,* 1998]. Reprinted with permission from MacMillan Publishers Ltd., *Nature*, *396*, 159–161, 1998.

2.1. *Wilcock's [2004] Work*

Wilcock [2004] showed that Darcian flow through a porous medium could propagate a temperature perturbation a distance of ~1 km in a few days, provided the permeability of the discharge zone is ~10^{-9} m^2. This is much higher than the usual range of 10^{-12}–10^{-13} m^2 inferred for high-temperature hydrothermal systems [*Lowell and Germanovich*, 1994; *Pascoe and Cann*, 1995; *Wilcock and McNabb*, 1996; *Cherkaoui et al.*, 1997; *Fisher*, 1998; *Wilcock and Fisher*, 2004]. *Wilcock* [2004] also showed that pressure anomalies and permeability variations at depth could generate the temperature variations observed at the seafloor. The delay observed at the seafloor depends on the depth of the pressure perturbation as well as the permeability. For a pressure anomaly at ~1 km depth, a permeability of ~10^{-14} m^2 or smaller is required to delay the response at the seafloor by a few days [*Wilcock*, 2004]. This is smaller than the aforementioned range of 10^{-12}–10^{-13} m^2 that is characteristic of high-temperature venting.

One of *Wilcock*'s [2004] interpretations of the March 1995 seismic activity at 9°50′N, East Pacific Rise, is a cracking event that opened up a new pathway to hot rock at the base of the hydrothermal system [*Fornari et al.*, 1998; *Sohn et al.*, 1999]. Following *Bodvarsson* [1969], *Wilcock* [2004] first considered the temperature distribution in a single planar crack embedded in an infinite medium with an oscillatory fluid temperature imposed at the base of the crack. To fit the observed temperature anomaly (Figure 1), he then assumed that the initial swarm was accompanied by a basal temperature increase of ~50°C that linearly decreased over 16 days, followed by a smaller temperature increase of 17°C at the time of a secondary swarm, which in turn decayed over 25 days. He obtained a good fit to the first 90 days of the data shown in Figure 1 with a crack 0.6 mm wide.

A similar approach to temperature perturbation at the base of hydrothermal system was considered by *Lowell* [1975, 1976]. His model considers flow in an isolated narrow fracture (or set of fractures) embedded in impermeable rock. In this case, the temperature change, ΔT_v, at the vent related to an instantaneous temperature step, ΔT_b, at depth, z, is given by

$$\frac{\Delta T_v}{\Delta T_b} = \operatorname{erfc}\left(\frac{\lambda_r z}{q c_f \sqrt{at}}\right), \tag{2.1}$$

where q is the mass flux per unit length of the crack, λ_r is the thermal conductivity of the rock, $a = \lambda_r/(c_f\rho_f)$ is the effective thermal diffusivity, c_f is the specific heat of water, and t is time. Table 1 summarizes all the symbols used in the formulation and their range of values. Given an assumed ΔT_b ~ 50°C, z = 1 km, reasonable values of thermal parameters (see Table 1), and observed values of ΔT_v and time t (~10°C and ~10 days, respectively, at Bio 9), the mass flux calculated from (2.1) is q ~ 1 kg/(m · s). Assuming buoyancy-driven flow through a single effective crack, the effective crack width d_c can be calculated from [e.g., *Bear*, 1972]

$$d_c = \left(\frac{12\nu q}{g\Delta\rho}\right)^{1/3}, \tag{2.2}$$

where ν is the kinematic viscosity of the fluid, Δρ ~ 0.4 kg/m^3 is the density difference between descending and ascending fluids, and g is the acceleration due to gravity. For typical parameters, equation (2.2) gives d_c ~ 1 mm. The subsequent temperature decay in Figure 1 could be modeled using (2.1) and by invoking a temperature decrease at the base of the crack.

Both of these temperature perturbation models fit reasonably well the rapid temperature increase and secular decay after an eruption; the initial temperature delay being related to conductive cooling along the flow path. Neither model, however, explains the secular rise in temperature observed over the ensuing years (Figure 1). Moreover, the imposed temperature perturbations do not have a firm physical basis; and the high effective permeability in these single crack models yields a fluid residence time in the upflow zone of less than 1 hour. Hence, we consider these single crack models to be somewhat unrealistic.

2.2. *Approach in This Paper*

During porous discharge, lateral heat losses and mixing in the shallow subsurface result in a thin thermal boundary layer near the seafloor. This boundary layer can be derived from the equation of steady state heat transfer

$$v_0 \frac{dT}{dz} = a\frac{d^2T}{dz^2}, \tag{2.3}$$

where v_0 is the initial Darcian velocity (specific discharge) in the discharge zone and a is the effective thermal diffusivity. The thickness of the boundary layer is δ ~ a/v_0, and because advection prevails in the rest of the upflow zone, the temperature profile is almost constant everywhere except in the boundary layer (Figure 2). In this case, neither the fluid flow nor the temperature response to the earthquake has to propagate through the entire discharge zone, but need only affect the movement of the boundary layer. For a typical value of thermal diffusivity a ~ 10^{-6} m^2/s and a Darcian velocity of v ~ 10^{-6} m/s, should the boundary layer be altered, the expected response time would be ~ δ/v ~ 10 days.

One can apply this simple model to the case of high-temperature venting by using a heat transfer boundary con-

Table 1. Symbol Definitions and Parameter Values

Symbol	Definition	Value
a	effective thermal diffusivity, $\lambda_r/(\rho_f c_f)$	~10^{-6} m^2/s
a_*	effective thermal diffusivity in equation (3.21)	
a_r	thermal diffusivity of rock, $\lambda_r/(\rho_r c_r)$	~10^{-6} m^2/s
A_1	cross-sectional area of the upper part of the upflow zone	10^4 m^2
A_2	cross-sectional area of the lower part of the upflow zone	10^4 m^2
A_3	cross-sectional area of the horizontal limb of the deep recharge	10^5 m^2
A_4	cross-sectional area of the horizontal limb of the shallow recharge	10^2–10^3 m^2
b	length of vent field along strike	100 m
b_*	parameter in equation (3.21)	
c_f	specific heat of water	4×10^3 J/(kg°C)
c_r	specific heat of rock	1×10^3 J/(kg°C)
d	basal thermal boundary layer thickness	2–20 m
d_c	effective crack width	
f	height of the junction between upflow and shallow recharge limb	1–10 m
g	acceleration due to gravity	9.8 m/s^2
H	length of the upflow zone	1.5 km
$H(x)$	Heaviside function	
h	depth of the shallow recharge	100 m
h_d	height of the dike emplacement	600 m
I	source of material produced by the shallow recharge	
k	(temperature-dependent) permeability	
k_0	initial permeability	10^{-13}–10^{-12} m^2
k_{res}	residual permeability	$10^{-2}k_0$
l	length of the horizontal limb of the shallow recharge	
L	length of the horizontal limb of the deep recharge	1 km
L_m	latent heat of crystallization	4×10^5 J/kg
P	pressure	
P_0	pressure at the seafloor	250 bars
P_H	pressure at the depth H below the seafloor	
Q_i	mass flux in the ith limb	
q	mass flux per unit length of the crack	
q_d	heat source density per unit time and unit volume of the dike	
q_m	basal heat flux (heat uptake rate)	
q_w	lateral heat transfer density per unit time and unit volume	
r_i	hydrodynamic resistance per unit area in the ith limb	
s	"curvilinear" spatial variable along the flow path (deep limb and upflow)	
t	time	
t_c	time of dike crystallization	
t_d	time of dike emplacement	
T	temperature in the upflow zone	
T_*	effective temperature in equation (3.21)	
$T_{diffuse}$	temperature monitored at sites of diffuse venting	20°C
T_{in}	initial steady-state temperature	
T_j	temperature of the fluid from the shallow recharge at junction point	
T_m	melting temperature of basalt	1200°C
T_{rech}	far-field temperature in the deep recharge zone	0–100°C
T_{sh}	far-field temperature in the recharge zone (for shallow recharge)	0–10°C
ΔT_b	temperature change at the base of the upflow zone	
ΔT_v	temperature change at the vent	
v	Darcian flow velocity (specific discharge)	
v_0	initial Darcian flow velocity	
w	dike width (opening)	1 m
w_u	width of vent field (upflow zone)	100 m

Table 1. (Continued)

Symbol	Definition	Value
z	depth	
α_r	linear thermal expansion coefficient of rock	2×10^{-5}/°C
α_f	coefficient of the thermal expansion of water	10^{-3}/°C
δ	thermal boundary layer thickness	a/v
$\delta(s)$	Dirac delta function	
ϕ_0	initial porosity of the upflow zone	0.1–10%
γ	ratio α_r/ϕ_0	10^{-4}–10^{-1}
η	(temperature dependent) dynamic viscosity of water	
λ	parameter related to the rate of dike freezing [*Germanovich et al.*, 2000]	1
ν	kinematic viscosity of water	η/ρ
λ_r	coefficient of thermal conductivity of rock	2.5 W/(m°C)
ρ_f	(temperature-dependent) density of water	
ρ_r	rock density	3×10^3 kg/m^3
ρ_0	initial density of water at the ocean floor	10^3 kg/m^3
$\Delta\rho$	density difference between descending and ascending fluids	~0.4 kg/m^3

dition at the upper boundary [*Germanovich et al.*, 2001]. This condition can be interpreted as representing the mixing between high-temperature discharge and overlying low-temperature fluids in the upper crust. Such a boundary condition can be used in the context of a single-pass model (Plates 1 and 2) as well as in a one-dimensional upflow model. An advantage of using this boundary condition is that the same model formulation can be used to address perturbations to both high-temperature systems, such as Bio9 vent, and low-temperature diffuse flow systems, such as those observed on the Juan de Fuca Ridge. Using this model, we have shown that an increase in permeability can sufficiently perturb the boundary layer to give the observed temperature increases at Bio9 at the time scale of a week [*Ramondenc et al.*, 2005]. Changes in the stress field resulting from earthquakes tend to create or reopen fractures thereby increasing permeability [*Brodsky et al.*, 2003; *Elkhoury et al.*, 2006; *Rojstaczer and Wolf*, 1992; *Rojstaczer et al.*, 1995]. It is also possible that permeability increases would appear in the case of dike intrusion, especially near the margins of the dike [*Delaney et al.*, 1986; *Germanovich et al.*, 2000; *Pollard*, 1987]. Our simulations [*Ramondenc et al.*, 2005] indicate that the fluid residence time in the discharge zone is on the order of years, which is consistent with the general scaling of the upflow zone [e.g., *Lowell and Germanovich*, 2004].

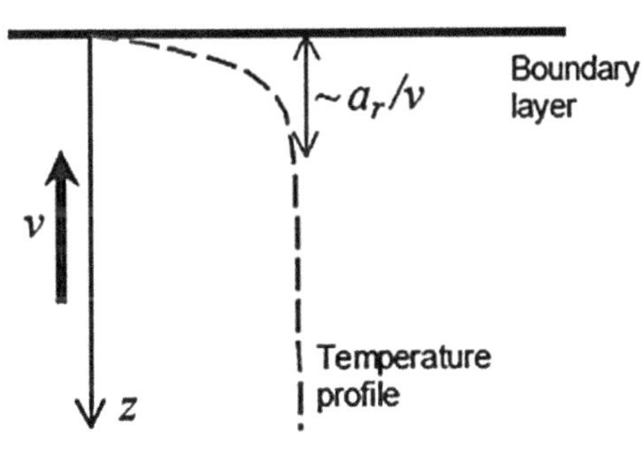

Figure 2. Steady-state temperature distribution in the discharge zone before the earthquake and the corresponding boundary layer.

Our initial approach [*Ramondenc et al.*, 2005] did not address the decay of the temperature spike, which occurs on a time scale of months, nor the longer-term temperature rise at Bio9, which occurs over a time scale of years (Figure 1). In this paper, however, we extend the idea of thermal boundary layer using a "double-loop" model (described in Section 3) to explain the observed temperature history within the first year after the seismic event at 9°50′N, East Pacific Rise. We also suggest a mechanism to explain the later rise of the observed temperature.

Lowell et al. [2007] have recently developed numerical models of single-pass convection in the presence of a high-permeability extrusive layer. Their results show that mixing between ascending hydrothermal fluid and seawater circulating in the extrusives can lead to either black smoker flow together with nearby diffuse flow or to a system exhibiting only low-temperature diffuse flow. The nature of the hydrothermal output depends mainly on the permeability ratio between the extrusives and the deep discharge channel, and secondarily on the thickness of the extrusives [see also *Lowell et al.*, 2008], although relatively small changes in permeability may affect system evolution toward low- or

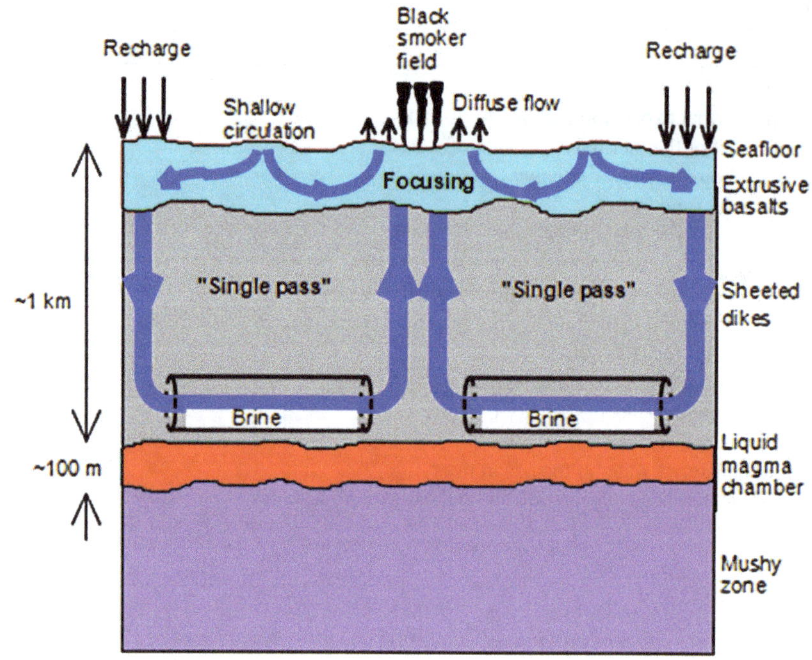

Plate 1. Cartoon of a single-pass hydrothermal circulation model at an ocean ridge crest at the local, ridge crest circulation scale. The fracture-like segment at the base of the sheeted dikes represents the region where chemical reactions and heat uptake from the magma chamber occur, and where brine may reside. The focused, high-temperature flow is thought to occur in the main single-pass limb. Diffuse flow may occur as a result of mixing of the deep circulation with shallow circulation in the shallow extrusive basalts. Chemical reactions occur throughout the circulation path. Biogeochemical processes also occur in the extrusives [from *Germanovich et al.,* 2001].

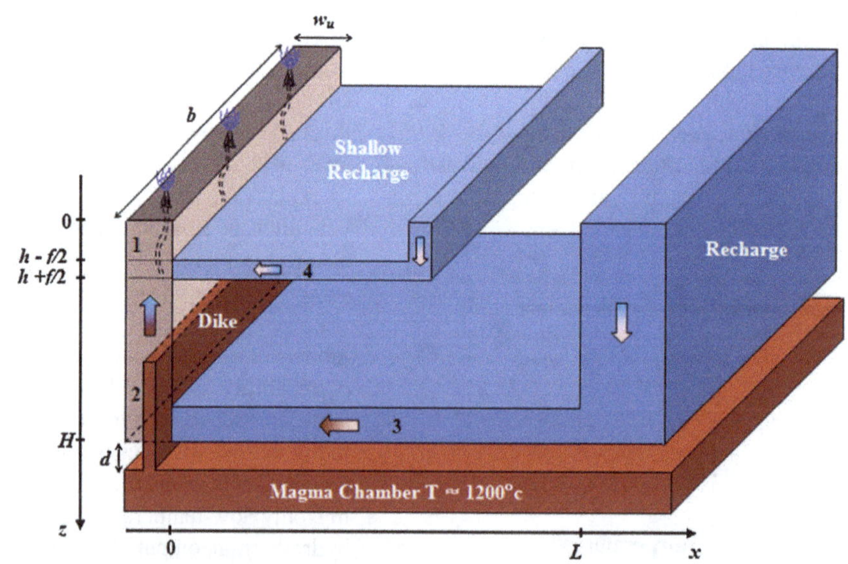

Plate 2. "Double-loop" single-pass model adapted from Plate 1. The region between $h - f/2$ and $h + f/2$ is the junction zone where high-temperature black smoker fluid from below mixes with seawater from the extrusives. Thermal boundary layers are located at the junction zone, the seafloor, and at the bottom of the system between liquid magma and the hydrothermal system, denoted by d.

high-temperature venting [*Germanovich et al.*, 2001]. However, *Lowell et al.* [2007] did not consider the effects of magmatic or tectonic events.

3. FORMULATION

3.1. Mixing in the Shallow Crust

At the seafloor, high-temperature venting occurs through discrete point sources such as sulfide chimneys, fissures, or flanges protruding from the walls of large sulfide edifices [*Delaney et al.*, 1992]. Low-temperature diffuse flow usually occurs over much larger areas of the seafloor [*Baker et al.*, 1993; *Bemis et al.*, 1993; *Rona and Trivett*, 1992]. The relationship between diffuse and focused flows remains puzzling. In general, focused venting constitutes only a fraction of the diffuse flow (Table 1 of *Ramondenc et al.* [2006]). At some vent sites only diffuse, low-temperature venting occurs (e.g., at the Galapagos Spreading Center; *Corliss et al.* [1979]; *Edmond et al.* [1979]), whereas at others black smokers and low-temperature diffuse flow coexist (e.g., the trans-Atlantic geotraverse vent field on the Mid-Atlantic Ridge [*Edmond et al.*, 1995; *Humphris and Tivey*, 2000; *James and Elderfield*, 1996], the Endeavour segment on the Juan de Fuca Ridge [*Schultz et al.*, 1992], or the 9°–10°N zone on the East Pacific Rise [*Scheirer et al.*, 2006; *Von Damm and Lilley*, 2004]). The V vent, at 9°47′N, East Pacific Rise, is as an example of diffuse flow that evolved toward high-temperature focused discharge in less than 10 years [*Von Damm*, 2000]. Similar geochemical changes occurred simultaneously in both diffuse and focused flows on the northern part of the 9°49–51′N area (also called "BioGeoTransect") on the East Pacific Rise [*Von Damm*, 2004], suggesting a coupling between focused and diffuse flows.

The chemistry of diffuse flow fluids sampled from the Transect on the East Pacific Rise between 1991 and 2000 implies some reaction at high temperature. *Von Damm and Lilley* [2004] argue that phase separation, requiring temperatures >389°C in this area, is the only consistent way to explain the chlorinity content in the sampled fluids, which is significantly lower than ambient seawater for most samples. The analysis of nonbioactive elements and the use of geochemical modeling show that the diffuse flow fluids in this area can be considered to be a nearly conservative mixture of seawater and high-temperature fluids, with additional conductive heating [*Von Damm and Lilley*, 2004]. Furthermore, Li isotope data [*Bray*, 2001] suggest that, once mixed, these fluids would have a short residence time (months or less) at a temperature less than 150°C. High-temperature vent fluid samples are generally thought to represent a mixture between end-member, Mg-free fluid and seawater that is entrained during sampling, but end-member fluids may mix below the seafloor prior to venting with small amounts of partially reacted seawater that has been heated to less than 150°C [*Ravizza et al.*, 2001].

Based on these observations, we conceptualize the upper part of the upflow zone as shown schematically in Figure 3. Following *Lowell et al.* [2003], we assume that sulfate-rich seawater circulating in the extrusive layer mixes with the hot sulfate-depleted ascending hydrothermal fluid. As a result, anhydrite is precipitated, thus forming low-permeability barriers that channel high-temperature fluids to their discharge points as black smokers. There appears to be little mixing of black smoker fluids with ambient seawater during their ascent [*Von Damm and Lilley*, 2004]. Because little conductive cooling occurs along the channels during rapid ascent, we consider black smokers to be "temperature probes" corresponding to the temperature at the base of the focusing zone, where the hot ascending fluid enters the focusing channels. In other words, black smoker fluid is a proxy of the high-temperature part of the upflow zone.

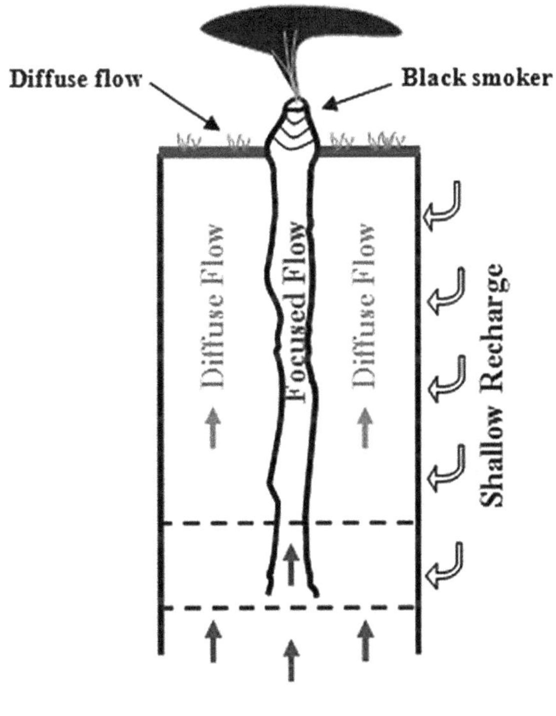

Figure 3. Schematic representation of focusing with adjacent diffuse flow in the upper part of the upflow zone. The dashed lines represent the junction at which mixing between hydrothermal fluid and seawater occurs.

In this model, black smoker fluid represents an end-member 0-Mg fluid mixed with a small amount of partially reacted seawater at the junction (Figure 3) as observed for Bio vent [*Ravizza et al.*, 2001]. In the extrusive layer outside the channels, low-temperature diffuse flow occurs. This fluid is a mixture of high-temperature fluid with the seawater circulating in the extrusives.

We will therefore add the effect of shallow circulation occurring in the shallow crust to the standard single-pass model [e.g., *Pascoe and Cann*, 1995; *Lowell et al.*, 2003] (Plate 2). This shallow circulation provides a mechanism of temperature regulation that can explain the variations observed following the temperature peak at Bio9 after the March 1995 event. In addition to facilitating the focusing mechanism, input of cold water from the shallow circulation system creates a thermal boundary layer in the junction zone, where the entrance to the focusing channels is located (Plate 2). In our model, the evolution of this boundary layer controls the black smoker temperature. When the temperature rises due to a perturbation at depth, the density decrease triggers an influx of cold fluid from the shallow circulation system. Temperature in the junction then drops, decreasing the input of cold fluid. The system gradually evolves toward a new steady state. Diffuse venting is affected by variations of another boundary layer that occurs at the seafloor. These variations arise from the interaction with cold seawater near the surface as well as from mixing between hot ascending fluid and cold recharge in the junction.

3.2. Mass Fluxes in the System and Conservation Laws

The single-pass model shown in Plate 2 is considered one-dimensional because the only dimension we are interested in is the distance from the discharge point at the seafloor. The z- and x-axes are here directed *against* the fluid flow in all the limbs of the model, that is, downward in the upflow zone and from left to right in the bottom horizontal limb. To write all the subsequent equations with respect to a unique coordinate along the fluid path in the deep horizontal limb and the upflow zone, we introduced a "curvilinear" coordinate s defined as

$$\begin{aligned} s &= z \quad \text{if } s \leq H \\ s &= z + x \quad \text{if } H < s \leq H + L \end{aligned} \tag{3.1}$$

For an easier formulation, we divided the fluid path into four main limbs, represented hereafter by the index i, where i takes values from 1 to 4 representing the limb (see Plate 2).

Darcy's law in the vertical limbs ($i = 1, 2$) can then be written as

$$v = -\frac{k(T)}{\eta(T)}\left[\frac{\partial P}{\partial s} - \rho(T)g\right]. \tag{3.2}$$

For the horizontal limbs ($i = 3, 4$), the same expression is used without the gravity term $\rho(T)g$. The temperature T is a function of time t and coordinate s. Hereafter, we use the subscript i (where $i = 1, 2, 3, 4$) when a physical value refers to a specific limb.

Let us first consider the case where the heat input from the shallow recharge is modeled as a concentrated (or point) source. At the junction point $s = h$, conservation of mass can be written as $Q_1 = Q_2 + Q_4$. By convention, we choose $Q_i > 0$, whereas $v_i < 0$ is imposed by the orientation of the z- and x-axes in (3.2). Moreover, we consider that at any time step, the velocity v_i is constant in the ith limb. This implies that the mass fluxes $Q_i = -\rho A_i v_i$ are also constant in each limb at any time step, because we invoke Boussinesq's approximation.

Rewriting Darcy's law in terms of Q and using the kinematic viscosity, $\nu(T) = \eta(T)/\rho(T)$, we obtain

$$\frac{\partial P}{\partial s} = \frac{Q_i}{A_i}\frac{\nu(T)}{k(T)} + \rho(T)g \qquad (i = 1, 2, 3, 4) \tag{3.3}$$

where the last term on the right-hand side is omitted for horizontal limbs ($i = 3, 4$). The hydrostatic pressures at the bottom of the shallow and deep recharge zones, respectively, can be written as

$$\begin{aligned} P(h+l) &= P_0 + g\int_0^h \rho(T_{sh})ds, \\ P(H+L) &= P_0 + g\int_0^H \rho(T_{rech})ds, \end{aligned} \tag{3.4}$$

where T_{sh} and T_{rech} are the far-field temperature in the shallow and deep recharge zones, respectively.

Also, because pressure is continuous at $s = h$ and $s = H$, we have:

$$P(h+l) - P_0 = \int_0^{h+l}\frac{\partial P}{\partial s}ds, \quad P(H+L) - P_0 = \int_0^{H+L}\frac{\partial P}{\partial s}ds. \tag{3.5}$$

We further introduce the hydrodynamic resistance

$$r = \int_{s_1}^{s_2}\frac{\nu(T)}{k(T)}\frac{ds}{A(s)}, \tag{3.6}$$

where $A(s)$ is the cross-sectional area between $s = s_1$ and s_2. Combining (3.3), (3.4), and (3.5), and using conservation of mass, we finally obtain

$$\begin{cases} Q_1 = Q_2 + Q_4 \\ Q_1 = -Q_4\dfrac{r_4}{r_1} + \dfrac{g}{r_1}\displaystyle\int_0^h [\rho(T_{sh}) - \rho(T)]\,ds \\ Q_2 = Q_3 \\ Q_4 = \dfrac{g}{(r_2+r_3)(r_1+r_4)+r_1r_4} \\ \quad \times \left\{(r_1+r_2+r_3)\displaystyle\int_0^h [\rho(T_{sh}) - \rho(T)]\,ds \right. \\ \quad \left. - r_1\displaystyle\int_0^H [\rho(T_{rech}) - \rho(T)]\,ds\right\} \end{cases} \tag{3.7}$$

where r_i is the resistance of the ith limb. Expressions (3.7) can also be obtained by considering the electric current analogy.

Lowell [1990] and *Germanovich and Lowell* [1992] showed that thermoelastic stresses induced by hot hydrothermal fluid flowing through cooler rocks may close some pathways and thus focus flow. Therefore, the permeability of the upflow zone is assumed to be temperature-dependent and represented by [*Germanovich et al.*, 2000, 2001]

$$k(T) = k_0\left[1 - \gamma(T - T_{in})\right]^3 H\left[1 - \gamma(T - T_{in})\right] + k_{res}, \tag{3.8}$$

where k_0 is the permeability of the upflow zone at the initial steady-state temperature T_{in}, $\gamma = \alpha_r/\phi_0$ is the ratio of the thermal expansion coefficient of the rock to the initial porosity, $H(x)$ is the Heaviside step function [$H(x) = 0$ if $x \le 0$ and $H(x) = 1$ if $x > 0$], and k_{res} is the residual permeability. The factor multiplying k_0 represents the effect of thermoelastic stresses, whereas k_{res} reflects the permeability remaining after the closure of the major fractures and asperities as temperature increases. Specifically, if $T < T_{in} + 1/\gamma$, $H[1 - \gamma(T - T_{in})] = 1$ in (3.8) and $k(T) = k_0[1 - \gamma(T - T_{in})]^3 + k_{res}$. Although typically $k_{res} \ll k_0$, k_{res} cannot be ignored here because the first term is reducing to zero with increasing temperature. Once $T \ge T_{in} + 1/\gamma$, $H[1 - \gamma(T - T_{in})] = 0$ and $k(T) = k_{res}$. The dependence of the permeability on pressure is negligible compared to that on temperature for the conditions considered in this study. The temperature dependence of fluid viscosity η is approximated as [*Germanovich et al.*, 2000]

$$\eta(T) = \frac{C_1}{T + C_2}, \tag{3.9}$$

where $C_1 = 0.032$ Pa · s°C, and $C_2 = 15.4$°C. Finally, the dependence of fluid density on temperature is given by

$$\rho_f = \rho_0\left[1 - \alpha_f(T - T_0)\right] \tag{3.10}$$

We now modify the formulation of *Germanovich et al.* [2000, 2001] to take into account recharge at the base of the extrusives and heat transfer from the magma body at the base of the hydrothermal system. Then, conservation of energy along the flow path 1-2-3 (Plate 2) can be written as

$$c_r\rho_r A\frac{\partial T}{\partial t} - c_f Q\frac{\partial T}{\partial s} = \lambda_r A\frac{\partial^2 T}{\partial s^2} - 2b(q_w + q_d) + 2c_f(T_j - T)AI + bq_m, \tag{3.11}$$

where the source term I represents the mass flux per unit volume delivered by the shallow recharge to the junction region $h - f/2 \le s \le h + f/2$ (Plate 2) and q_m represents basal heat transfer from the magma body to the base of the hydrothermal system. The quantity $2I(T_j - T)$ represents cooling of the system, where T_j denotes the temperature of the fluid from the shallow recharge at the junction point (Plate 2), and $T > T_j$. Here, $I = [Q_4/(A_4w_u)]\delta(s - h)$, where $\delta(s - h)$ is the Dirac delta function that models the concentrated influx of the shallow recharge into the upflow zone (corresponding to $f \to 0$ in Plate 2). In (3.11), q_w represents lateral heat transfer through the walls of the upflow zone, and q_d the heat from the freezing and cooling of a dike (when dike emplacement is modeled, as in Section 5). Their asymptotic approximations are discussed by *Germanovich et al.* [2000] and can be written as follows:

$$q_w = -\frac{2}{\sqrt{\pi}}\frac{\lambda_r(T - T_{in})}{\sqrt{a_r t}}, \tag{3.12}$$

$$q_d(z,t) = \begin{cases} -\dfrac{\lambda_r}{\sqrt{\pi}\,\mathrm{erf}(\lambda)}\dfrac{T_m - T}{\sqrt{a_r t}} & \text{if } 0 < t < t_c \\ -\dfrac{\lambda_r(T_m - T)}{\sqrt{\pi}\,\mathrm{erf}(\lambda)\sqrt{a_r t_c}} * \exp\left(-\dfrac{\pi^2 a_r(t - t_c)}{w^2}\right) & \text{if } t \ge t_c \end{cases} \tag{3.13}$$

The upper and lower expressions in (3.13) correspond to the freezing and cooling stages after dike emplacement, $t_c \approx w^2(16a_r\lambda)^{-1}$ is the time for dike solidification, w is the dike width, and T_m is the temperature of the basalt melt emplaced by the dike. In (3.13), λ is the root of equation $\lambda \exp(\lambda^2)\mathrm{erf}(\lambda) = c_r(T_m - T)(L_m\pi^{1/2})^{-1}$ [*Carlsaw and Jaeger*, 1986]. For typical parameter values (Table 1) and $T_m - T \sim 10^3$°C, $\lambda \approx 1$ [*Germanovich et al.*, 2000]. Then, for $a_r = 10^{-6}$ m²/s and $w = 1$ m, $t_c \sim 1$ day.

When a dike is suddenly emplaced in fluid-saturated country rock, rapid heating will tend to pressurize the adjacent fluid and drive flow away from the dike wall [*Delaney*, 1982]. In turn, pressurization of the fluid may lead to fracturing of the adjacent rock [e.g., *Germanovich and Lowell*, 1995]. Local boiling and phase separation may also occur. For seafloor systems with high permeability, and ambient pressure of ~30 MPa, the pressure increment is negligible [*Delaney*, 1982] and so is the corresponding increase of the sizes of preexisting fractures [*Germanovich and Lowell*, 1995]. If an increase of permeability results from fluid pressurization, it is incorporated into the permeability increase that we assume results from dike emplacement (see Section 5.2). The width of the two-phase zone is only ~10 cm [*Lewis and Lowell*, 2004], and hence we neglect it. As noted by *Wilcock* [2004] and *Lowell and Germanovich* [1995], heating adjacent to the dike may lead to large pressure increases but only within a narrow region.

In (3.11), $q_m = 0$ everywhere except along the horizontal limb 3, where heat is conducted without heat loss through an impermeable thermal boundary layer of thickness d. If T_m is the temperature of the magma body, then asymptotically [*Germanovich et al.*, 2001; *Lowell and Germanovich*, 2004]

$$q_m = \frac{\lambda_r(T_m - T)}{d}. \qquad (3.14)$$

Heat input from the shallow recharge modeled as a concentrated (or point) source has an important limitation. If T_j is the temperature of the fluid coming from the shallow recharge, the heat balance yields $Q_2[T(h^+) - T_j] = Q_1[T(h^-) - T_j]$. This expression can be satisfied only if $Q_1 = Q_2$, which does not apply in our case. Therefore, in the point source model, the temperature has a discontinuity at the junction. To avoid computational inconvenience, we model the input from the shallow recharge over a finite area between the depths $z = s = h - f/2$ and $z = s = h + f/2$, f being the vertical dimension of junction zone (Plate 2). Then, in (3.11), $I = Q_4/(A_4 w_u)$ is assumed to be constant within the junction zone, so that

$$I(z) = \begin{cases} \dfrac{Q_4}{A_4 w_u} & \text{if } h - f/2 \le s \le h + f/2 \\ 0 & \text{otherwise} \end{cases} \qquad (3.15)$$

Within the junction interval $h - f/2 \le s \le h + f/2$, $\partial Q/\partial s = -AI$, where $A = A_1 = A_2$, and the mass flux

$$Q(s,t) = Q_2(t) - \left(s - h - \frac{f}{2}\right) AI \qquad (3.16)$$

is linearly distributed along this interval. In the numerical calculations described below, we adjusted Q in the energy balance expression (3.11) based on (3.16); however, the correction to the mass balance expression (3.7) proved to be negligible because $f << h$. Because the geochemical analyses of *Bray* [2001] suggest that the residence time of the diffuse flow fluid to be on the order of a few months, we assumed that focusing occurred in the upper 100 m of the crust and placed the junction zone at that depth.

3.3. Boundary Conditions

In previous work with the single-pass model [*Germanovich et al.*, 2000, 2001], we did not explicitly include the deep horizontal limb (limb 3 in Plate 2), which required specifying a separate boundary condition at the bottom of the upflow zone [*Germanovich et al.*, 2001, equation (14)]. Here, this horizontal limb and its effect are explicitly included, and we now have the following boundary conditions:

$$T = T_{diffuse} \qquad \text{if } s = 0,\ t \ge 0 \qquad (3.17)$$

and

$$T = T_{rech}(H) \qquad \text{if } s = H + L,\ t \ge 0 \qquad (3.18)$$

where $T_{diffuse}$ is the temperature monitored at sites of diffuse venting. To limit the number of unknowns, we model the recharge zone by a linear temperature distribution (as also discussed in Section 4).

To solve equations (3.7), we could estimate the hydrodynamic resistance of the shallow limb 4 directly from (3.6), but if the temperature in horizontal limb 4 is constant and equal to the temperature of the far-field recharge at this depth, then

$$r_4 \approx \frac{\nu_4 l}{k_4 A_4}, \qquad (3.19)$$

where the subscript 4 refers to limb 4, and l is the length of the horizontal limb of the shallow recharge. Because l and A_4 in (3.19) are poorly constrained, we use r_4 as an independent parameter in our simulations. The permeability and the recharge fluid viscosity are calculated using (3.8) and (3.9), respectively.

3.4. Initial Conditions

We define the initial conditions as the steady-state of the system prior to any perturbation. In steady-state, equation (3.11) becomes

$$-c_f Q \frac{dT}{ds} = \lambda_r A \frac{d^2T}{ds^2} + 2c_f(T_j - T)AI + bq_m, \qquad (3.20)$$

so that the initial condition is given by the solution of (3.20) with boundary conditions (3.17) and (3.18). Given (3.14), equation (3.20) can be written as

$$a_* \frac{d^2(T - T_*)}{ds^2} - Q\frac{d(T - T_*)}{ds} - b_*(T - T_*) = 0\,, \qquad (3.21)$$

where $a_* = \lambda_r A/(-c_f)$, $T_* = T_m$ in the deep horizontal limb 3, $T_* = T_j$ in the junction of the upflow zone with the shallow recharge, and $T_* = 0$ everywhere else; and b_* is defined as either $b_* = -(\lambda_r b)/(c_f d)$ in limb 3, $b_* = -2AI$ in the junction, or $b_* = 0$ elsewhere. The general solution of the linear, homogeneous ordinary differential equation (3.21) is

$$T = T_* + C_1 \exp\left(\frac{Q + \sqrt{Q^2 + 4a_*b_*}}{2a_*}s\right) + C_2 \exp\left(\frac{Q - \sqrt{Q^2 + 4a_*b_*}}{2a_*}s\right), \qquad (3.22)$$

where constants C_1 and C_2 are different for limbs 1, 2, 3, and the junction zone. They are determined from the boundary conditions (3.17), (3.18), and the continuity of temperature and heat flow between intervals.

One obtains these constants as functions of the corresponding mass fluxes. Then, (3.22) is substituted into (3.7) and the resulting algebraic equations are solved to obtain the mass fluxes Q_i in the steady-state configuration. Alternatively, equation (3.20) can be solved numerically to obtain the steady-state solution using the numerical technique described below for solving transient problems.

4. NUMERICAL RESULTS

We chose an implicit numerical scheme that uses a finite difference discretization with a zeroth-order approximation with respect to time. At each time step, the mass fluxes (and velocities) are considered to be constant. This allows solving the system of equations using the standard elimination method [*Samarskii*, 2001]. The mass fluxes and velocities are then recalculated, along with values of density, permeability, and viscosity, before the next step. We consider the flow path from the bottom of the recharge zone to the top of the upflow zone. The bottom boundary condition corresponds to the temperature at the bottom of the deep recharge. This temperature is not well constrained, but test computations showed that the results were not sensitive to temperatures ranging between 0° and 100°C. For results presented here, we assumed a linear gradient in the recharge zone with 0°C at the seafloor and 100°C at the bottom (i.e., at the depth of 1.5 km).

In all the calculations, we chose a uniform grid density with a grid spacing of 1 cm. We checked the accuracy of the computations by comparing the results with different grid spacing and time steps. The difference between the grids with a 10- and a 1-cm spacing did not exceed 1%. Similarly, we decreased the time step until it did not affect the results by more than 1%. The program was implemented as a FORTRAN 90 [*Press et al.*, 1992] code. Our typical simulation for the 2.5-km-long spatial domain (upflow zone plus deep horizontal limb) discretized at 250,000 grid points required 5 min for 10^3 time steps on a personal computer running with 3 GB of RAM and a 2-GHz processor.

Figure 4a shows how the steady-state temperature profile varies with the thickness d of the impermeable thermal boundary layer situated between the top of the magma chamber and the base of the hydrothermal system. Note that for the results presented from this point on, the term "distance" represents the (curvilinear) distance s from the seafloor along the flow path 1-2-3. Therefore, in Figure 4 and following, the upper part of the upflow zone, representing the diffuse flux, corresponds to distance 0–100 m; the lower part of the upflow zone corresponds to distance 100–1500 m, with conditions at 100 m representing the focused flux; and the deep horizontal limb corresponds to distance 1500–2500 m. For a fixed mass flux in the horizontal limb, as d gets thinner, the amount of heat that is conducted into the fluids increases and the fluid temperature rises. In the example presented here, a difference of 1 m generates a shift of ~20°–30°C in the lower part of the upflow zone. At the junction, however (Figure 4b), all the profiles become identical. This is a result of the long-term regulating role of the cold shallow recharge. As can be seen in Figure 4c, the boundary layer at the seafloor remains the same in all the cases.

The hydrodynamic resistances for the simulations presented in Figure 4 ($r_3 = 3.6 \times 10^5\ (\mathrm{ms})^{-1}$, $r_4 = 5 \times 10^2\ (\mathrm{ms})^{-1}$) are such that the mass fluxes are moderate ($Q_1 = 37$ kg/s, $Q_2 = 14$ kg/s, and $Q_4 = 23$ kg/s). Figure 5 presents similar

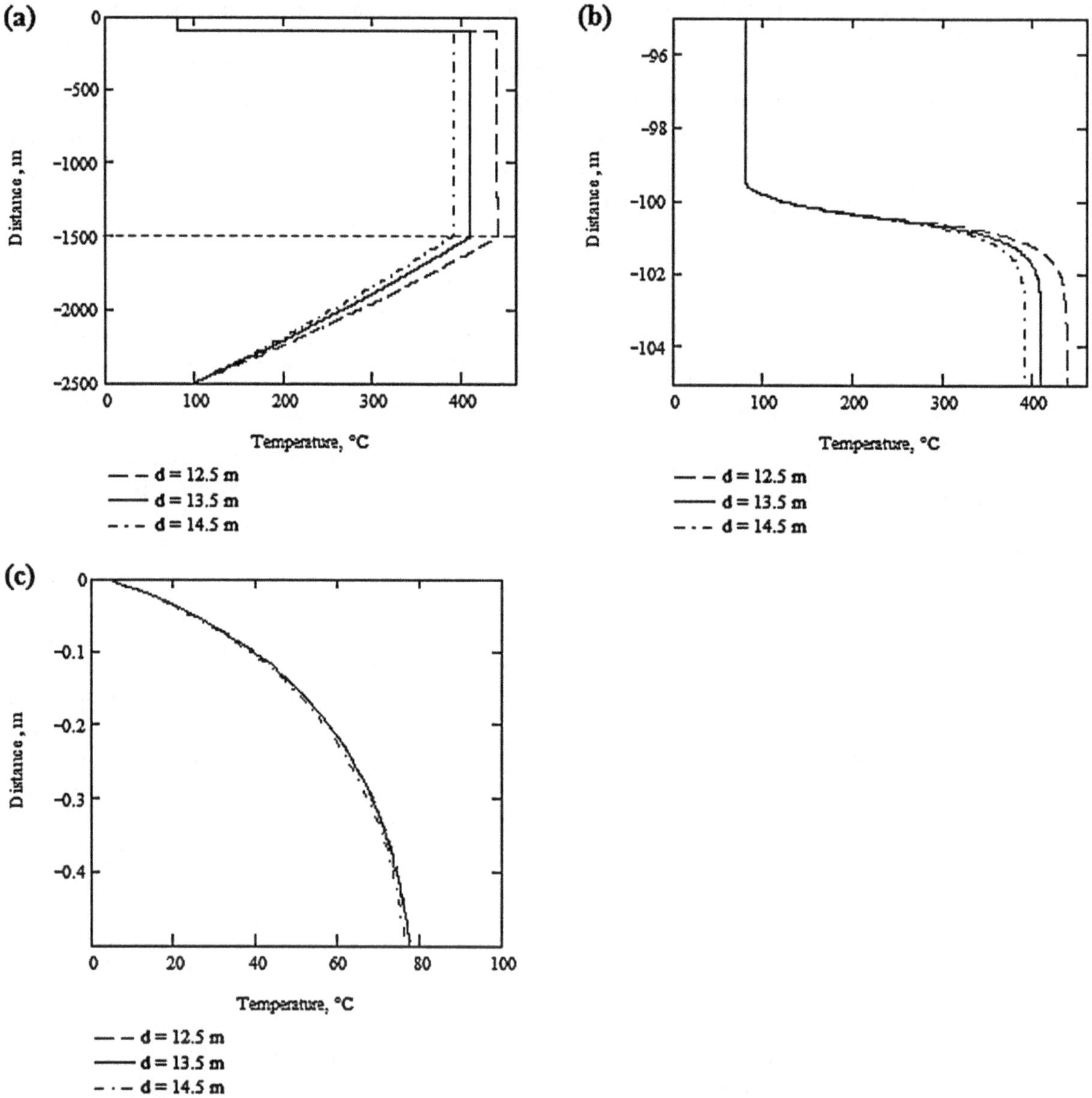

Figure 4. (a) Variation of the steady-state temperature profile with the thickness d of the impermeable thermal boundary layer situated between the magma chamber and the base of the hydrothermal system. The dashed line corresponds to $d = 12.5$ m, the solid line to $d = 13.5$ m, and the dash-dot line to $d = 14.5$ m. The dashed horizontal line marks the base of the upflow zone. (b) Detail of the profile between $s = -95$ m and $s = -105$ m. (c) Detail of the profile in the boundary layer near the seafloor.

results but for smaller resistances, and thus higher fluxes of $Q_1 = 53$ kg/s, $Q_2 = 26$ kg/s, and $Q_4 = 27$ kg/s. The latter case requires smaller values of d, on the order of 7–8 m, to obtain temperatures representative of East Pacific Rise conditions. Also, smaller variations of d are required to generate shifts of the temperature profile in the lower part of the upflow zone.

We then tested the transient response of the system to an instantaneous change in the thickness d of the boundary layer (Figures 6 and 7). Instantaneous thinning of this boundary layer corresponds to a downward cracking event and the formation of new permeability near the top of the magma chamber. For this example, the mass fluxes are the same as for Figure 4. After 1 year, a slight change in the profile of

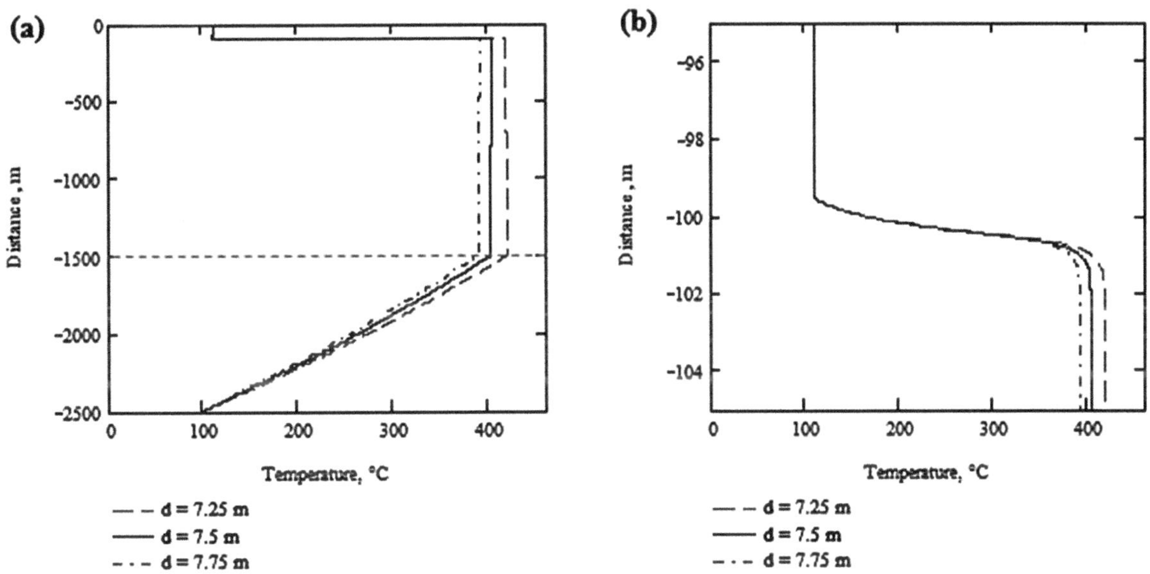

Figure 5. (a) Similar to Figure 4 but for somewhat larger mass fluxes (see text for details). The dashed horizontal line marks the base of the upflow zone. (b) Detail of the profile between $s = -95$ m and $s = -105$ m.

the horizontal limb can be observed, whereas the thermal boundary layers at the junction zone (Figure 6b) and beneath the seafloor are unchanged. After 5 years, the perturbation has started to propagate upward in the upflow region, but the junction zone is unchanged (Figure 7b).

These simulations illustrate that any modification of the thickness of the bottom boundary layer d would have only long-term effects. It would take many years before the flow regime is altered enough to modify the boundary layers at the junction zone and at the seafloor. Because the conductive

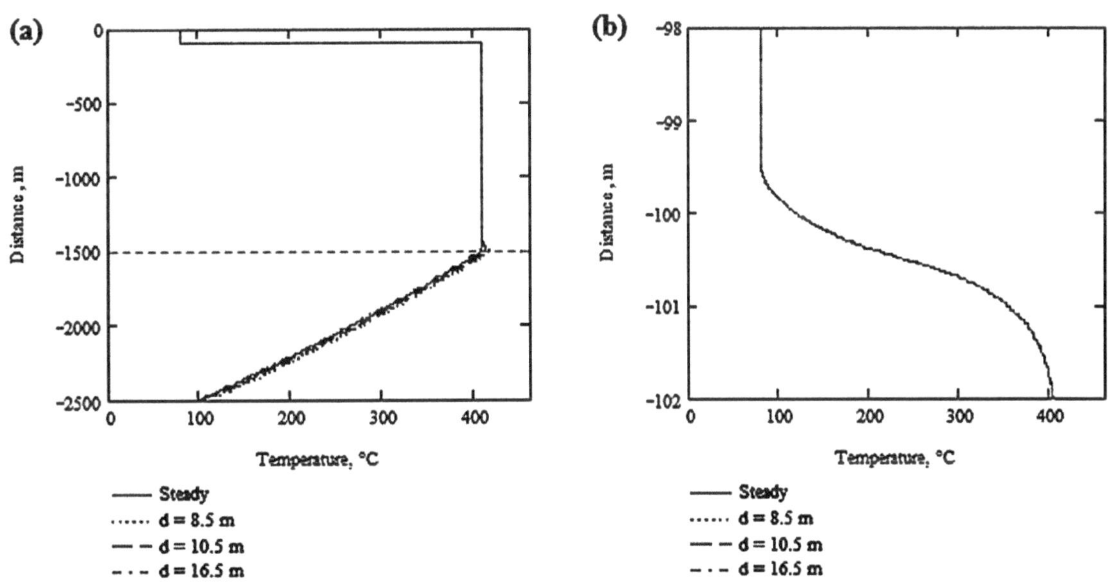

Figure 6. (a) Temperature profile 1 year after changing the thickness d of the impermeable thermal boundary layer. Steady-state corresponds to $d = 13.5$ m, and the other curves correspond to values of d after perturbation. The dashed horizontal line marks the base of the upflow zone. (b) Detail of the profile at the junction zone, between $s = -98$ m and $s = -102$ m.

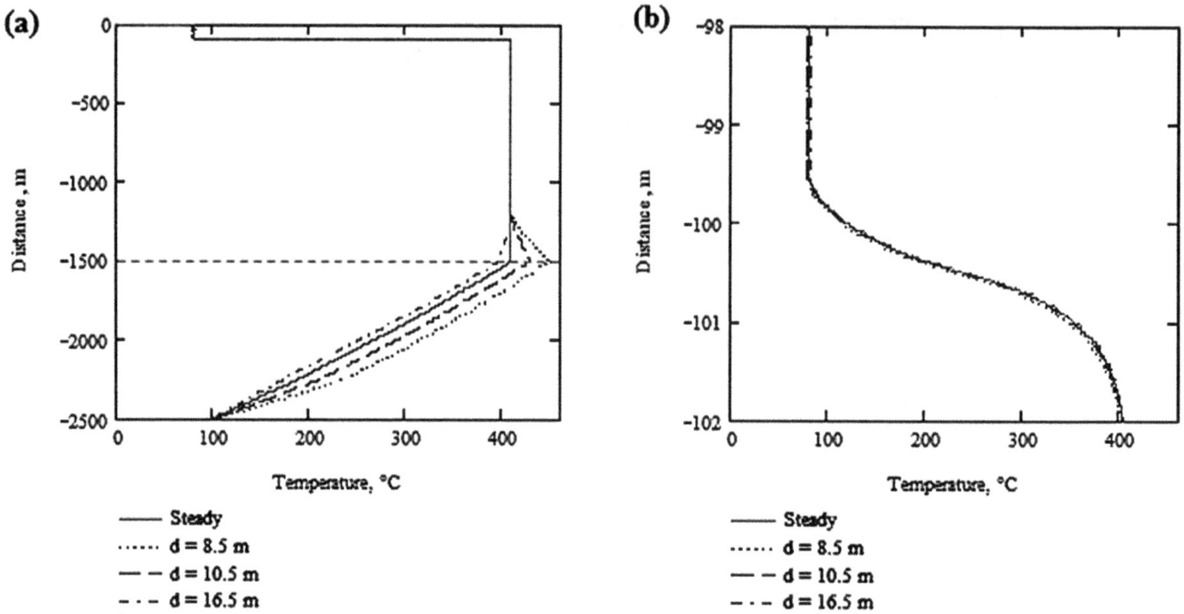

Figure 7. (a) Temperature profile 5 years after changing the thickness d of the impermeable thermal boundary layer. Steady-state corresponds to $d = 13.5$ m, and the other curves correspond to the values of d after perturbation. The dashed horizontal line marks the base of the upflow zone. (b) Detail of the profile at the junction zone, between $s = -98$ m and $s = -102$ m.

boundary layer d is already quite thin, the new permeability generated by a deep cracking event that further thins the boundary layer is not likely to cause a rapid perturbation in the vent temperature.

5. MODELING THE MARCH 1995 EAST PACIFIC RISE EVENT

5.1. A Possibility of Diking Event

The March 1995 East Pacific Rise swarm is typically explained as a thermal cracking episode, enabling hydrothermal fluid to mine heat deeper in the reaction zone [*Fornari et al.*, 1998; *Sohn et al.*, 1998, 1999]. The observed seismic pattern (Figure 8a) was interpreted as a slip along a vertical fault generated by the release of thermal stresses at the base of the hydrothermal system [*Sinha and Evans*, 2004; *Sohn et al.*, 1999]. The slip would have rapidly propagated downward, as the axial magma chamber lid subsided, opening pathways for the fluid that presumably reached the seafloor and caused the temperature anomaly four days later [*Fornari et al.*, 1998; *Sohn et al.*, 1998, 1999]. Here, we suggest an alternative interpretation based on the fact that the microearthquakes are located well above the margins of the magma lens (Figure 8a). (See also Germanovich et al., Seismic and hydrothermal evidence for a diking event on the East Pacific Rise at 9°50′N, *J. Geophys. Res.*, submitted, 2008, hereinafter referred to as Germanovich et al., submitted, 2008).

As shown by *Sim* [2004] and *Sim et al.* [2004], the pressurization of a lens-shaped magma chamber results in the tensile stress concentration near the tips of the lens, whereas the rest of the host rock is in compression. This implies that an episode of pressurization is likely to result in dikes initiating near the lens tips. Finite element calculations [*Sim*, 2004; *Sim et al.*, 2004] show that the dike then propagates almost vertically toward the seafloor from the lens tips (Figure 8b). The hypocentral pattern in Figure 8a could thus be the expression of dike emplacement initiated at the tip of the axial magma chamber lying beneath the ridge. As noted by *Sohn et al.* [1999], the events shown in Figure 8a are compatible with dike-induced seismicity.

Although we prefer the diking explanation of the March 1995 seismic swarm (Germanovich et al., submitted, 2008), fault slip and dike propagation are likely to enhance the permeability in the upflow zone [*Brodsky et al.*, 2003; *Delaney et al.*, 1986; *Germanovich et al.*, 2000; *Pollard*, 1987; *Rojstaczer and Wolf*, 1992; *Rojstaczer et al.*, 1995]. Although *Sohn et al.* [1999] do not describe this explicitly, their model

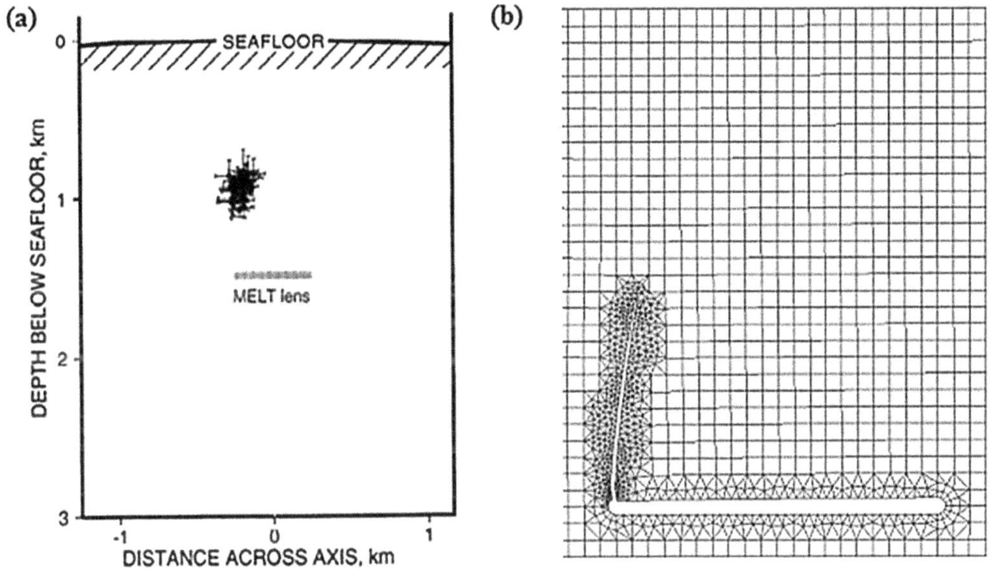

Figure 8. (a) Ridge-normal cross section of relocated north group events with error bars (modified from *Sohn et al.* [1999]). (b) Finite element simulation of dike propagation from magma chamber suggested in (a). We used a pressurized 1.5-km-deep, 1-km-wide, and 50-m-thick magma lens, and propagated the dike 600 m in order to match the hypocentral pattern shown in (a).

can be interpreted as the slip of a vertical fault enhancing permeability in the area around the hypocenters and below (Figure 8a). In this section, we consider the hydrothermal implications of the change of permeability above the west margin of the axial magma chamber when the seismic swarm was observed. Mechanical implications are discussed in more detail by Germanovich et al. (submitted 2008).

Before studying the impact of permeability enhancement in the upflow zone, we first consider the influence of heat input in the case of a dike intrusion. To test this hypothesis, the heat flux q_d from dike emplacement is accounted for when solving (3.11), and an emplacement height of $h_d = 600$ m is invoked to match the hypocentral pattern in Figure 8a (as sketched in Plate 2). Figure 9 shows the modified temperature profile for various dike thicknesses, 1 year after the emplacement. For this simulation, we chose the large mass fluxes used in the previous section. Only the temperature of the fluid situated around the dike was significantly altered (Figure 9a). The additional heat results in a slight increase of the mass flux and, consequently, a moderate upward displacement in the boundary layers at the junction zone and at the seafloor. This increase is then followed by a gradual decrease of the mass fluxes, which results in a downward displacement of the boundary layers, and a decrease in temperature at any fixed point in the boundary layers. The system eventually evolves toward a new steady state. The temperature behavior is shown in Figure 9b. This simulation demonstrates that shallow recharge regulates the outlet temperatures of the system. The rapid decrease of fluid density in the upflow zone increases for wider dikes because the heat input from wider dikes is larger than for smaller ones. The increased buoyancy in turn drives an enhanced response of the shallow recharge, which explains why the subsequent temperature decrease is amplified for thicker dikes. The same simulation conducted with smaller steady-state mass fluxes showed a similar response, although more attenuated [*Ramondenc et al.*, 2005]. These simulations suggest that unless the dike is exceptionally wide (~10 m or wider), additional heat from the dike alone does not result in the observed changes in black smoker temperatures.

5.2. Fitting the Observed Data

Although the simulation described above shows that heat alone cannot induce temperature variations such as those observed at East Pacific Rise during the March 1995 swarm [*Fornari et al.*, 1998], it is likely that dike emplacement is accompanied by permeability enhancement around the

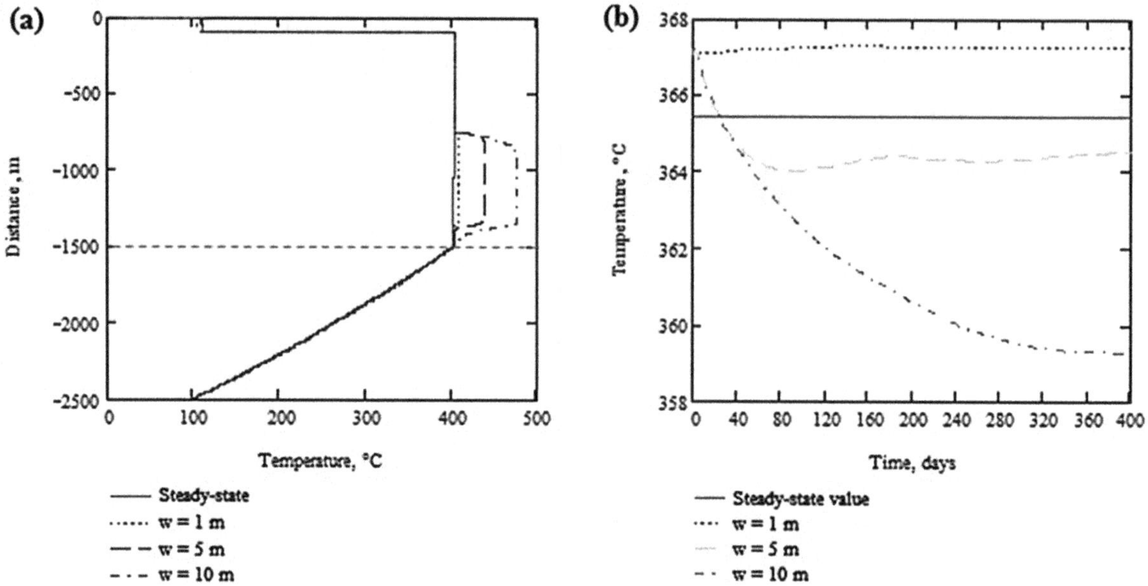

Figure 9. (a) Temperature profile 1 year after the emplacement of a dike. The steady-state corresponds to $d = 7.5$ m. The other curves correspond to different dike widths w. The dashed horizontal line marks the base of the upflow zone. (b) Temperature evolution in the junction zone at a fixed point of depth $s = -100.73$ m for different dike widths w.

margins of the dike [*Delaney et al.*, 1986; *Germanovich et al.*, 2000; *Pollard*, 1987]. Figure 10 shows the evolution of the temperature profile over time for the same parameters as Figure 4, when both added heating and enhanced permeability are taken into account. The dike thickness was chosen as $w = 1$ m, based on our mechanical computations (Figure 8b). Test simulations showed that for mass fluxes of the order of 15 kg/s in the upflow zone, the system response does not result in temperature decreases after the initial increase. Therefore, a thin thermal boundary layer at the top of the magma chamber ($d = 7.5$ m) and relatively small hydrodynamic resistances in the horizontal limbs were used in order to achieve the higher steady-state mass fluxes described in Section 4 (i.e., $Q_1 = 53$ kg/s, $Q_2 = 26$ kg/s, and $Q_4 = 27$ kg/s, with r_4 chosen to be $5 \times 10^2 (\text{ms})^{-1}$ and r_3 found to be 3.4×10^4 $(\text{ms})^{-1}$). Finally, the permeability was multiplied by a factor of 1.5 in the 600-m section of the upflow zone where the dike was emplaced, and the height of the junction with the shallow recharge of cold fluid was taken to be 5 m, rather than 1 m as in the previous examples. These values provide the best fit for the temperature variations at Bio9 following the earthquake swarm.

Figure 10a shows the resulting evolution of the temperature profile at the bottom of the upflow zone. Temperature increases for approximately 7 days after the emplacement of the dike then gradually decreases as the fluid flows upward. Figure 10b shows the evolution of temperature at the junction with shallow recharge, representing the behavior of black smokers. Temperature immediately increases in the junction zone after the dike is emplaced, before decreasing approximately 1 week later. This decrease results from the input of cold fluid from shallow recharge.

Figure 11a shows temperature as a function of time at the point near the junction zone that best matches the temperatures observed at Bio9 before and during the event (at depth $z = 102.40$ m). Our model matches the different variations over the first year that have been observed in the temperature time series (Figure 11a). The large-scale oscillations in the model are related to the interplay between cold recharge from the shallow part of the system and hot discharge from below as the system evolves to a new steady state. The fit is further improved after translating the record after November 1995 by 3°C (Figure 11b), time at which the temperature probe (or hobo) was replaced [*Fornari et al.*, 1998; Fornari, personal communication, 2006].

5.3. Discussion

In the model presented in this work, we assumed that black smokers represent "temperature probes" (proxies) of the base of the focusing zone. In turn, we assumed that focusing occurs in the narrow junction where the upflow meets the

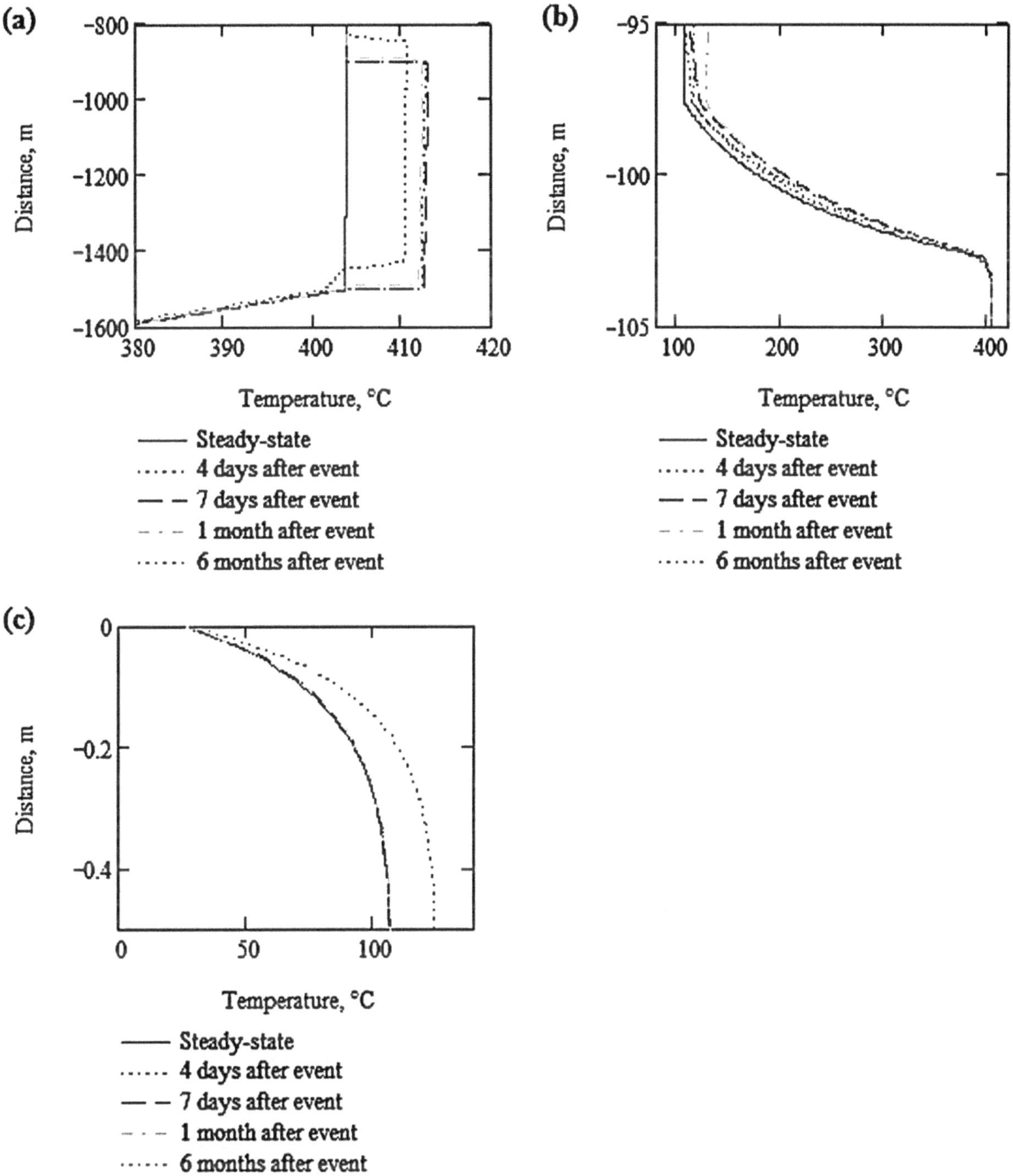

Figure 10. Evolution of temperature profile (a) in the bottom of the upflow zone, (b) at the junction with the shallow recharge, and (c) in the first 50 cm below the seafloor (d = 7.5 m, f = 5 m, w = 1 m, and $k = 1.5 \times k_0$ around the dike margins). Note that the black smoker temperature is represented by the temperature at some fixed point within the junction zone shown in (b), whereas (c) corresponds to low-temperature diffuse flow.

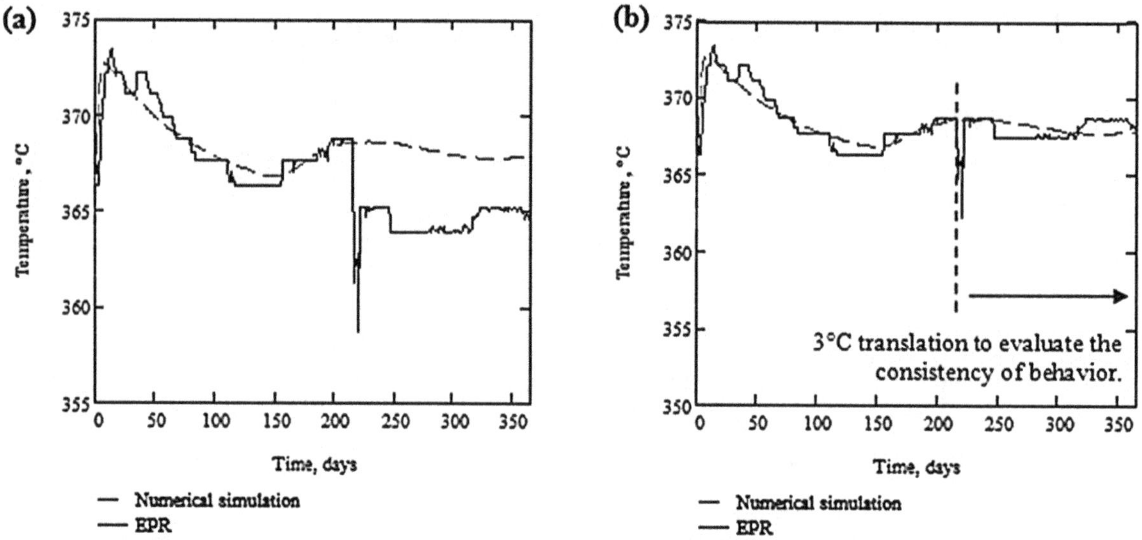

Figure 11. Comparison of the calculated response to dike emplacement with the observed response at Bio9 after the March 1995 East Pacific Rise event. In (a), the monitored response is unchanged while in (b) we applied a 3°C translation to the records at the time of temperature probe replacement. For this simulation, Q_1 = 53 kg/s, Q_2 = 26 kg/ls, Q_4 = 27 kg/ls, $r_3 = 10^4(\text{ms})^{-1}$, $P_4 = 10^2(\text{ms})^{-1}$. ($d$ = 7.5 m, f = 5 m, w = 1 m, and $k = 1.5 \times k_0$ around the dike margins)

shallow recharge of cold seawater (Plate 2). One can argue, however, that this recharge may occur over a much larger portion of the extrusives (Figure 3). In essence, our approach consists in using an "equivalent" shallow recharge, delivering the same amount of cooling that may occur in reality. From this point of view, our model captures relatively well the physics of shallow recharge. In this context, the depth of the junction scales with the depth of the shallow recharge zone and its exact value is relatively unimportant. One way to obtain a better estimate of the position of shallow recharge would be to compare both the diffuse low-temperature and focused high-temperature responses to seismic activity. Unfortunately, there was no temperature sensor deployed in the diffuse venting zone of Bio9 area during the 1995 activity [*Scheirer et al.*, 2006].

Furthermore, using diffuse flow temperature records for modeling similar to that presented here may be less reliable than using black smoker temperatures. Diffuse flow is typically much broader and is not easily defined [e.g., *Ramondenc et al.*, 2006]. In any patch of diffuse venting, the temperature may range from nearly ambient seawater temperature to tens of degrees. Moreover, the temperature of diffuse venting is often measured at only a few selected points. Consequently, the average temperature of diffuse venting, which is determined in our model, as well as in most of other models, is not well constrained. This is why we did not attempt to model the June 1999 Endeavour event at the Juan de Fuca Ridge [*Johnson et al.*, 2000; *Seyfried et al.*, 2003]. In that case, only diffuse venting records at several points are available and no black smoker temperature was recorded.

Nevertheless, in principle our model can be used to simulate the temperature-averaged behavior of diffuse venting. Figure 10c represents the temperature evolution in the first 50 cm below the ocean floor. Displacements of the boundary layer are nearly nonexistent in the first couple of months after the event, whereas the response after 6 months is larger. This can be explained by the arrival of the temperature spike from the junction zone that resulted from dike emplacement at depth. As is the case in the lower part of the upflow zone, this spike is gradually advected toward the seafloor at the same time as the fluid loses heat through the walls of the upflow zone. Although there are no data on diffuse venting from Bio9 vent, temperature time series are available for the Bio9 Riftia community starting November 1995 [*Scheirer et al.*, 2006]. These records indicate a sudden temperature increase at the beginning of 1996, followed by a gradual decrease over several years. This behavior is consistent with the delayed response that our model suggests for sites of diffuse venting.

The main advantage of our model is that, despite the apparently large number of parameters, essentially only

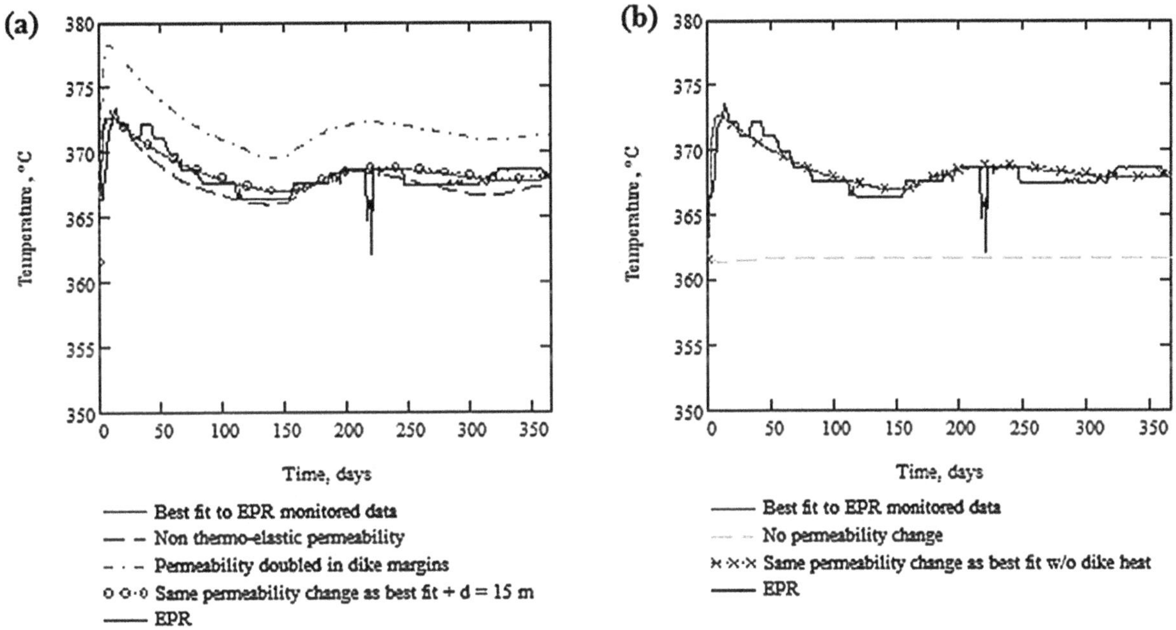

Figure 12. Comparison of the best fit to the March 1995 East Pacific Rise event ($d = 7.5$ m, $f = 5$ m, $w = 1$ m, and $k = 1.5 \times k_0$ around the dike margins) with other scenarios; only one parameter is modified at a time, all the others remaining the same. (a) Comparison of two types of permeability changes; doubling the thickness of the thermal boundary layer d does not affect the results. (b) Comparison of cases where heat alone is taken into account (without any permeability change), and only permeability change is considered (without any heat).

two control the behavior of the system: the thickness d of the thermal boundary layer between the top of the magma chamber and the hydrothermal system, and the permeability k of the upflow zone. Figure 12 shows that at the time scale of a year, permeability is the main controlling parameter, whereas for longer time scales, mainly the thickness, d, of the basal boundary layer affects the evolution of the system. Even doubling this thickness has little effect over time scales shorter than a year (Figure 12a). In contrast, our best fit requires multiplying the steady-state permeability by 1.5, and increasing this to a factor of 2 has a large effect. In comparison, the permeability of on-land hydrological systems in Southern California has been shown to undergo increases by a factor as high as 3 at the time of earthquakes [*Elkhoury et al.*, 2006]. The sensitivity to permeability changes would become greater if changes occur over a larger portion of the upflow zone.

Figure 12b further illustrates the fact that heat alone cannot generate the observed temperature variations; however, permeability enhancement without heating could produce satisfactory results. This shows that permeability is the main controlling parameter, at least when the dike thickness w is less than 10 m (Figure 9b). If the dike were considerably thicker than $w = 1$–10 m assumed here (Figure 9b), the temperature changes in the junction zone may be large enough to counterbalance the response to the permeability change. Thus, dike thickness has some influence on the response of the system.

To address the longer-term (>1 year) temperature increase (Figure 1), our model would need to be complemented by some other factors. A gradual decrease in permeability either deep in the system or at the mixing junction would lead to a gradual increase in vent temperature. Our model does not consider mineral precipitation. In the junction zone, the temperature fluctuations that result from mixing may lead to clogging by precipitation of anhydrite [*Lowell et al.*, 2003] or some other mineral. This would reduce the input of cold shallow recharge, thus leading to a slow increase of temperature at black smokers. In addition to inorganic mineral precipitation, subsurface microbial activity may clog the pore space, although recent modeling of snowblower vents at 9°50′N on the East Pacific Rise suggests that this effect may be small [*Lowell et al.*, 2008; Crowell *et al.*, 2008]. Mineral precipitation may also occur deep in the crust and eventually clog the permeability that was enhanced as a result of the hypothesized dike emplacement [*Lowell et al.*, 1993; *Martin and Lowell*, 2000].

6. CONCLUSION

The mechanisms of hydrothermal response of mid-ocean ridge venting systems to seismic and/or magmatic activity remain a matter of debate. Our model of hydrothermal circulation within the crust offers a way to study different types of perturbations while invoking relatively simple circulation pathways. It allows simultaneous consideration of both low-temperature diffuse and high-temperature focused venting through the introduction of shallow recharge of cold fluid within the shallow crust. The most important result of this work is that a thermal perturbation at depth need not be advected along the entire discharge zone; it only needs to affect thermal boundary layers near the seafloor and at the junction with the shallow recharge to generate the observed temperature variations. Moreover, our results show that a deep cracking event alone that generates new permeability near the base of the system is not likely to cause a rapid perturbation in the vent temperature. If modest permeability increases resulting from seismic activity or dike emplacement above the west margin of the axial magma chamber where the seismic swarm was observed in March 1995 at the East Pacific Rise occur, our model yields results that are in good agreement with the observed temperature behavior there.

Acknowledgments. We thank Robert Reves-Sohn, Maya Tolstoy, and the Editor Jeff Seewald for their thorough and thoughtful reviews of this manuscript. We are particularly grateful to the anonymous reviewer not only for his technical comments but also for the thorough copy editing of the original manuscript. We are indebted to Karen Von Damm for insightful discussions and the opportunity to join her 2004 and 2006 cruises to the 9°50′N area. We thank Dan Fornari for generously sharing information about his data set and the March 1995 event. We are also grateful to Marv Lilley for his helpful comments and suggestions. This work was supported in part by NSF Grants OCE-0221974 and OCE-0527208 to R.P.L and L.N.G. The authors appreciate Schlumberger supporting color figures in this paper.

REFERENCES

Baker, E. T. (1998), Patterns of event and chronic hydrothermal venting following a magmatic intrusion: New perspectives from the 1996 Gorda Ridge eruption, *Deep Sea Res. II*, *45*, 2599–2618.

Baker, E. T., C. G. Fox, and J. P. Cowen (1999), In situ observations of the onset of hydrothermal discharge during the 1998 submarine eruption of Axial Volcano, Juan de Fuca Ridge, *Geophys. Res. Lett.*, *26*(23), 3445–3448.

Baker, E. T., G. J. Massoth, and R. A. Feely (1987), Cataclysmic hydrothermal venting on the Juan de Fuca Ridge, *Nature*, *329*, 149–151.

Baker, E. T., G. J. Massoth, R. A. Feely, G. A. Cannon, and R. E. Thomson (1998), The rise and fall of the CoAxial hydrothermal site, 1993–1996, *J. Geophys. Res.*, *103*(B5), 9791–9806.

Baker, E. T., G. J. Massoth, R. A. Feely, R. W. Embley, R. E. Thomson, and B. J. Burd (1995), Hydrothermal event plumes from the CoAxial seafloor eruption site, Juan de Fuca Ridge, *Geophys. Res. Lett.*, *22*(2), 147–150.

Baker, E. T., G. J. Massoth, S. L. Walker, and R. W. Embley (1993), A method for quantitatively estimating diffuse and discrete hydrothermal discharge, *Earth Planet. Sci. Lett.*, *118*, 235–249.

Bear, J. (1972), *Dynamics of Fluids in Porous Materials,* Elsevier, New York, 1972.

Bemis, K. G., R. P. Von Herzen, and M. J. Mottl (1993), Geothermal heat flux from hydrothermal plumes on the Juan de Fuca Ridge, *J. Geophys. Res.*, *98*(B4), 6351–6365.

Bodvarsson, G. (1969), On temperature of water flowing through fractures, *J. Geophys. Res.*, *74*(8), 1987–1992.

Bohnenstiehl, D. R., R. P. Dziak, M. Tolstoy, C. G. Fox, and M. Fowler (2004), Temporal and spatial history of the 1999–2000 Endeavour Segment seismic series, Juan de Fuca Ridge, *Geochem. Geophys. Geosyst.*, *5*, Q09003, doi:10.1029/2004GC000735.

Bray, A. M. (2001), The geochemistry of boron and lithium in mid-ocean ridge hydrothermal vent fluids, Ph.D. thesis, 125 pp., University of New Hampshire, Durham.

Brodsky, E. E., E. Roeloffs, D. Woodcock, I. Gall, and M. Manga (2003), A mechanism for sustained groundwater pressure changes induced by distant earthquakes, *J. Geophys. Res.*, *108*(B8), 2390, doi:10.1029/2002JB002321.

Carlsaw, H. S., and J. C. Jaeger (1986), *Conduction of Heat in Solids*, 2nd ed., Clarendon Press, Oxford, 510 pp.

Cherkaoui, A. S. M., W. S. D. Wilcock, and E. T. Baker (1997), Thermal fluxes associated with the 1993 diking event on the CoAxial segment, Juan de Fuca Ridge: A model for the convective cooling of a dike, *J. Geophys. Res.*, *102*(B11), 24,887–24,902.

Corliss, J. B., et al. (1979), Submarine thermal springs on the Galápagos Rift, *Science*, *203*, 1073–1083.

Cowen, J. P., E. T. Baker, and R. W. Embley (2004), Detection of and response to mid-ocean ridge magmatic events: Implications for the subsurface biosphere, in *The Subseafloor Biosphere at Mid-Ocean Ridges*, *Geophys. Monogr. Ser.*, vol. 144, edited by W. S. D. Wilcock, et al., pp. 227–243, AGU, Washington, D. C.

Crowell et al. (2008), A model for the production of sulfur floc and "Snowblower" vents at mid-ocean ridges, *Geochemistry*, *Geophysics*, *Geosystems*, in press.

Davis, E., K. Becker, R. Dziak, J. Cassidy, K. Wang, and M. Lilley (2004), Hydrological response to a seafloor spreading episode on the Juan de Fuca ridge, *Nature*, *430*, 335–338, doi:10.1038/nature02755.

Davis, E. E., K. Wang, R. E. Thomson, K. Becker, and J. F. Cassidy (2001), An episode of seafloor spreading and associated plate deformation inferred from crustal fluid pressure transients, *J. Geophys. Res.*, *106*(B10), 21,953–21,963.

Delaney, J. R., V. Robigou, and R. E. McDuff (1992), Geology of a vigorous hydrothermal system on the Endeavour Segment, Juan de Fuca Ridge, *J. Geophys. Res.*, *97*(B13), 19,663–19,682.

Delaney, P. T. (1982), Rapid intrusion of magma into wet rock: Groundwater flow due to pore pressure increases, *J. Geophys. Res.*, *87*, 7739–7756.

Delaney, P. T., D. D. Pollard, J. I. Ziony, and E. H. Mckee (1986), Field relations between dikes and joints: Emplacement processes and paleostress analysis, *J. Geophys. Res.*, *91*(B5), 4920–4938.

Detrick, R. S., S. M. Carbotte, E. van Ark, J. P. Canales, G. M. Kent, A. Harding, J. B. Diebold, and M. R. Nedimović (2002), New multi-channel seismic constraints on the crustal structure of the Endeavour Segment, Juan de Fuca Ridge: Evidence for a crustal magma chamber, *Eos Trans. AGU*, *83*(47), Abstract T21B-1316.

Dziak, R., B. Chadwick, J. Cowen, E. Baker, R. Embley, D. Bohnenstiehl, and J. Resing (2006), Detecting Volcanic Events in the Northeast Pacific, *Eos Trans. AGU*, *87*(4), 37.

Dziak, R. P., and C. G. Fox (1999), The January 1998 earthquake swarm at Axial Volcano, Juan de Fuca Ridge: Hydroacoustic evidence of seafloor volcanic activity, *Geophys. Res. Lett.*, *26*(23), 3429–3432.

Dziak, R. P., C. G. Fox, and A. E. Schreiner (1995), The June-July 1993 seismo-acoustic event at CoAxial segment, Juan de Fuca Ridge: Evidence for a lateral dike injection, *Geophys. Res. Lett.*, *22*(2), 135–138.

Dziak, R. P., and H. P. Johnson (2002), Stirring the oceanic incubator, *Science*, *296*, 1406–1407.

Dziak, R. P., D. K. Smith, D. R. Bohnenstiehl, C. G. Fox, D. Desbruyeres, H. Matsumoto, M. Tolstoy, and D. J. Fornari (2004), Evidence of a recent magma dike intrusion at the slow spreading Lucky Strike segment, Mid-Atlantic Ridge, *J. Geophys. Res.*, *109*, B12102, doi:10.1029/2004JB003141.

Edmond, J. M., A. C. Campbell, M. R. Palmer, G. P. Klinkhammer, C. R. German, H. N. Edmonds, H. Elderfield, G. Thompson, and P. A. Rona (1995), Time series studies of vent fluids from the TAG and MARK sites (1986, 1990) Mid-Atlantic Ridge: A new solution chemistry model and a mechanism for Cu/Zn zonation in massive sulphide ore bodies, in *Hydrothermal Vents and Processes, Geol. Soc. Spec. Publ.*, vol. 87, edited by L. M. Parson, et al., pp. 77–86, The Geological Society, London, U.K.

Edmond, J. M., C. Measures, R. E. McDuff, L. H. Chan, R. Collier, B. Grant, L. I. Gordon, and J. B. Corliss (1979), Ridge crest hydrothermal activity and the balances of the major and minor elements in the ocean: The Galapagos data, *Earth Planet. Sci. Lett.*, *46*, 1–18.

Elkhoury, J. E., E. E. Brodsky, and D. C. Agnew (2006), Seismic waves increase permeability, *Nature*, *441*, 1135–1138, doi:10.1038/nature04798.

Embley, R., and E. Baker (1999), Interdisciplinary group explores seafloor eruption with remotely operated vehicle, *Eos Trans. AGU*, *80*(19), 213.

Embley, R. W., W. W. Chadwick Jr., I. R. Jonasson, D. A. Butterfield, and E. T. Baker (1995), Initial results of the rapid response to the 1993 CoAxial event: Relationships between hydrothermal and volcanic processes, *Geophys. Res. Lett.*, *22*(2), 143–146.

Embley, R. W., and J. E. Lupton (2004), Diking, Event Plumes, and the Subsurface Biosphere at Mid-Ocean Ridges, in *The Subseafloor Biosphere at Mid-Ocean Ridges, Geophys. Monogr. Ser.*, vol. 144, edited by W. S. D. Wilcock, et al., pp. 75–97, AGU, Washington, D. C.

Fisher, A. T. (1998), Permeability within basaltic oceanic crust, *Rev. Geophys.*, *36*(2), 143–182.

Fornari, D. J., T. Shank, K. L. Von Damm, T. K. P. Gregg, M. Lilley, G. Levai, A. Bray, R. M. Haymon, M. R. Perfit, and R. Lutz (1998), Time-series temperature measurements at high-temperature hydrothermal vents, East Pacific Rise 9°49′–51′N: Evidence for monitoring a crustal cracking event, *Earth Planet. Sci. Lett.*, *160*, 419–431.

Fox, C. G. (1995), Special collection on the June 1993 volcanic eruption on the Coaxial Segment, Juan de Fuca Ridge, *Geophys. Res. Lett.*, *22*(2), 129–130.

Fox, C. G., and R. P. Dziak (1998), Hydroacoustic detection of volcanic activity on the Gorda Ridge, February-March 1996, *Deep Sea Res. II*, *45*, 2513–2530.

Fox, C. G., R. P. Dziak, H. Matsumoto, and A. E. Schreiner (1993/1994), Potential for monitoring low-level seismicity on the Juan de Fuca Ridge using military hydrophone arrays, *Mar. Technol. Soc. J.*, *27*, 22–30.

Fox, C. G., H. Matsumoto, and T. K. A. Lau (2001), Monitoring Pacific Ocean seismicity from an autonomous hydrophone array, *J. Geophys. Res.*, *106*(B3), 4183–4206.

Fox, C. G., W. E. Radford, R. P. Dziak, T. K. Lau, H. Matsumoto, and A. E. Schreiner (1995), Acoustic detection of a seafloor spreading episode on the Juan de Fuca Ridge using military hydrophone arrays, *Geophys. Res. Lett.*, *22*(2), 131–134.

Germanovich, L. N., and R. P. Lowell (1992), Percolation theory, thermoelasticity, and discrete hydrothermal venting in the earth's crust, *Science*, *255*, 1564–1567.

Germanovich, L. N. and R. P. Lowell (1995), The mechanism of phreatic eruptions, *J. Geophys. Res.*, *100*, 8417–8434.

Germanovich, L. N., R. P. Lowell, and D. K. Astakhov (2000), Stress-dependent permeability and the formation of seafloor event plumes, *J. Geophys. Res.*, *105*(B4), 8341–8354.

Germanovich, L. N., R. P. Lowell, and D. K. Astakhov (2001), Temperature-dependent permeability and bifurcations in hydrothermal flow, *J. Geophys. Res.*, *106*(B1), 473–495.

Haymon, R. M., et al. (1993), Volcanic eruption of the mid-ocean ridge along the East Pacific Rise crest at 9°45–52′N: Direct submersible observations of seafloor phenomena associated with an eruption event in April, 1991, *Earth Planet. Sci. Lett.*, *119*, 85–101.

Hill, D. P., et al. (1993), Seismicity remotely triggered by the magnitude 7.3 Landers, California, Earthquake, *Science*, *260*, 1617–1623.

Humphris, S. E., and M. K. Tivey (2000), A synthesis of geological and geochemical investigations of the TAG hydrothermal fluid: Insights into fluid flow and mixing processes in a hydrothermal system, in *Ophiolites and oceanic crust: New insights from field studies and the Ocean Drilling Program, Spec. Pap.*, vol. 349, edited by Y. Dilek, et al., pp. 213–235, Geological Society of America, Boulder, CO.

James, R. H., and H. Elderfield (1996), Chemistry of ore-forming fluids and mineral formation rates in an active hydrothermal sulfide deposit on the Mid-Atlantic Ridge, *Geology*, *24*, 1147–1150.

Johnson, H. P., R. P. Dziak, C. R. Fisher, C. G. Fox, and M. J. Pruis (2001), Earthquakes' impact on hydrothermal systems may be far-reaching, *Eos Trans. AGU*, *82*(21), 233.

Johnson, H. P., M. Hutnak, R. P. Dziak, C. G. Fox, I. Urcuyo, J. P. Cowen, J. Nabelek, and C. Fisher (2000), Earthquake-induced changes in a hydrothermal system on the Juan de Fuca mid-ocean ridge, *Nature*, *407*, 174–177.

Juniper, S. K., P. Martineu, J. Sarrazin, and Y. Gelinas (1995), Microbial-mineral floc associated with nascent hydrothermal activity on CoAxial segment, Juan de Fuca Ridge, *Geophys. Res. Lett.*, *22*(2), 179–182.

Kadko, D., and W. Moore (1988), Radiochemical constraints on the crustal residence time of submarine hydrothermal fluids: Endeavour Ridge, *Geochim. Cosmochim. Acta*, *52*, 659–668.

Kent, G. M., A. J. Harding, and J. A. Orcutt (1993), Distribution of magma beneath the East Pacific Rise between the Clipperton transform and the 9°17′N deval from forward modeling of common depth point data, *J. Geophys. Res.*, *98*(B8), 13,945–13,969.

Lewis, K. C. and R. P. Lowell (2004), Mathematical modeling of phase separation of seawater near an igneous dike, *Geofluids*, *4*, 197–209.

Lilley, M. D., D. A. Butterfield, J. E. Lupton, and E. J. Olson (2003), Magmatic events can produce rapid changes in hydrothermal vent chemistry, *Nature*, *422*, 878–881, doi:10.1038/nature01569.

Lowell, R. P. (1975), Circulation in fractures, hot springs, and convective heat transport on mid-ocean ridge crests, *Geophys. J. R. Astron. Soc.*, *40*, 351–365.

Lowell, R. P. (1976), Comments on 'Theory of heat extraction from fractured hot dry rock' by A. C. Gringarten, P. A. Witherspoon, and Yuzo Ohnishi, *J. Geophys. Res.*, *81*(2), 359.

Lowell, R. P. (1990), Thermoelasticity and the formation of black smokers, *Geophys. Res. Lett.*, *17*(6), 709–712.

Lowell, R. P., B. W. Crowell, K. Lewis, and L. Liu (2008), Modeling multiphase, multicomponent processes at oceanic spreading centers, this volume.

Lowell, R. P., S. R. Gosnell, and Y. Yang (2007), Numerical simulations of single-pass hydrothermal convection at mid-ocean ridges: Effects of the extrusive layer and temperature dependent permeability, *Geochem. Geophys. Geosyst.*, *8*, Q10011, doi:10.1029/2007GC001653.

Lowell, R. P., and L. N. Germanovich (1994), On the temporal evolution of high-temperature hydrothermal systems at ocean ridge crests, *J. Geophys. Res.*, *99*(B1), 565–575.

Lowell, R. P., and L. N. Germanovich (2004), Hydrothermal Processes at Mid-Ocean Ridges: Results From Scale Analysis and Single-Pass Models, in *Mid-Ocean Ridges: Hydrothermal Interactions Between the Lithosphere and Oceans, Geophys. Monogr. Ser.*, vol. 148, edited by C. R. German, et al., pp. 219–244, AGU, Washington, D. C.

Lowell, R. P., P. Van Cappellen, and L. N. Germanovich (1993), Silica Precipitation in Fractures and the Evolution of Permeability in Hydrothermal Upflow Zones, *Science*, *260*, 192–194.

Lowell, R. P., Y. Yao, and L. N. Germanovich (2003), Anhydrite precipitation and the relationship between focused and diffuse flow in seafloor hydrothermal systems, *J. Geophys. Res.*, *108*(B9), 2424, doi:10.1029/2002JB002371.

Martin, J. T., and R. P. Lowell (2000), Precipitation of quartz during high-temperature, fracture controlled hydrothermal upflow at ocean ridges: Equilibrium versus linear kinetics, *J. Geophys. Res.*, *105*(B1), 869–882.

Pascoe, A. R., and J. R. Cann (1995), Modeling diffuse hydrothermal flow in black smoker vent fields, in *Hydrothermal Vents and Processes*, *Geol. Soc. Spec. Publ.*, vol. 87, edited by L. M. Parson, et al., pp. 159–173, The Geological Society, London, U.K.

Pollard, D. D. (1987), Elementary fracture mechanics applied to the structural interpretation of dikes, in *Mafic Dyke Swarms*, *Spec. Pap. Geol. Assoc. Can.*, edited by H. C. Halls and W. F. Fahrig, pp. 5–24, Geological Association of Canada, St. Johns, Newfoundland, Canada.

Press, W. H., B. P. Flannery, S. A. Teukolsky, and W. T. Vetterling (1992), *Numerical Recipes in FORTRAN: The Art of Scientific Computing*, 2nd ed., 992 pp., Cambridge University Press, New York.

Ramondenc, P., L. N. Germanovich, and R. P. Lowell (2005), Modeling hydrothermal response to earthquakes at mid-ocean ridges: EPR, *Eos Trans. AGU*, *86*(52), Abstract T31B-0500.

Ramondenc, P., L. N. Germanovich, K. L. Von Damm, and R. P. Lowell (2006), The first measurements of hydrothermal heat output at 9°50′N, East Pacific Rise, *Earth Planet. Sci. Lett.*, *245*, 487–497, doi:10.1016/j.epsl.2006.03.023.

Ravizza, G., J. Blusztain, K. L. Von Damm, A. M. Bray, W. Bach, and S. R. Hart (2001), Sr isotope variations in vent fluids form 9°46′–9°54′N East Pacific Rise: Evidence of a non-zero-Mg fluid component, *Geochim. Cosmchim. Acta*, *65*, 729–739.

Roeloffs, E. (1996), Poroelastic techniques in the study of earthquake-related hydrologic phenomena, *Adv. Geophys.*, *37*, 135–195.

Rojstaczer, S., and S. Wolf (1992), Permeability changes associated with large earthquakes: An example from Loma Prieta, California, *Geology*, *20*, 211–214.

Rojstaczer, S., S. Wolf, and R. Michel (1995), Permeability enhancement in the shallow crust as a cause of earthquake-induced hydrological changes, *Nature*, *373*, 237–239.

Rona, P. A., and D. A. Trivett (1992), Discrete and diffuse heat transfer at ASHES vent field, Axial Volcano, Juan de Fuca Ridge, *Earth Planet. Sci. Lett.*, *109*, 57–71.

Samarskii, A. A. (2001), *The Theory of Difference Schemes*, 786 pp., Marcel Dekker Inc., New York.

Scheirer, D. S., T. M. Shank, and D. J. Fornari (2006), Temperature variations at diffuse and focused flow hydrothermal vent sites along the northern East Pacific Rise, *Geochem. Geophys. Geosyst.*, *7*, Q03002, doi:10.1029/2005GC001094.

Schultz, A., J. R. Delaney, and R. E. McDuff (1992), On the partitioning of heat flux between diffuse and point source seafloor venting, *J. Geophys. Res.*, *97*(B9), 12,299–12,314.

Seewald, J. S., A. Cruse, and P. J. Saccocia (2003), Aqueous volatiles in hydrothermal fluids from the Main Endeavour Field, northern Juan de Fuca Ridge: Temporal variability following earthquake activity, *Earth Planet. Sci. Lett.*, *216*, 575–590.

Seyfried, W. E., Jr., J. S. Seewald, M. E. Berndt, K. Ding, and D. I. Foustoukos (2003), Chemistry of hydrothermal vent fluids from the Main Endeavour Field, northern Juan de Fuca Ridge: Geochemical controls in the aftermath of June 1999 seismic events, *J. Geophys. Res.*, *108*(B9), 2429, doi:10.1029/2002JB001957.

Sim, Y. (2004), Mechanics of complex hydraulic fractures in the earth's crust, Ph.D. thesis, 324 pp., Georgia Institute of Technology, Atlanta.

Sim, Y., L. N. Germanovich, R. P. Lowell, and P. Ramondenc (2004), Diking, magma lenses, and location of hydrothermal sites at mid-ocean ridges, *Eos Trans. AGU*, *85*(47), Abstract B13A-0206.

Sinha, M. C., and R. L. Evans (2004), Geophysical constraints upon the thermal regime of the ocean crust, in *Mid-Ocean Ridges: Hydrothermal Interactions Between the Lithosphere and Oceans*, *Geophys. Monogr. Ser.*, vol. 148, edited by C. R. German, et al., pp. 19–62, AGU, Washington, D. C.

Smith, D. K., M. Tolstoy, C. G. Fox, D. R. Bohnenstiehl, H. Matsumoto, and M. J. Fowler (2002), Hydroacoustic monitoring of seismicity at the slow-spreading Mid-Atlantic Ridge, *Geophys. Res. Lett.*, *29*(11), 1518, doi:10.1029/2001GL013912.

Sohn, R. A., D. J. Fornari, K. L. Von Damm, J. A. Hildebrand, and S. C. Webb (1998), Seismic and hydrothermal evidence for a cracking event on the East Pacific Rise crest at 9°50′N, *Nature*, *396*, 159–161.

Sohn, R. A., J. A. Hildebrand, and S. C. Webb (1999), A microearthquake survey of the high-temperature vent fields on the volcanically active East Pacific Rise (9°50′N), *J. Geophys. Res.*, *104*(B11), 25,367–25,377.

Tolstoy, M. (2008), Seismological constraints on magmatic and hydrothermal processes at ridges, this volume.

Tolstoy, M., et al. (2006), A Sea-Floor Spreading Event Captured by Seismometers, *Science*, *314*, 1920–1922, doi:10.1126/science.1133950.

Von Damm, K. L. (2000), Chemistry of hydrothermal vent fluids from 9°–10°N, East Pacific Rise: "Time zero," the immediate posteruptive period, *J. Geophys. Res.*, *105*(B5), 11,203–11,222.

Von Damm, K. L. (2004), Evolution of the hydrothermal system at East Pacific Rise 9°50′N: Geochemical evidence for changes in the upper oceanic crust, in *Mid-Ocean Ridges: Hydrothermal Interactions Between the Lithosphere and Oceans*, *Geophys. Monogr. Ser.*, vol. 148, edited by C. R. German, et al., pp. 285–304, AGU, Washington, D. C.

Von Damm, K. L., and M. D. Lilley (2004), Diffuse flow hydrothermal fluids from 9°50′N East Pacific Rise: Origin, evolution and biogeochemical controls, in *The Subseafloor Biosphere at Mid-Ocean Ridges*, *Geophys. Monogr. Ser.*, vol. 144, edited by W. S. D. Wilcock, et al., pp. 245–268, AGU, Washington, D. C.

Wilcock, W. S. D. (2004), Physical response of mid-ocean ridge hydrothermal systems to local earthquakes, *Geochem. Geophys. Geosyst.*, *5*, Q11009, doi:10.1029/2004GC000701.

Wilcock, W. S. D., and A. T. Fisher (2004), Geophysical constraints on the subseafloor environment near mid-ocean ridges, in *The Subseafloor Biosphere at Mid-Ocean Ridges*, *Geophys. Monogr. Ser.*, vol. 144, edited by W. S. D. Wilcock, et al., pp. 51–74, AGU, Washington, D. C.

Wilcock, W. S. D., and A. McNabb (1996), Estimates of crustal permeability on the Endeavour segment of the Juan de Fuca mid-ocean ridge, *Earth Planet. Sci. Lett.*, *138*, 83–91.

R. P. Lowell (corresponding author), Virginia Polytechnic Institute & State University, Department of Geosciences, Blacksburg, Virginia 24061-0420, USA. (rlowell@vt.edu)

P. Ramondenc, Schlumberger, Integrated Productivity and Conveyance Centre, Sugar Land, Texas 77478-3154, USA.

L. N. Germanovich, Georgia Institute of Technology, School of Civil and Environmental Engineering, Atlanta, Georgia 30332-0355, USA.

The Chemistry of Diffuse-Flow Vent Fluids on the Galapagos Rift (86°W): Temporal Variability and Subseafloor Phase Equilibria Controls

N. J. Pester,[1] D. A. Butterfield,[2] D. I. Foustoukos,[1,4] K. K. Roe,[2] K. Ding,[1] T. M. Shank,[3] and W. E. Seyfried Jr.[1]

Recent (2002, 2005) expeditions to the Galapagos Rift, 86°W, have conducted integrated experiments to assess temporal and spatial changes in both the chemistry and associated biology of diffuse-flow hydrothermal vents. The chemical data add significantly to time series observations of hydrothermal processes at the Galapagos Spreading Center (GSC), where seafloor venting was first discovered in 1977. Assuming quartz-fluid equilibrium (0.55 NaCl solution), end-member values for dissolved silica in 2002 and 2005 indicate temperatures ~50°–100°C lower than in 1977 (~400°C). The higher predicted temperature in 1977 is consistent with observed chloride depletion in fluids issuing from Oyster Beds vent, indicating fluid phase separation, whereas all fluid samples from 2002 and 2005 are at or above seawater concentration. Recent data for dissolved Mn further suggest a lower temperature hydrothermal reaction zone relative to three decades ago. The Li/Si ratios in 1977, however, do not differ significantly from analogous data in 2002 and 2005. This is in contrast with ratios for species known to be temperature-dependent, suggesting that other factors (e.g., fluid/rock ratio, protolith composition) may exert a greater influence on dissolved Li in the hydrothermal end-member. Geologic evidence indicates that between 1990 and 2002 a volcanic eruption covered the historical Rose Garden vent field, ending years of biological observations. Although the precise age of the lava flow is uncertain, the lack of chloride depletion and low H_2S/heat ratios in the 2002 and 2005 vent fluids suggest that the 2002 expedition arrived well after the eruption based on observations from other diffuse-flow

[1]Department of Geology and Geophysics, University of Minnesota, Minneapolis, MN 55455, USA

[2]Joint Institute for the Study of Atmosphere and Ocean, University of Washington, Seattle, WA 98195, USA

[3]Biology Department, Woods Hole Oceanographic Institution, Woods Hole, Massachusetts, 02543, USA

[4]Now at the Carnegie Institute Geophysical Laboratory, Washington, D. C., USA.

Magma to Microbe: Modeling Hydrothermal Processes at Ocean Spreading Centers
Geophysical Monograph Series 178

10.1029/178GM07

systems perturbed by volcanic activity. Lower dissolved silica, Mn, and H_2S in the 2005 vent fluids relative to 2002, however, may indicate continued cooling at depth in the aftermath of the recent eruption. Geochemical modeling results of evolved seawater in equilibrium with a moderately oxidizing mineral assemblage are consistent with the chemistry of recent GSC fluid samples, predicting $H_2S/Fe > 1$, with little Fe in fluids diluted to the extent typical of 86°W vents. In situ chemical sensor measurements indicate that at temperatures below ~10°C, dissolved sulfide species change in a complex way with temperature, suggesting the involvement of processes other than simple dilution, including, perhaps, the effects of microbial metabolism, recognized in other diffuse-flow vent systems.

1. INTRODUCTION

The discovery of warm springs issuing from cracks in the ocean floor on the Galapagos Spreading Center (GSC), 86°W, in 1977 [*Corliss et al.*, 1979; *Edmond et al.*, 1979b] marks a significant event in the study of subseafloor hydrothermal processes. Fluid-mineral interactions, occurring at high pressures and temperatures in the oceanic crust, were shown to play a substantial role in the chemical evolution of major and minor element concentrations in seawater. At the same time, this landmark discovery highlighted numerous biological processes that could be linked in fundamental ways to vent fluid chemistry [*Jannasch and Mottl*, 1985]. The hydrothermal activity at 86°W, with fluids then emanating at temperatures ~3°–13°C [*Edmond et al.*, 1979b] is now characterized as diffuse or "diffuse-flow" venting ($T < \sim 50°C$) to differentiate from the high-temperature "black smoker" style of venting discovered shortly after on the East Pacific Rise (EPR), 21°N [*Spiess et al.*, 1980].

The chemistry of the vent fluids sampled from 86°W in 1977 [*Edmond et al.*, 1979a; *Edmond et al.*, 1979b; *Lilley et al.*, 1983] is consistent with what is now considered a typical flux of chemical species attending the metamorphism of basalt into upper greenschist and lower amphibolite facies mineral assemblages [*Bowers et al.*, 1985; *Gillis*, 1995]. Thus, relative to seawater, the vent fluid compositions exhibited enrichment in dissolved silica and selected cations (Li, K, Rb, Ca, Sr, Ba, Mn), moderately reducing source conditions, and depletion in dissolved Mg, O_2, SO_4^{2-}, and alkalinity. Although the seafloor expressions of the source fluids were greatly diluted, high-precision analytical techniques still allowed a faint hydrothermal signal to be recognized and quantified [*Edmond et al.*, 1979a, 1979b]. Dissolved silica concentrations in the sample set followed a linear relationship with respect to temperature and the fluids were subsequently interpreted as a subsurface dilution of a single hydrothermal end-member by cold seawater circulating in the upper crust [*Edmond et al.*, 1979a, 1979b]. This simple mixing model was soon supported by the discovery of high-temperature vent fluids (~350°C) at EPR, 21°N [*Von Damm et al.*, 1985], where end-member concentrations were remarkably similar to those estimated for the diffuse fluids of the GSC. Although common end-member concentrations were noted for many species (e.g., silica, chloride, Mg, Mn), others, such as Fe and additional transition metals, proved to be highly variable with extent of mixing. As the more conservative elements can supply clues to hydrothermal reaction conditions at depth, the latter components are likely affected by near-surface biogeochemical processes and/or mineral precipitation induced by cooling [*Edmond et al.*, 1979a; *Jannasch and Mottl*, 1985; *Luther III et al.*, 2001].

When initially discovered in 1977, and in some years to follow [*Campbell et al.*, 1988], the chemistry of hydrothermal fluids was generally interpreted as stable on the order of decades, with temperature maxima that effectively precluded phase separation in the $NaCl\text{-}H_2O$ system. These once strongly held tenets have largely been replaced by more dynamic models, usually involving tectonic/magmatic processes that allow temperatures to intersect the two-phase boundary of seawater, with concomitant effects on reaction conditions and heterogeneous phase relations. Integrated studies, such as those on the "Bio-Geo-Transect" of EPR, 9°50′N [*Shank et al.*, 1998; *Von Damm and Lilley*, 2004], and at Axial Volcano, Juan de Fuca Ridge (JdFR) [*Butterfield et al.*, 2004], have become increasingly important in developing a more thorough understanding of evolutionary change in seafloor hydrothermal systems. The comprehensive nature of such studies includes a wide range of time series data for both dissolved organic/inorganic species and vent-associated faunal assemblages. After the initial discovery in 1977, 86°W has been revisited on several occasions, although the return visits were largely focused on clarifying

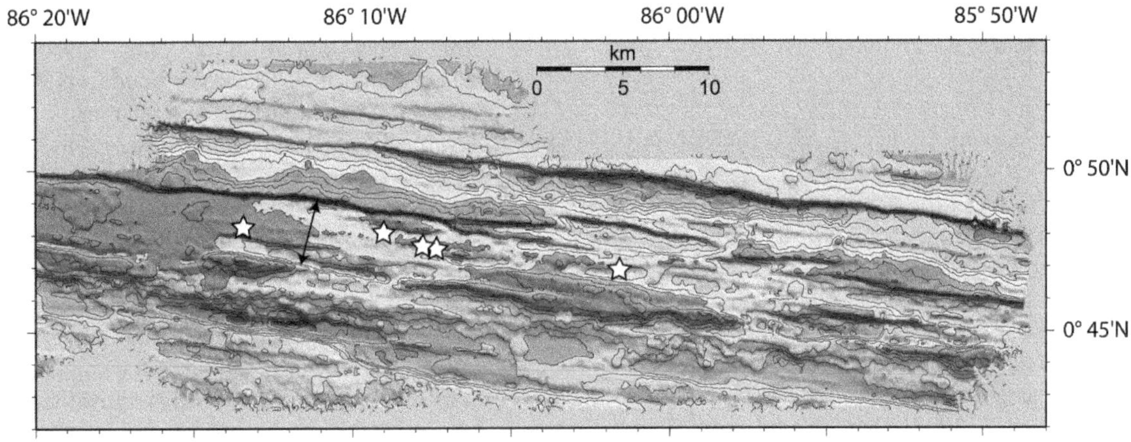

Figure 1. Multibeam bathymetric map of the 86°W study area (modified after *Shank et al.* [2003] and *Ballard et al.* [1982]) located on the Galapagos Spreading Center. Open stars mark both current and historical vent sites (from left to right): Rose Garden/Rosebud (~2440 m), Clambake, Oyster Beds, Garden of Eden, and East of Eden. Arrows depict the width of the rift valley near Rosebud, which subsequently narrows to the east (see text).

the role of dissolved redox species as an energy source for chemoautotrophic organisms [*Jannasch and Mottl*, 1985; *Johnson et al.*, 1988; *Lilley et al.*, 1983].

Coinciding with the 25th anniversary of the discovery expedition, a NOAA Dive and Discover program was initiated in 2002 to assess the continued evolution of 86°W in a more integrated context. The main objective was further detailed observations in the Rose Garden field (~86°13.5′W, Figure 1), including the deployment of in situ chemical sensors/data loggers and full chemical analysis of collected fluid samples. These planned multidisciplinary studies, however, took an unexpected turn as the first few dives revealed that Rose Garden had been paved over by lava flows [*Shank et al.*, 2003] subsequent to the last visit in 1990. Although this eruption seemed to serve as a premature end to the time series biological studies, the concurrent discovery of nascent biological communities and associated venting ~300 m northwest of the now inactive Rose Garden allowed the intended program to be carried out at a site in its earliest stage of biological succession. Previously reported results from a wide range of time series studies at seafloor hydrothermal vents attest to the dynamic chemical/biological response after eruptions and subseafloor diking events [*Butterfield et al.*, 2004; *Delaney et al.*, 1998; *Embley et al.*, 1998; *Fornari et al.*, 1998; *Lutz et al.*, 1994; *Mullineaux et al.*, 2000; *Shank et al.*, 1998; *Tsurumi and Tunnicliffe*, 2001; *Tunnicliffe et al.*, 1997]. Thus, vent fluids were sampled, while at the same time, type and distribution of biological communities were extensively mapped and sampled [*Shank et al.*, 2003]. Preliminary observations from this newly discovered site (Rosebud) provided sufficient justification for a follow-up expedition, conducted in 2005, which included a visit to Garden of Eden (~86°07.5′W), a site sampled in 1977. In general, the temperatures of the vent fluids sampled in 2002 and 2005 ranged from ~6° to 17°C, indicating a continued influence of subsurface mixing on seafloor expressions of the hydrothermal component.

We report the concentrations of select dissolved components in the vent fluids sampled from 86°W in 2002 and 2005. We note that the high dilution factors (~20–30×) associated with diffuse fluids introduce an increased challenge to data interpretation by mitigating hydrothermal component concentrations and heightening the risk of nonconservative behavior for some species upon mixing/cooling. Although these complications cannot be disregarded, certain chemical constituents can be analyzed with great precision and, in basalt-hosted systems, appear to exhibit a high degree of conservative behavior. Such elements can still allow meaningful interpretations regarding the dominant reaction conditions at depth. Thus, the data we report are not comprehensive, but rather we focus on species (Mg, Si, Li, Mn, H_2S, and Cl) likely to reveal the deep-seated, high-temperature conditions that result in the pure hydrothermal component of these hot springs.

To broaden the time series implications of this study, we also make an effort to compare the present data with those previously reported. In particular, routine use is made of the dataset from the 1977 expedition to 86°W [*Corliss et al.*, 1979; *Edmond et al.*, 1979a, 1979b]. Although differences in the specific location of vent sites between 1977 and 2002/2005 makes quantitative interpretations less certain, such comparisons are useful nevertheless, especially if the earlier data is viewed historically, taking advantage of more than 25 years spent researching causes/effects of compositional change in hydrothermal fluids [*Butterfield et al.*, 1997; *Butterfield and Massoth*, 1994; *Lilley et al.*,

2003; *Seyfried and Ding*, 1995; *Von Damm*, 1988, 1995; *Von Damm and Lilley*, 2004].

2. GEOLOGIC SETTING AND HYDROTHERMAL ACTIVITY

Since the initial discovery of hydrothermal venting, follow-up studies have demonstrated that the spreading ridge at 86°W represents an unusually complex geological and geophysical setting. The Galapagos Rift is an east-west trending ridge that exhibits systematic changes in both axial morphology and depth with distance from the 91°W transform [*Canales et al.*, 1997]. Here, the influence of the Galapagos Hotspot induces a bathymetric high of ~1500 m. 86°W falls on the deepest (~2500–2800 m), easternmost, and fastest spreading (~64 mm/yr, full rate) segment of the ridge, bound to the west by the 87°W propagator/overlapping spreading center and to the east by the Inca Fracture Zone at 85.5°W [*Christie et al.*, 2005; *Perram and Macdonald*, 1994]. This locale exhibits transitional axial morphology (TAM) [*Canales et al.*, 1997] consisting of a shallow rift valley ~3–4 km wide and ~200–250 m deep that hosts a small volcanic ridge rising ~20 m from the valley floor [*Corliss et al.*, 1979]. Estimates derived from thermal modeling suggest the persistence of a steady-state axial magma chamber (AMC) or seismic reflector for ridges spreading at a rate equivalent to GSC, 86°W [*Morgan and Chen*, 1993b; *Sleep*, 1975], whereas the depths of seismic reflectors show consistent variability with axial morphology and spreading rate [*Chen and Morgan*, 1996; *Morgan and Chen*, 1993a; *Sinton and Detrick*, 1992]. Although seismic reflection studies have not yet been performed in the vicinity of 86°W, the TAM and evidence of recent (< ~10 years) volcanism indicate the likely presence of an AMC ~2.5 km below the ridge axis. This inference is based on recent seismic observations from both the JdFR [*Carbotte et al.*, 2006; *Van Ark et al.*, 2007] and the GSC west of the 91°W transform [*Blacic et al.*, 2004; *Canales et al.*, 1997; *Chen and Lin*, 2004]. Furthermore, the composition of fresh lava collected from Rosebud in 2002 is enriched in MgO, a trait also characteristic of samples collected from Rose Garden in 1985. This not only suggests that source conditions have remained stable over several decades, but also underscores the magmatically robust nature of this site relative to more easterly areas of the ridge axis (M. Perfit, personal communication, 2007). Clearly, episodic magmatism is the dominant factor influencing the evolution of fluid chemistry on fast/intermediate spreading ridges [*Butterfield et al.*, 1997; *Butterfield and Massoth*, 1994; *Lilley et al.*, 2003; *Seewald et al.*, 2003; *Seyfried et al.*, 2003; *Von Damm*, 2000]. The area surrounding Rose Garden/Rosebud (86°13.5′W) shows evidence of regular sheet flow events [*Ballard et al.*, 1982; *Shank et al.*, 2003]; yet when approaching Garden of Eden from the west, the ridge appears in transition to a more magmatically starved [*Kennish and Lutz*, 1998; *Macdonald*, 1998] setting dominated by older pillow flows and extensive fissuring [*Ballard et al.*, 1982]. This shift, accompanied by an approximately two-fold decrease in the width of the axial valley (Figure 1), suggests that the Garden of Eden site may be the eastern limit of the area influenced by the prevailing heat source at depth [*Carbotte et al.*, 2000]. Additional support for this interpretation is based on its setting in transitional sheet-pillow basalts [*Corliss et al.*, 1979]. This transition in geological and geophysical characteristics should influence the intensity of hydrothermal activity and may be due to factors such as increasing proximity of the Inca Fracture Zone, a small-scale discontinuity in the ridge axis [*Macdonald*, 1998] or relative distance from the Galapagos Hotspot [*Chen and Lin*, 2004].

An inherent component of both the 2002 and 2005 expeditions was the search for high-temperature venting, but this phenomenon was not observed in the vicinity of 86°W. Although both active, high-temperature vents and relict sulfide chimneys have been identified west of the 91°W transform (R. Haymon, personal communication, 2007), current near-bottom magnetic data for 86°W do not contain low magnetization anomalies, which have proven indicative of recent black smoker activity in the Main Endeavour Field (MEF), JdFR [*Tivey and Johnson*, 2002]. These data, coupled with the lack of any visual/physical markers, indicate that this segment of the GSC has likely remained a diffuse-flow hydrothermal regime since it was first explored in 1977. Active hydrothermal venting at ~86°00–86°15′W (Figure 1) appears limited to within ~50 m of an eruptive fissure [*Ballard et al.*, 1982] due to its association with the most recent pillow flows. Furthermore, upon initial discovery of venting on the GSC, *Corliss et al.* [1979] noted that all vent fluids were seeping from cracks between basalt pillows, near the contact zone of these recent flows. This may indicate that the contact is one of the primary conduits of up-flow, and/or that a new extrusive layer provides a temporary cap of lower permeability, leading to an increase in the extent of subseafloor mixing and, therefore, possibly contributing to the lack of high-temperature venting at 86°W. Most black smoker sulfide chimneys appear to have developed along fault lines [*Butterfield et al.*, 1994; *Fornari and Embley*, 1995; *Tivey and Delaney*, 1986], suggesting the onset of high-temperature venting results from the availability of a high-permeability conduit between deep source fluids (reaction zone) and the seafloor.

The insulating effect of marginal anhydrite formation in hydrothermal up-flow zones within the crust undoubtedly plays a key role in the development of focused (high-temperature) venting [*Tivey and Delaney*, 1986]. At 86°W, magma emplacement and associated tectonic processes in the shallow subsurface may sufficiently enhance permeability to preclude temperatures high enough for anhydrite formation, in keeping with elements of the model advanced earlier by *Lowell et al.* [2003]. Moreover, the apparent existence of a relatively thick layer 2A at 86°W [*Blacic et al.*, 2004; *Buck et al.*, 1997; *Carbotte et al.*, 2000] enhances the likelihood for ingression of cold seawater all along the up-flow zone, further inhibiting anhydrite formation and thereby maintaining permeability that manifests in low vent exit temperatures. Single-pass models of hydrothermal convection cells suggest that diffuse venting in the absence of black smoker activity is likely to result if the permeability of the upper extrusive layer is several orders-of-magnitude greater than that of the deeper circulation pathways [*Lowell et al.*, this volume, 2007, 2003]. This effect is further enhanced with increasing thickness of the extrusives. High-temperature vent fluids have been reported near the 91°W transform (GSC), although there, the extrusive layer is relatively thin [*Blacic et al.*, 2004], which may be conducive to sealing of up-flow zones with subsurface mineral precipitation. Magmatic hydrothermal systems associated with intermediate spreading ridges seem to exhibit characteristics of both fast- and slow-spreading centers [*Van Ark et al.*, 2007; *Wilcock and Delaney*, 1996]. The prevailing hydrothermal regimes along the Galapagos Rift, both spatially and temporally, are therefore the result of complex/variable interplay between magmatism and extentional faulting.

3. SAMPLE COLLECTION AND ANALYSIS

Sampling efforts, conducted from DSV *Alvin*, were primarily focused at the Rosebud vent field in both 2002 and 2005. As noted earlier, in 2005, vent fluids were also retrieved from the still active Garden of Eden, first observed in 1977 [*Corliss et al.*, 1979]. Lettered markers (e.g., marker B) placed during the 2002 dives facilitated the location of Rosebud during the return expedition, whereas markers N and I were deployed in 2005 to note additional sites sampled in that year. Garden of Eden was recognized by the existence of relict biological experiments and *Alvin* dive weights bearing a number corresponding to the second visit to the site in 1979. Data from ALR Vent in 2002 and a new site in 2005, located ~900 m north and ~150 m southwest of Rosebud, respectively, are also included in the sample set.

The 2005 vent fluid samples were acquired with titanium major-pair samplers [*Von Damm et al.*, 1985], whereas, in 2002, one major pair was taken (dive 3789) and all remaining samples were obtained by the NOAA manifold sampler [*Massoth et al.*, 1989; *Von Damm and Lilley*, 2004] modified with a PEEK plastic intake hose. The manifold device is generally better suited for sampling diffuse-flow fluids because the intake pump can be adjusted to low speeds, helping to minimize the entrainment of ambient seawater by sampling at a rate comparable to or lower than the outflow of vent fluid. Moreover, temperature can be monitored at both the inlet nozzle and before diversion into collection bottles, providing in-line temperature measurement of the sampled fluid that permits monitoring for when the sampling line has been flushed with fluids maintaining a stable temperature.

3.1. Analytical Procedures

Vent fluid samples were processed immediately after recovery of the submersible, with priority given to the most labile components. Unless noted, analytical methods were the same in both 2002 and 2005. In 2002, a sample aliquot was expressed from the titanium samplers through silicone tubing into a syringe with zero headspace and then analyzed in replicate for total dissolved sulfide (S_T) by the methylene blue colorimetric method (modified after *Cline* [1969]). A gravimetrically prepared, iodometrically standardized Na_2S standard solution was maintained in a tightly stoppered septum bottle with nitrogen overpressure and diluted to make daily working standards. In 2005, replicate samples (~5-mL aliquots) were taken from the titanium sampler, diluted (1:1) with a high-pH antioxidant buffer, and directly measured with a sulfide-specific electrode. The electrode was calibrated by $Pb(NO_3)_2$ titrations of sequentially diluted Na_2S standards, which had been prepared and precalibrated prior to the cruise by spectrophotometry and gravimetric analysis. Results of replicate standard/sample measurement indicate an uncertainty of 7 and 15% (2σ) for the 2002 and 2005 values of S_T, respectively (Tables 1 and 2, values reported as H_2S).

After H_2S analysis, pH and alkalinity measurements were made using potentiometry and Gran titration, respectively. Filtered, diluted, and acidified aliquots of vent sample source fluid were also sealed in preweighed and acid-cleaned high- (HDPE) or low-density polyethylene (LDPE) bottles for shore-based analysis at the University of Washington (2002) or the University of Minnesota (2005). Dissolved chloride and Mg were determined by $AgNO_3$ and ethylenediaminetetraacetic acid/ethylene glycol tetraacetic acid (EDTA/EGTA) titration, respectively, using undiluted fractions of source fluid. In both cases, standardization was performed using IAPSO (International Association for Physical Sciences of the Oceans) standard seawater and all samples were titrated a minimum of three times to ensure precision. Dissolved silica was determined spectrophotometrically as reduced silicomolybdate dye and, in 2005, also measured by ICP-MS on acidified aliquots with a 30-fold dilution.

Furthermore, ICP-MS measurements of silica utilized internal scandium and rhodium standards to minimize uncertainties. In 2002, Li and Mn were determined by flame atomic emission and graphite furnace absorption spectroscopy, respectively, whereas, in 2005, Li and Mn were determined on replicate measurements using inductively coupled plasma–mass spectrometry (ICP-MS) programmed for standard and blank (NaCl) correction between individual determinations. Li was also determined by using ion chromatography (IC) in 2005 (Table 2), wherein this technique proved to achieve a higher level of precision than that of ICP-MS with the standardization procedures employed. A summary of the relevant analytical techniques, along with associated uncertainties, is presented in Table 1. Although not reported here, all fluids were determined for major cations and anions by a combination of ICP-MS, IC and/or spectroscopic techniques.

3.2. The 2002 and 2005 Vent Fluid Data

Fluids venting during the 2002 and 2005 expeditions reveal similar enrichments and depletions in equivalent species relative to their respective measured ambient bottom seawater (ABSW) values. In agreement with the silica value reported by *Edmond et al.* [1979b], an ABSW value of 165 μmol/kg was again measured for both the 2002 and 2005 studies. There is, however, an observed offset in expected seawater values for several species including Mg, chloride, and Li. Such offsets are best accounted for by differing standardization parameters with similar techniques or uncertainties intrinsic to the different analytical methods employed (see above). Thus, to appropriately compare these two recent datasets, the dissolved Mg and chloride concentrations of 2002 were normalized to the 2005 data. This normalization procedure was also adopted for Li values measured in the 2005 samples because a similar offset was observed. Normalizing these data essentially corresponds to correcting the calibration of the IC such that both datasets run through a common seawater end-member. In all cases, any adjustments made were within acceptable uncertainties imposed by the analytical methods used to process the fluids from the two different sampling intervals. These combined data are presented in Table 2.

Species concentrations in the pure hydrothermal components of 2002 and 2005 (Table 2) were estimated by using a standard linear regression of the data, assuming zero-Mg and ABSW end-members [*Seyfried et al.*, 2003; *Von Damm et al.*, 1985]. For silica and Mn in 2005, regressions were weighted utilizing the individual standard deviations recorded by the ICP-MS. The reported estimates of standard error for all end-members were calculated at the 95% confidence interval using a bootstrapping algorithm, which performed 1000 iterations. Hydrothermal end-members for chloride and H_2S are not reported due to accuracy limitations and potential nonconservative mixing effects, respectively (see below). Furthermore, our discussion deals largely with reaction temperature constraints and although Mg regressions are better suited for deriving end-member concentrations, dissolved silica serves as the best temperature proxy [*Edmond et al.*, 1979b]. Thus, for ease of comparison with historical data and to better portray species-temperature relationships, data for Li, Mn, and H_2S are graphically depicted relative to silica.

Table 1. Analytical methods for fluid samples from GSC, 86°W, 2002/2005.

	Method	% RSD (2σ)
2002 Samples[a]		
Cl	Ag NO_3 titration	0.1%
Mg	EDTA/EGTA titration	0.1%
Si	Silicomolybdate, colorimetric	1%
Mn	Graphite furnace atomic absorption spectroscopy	5%
H_2S	Methylene blue	7%
Li	Flame atomic emission spectroscopy	5%
2005 Samples[b]		
Cl	Ag NO_3 titration	0.1%
Mg	EDTA/EGTA titration	0.3%
Si	ICP-MS with internal standard and silicomolybdate, colorimetric	<6.5%
Mn	ICP-MS	<4%
H_2S	$Pb(NO_3)_2$ titration	15%
Li	Ion Chromatography	5%

[a]Analyses performed by the University of Washington.
[b]Analyses performed by the University of Minnesota.

Table 2. Vent fluid compositions[a] from GSC, 86°W, 2002/2005.

Site	Sample[b]	*T* (avg), °C	Cl, mm	Mg, mm	Si, μm	Mn, μm	H_2S, μm	Li, μm
	2005[c]							
Rosebud	B-M4114-23		541.6	51.08	499	2.3	5	34.1
(~86°13.5′W)	B-M4114-25		540.8	50.52	735	5.0	21	50.4
	B-M4114-26		541.4	51.39	522	3.0	7	38.7
	B-M4115-23		540.3	50.45	761	5.3	25	52.1
	B-M4115-24		541.8	51.31	378	2.4	2	36.3
	B-M4115-25		540.9	50.60	700	5.0	19	49.9
	B-M4115-26		540.7	50.59	615	4.2	9	46.1
	B-M4124-25		540.5	50.68	748	5.3	21	52.7
	B-M4124-26	16	541.2	51.04	408	2.2	0	37.0
	N-M4116-23		540.8	50.39	746	5.1	18	50.2
	N-M4116-24		540.6	50.29	727	5.0	16	51.6
	N-M4116-25		541.0	49.87	734	4.8	14	47.1
	N-M4116-26	16	540.9	49.88	750	5.0	14	48.0
	I-M4117-23		542.0	50.38	756	4.1	23	49.8
	I-M4117-24		541.5	50.24	749	4.1	19	49.7
	I-M4117-25		541.8	50.17	694	4.1	21	51.6
	I-M4117-26		541.9	50.33	659	4.0	18	49.1
	I-M4122-23		542.1	50.39	714	4.0	16	45.5
	I-M4122-24	14	541.9	50.96	513	2.3	11	38.5
Garden of Eden	Q-M4120-25		544.7	51.01	686	5.8	27	46.8
(~86°07.5′W)	Q-M4120-26		544.0	50.89	703	5.7	26	46.4
	Q-M4121-23		543.2	50.72	655	4.6	16	45.5
	Q-M4121-24		542.9	50.24	664	4.3	17	43.8
	Q-M4121-25		543.0	50.56	660	5.1	17	46.4
	Q-M4121-26	17	543.3	50.53	622	4.8	18	44.0
New Site	M4124-23		542.7	51.76	360	1.9	0	34.2
(~86°13.5′W)	M4124-24	8	542.6	51.66	414	2.0	0	37.1
	End-members		x	0	12.2[d]	100	x	557
	2002							
Rosebud	B-M3789m21		540.2	52.15	268	2.5	1	29.5
(~86°13.5′W)	B-M3789m22		541.6	51.48	695	5.8	26	46.3
	B-man3789m10c		540.4	51.25	552	4.4	15	42.0
	B-man3789m1c		540.7	52.25	391	5.1	1	34.4
	B-man3789m20	18	541.5	51.54	439	3.0	12	36.9
	F-man3790m10c		540.9	51.38	605	3.7	36	41.5
	F-man-3790m1c		541.5	51.33	571	4.1	33	44.2
	F-man3790m20	14	539.9	51.48	569	4.1	42	42.0
ALR Vent	E-man3791m1c		541.5	52.17	353	0.7	0	31.7
(~86°13.5′W)	E-man3791m20	10	539.2	51.77	331	0.6	0	31.1
	End-members		x	0	16.9[d]	169	x	722
ABSW		2	541.1	52.65	165	0.1	0	25.1

[a]Chemical compositions normalized to a common ABSW for Cl, Mg and Li; mm, mmol/kg fluid; μm, μmol/kg fluid.

[b]Sample prefix: first letter indicates site marker; "M" or "man" denote acquisition with the major-pair or manifold sampler, respectively.

[c]Sample B-M4114-24 not reported as this half of the sampler did not fire properly.

[d]Expressed in mmol/kg.

4. RESULTS AND DISCUSSION

Vent fluid chemical data from the 2002 and 2005 cruises to GSC add significantly to a relatively small time series database [*Johnson et al.*, 1988; *Lilley et al.*, 1983] developed subsequent to the discovery dive results of 1977 [*Corliss et al.*, 1979; *Edmond et al.*, 1979b]. Because the 1977–1979 dive programs primarily encompassed studies along axis between Rose Garden (~86°13.5′W) and Garden of Eden (~86°07.5′W), the present report emphasizes results from these two locales (Rosebud being roughly equivalent in location to Rose Garden). These two sites are of unique importance, because other sites visited originally in 1977–1979 are now believed inactive.

Before fluid samples were obtained in 2002 and 2005, vent fluid temperatures were measured with the *Alvin* low-temperature probe. In general, the highest fluid temperatures (~18°C) were associated with the base of tubeworm colonies, whereas lower temperatures were typically recorded at the base of mussel beds (~6°–10°C). Owing to constraints imposed by the respective sampling techniques, the relationship between fluid samples and corresponding temperature is uncertain. This problem is especially common when attempting to obtain fluid issuing from cracks at low rates of flow. Accordingly, species known to be sensitive to temperature and believed conservative with seawater mixing will best serve to establish temperature of last equilibration in subseafloor reaction zones. For Galapagos, as with many other hydrothermal systems, the component that best serves this purpose is dissolved silica [*Edmond et al.*, 1979b; *Johnson et al.*, 1988; *Von Damm et al.*, 2003]. For example, the chemical fluxes originally calculated from the 1977 data used silica as a temperature proxy to further decrease uncertainties in extrapolation [*Edmond et al.*, 1979b]. In addition, anomalous concentrations of dissolved chloride relative to seawater can provide clues to the temperature-dependent history of hydrothermal vent fluids [*German and Von Damm*, 2003; *Sohn et al.*, 1999; *Von Damm*, 2000]. Thus, the approach we take here is to combine such chemical proxies for temperature with other indicators of hydrothermal alteration to better constrain subseafloor mass transfer processes for the 2002 and 2005 samples in comparison with available historical data.

4.1. Constraints on Subseafloor Reaction Temperature at 86°W

4.1.1. Dissolved silica systematics. The extent of seawater dilution in hydrothermal fluid samples is most often indicated by the relative concentration of Mg in the mixture. For the Rosebud vent fluids sampled in both 2002 and 2005, a clear linear relationship is indicated between dissolved Mg and Si (Figure 2). In 2005, vent fluids from the so-called "new site" and Garden of Eden also appear consistent with a common source (Figure 2). In contrast with the relatively shallow slope depicted by the 2005 Mg-Si regression, data reported by *Edmond et al.* [1979b] for fluids collected in 1977 exhibit a considerably steeper slope (dotted line, Figure 2). Extrapolation of silica in the 2005 and 1977 fluids to zero Mg indicates sharply contrasting end-member concentrations of 12.2 and 21.6 mmol/kg, respectively. The data of 2002 exhibit an intermediate trend, having an end-member of 16.9 mmol/kg.

Numerous experimental and theoretical studies have been conducted to constrain quartz solubility in pure water and NaCl-bearing fluids for a wide range of pressure and temperature (*P-T*) conditions [*Fournier*, 1983; *Fournier and Potter*, 1983; *Foustoukos and Seyfried*, 2007a; *Von Damm et al.*, 1991]. Utilizing quartz solubility data from *Von Damm et al.* [1991], coupled with activity-concentration relations for dissolved silica in NaCl-bearing fluids [*Foustoukos and Seyfried*, 2007a], isobars for silica in equilibrium with quartz can be calculated as a function of temperature (Figure 3).

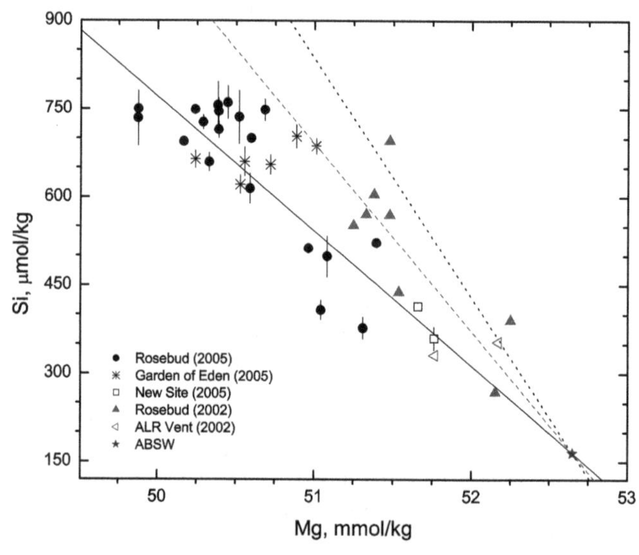

Figure 2. Mg versus Si data for 2002/2005 (Table 2). Individual error bars (ICP-MS) are shown for the 2005 Si data. The solid line and dashed line represent linear regressions for the 2005 and 2002 data, respectively. Regressions were used to calculate estimated values of the hydrothermal component. The 2005 and 2002 data extrapolate to zero-Mg at Si = 12.2 (±0.1) and 16.9 (±0.5) mmol/kg, respectively. The dotted line is the Mg-Si regression for the 1977 data [*Edmond et al.*, 1979b], which predicts a significantly higher Si end-member of 21.6 mmol/kg.

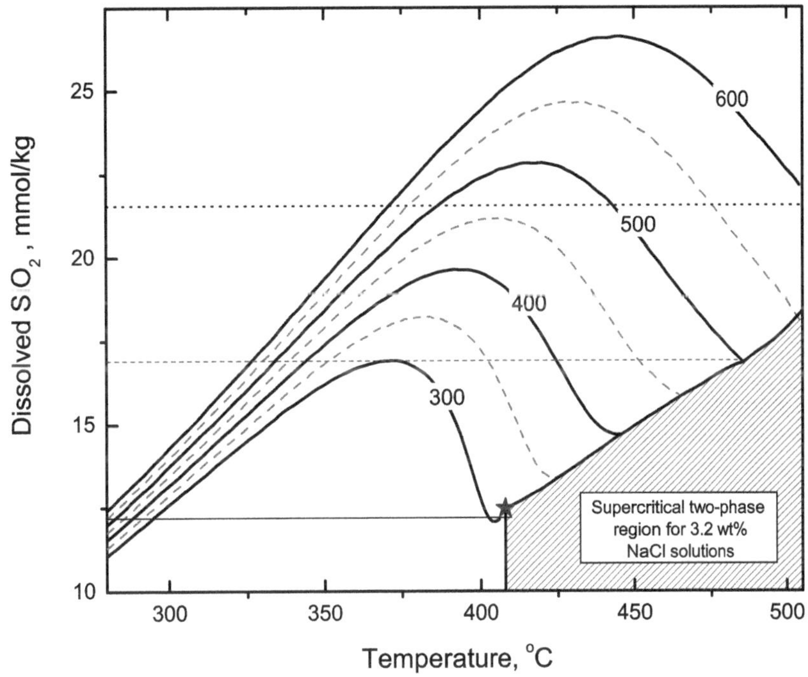

Figure 3. Isobaric representations (in bars) of dissolved silica concentrations with respect to temperature, assuming a fluid of seawater salinity in equilibrium with quartz (dashed lines are intermediate isobars, 350, etc.), generated with data from *Von Damm et al.* [1991] and *Foustoukos and Seyfried* [2007a]. Relevant temperatures and pressures for the two-phase boundary of seawater are calculated from *Bischoff and Rosenbauer* [1988]. Dotted line is $y = 21.6$, dashed line is $y = 16.9$, solid line is $y = 12.2$, star symbol is the critical point of seawater.

These studies were conducted with specific relevance to inferred hydrothermal alteration conditions/processes in subseafloor reaction zones. A diagram similar to this, but for pure water, was constructed by *Edmond et al.* [1979b] to illustrate the relatively high temperature and pressure required to produce the hydrothermal component in 1977. Although technically more relevant for natural systems, use of solubility data in NaCl solutions do little to change this interpretation. The difficulty in constraining subseafloor pressure precludes unambiguous determination of temperature, but those in the vicinity of 400°C are likely if subseafloor pressure is no greater than ~500 bars.

An added constraint on the *P-T* conditions in the reaction zone beneath 86°W in 1977 comes from the Si-T correlation [*Edmond et al.*, 1979b]. Assuming quartz-fluid equilibria, these data suggest a temperature of ~345°C, but this requires a pressure in excess of 700 bars, which seems unreasonably high. Magma chambers associated with active hydrothermal venting seldom exceed 2.5–3 km in depth [*Van Ark et al.*, 2007]. It is more likely that uncertainties inherent in the temperature measurements (±0.5°C) over the range 2°–13°C [*Edmond et al.*, 1979b] allow higher temperatures with correspondingly lower pressures to come in to play (Figure 3). Although the Si-T regression is imperfect, the correlation that does exist provides strong support for the conservative tendency of dissolved silica in the near-seafloor mixing environment.

Application of constraints imposed by quartz solubility to the silica end-members of 2002 and 2005 indicate very different reaction conditions relative to 1977. The relatively low silica concentrations (Figure 2, Table 2) suggest correspondingly low temperatures (no greater than ~325°C). Unlike higher temperature conditions where pressure is a critical constraint on quartz solubility, such is not the case here as indicated by the coalescence of isobars on the solubility diagram as temperature decreases (Figure 3). While the silica end-member of 2002 does not preclude high temperatures (>400°C), low concentrations of other coexisting elements (e.g., Mn; see below) make this highly unlikely. Given the evidence of recent volcanism, the lower silica in 2005 relative to 2002, however, may represent continued cooling at depth in the aftermath of the seafloor eruption. Regardless, a low reaction zone temperature is almost certain in 2005 and, if we assume a similar quartz equilibration pressure/depth as the hydrothermal component of 1977, the contrast in silica suggests a temperature difference on the order of ~100°C.

4.1.2. Dissolved chloride systematics. The high reaction zone temperature inferred by the silica end-member in 1977 is consistent with the variability in dissolved chloride, suggesting phase separation at depth. Due to extensive mixing and, therefore, the predominant seawater composition of the fluids, end-member chloride values were calculated from salinity regressions [*Edmond et al.*, 1979b] and ranged from moderate enrichment to significant depletion relative to ABSW. *Edmond et al.* [1979b] reported values of ~595 and 322 mmol/kg for Clambake and Oyster Beds, respectively, but little or no deviation (~543) from ABSW (541.1) for Garden of Eden. The chloride variability of the vents in 1977, relative to location, may be a result of the local permeability structure. *Corliss et al.* [1979] noted that fluids of Oyster Beds (chloride-depleted) were issuing from an atypical, high-relief terrain produced by normal faulting, which could imply deeper fluid penetration. The greater extent of faulting and fissuring in this previously noted transition zone may also, in part, explain the cessation of hydrothermal venting at sites such as Oyster Beds and Clambake due to increased efficiency in cooling [*Macdonald*, 1982].

In light of the high pressure (~500 bars) necessary to account for the corresponding silica, unusually high temperatures (Figure 3) would have been required for a fluid of seawater composition to intersect the two-phase boundary [*Anderko and Pitzer*, 1993; *Berndt et al.*, 2001; *Bischoff and Pitzer*, 1989; *Bischoff and Rosenbauer*, 1985, 1988], accounting for the chloride depletion at Oyster Beds. Such temperatures, however, are too high to be consistent with silica values approaching 20 mmol/kg. A probable explanation for this disparity, first proposed by *Welhan and Craig* [1979], involves vapor-rich fluids produced at high *P-T* mixing into a seawater-dominated hydrothermal reaction zone at conditions more consistent with constraints imposed by quartz solubility (~400°C, 500 bars). Similar models have been proposed for MEF [*Butterfield et al.*, 1994] and, more recently, for 21°S, EPR [*Foustoukos and Seyfried*, 2007b] as the use of experimentally derived coefficients for trace element partitioning during phase separation [*Berndt and Seyfried*, 1990, 1997; *Foustoukos and Seyfried*, 2007b] suggest the presence of a much higher temperature component than can be accounted for by contemporaneous silica-chloride systematics. Temperatures (at relevant pressures) resulting in dilute vapor fractions are well within the range of those thought achieved by some fluids circulating in the crust based on fluid inclusion studies. Hydrothermally altered breccias dredged from Hess Deep in the EPR and the Kane Fracture Zone, Mid-Atlantic Ridge (MARK) [*Saccocia and Gillis*, 1995], exhibit phase separation effects as high as 500° and 625°C, respectively, whereas altered gabbros, also recovered near MARK [*Kelley and Delaney*, 1987], suggest temperatures in excess of 700°C. A magmatic source, however, has been theorized for the latter [*Kelley and Delaney*, 1987]. Thus, the model we propose for 1977 involves mass transfer associated with at least two stages of hydrothermal reaction: the first involves phase separation and vapor release at extreme temperatures, whereas the second entails heterogeneous phase equilibria at more moderate *P-T* (~400°C, 500 bars). It is during this second hydrothermal stage where deeply circulating seawater incorporates rising vapor-like fluids (if available), while at the same time reacting with unstable minerals in basalt. Here, pH and bulk fluid chemistry are buffered by mineral transformations, producing the hydrothermal end-member that, in turn, rises to become diluted by cold seawater in the shallow crust.

In contrast with the 1977 data, the measured chloride concentrations in fluids from Rosebud, 2002/2005, do not vary significantly from ABSW and average 541.1 (±0.6) mmol/kg (Figure 4). The Garden of Eden samples, however, show slight but statistically meaningful chloride enrichment, possibly suggesting seawater mixing with stored brine. Curiously, there is no evidence in the chloride data for phase separation in spite of the fact that the geologic observations at Rose Garden in 2002 clearly indicate the area was recently

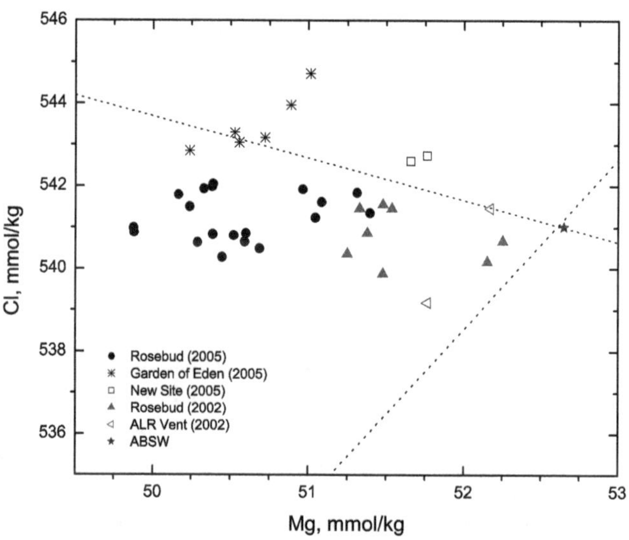

Figure 4. Mg versus Cl data for 2002/2005 (Table 2). Dotted lines are the linear regressions of Clambake (chloride-enriched, 1977) and Oyster Beds (chloride-depleted, 1977); data from *Edmond et al.* [1979b]. Although the current samples from Garden of Eden (and the new site) appear comparably chloride-enriched relative to Clambake in 1977, most likely indicative of stored brine entrainment, the 2002/2005 data show no evidence of chloride depletion relative to ABSW and hence, no contemporary phase separation. Conversely, Oyster Beds, 1977, fluids are significantly depleted in chloride (~325 mmol/kg).

covered by a lava flow. Where the time of a volcanic eruption can be constrained, coexisting vent fluids are typically chloride depleted for up to several years [*Butterfield et al.*, 1997, 1994; *Von Damm*, 2000]. As noted, the variability in chloride reported by *Edmond et al.* [1979b] was unambiguously recognized from accurate measurements of salinity. If the magnitude of chloride depletions rivaled those observed in 1977 (e.g., Oyster Beds), however, then such a trend would be resolved in the current data. The lack of vapor enrichment suggests that the heat supply fueling hydrothermal circulation in recent years is either less robust at equivalent depths than in 1977, or other factors have developed that inhibit fluid penetration into rock sufficiently hot to induce formation of vapor-phase fluids. Regardless, the reported chloride values (Table 2) are consistent with the low equilibration temperatures suggested by quartz geothermometry. High possible temperatures for 2002 would not necessarily induce fluid phase separation (Figure 3). Such temperatures, however, are still unlikely for reasons noted earlier. Whether these cooler systems are a result of magmatic or tectonic factors is not clear. Gradual quenching of the heat supply, however, could be expected to induce quartz precipitation, which in the absence of seismic events, might contribute to a loss of permeability needed for fluid penetration into deeper and hotter portions of the crust [*Martin and Lowell*, 2000].

It is now well recognized that the magmatic heat source driving any given hydrothermal system can vary profoundly with time [*Butterfield et al.*, 1997; *German and Lin*, 2004; *Sohn et al.*, 1998; *Von Damm*, 2000, 2004], consequentially influencing the observed chemistry and temperature of the vent fluids. The H_2S/heat ratios in these recent fluids are clearly not elevated relative to previous measurements at 86°W (see below) or other hydrothermal systems observed in post-eruptive development [*Butterfield et al.*, 1997; *Butterfield and Massoth*, 1994]. After the 1998 eruption at Axial Volcano, JdFR, initially high H_2S/heat ratios in the diffuse fluids quickly declined to pre-eruption levels after ~2 years [*Butterfield et al.*, 2004]. Thus, a similar conclusion that may be drawn from the 2002/2005 chloride and H_2S data is that the 2002 expedition arrived several years after the flows that covered Rose Garden. Furthermore, the slightly lower H_2S concentrations of 2005 relative to 2002 may, again, represent continued cooling of this system, although we cannot rule out the possibility that this is an artifact of the different sampling methods employed.

The fluid chemistry before the post-1990 eruption at 86°W is difficult to constrain with respect to the evolutionary stage of the ridge. Because significant geochemical variables (e.g., Mg, chloride) are not accounted for in samples from 1979 and 1985, differentiation between the prevailing reaction conditions in 1977–1979 and 1985 is not feasible. In contrast with the recent samples from Rosebud, however, samples from Rose Garden, 1979 and 1985, show similar H_2S enrichment to Oyster Beds, 1977, which, in turn, exhibits significant chloride depletion (Figure 4). Although dissolved silica and chloride end-members cannot be calculated for Rose Garden, the high H_2S, whether related to vapor induced fractionation or not, indicates higher temperatures of equilibration than the 2002 and 2005 Rosebud samples.

4.1.3. Dissolved lithium. Dissolved Li concentrations in the 86°W vent fluids from 2002 and 2005 fall along a common line when plotted against silica (Figure 5), despite having somewhat different hydrothermal end-member concentrations of 722 and 557 μmol/kg, respectively. Thus, the 2005 concentration is less than 50% of those reported by *Edmond et al.* [1979b], with the exception of the chloride-depleted Oyster Beds fluids. Furthermore, on a chloride-normalized basis, all of the vents in 1977 indicate end-members exceeding 1100 μmol/kg. Because it is well known that Li is highly soluble during basalt alteration at elevated *P-T* [*Seyfried*,

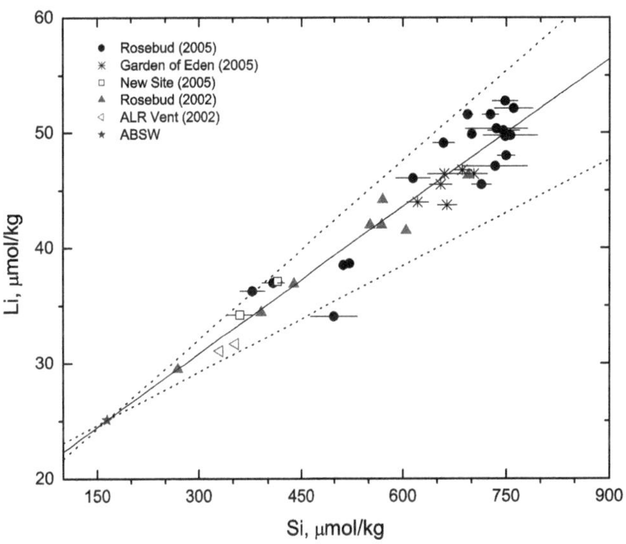

Figure 5. Si versus Li data for 2002/2005 (Table 2). Individual error bars (ICP-MS) are shown for the 2005 Si data. Li end-members calculated for 2005 and 2002 are 557 (±2) and 722 (±8) μmol/kg, respectively. Data from 1977 [*Edmond et al.*, 1979b] predict Li end-members of ~1142 and 689 μmol/kg for Clambake and Oyster Beds, respectively. The lower value for Oyster Beds is likely due to the observed chloride depletion in these fluids. The slopes of the 1977 regressions (dotted) are only slightly different than that of the 2002/2005 data. Furthermore, despite differing silica end-members for 2002 and 2005, the data essentially fall on the same line and were therefore not regressed separately here (solid line). Comparison of current and historical data suggests a lack in phase equilibria control on dissolved Li concentrations (see text for details).

1987; *Seyfried et al.*, 1984], one interpretation of these data is a current decrease in the extent and/or effectiveness of fluid-rock interaction relative to three decades ago. If the higher temperature fluids of 1977 are indicative of circulation in deeper portions of the crust, the attending decrease in the water/rock ratio may account for some portion of the higher Li [*Seyfried et al.*, 1984]. Considering the significant temperature decrease of subseafloor reactions, as indicated by the respective silica values between 1977, 2002, and 2005, mineral precipitation in response to this change may also have caused a resultant decrease in permeability, limiting fluid access to unaltered basalt, thereby constraining the hydrothermal flux of Li.

Considering that dissolved silica is an explicit function of temperature, the generally similar Si-Li correlation for earlier (1977) and more recent (2002 and 2005) fluid data suggest the effect of temperature on Li release is also similar for all periods. This is surprising if the solubility of Li is governed by fluid-mineral equilibria [*Chan et al.*, 2002, 1993; *Seyfried et al.*, 1984; *Von Damm*, 2000]. Although it is not possible to completely discount an aspect of phase equilibria control, it may be that above a threshold temperature (e.g., ~300°C; *James et al.* [2003]), leaching of Li during dissolution of primary minerals in basalt is not reversible. In fact, it is for this reason that Li has been effectively used to qualitatively indicate the relative abundance of fluid and fresh rock in subseafloor reaction zones [*Foustoukos et al.*, 2004]. At some point, dissolved Li concentrations are sufficient such that the drive for additional release lessens, allowing some residual Li to be retained by the rock (fresh or altered). Basalt alteration experiments at elevated *P-T* have confirmed the rapid dissolution of Li during incipient reaction followed by steady-state concentrations at a range of temperatures. The Li concentration achieved at steady state tends to be equivalent to ~60% of that initially in the rock [*Seyfried and Ding*, 1995; *Seyfried et al.*, 1984] with fluid/rock ratios of ~1. Most importantly, however, is the observation that the Li concentrations remain essentially unchanged (Figure 6), even during temperature cycling between 400° and 300°C [*Seewald and Seyfried*, 1990]. The data from such studies argue strongly against the influence of reversible phase equilibria on Li in seafloor hydrothermal systems, while also helping to account for the relatively similar slopes in Si-Li data from GSC, despite large changes in reaction zone temperature.

4.1.4. Dissolved manganese. Dissolved Mn concentrations of vent fluids at 86°W correlate well with silica, as might be expected with two species known for conservative behavior during mixing/cooling [*Von Damm and Lilley*, 2004] (Figure 7). In comparison with the vents in 1977 [*Edmond et al.*, 1979b], the 2002 and 2005 datasets are characterized by relatively low Mn/Si ratios, indicating Mn concentrations distinctly lower than the fluids of both 1977 and 1985 (Figure 7). The variability in the individual regression lines for the 1977 samples is best accounted for by the previously described variability in chloride that characterizes each of the vent sites [*Edmond et al.*, 1979b]. Interestingly, the 2005 Garden of Eden (and new site) samples, which have slightly elevated chloride (Table 2, Figure 7), fall slightly above the regression line, suggesting a similar effect, but to a lesser degree. There is no question that an increase in chloride will increase levels of Mn due to formation of Mn-Cl complexes. In addition, increasing temperatures lower the dielectric constant of water and increase the strength of aqueous complex formation, reinforcing the effect of chloride on Mn solubility [*Sverjensky et al.*, 1997].

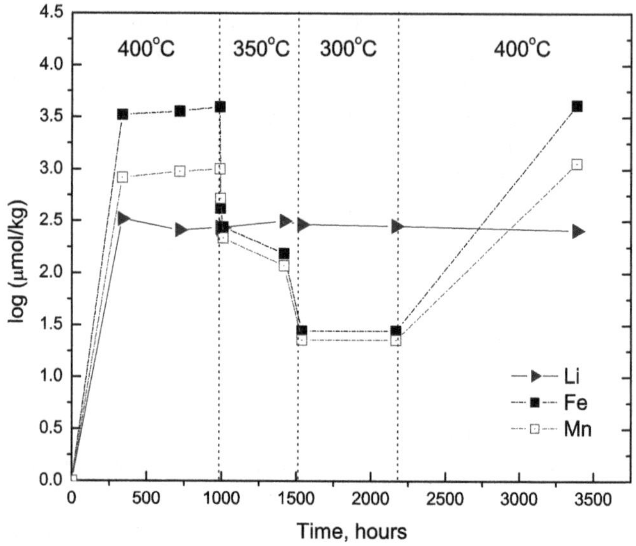

Figure 6. Measured concentrations of Fe, Mn, and Li during experimental alteration of crystalline basalt and T = 400°, 350°, and 300°C, and 400 bars. Initial fluid/rock mass ratio = 1 [see *Seewald and Seyfried,* 1990]. Li maintains a stable concentration through temperature swings between 400° and 300°C. Species such as Fe and Mn, conversely, appear strongly temperature-dependent, which is more typical of solubility/phase-equilibria control. At 400°C, the Fe/Mn ratio is ~3 but approaches unity at lower temperatures (see text for discussion). Data from *Seewald and Seyfried* [1990].

Regression data reported by *Edmond et al.* [1979b] indicate Mn concentrations for hydrothermal end-members that range from ~360 to 1140 µmol/kg. Analogous data for the 2002 and 2005 samples extrapolate to 169 and 100 µmol/kg, respectively. A separate regression for the samples that appear to have elevated chloride (i.e., Garden of Eden and the new site), however, indicate a slightly higher concentration (132 µmol/kg). Unfortunately, it is not possible to directly

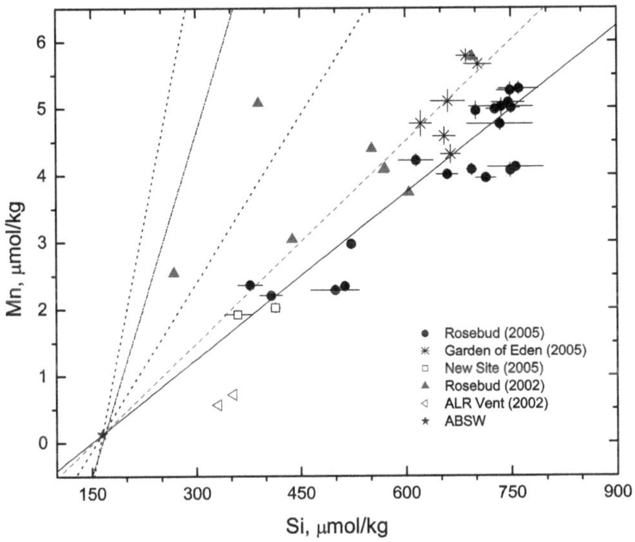

Figure 7. Si versus Mn data for 2002/2005 (Table 2). Individual error bars (ICP-MS) are shown for the 2005 Si and Mn data. Mn end-members calculated for 2005 and 2002 are 100 (±1) and 169 (±3) μmol/kg, respectively. Dotted regression lines from 1977 (left is Clambake, end-member = 1140 μmol/kg and right is Oyster Beds, end-member = 360 μmol/kg) pose as boundaries between which all of the 1977 data fall [*Edmond et al.,* 1979]. Si-Mn data from Rose Garden in 1985 [*Johnson et al.,* 1988] also fall within this region (dot-dash line).

apply phase equilibria to constrain Mn concentrations because there is a dearth of thermodynamic data for Mn-bearing phases in basalt and alteration products at elevated *P-T*. It is helpful, however, to compare these data with results of basalt alteration experiments. For example, *Seewald and Seyfried* [1990] reported Mn concentrations in Cl-bearing fluids (0.55 mmol/kg) coexisting with basalt and alteration products between 300° and 400°C, 400 bars (Figure 6). The results indicate apparent equilibrium Mn concentrations of ~1145 μmol/kg at 400°C, decreasing to ~118 and 23 μmol/kg at 350° and 300°C, respectively. Although variables other than temperature may, in part, account for these data (e.g., pH, redox), temperature appears to be the dominant influence. Relation of these data to the GSC samples retrieved in 1977, 2002, and 2005 is generally consistent with the previously discussed temperature constraints imposed by quartz geothermometry.

4.2. Modeling Subseafloor Iron/Redox Controls

4.2.1. Redox and temperature effects on Fe and H_2S mobility. Experimental data indicate a close correspondence between dissolved Fe and Mn [*Mottl et al.*, 1979; *Seewald and Seyfried*, 1990] with an increase in the Fe/Mn ratio from ~1 to 3 between 300° and 400°C, respectively (Figure 6). Although empirical, these data permit estimates to be made for Fe in the hydrothermal reaction zone at GSC, but to examine this further requires independent data that can be calculated with geochemical reaction path modeling codes. Thus, we performed a series of equilibrium and subsequent mixing calculations using Geochemist's Workbench (GWB) [*Bethke*, 1996], which was supported by thermodynamic data from 2° to 400°C, 500 bars generated by SUPCRT92 [*Johnson et al.*, 1992] including recent database updates [*Shock and Koretsky*, 1993; *Shock et al.*, 1997; *Sverjensky et al.*, 1997]. These data were further supplemented with experimental data that is well suited to enhance the accuracy of model results for moderately acidic and H_2S-bearing fluids at elevated *P-T* [*Ho et al.*, 2001, 1994; *Kishima*, 1989]. The reader is referred to *Alt-Epping and Diamond* [this volume] for a current review of reaction path modeling.

Calculations were performed between 300° and 400°C, 500 bars to simulate hydrothermal mass transfer processes at GSC in 2005 and 1977, respectively. In these models, dissolved Fe and H_2S concentrations were predicted in an evolved-seawater fluid [*Mottl*, 1983] where imposed redox conditions consisted of both a relatively oxidizing mineral system [hematite-magnetite-pyrite (HMP)] and a more reducing system [pyrite-pyrrhotite-magnetite (PPM)]. An activity of unity was assumed for all minerals originally present in the solutions. Results of experimental, theoretical, and field studies have indicated that redox conditions buffering vent fluid chemistry tend to range between those imposed by these two mineral assemblages [*Lilley et al.*, 2003; *Seyfried and Ding*, 1995; *Wetzel and Shock*, 2000]. Similar studies have also demonstrated that the pH of fluids in basalt-hosted systems is typically within one-half log unit of neutrality [*Ding and Seyfried*, 1996, 2007; *Ding et al.*, 2005; *Seyfried and Ding*, 1995]. Accordingly, we assume a pH of 5 in the model calculations. The extent to which the Galapagos fluids have mixed with seawater prior to sampling effectively precludes using the measured pH to define $pH_{(P,T)}$ (pH at elevated *P-T*). Although the model calculations are qualitative at best, they are still successful in demonstrating the relative influence of redox and temperature on resulting fluid chemistry.

Due to the significantly higher $fH_{2(g)}$ and $fH_2S_{(g)}$ when coexisting with PPM, dissolved Fe and H_2S concentrations are considerably greater than with HMP-buffered systems at all temperatures (500 bars) (Figure 8). It should also be noted that, similar to Mn, Fe will also vary with chloride concentration because of the temperature-dependent stability of $FeCl_{2(aq)}$, which is the dominant form of dissolved Fe in acidic, chloride-bearing solutions under the chemical and

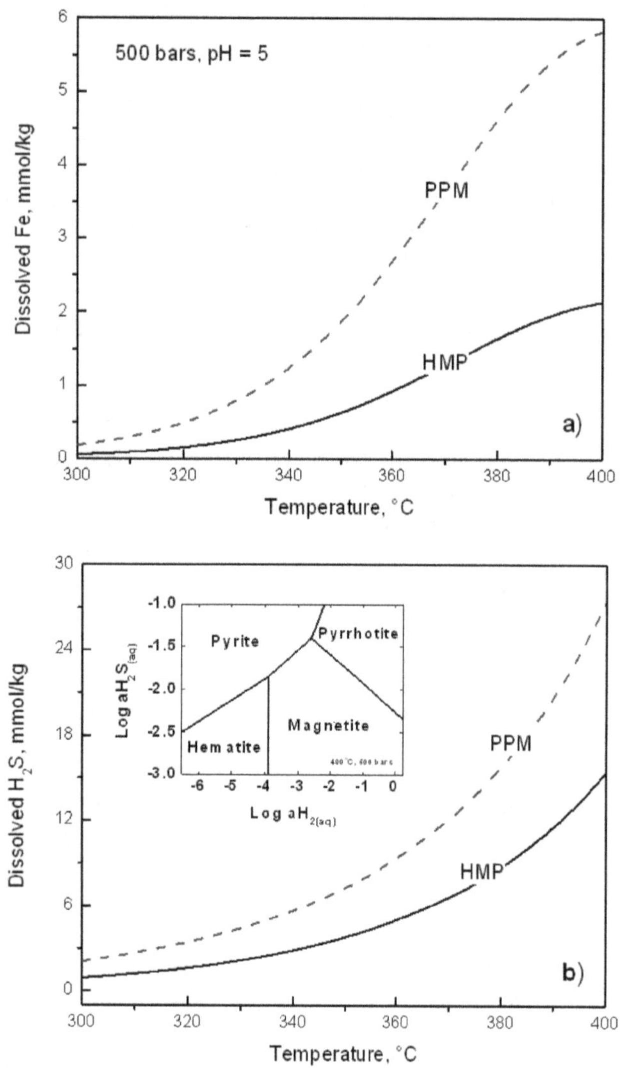

Figure 8. (a) Model dissolved Fe concentrations in equilibrium with either pyrite-pyrrhotite-magnetite (PPM, reducing) or hematite-magnetite-pyrite (HMP, oxidizing) over the range T = 300°–400°C with an inferred pH of 5 at P = 500 bars. (b) Similar to panel (a) but for dissolved H_2S. Inset: activity-phase diagram (H_2 versus H_2S) for 400°C, 500 bars depicting relevant mineral assemblages (for reference).

physical conditions considered [*Ding and Seyfried*, 1992; *Seyfried and Ding*, 1995].

Thus, PPM and HMP-fluid equilibria (pH = 5) at 400°C suggest end-member Fe concentrations of approximately 6 and 2.2 mmol/kg, respectively (Figure 8a). The relatively low chloride concentrations indicated by the 1977 Oyster Beds fluids [*Edmond et al.*, 1979a, 1979b] could be anticipated to cause a corresponding decrease in Fe relative to these estimates. Although this model range of Fe concentrations cannot be directly related to the GSC vent fluids due to apparent reaction (precipitation) of Fe during seawater mixing [*Edmond et al.*, 1979a], the Mn data, as noted, can provide a valuable constraint. Accordingly, from the hydrothermal Mn concentrations of 360 and 1140 µmol/kg, end-member Fe concentrations can be estimated to range from 1.2 to 3.3 mmol/kg, respectively, assuming Fe/Mn ≈ 3. If chloride effects are assumed, the estimated values of Fe are just slightly greater than for HMP-fluid equilibria. This agrees well with both H_2 (redox) measured in more steady-state high-temperature vents (i.e., those not recently perturbed by magmatic activity) [*Ding et al.*, 2001; *Lilley et al.*, 2003] and with constraints imposed by full equilibria involving minerals and fluids typical of basalt-hosted hydrothermal systems [*Foustoukos and Seyfried*, 2005; *Seewald and Seyfried*, 1990; *Seyfried and Ding*, 1995].

HMP-fluid equilibria (pH = 5) suggest a dissolved Fe concentration of ~90 µmol/kg at 300°C, the likely reaction zone temperature for the 86°W fluids in 2005 (Figure 8a). In this case, the lower reaction temperature suggests a Fe/Mn ratio of unity [*Mottl et al.*, 1979; *Seewald and Seyfried*, 1990], giving a hydrothermal end-member Mn value close to that estimated (~100 µmol/kg) from the 2005 GSC data (Figure 7). Although not reported here, Fe concentrations measured for the 2002 and 2005 samples indicate variably low and inconsistent data. Thus, constraints imposed by Mn on dissolved Fe are consistent with the existence of a relatively oxidizing and moderately low temperature reaction zone for 86°W in 2002 and 2005.

4.2.2. Subsurface mixing and the availability of H_2S. Assuming that the hydrothermal reaction zone beneath 86°W was buffered by HMP in both 1977 and 2005, the effective temperature difference (~100°C) would then be expected to correspondingly affect the H_2S concentration in venting fluids. To illustrate this, the buffered (HMP) fluid predicted at both 300° and 400°C (Figure 8) was progressively reacted/mixed with a fluid of seawater composition (2°C, pH = 8). For both temperatures, quartz-fluid equilibria were also assumed for the hydrothermal component prior to mixing. It is important to emphasize that in these model calculations, the hydrothermal source fluid has been separated from the buffer assemblage prior to seawater mixing. Although minerals (except quartz) were allowed to precipitate with reaction (mixing) progress, only the two fluid end-members were initially present, with temperature then changing in accordance with the proportion of seawater added. Experimental and field studies have confirmed sulfur redox disequilibria for homogeneous/heterogeneous reactions in the FeO-Fe_2O_3-H_2S-H_2O system, especially at moderate to low temperatures [*Janecky and Seyfried*, 1984; *Janecky et al.*, 1986; *Ohmoto* *and Lasaga*, 1982]. Furthermore, recent observations have indicated significant dissolved H_2 concentrations during such mixing processes [*Houghton*, 2003; *Houghton and Seyfried*, 2000; *Seewald et al.*, 2006; *Von Damm and Lilley*, 2004]. To be consistent with these observations, redox equilibria were not permitted for the SO_4^{2-}-HS^- or H_2-O_2 couples. As the hydrothermal components were quartz saturated, the mixing results are presented using silica as a temperature proxy in order to more accurately compare them with measured fluid chemistry from the GSC [*Johnson et al.*, 1988; *Von Damm and Lilley*, 2004]. It should be noted that the mixing calculations neglect possible conductive heat loss/gain and changes in heat capacity of the solution as a function of temperature. The error introduced by this simplification is, however, believed to be negligible [*Janecky and Seyfried*, 1984].

Dissolved H_2S is predicted to decrease sharply with sufficient seawater dilution to achieve temperatures typical of 86°W vents (~5°–20°C) (Figures 8 and 9). As expected, the H_2S delivered by the 400°C source fluid results in a relatively steep H_2S-Si profile, consistent with much of the data reported from 1977–1979 and 1985. Differences do exist, however, suggesting that factors other than simple mixing and mineralization need to be considered. Seawater dilution of the 300°C fluid predicts H_2S-Si that simulates reasonably well the compositional trends of the recently sampled GSC fluids (Figure 9). Time series data spanning the past 30 years at 86°W suggest GSC fluids are more typical of seafloor vents, having source fluids with H_2S/Fe ratios greater than unity [*Von Damm*, 1995]. One exception among GSC vent sites is Clambake, 1977, where fluids had H_2S/Fe ratios as low as 0.5 [*Edmond et al.*, 1979a]. This, however, is undoubtedly the result of chloride effects. The former is consistent with constraints imposed by HMP-fluid equilibria at both 400° and 300°C (Figures 8 and 9) and generally occurs where equilibrating fluids have chloride concentrations at or below seawater values. Although this model reveals that sulfide may mix conservatively under such circumstances, Fe and pH are expected to react differently. Changes in these species reveal highly nonlinear patterns with increased mixing (Figure 10). The pH minima between ~200° and 300°C are caused by the precipitation of pyrite. Continued mixing, however, substantially increases the predicted pH as the seawater end-member is approached. The high pH, in turn, is expected to render Fe insoluble (Fig. 10). Low Fe is a common observation in diffuse fluids [*Butterfield et al.*, 2004; *Edmond et al.*, 1979a; *Le Bris et al.*, 2006; *Von Damm and Lilley*, 2004]. Thus, it is not surprising that the concentration in GSC vents is either near/below detection

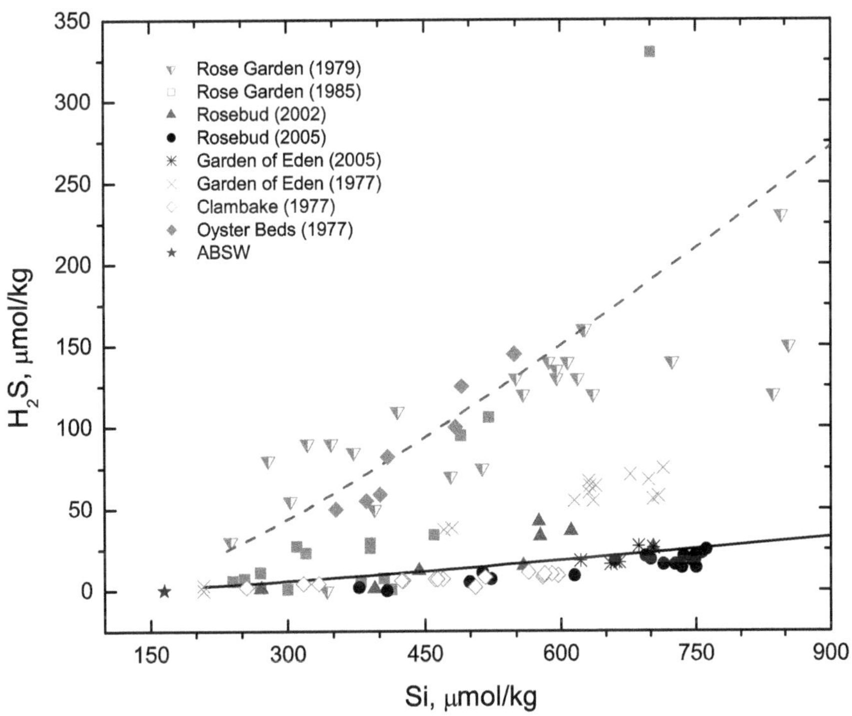

Figure 9. Time series Si versus H_2S for discreet samples taken from the GSC, 86°W, 1977–2005. Superimposed are modeled mixing lines between ABSW and an HMP-derived hydrothermal component equilibrated at both 400°C (dashed) and 300°C (solid). See text for discussion. Supplementary data; 1977–1979 [*Lilley et al.*, 1983] and 1985 [*Johnson et al.*, 1988].

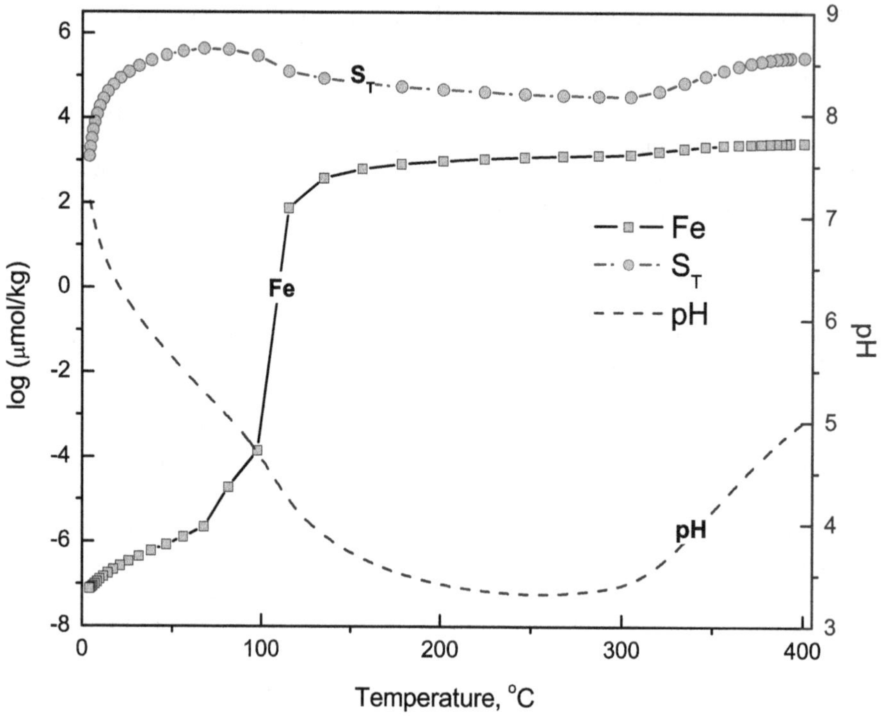

Figure 10. Model mixing calculations illustrating progressive changes in pH (dashed line, right axis), and total dissolved sulfur (S_T) and Fe with increasing seawater (2°C) dilution for a hydrothermal source fluid having been in equilibrium with HMP at 400°C, 500 bars. Note: hydrothermal component is no longer buffered by HMP, temperature change is solely the result of seawater mixing.

levels (2002/2005), or highly nonlinear against silica [*Edmond et al.*, 1979a]. Although an exact model accounting for the full effects of thermal/chemical evolution from hydrothermal source fluid to diffuse effluent is not yet at hand, these simulations still provide a useful frame of reference for comparison with natural systems, as long as exceptions are recognized when more complex kinetic/thermodynamic processes prevail.

4.2.3. Insight from in situ fluid chemical studies. Results from in situ chemical studies provide crucial insight into the effect of fluid temperature on the distribution of aqueous species in diffuse fluids [*Johnson et al.*, 1988; *Le Bris et al.*, 2006, 2000, 2001; *Luther et al.*, 2001]. For example, a study conducted in Rose Garden in 1985 examined concentrations of important redox active species and incorporated in situ measurements of silica, sulfide, oxygen, and temperature [*Johnson et al.*, 1988]. The results of this study stand in confirmation that microorganisms can have a profound effect on vent fluid chemistry, especially oxygen, sulfide, and nitrate. An element of nonlinearity was noted between silica and temperature, but this was largely restricted to conditions characterized by unusually high degrees of seafloor mixing. More recently, in situ chemical/temperature data from Tica and Biovent of EPR, 9°50′N [*Le Bris et al.*, 2006] indicate that sulfide and temperature in these diffuse vents are largely consistent with a conservative mixing model. Other sites in the vicinity (e.g., Riftia Field), however, reveal significant departures, likely due to the lower H_2S/Fe ratio in the source fluid, which, in turn, contributes to the unusually low pH. The acidity of these fluids may reflect Fe oxide mineralization and could also account for the observed difference in macrofaunal abundance/distribution via change in sulfide speciation.

A key objective of the integrated 2005 dive program was deployment of in situ chemical sensors to better understand the feedback between hydrothermal and biologic systems. Previous studies of this sort were often limited in duration to the time span of an individual dive. In 2005, however, an autonomous electrochemical sensor system [*Ding and Seyfried*, 2007; *Ding et al.*, 2004] was off-loaded by *Alvin* to provide a continuous 24-h record of chemical (sulfide) and temperature data at marker B of Rosebud (Figure 11a). These data (Figures 11b and 11c) reveal highly nonlinear correlations, wherein relatively small changes in temperature can result in sharp changes in dissolved sulfide. This is especially prevalent near the midpoint (~12 h) of the deployment. Clearly, changes in temperature will reflect changes in the flux of hydrothermal fluid, but also, bottom currents can variably affect the amount of seawater mixing with the monitored effluent. Nevertheless, the sulfide data reported by the in situ sensor are roughly equivalent to concentrations measured in discrete fluid samples when temperature effects are explicitly accounted for.

It is of interest to note that in 2002 and 2005, all sampled vents at 86°W having exit temperatures lower than 10°C lacked H_2S. This is consistent with the in situ sensor measurements at Rosebud, 2005 (Figures 11b and 11c). This cannot be accounted for by simple dilution and may suggest chemical and/or microbiological effects that dominate once certain thresholds are exceeded. For example, an increased influx of ambient seawater might be sufficient to induce microbially mediated oxidation effects, thereby removing a large fraction of the sulfide. *Butterfield et al.* [2004] note that H_2S/heat ratios associated with Axial Volcano, JdFR, appear to decrease with vent exit temperature yet microbial cell counts increase. This is compelling evidence of nonconservative mixing for H_2S and microbial involvement in the temporal evolution of the fluids. At Axial Volcano, subseafloor microbial reactions are quite certain and the oxidation of H_2S is likely the primary energy source for resident microorganisms. These effects may not be revealed by previous in situ studies because effluent was not monitored continuously at the exit point near the rock. Although largely speculative at this point, additional sensor deployments, such as those conducted in 2005, might provide the kinetic data needed to link geochemical and microbial processes in real time. There can be no question, however, that a better understanding of diffuse-flow hydrothermal systems will require future integration of kinetic as well as thermodynamic processes.

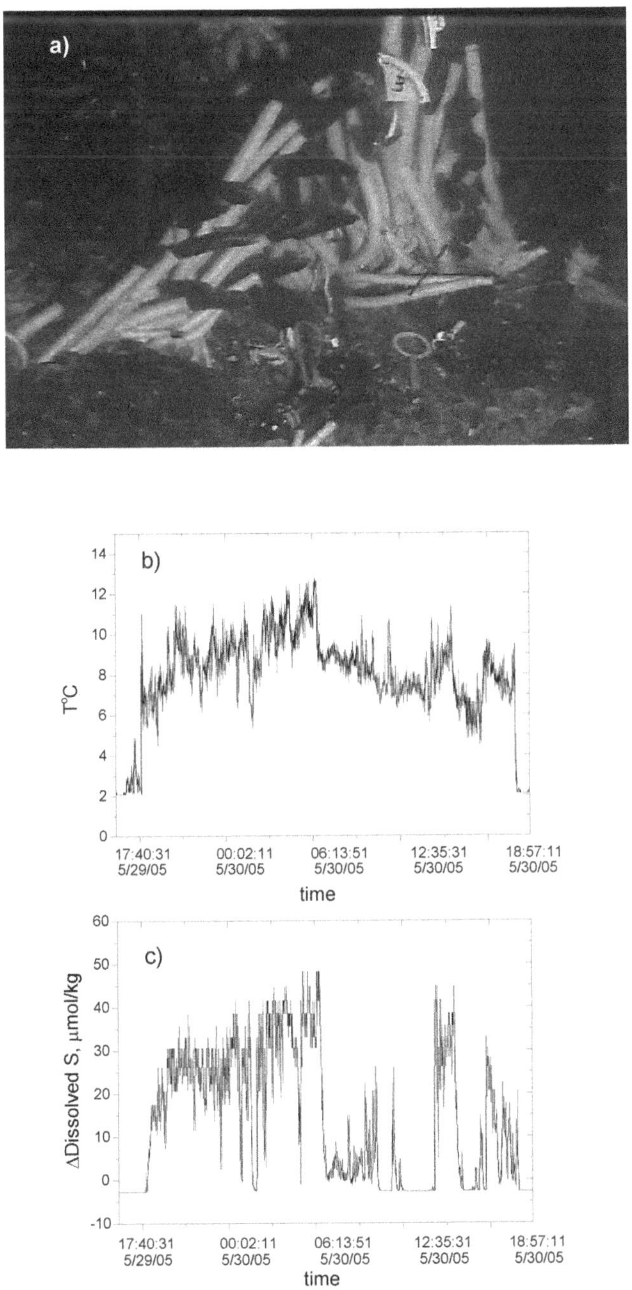

5. CONCLUSIONS

Hydrothermal activity on the GSC, 86°W appears to have remained a diffuse-flow regime since the discovery of seafloor venting in 1977. Measured concentrations of species believed conservative during seawater mixing indicate lower temperatures for the 2002 and 2005 hydrothermal components relative to 1977. The hydrothermal end-members regressed from the 1977 data provide clear evidence for relatively high temperatures and phase separation effects. The *P-T* conditions indicated by quartz geothermometry (~500 bars, 400°C), however, are inconsistent with the observed chloride variability, where temperatures in excess of ~480°C (assuming similar pressure) are needed to intersect the two-phase boundary of seawater (3.2% NaCl). Thus,

Figure 11. (a) In situ chemical/temperature probe and accompanying data logger situated near the base of a tubeworm patch in Rosebud vent field, 2005. The solid-state sensors in the tip of the probe were used to gather time series data including temperature and S_T from the diffuse-flow vents of 86°W. In situ measurements of (b) temperature and (c) S_T in the fluids issuing from marker B, Rosebud, were compiled for a period of 24 h. Simultaneously monitored, these data show a clear correspondence between temperature and S_T. This relationship, however, appears to be highly nonlinear. It is important to note that the S_T data is presented relative to an internal reference temperature (here, ~5.5°C) corresponding to the reliable detection limit of S_T (~10 μmol/kg) in the diffuse fluids. This is to minimize potential uncertainties involving laboratory calibration of the sensor.

these data suggest the likely existence of multiple reaction stages wherein deeply penetrating fluids achieve *P-T* conditions sufficient to evolve a highly dilute (chloride-depleted) vapor phase, which then rises to mix with fluids equilibrating near the conditions predicted by quartz-fluid phase equilibria. The moderately high pressure that characterizes subseafloor hydrothermal processes at GSC more easily permits multiple reaction stages to be resolved.

In contrast, quartz geothermometry suggests that temperatures likely did not exceed ~325°C for the fluids sampled from 86°W in 2002 and 2005. Relatively low Mn and H_2S concentrations and the lack of chloride depletion further support this conclusion. Despite evidence of a recent eruption, these data are not consistent with the subsequent period of high heat flux observed in other hydrothermal systems. The most recent magmatic activity, however, was likely the eruption that covered Rose Garden a maximum of ~10 years before 2002. Assuming this is the case, quiescent periods lacking significant volcanism/tectonism may encourage mineral precipitation, obstructing fluid pathways above the magma-hydrothermal interface. Decreasing permeability would then begin to limit the temperature maxima of hydrothermal fluids. Unfortunately, the age of the most recent flow is uncertain on the order of a decade, with chloride and H_2S data suggesting that the 2002 expedition arrived too late to observe the expected high-temperature aftermath of such an event. It is possible, however, that the lower slopes in the 2005 Mg-Si and Si-Mn regressions represent further cooling relative to 2002.

Basalt alteration experiments at elevated *P-T* have consistently demonstrated the sensitivity of dissolved silica, Mn, Fe, and H_2S to temperature, because all of these species are controlled largely by heterogeneous phase equilibria. The same is not true for dissolved Li, at least to the extent demonstrated by these other species. The remarkably similar correlation between dissolved silica and Li for the 1977, 2002, and 2005 datasets is therefore consistent with Li release being largely temperature-independent above a certain threshold. Still consistent with higher temperatures, high Li concentrations may therefore be indicative of deeper fluid circulation through fresh rock at lower water/rock ratios. The diminished permeability suggested to account for the cooler systems in 2002 and 2005 might then restrict hydrothermal circulation to previously altered portions of the crust, resulting in lower Li concentrations in the end-member fluids.

Assuming fluid equilibration with a moderately oxidizing mineral assemblage, model calculations predict dissolved H_2S/Fe ratios greater than unity in GSC hydrothermal components, a result in keeping with time series H_2S data from 86°W. Furthermore, the estimated temperature difference of ~100°C between 1977 and 2005 should have a profound effect on the mobility of Fe and H_2S, consequently effecting the availability of H_2S to vent biological communities. After the essentially complete removal of Fe in fluids diluted to the extent of those at 86°W, model calculations indicate dilution dominated concentration changes. Data from in situ chemical/temperature sensors deployed at Rosebud, however, reveal nonlinear behavior of total dissolved sulfide relative to temperature, especially below 10°C. These data may indicate the effects of microbial metabolism. In situ sensor deployments, possibly on greater time scales, may be the key to better understanding these processes, a requisite first step for developing more quantitative models of seafloor mixing environments.

Acknowledgments. We wish to thank the ALVIN Group and Master, officers, and crew of the R/V *Atlantis*. Without their help this research would not have been possible. We greatly appreciate the counsel of Rachel Haymon, Ken Macdonald, and Michael Perfit, all of whom provided important insight into the geological and geophysical characteristics of the GSC. We also thank Rick Knurr and Mikaella Rough for their careful and diligent chemical analysis of the 2005 fluid samples, and Justin Revenaugh for creating the algorithm used for the end-member extrapolations. Finally, thoughtful reviews by Wolfgang Bach, Mike Mottl, and Bob Lowell helped to greatly improve the clarity and content of this manuscript. Research funding was largely provided (to T.S.) by NOAA's Office of Ocean Exploration (NA04OAR4600084), the Cooperative Institute for Climate and Ocean Research (NA17RJ1223), and WHOI through the Vetlesen and Richards Foundation Grants and a Fellowship through the Deep-Ocean Exploration Institute. This publication is also partially funded by the Joint Institute for the Study of the Atmosphere and Ocean (JISAO) under NOAA Cooperative Agreement No. NA17RJ1232, contribution no. 1418, PMEL contribution no. 3088 and through NSF grants OCE-0351069 (W.E.S.), OCE-0525907 (K.D.), and EAR-0649044 (M.H.).

REFERENCES

Alt-Epping, P., and L. W. Diamond (2008), Reactive transport and numerical modeling of seafloor hydrothermal systems: A review, this volume.

Anderko, A., and K. S. Pitzer (1993), Equation of state representation of phase equilibria and volumetric properties of NaCl-H_2O above 573°C, *Geochim. Cosmochim. Acta*, *57*, 1657–1680.

Ballard, R. D., T. H. Van Andel, and R. T. Holcomb (1982), The Galapagos Rift at 86°W: 5. Variations in volcanism, structure, and hydrothermal activity along a 30-kilometer segment of the rift valley, *J. Geophys. Res.*, *87*, 1149–1161.

Berndt, M. E., and W. E. Seyfried Jr. (1990), Boron, bromine, and other trace elements as clues to the fate of chlorine in mid-ocean ridge vent fluids, *Geochim. Cosmochim. Acta*, *54*, 2235–2245.

Berndt, M. E., and W. E. Seyfried Jr. (1997), Calibration of Br/Cl fractionation during subcritical phase separation of seawater;

possible halite at 9 to 10°N East Pacific Rise, *Geochim. Cosmochim. Acta, 61*, 2849–2854.

Berndt, M. E., M. E. Person, and W. E. Seyfried Jr. (2001), Phase separation and two-phase flow in seafloor hydrothermal systems: Geophysical modeling in the NaCl-H_2O system, paper presented at Eleventh Annual V. M. Goldschmidt Conference, Geochemical Society, Hot Springs, VA, 20–24 May.

Bethke, C. (1996), *Geochemical Reaction Modeling: Concepts and Applications*, 397 pp., Oxford University Press, New York.

Bischoff, J. L., and K. S. Pitzer (1989), Liquid-vapor relations for the system NaCl-H_2O: Summary of the *P-T-x* surface from 300°C to 500°C, *Am. J. Sci., 289*, 217–248.

Bischoff, J. L., and R. J. Rosenbauer (1985), An empirical equation of state for hydrothermal seawater (3.2% NaCl), *Am. J. Sci., 285*, 725–763.

Bischoff, J. L., and R. J. Rosenbauer (1988), Liquid-vapor relations in the critical region of the system NaCl-H_2O from 380 to 415°C: A refined determination of the critical point and two-phase boundary of seawater, *Geochim. Cosmochim. Acta, 52*, 2121–2126.

Blacic, T. M., G. Ito, J. P. Canales, R. S. Detrick, and J. Sinton (2004), Constructing the crust along the Galapagos Spreading Center 91.3°–95.5°W: Correlation of seismic layer 2A with axial magma lens and topographic characteristics, *J. Geophys. Res., 109*, B10310, doi: 10.1029/2004JB003066.

Bowers, T. S., K. L. Von Damm, and J. M. Edmond (1985), Chemical evolution of mid-ocean ridge hot springs, *Geochim. Cosmochim. Acta, 49*, 2239–2252.

Buck, W. R., S. M. Carbotte, and C. Mutter (1997), Controls on extrusion at mid-ocean ridges, *Geology, 25*, 935–938.

Butterfield, D. A., and G. J. Massoth (1994), Geochemistry of north Cleft segment vent fluids: Temporal changes in chlorinity and their possible relation to recent volcanism, *J. Geophys. Res., 99*, 4951–4968.

Butterfield, D. A., R. E. McDuff, M. J. Mottl, M. D. Lilley, J. E. Lupton, and G. J. Massoth (1994), Gradients in the composition of hydrothermal fluids from the Endeavor Segment vent field: Phase separation and brine loss, *J. Geophys. Res., 99*, 9561–9583.

Butterfield, D. A., I. R. Jonasson, G. J. Massoth, R. A. Feely, K. K. Roe, R. W. Embley, J. F. Holden, R. E. McDuff, M. D. Lilley, and J. R. Delaney (1997), Seafloor eruptions and evolution of hydrothermal fluid chemistry, *Philos. Trans. R. Soc. Lond. A, 355*, 369–386.

Butterfield, D. A., K. K. Roe, M. D. Lilley, J. A. Huber, J. A. Baross, R. W. Embley, and G. J. Massoth (2004), Mixing, reaction and microbial activity in the sub-seafloor revealed by temporal and spatial variation in diffuse flow vents at Axial Volcano, in *The Subseafloor Biosphere at Mid-Ocean Ridges, Geophys. Monogr. Ser.*, vol. 144, edited by W. S. D. Wilcock, et al., pp. 269–289, AGU, Washington, D. C.

Campbell, A. C., et al. (1988), Chemistry of hot springs on the Mid-Atlantic Ridge, *Nature, 335*, 514–519.

Canales, J. P., J. J. Dañobeitia, R. S. Detrick, E. E. E. Hooft, R. Bartolome, and D. F. Naar (1997), Variations in axial morphology along the Galápagos Spreading Center and the influence of the Galápagos Hotspot, *J. Geophys. Res., 102*, 27,341–27,354.

Carbotte, S. M., A. Solomon, and G. Ponce-Correa (2000), Evaluation of morphological indicators of magma supply and segmentation from a seismic reflection study of the East Pacific Rise 15°30′–17°N, *J. Geophys. Res., 105*, 2737–2759.

Carbotte, S. M., R. S. Detrick, A. Harding, J. P. Canales, J. Babcock, G. Kent, E. Van Ark, M. Nedimovic, and J. Diebold (2006), Rift topography linked to magmatism at the intermediate spreading Juan de Fuca Ridge, *Geology, 34*, 209–212.

Chan, L. H., J. M. Edmond, and G. Thompson (1993), A lithium study of hot springs and metabasalts from mid-ocean ridge hydrothermal systems, *J. Geophys. Res., 98*, 9653–9659.

Chan, L. H., J. C. Alt, and D. A. H. Teagle (2002), Lithium and lithium isotope profiles through the upper oceanic crust: A study of seawater-basalt exchange at ODP Sites 504B and 896A, *Earth Planet. Sci. Lett., 201*, 187–201.

Chen, Y. S. J., and J. Lin (2004), High sensitivity of ocean ridge thermal structure to changes in magma supply: The Galapagos Spreading Center, *Earth Planet. Sci. Lett., 221*, 263–273.

Chen, Y. J., and J. P. Morgan (1996), The effects of spreading rate, the magma budget, and the geometry of magma emplacement on the axial heat flux at mid-ocean ridges, *J. Geophys. Res., 101*, 11,475–11,482.

Christie, D. M., R. Werner, F. Hauff, K. Hoernle, and B. B. Hanan (2005), Morphological and geochemical variations along the eastern Galapagos Spreading Center, *Geochem. Geophys. Geosyst., 6*, Q01006, doi: 10.1029/2004GC000714.

Cline, J. D. (1969), Spectrophotometric determination of hydrogen sulfide in natural waters, *Limnol. Oceanogr., 14*, 454–458.

Corliss, J. B., J. Dymond, L. I. Gordon, J. M. Edmond, R. P. V. Herzen, R. D. Ballard, K. Green, D. Williams, A. Bainbridge, K. Crane, and T. H. Vanandel (1979), Submarine thermal springs on the Galapagos Rift, *Science, 203*, 1073–1083.

Delaney, J. R., D. S. Kelley, M. D. Lilley, D. A. Butterfield, J. A. Baross, W. S. D. Wilcock, R. W. Embley, and M. Summit (1998), The quantum event of oceanic crustal accretion: Impacts of diking at mid-ocean ridges, *Science, 281*, 222–230.

Ding, K., and W. E. Seyfried Jr. (1992), Determination of Fe-Cl complexing in the low pressure supercritical region (NaCl fluid): Iron solubility constraints on pH of subseafloor hydrothermal fluids, *Geochim. Cosmochim. Acta, 56*, 3681–3692.

Ding, K., and W. E. Seyfried Jr. (1996), Direct pH measurement of NaCl-bearing fluid with an in-situ sensor at 400°C and 40 megapascals, *Science, 272*, 1634–1636.

Ding, K., and W. E. Seyfried Jr. (2007), In-situ measurement of pH and dissolved H_2 in mid-ocean ridge hydrothermal fluids at elevated temperatures and pressures, *Chem. Rev., 107*, 601–622.

Ding, K., W. E. Seyfried Jr., M. K. Tivey, and A. M. Bradley (2001), In-situ measurement of dissolved H_2 and H_2S in high-temperature hydrothermal vent fluids at the Main Endeavour Field, Juan de Fuca Ridge, *Earth Planet. Sci. Lett., 186*, 417–425.

Ding, K., W. E. Seyfried Jr., Z. Zhang, C. Yang, and D. I. Foustoukos (2004), Time-series deployment of chemical sensor datalogger in a Riftia patch at Tica vent (EPR 9°50′N), *Eos Trans. AGU, 85*, Fall Meet. Suppl., Abstract B13A-0170.

Ding, K., W. E. Seyfried Jr., M. K. Tivey, K. L. Von Damm, A. M. Bradley, and Z. Zhang (2005), In-situ pH measurement of hydrothermal fluids at mid-ocean ridges, *Earth Planet. Sci. Lett., 237*, 167–174.

Edmond, J. M., C. Measures, B. Mangum, B. Grant, F. R. Sclater, R. Collier, A. Hudson, L. I. Gordon, and J. B. Corliss (1979a), On the formation of metal-rich deposits at ridge crests, *Earth Planet. Sci. Lett.*, *46*, 19–30.

Edmond, J. M., C. Measures, R. E. McDuff, L. Chan, R. Collier, B. Grant, L. I. Gordon, and J. Corliss (1979b), Ridge crest hydrothermal activity and the balances of the major and minor elements in the ocean: The Galapagos data, *Earth Planet. Sci. Lett.*, *46*, 1–18.

Embley, R. W., J. E. Lupton, G. Massoth, T. Urabe, V. Tunnicliffe, D. A. Butterfield, T. Shibata, O. Okano, M. Kinoshita, and K. Fujioka (1998), Geological, chemical, and biological evidence for recent volcanism at 17.5°S: East Pacific Rise, *Earth Planet. Sci. Lett.*, *163*, 131–147.

Fornari, D. J., and R. W. Embley (1995), Tectonic and volcanic controls on hydrothermal processes at the mid-ocean ridge: An overview based on near-bottom and submersible studies, in *Seafloor Hydrothermal Systems: Physical, Chemical, Biological, and Geological Interactions*, *Geophys. Monogr. Ser.*, vol. 91, edited by S. E. Humphris, et al., pp. 1–46, AGU, Washington, D. C.

Fornari, D. J., T. Shank, K. L. Von Damm, T. K. P. Gregg, M. Lilley, G. Levai, A. Bray, R. M. Haymon, M. R. Perfit, and R. Lutz (1998), Time-series temperature measurements at high-temperature hydrothermal vents, East Pacific Rise 9°49′–51′N: Evidence for monitoring a crustal cracking event, *Earth Planet. Sci. Lett.*, *160*, 419–431.

Fournier, R. O. (1983), A method of calculating quartz solubilities in aqueous sodium chloride solutions, *Geochim. Cosmochim. Acta*, *47*, 579–586.

Fournier, R. O., and P. W. Potter (1983), An equation correlating the solubility of quartz in water from 25° to 900°C at pressures up to 10,000 bars, *Geochim. Cosmochim. Acta*, *46*, 1969–1973.

Foustoukos, D. I., and W. E. Seyfried Jr. (2005), Redox and pH constraints in the subseafloor root zone of the TAG hydrothermal system, 26°N Mid-Atlantic Ridge, *Earth Planet. Sci. Lett.*, *235*, 497–510.

Foustoukos, D. I., and W. E. Seyfried Jr. (2007a), Quartz solubility in the two-phase and critical region of the NaCl-KCl-H_2O system: Implications for submarine hydrothermal vent systems at 9°50′N East Pacific Rise, *Geochim. Cosmochim. Acta*, *71*, 186–201.

Foustoukos, D. I., and W. E. Seyfried Jr. (2007b), Trace element partitioning between vapor, brine and halite under extreme phase separation conditions, *Geochim. Cosmochim. Acta*, *71*, 2056–2071.

Foustoukos, D. I., R. H. James, M. E. Berndt, and W. E. Seyfried Jr. (2004), Lithium isotopic systematics of hydrothermal vent fluids at the Main Endeavour Field, northern Juan de Fuca Ridge, *Chem. Geol.*, *212*, 17–26.

German, C. R., and J. Lin (2004), The thermal structure of the oceanic crust, ridge spreading and hydrothermal circulation: How well do we understand their inter-connections?, in *Mid-Ocean Ridges: Hydrothermal Interactions Between the Lithosphere and Ocean*, *Geophys. Monogr. Ser.*, vol. 148, edited by C. R. German, J. Lin, and L. M. Parsons, pp. 1–19, AGU, Washington, D. C.

German, C. R., and K. L. Von Damm (2003), Hydrothermal processes, in *Treatise on Geochemistry, vol. 6: The Oceans and Marine Geochemistry*, edited by H. Elderfield, pp. 181–222, Elsevier Pergamon, San Diego.

Gillis, K. M. (1995), Controls on hydrothermal alteration in a section of fast-spreading oceanic crust, *Earth Planet. Sci. Lett.*, *134*, 473–489.

Ho, P. C., D. A. Palmer, and R. E. Mesmer (1994), Electrical conductivity measurements of aqueous sodium chloride solutions to 600°C and 300 MPa, *J. Sol. Chem.*, *23*, 997–1018.

Ho, P. C., D. A. Palmer, and M. S. Gruszkiewicz (2001), Conductivity measurements of dilute aqueous HCl solutions to high temperatures and pressures using a flow-through cell, *J. Phys. Chem. B*, *105*, 1260–1266.

Houghton, J. L. (2003), Biogeochemistry of seafloor hydrothermal vent systems: An experimental study conducted at in-situ conditions, Ph.D. thesis, 170 pp, Univ. of Minnesota, Minneapolis, Minn.

Houghton, J. L., and W. E. Seyfried Jr. (2000), Kinetic inhibition of heterogeneous Fe and S reactions: Experimental and theoretical constraints on microbial energy sources in a subseafloor biosphere, paper presented at Fifth RIDGE Theoretical Institute: The subsurface biosphere at mid-ocean ridges, Big Sky, MT, 28 July–1 Aug.

James, R. H., D. E. Allen, and W. E. Seyfried Jr. (2003), An experimental study of alteration of oceanic crust and terrigenous sediments at moderate temperatures (51 to 350°C): Insights as to chemical processes in near-shore ridge-flank hydrothermal systems, *Geochim. Cosmochim. Acta*, *67*, 681–691.

Janecky, D. R., and W. E. Seyfried Jr. (1984), Formation of massive sulfide deposits on oceanic ridge crests: Incremental reaction models for mixing between hydrothermal solutions and seawater, *Geochim. Cosmochim. Acta*, *48*, 2723–2738.

Janecky, D. R., W. E. Seyfried Jr., and M. E. Berndt (1986), Fe-O-S redox reactions and kinetics in hydrothermal systems, in *Fifth International Symposium on Water-Rock Interaction*, Reykjavik, Iceland, 5, edited by B. Hitchon, pp. 282–285.

Jannasch, H. W., and M. J. Mottl (1985), Geomicrobiology of deep-sea hydrothermal vents, *Science*, *229*, 717–725.

Johnson, J. W., E. H. Oelkers, and H. C. Helgeson (1992), SUPCRT92: A software package for calculating the standard molal thermodynamic properties of minerals, gases, aqueous species, and reactions from 1 to 5000 bar and 0 to 1000°C, *Comput. Geosci.*, *18*, 899–947.

Johnson, K. S., J. J. Childress, R. R. Hessler, C. M. Sakamotoarnold, and C. L. Beehler (1988), Chemical and biological interactions in the Rose Garden hydrothermal vent field, Galapagos Spreading Center, *Deep-Sea Res. A*, *35*, 1723–1744.

Kelley, D. S., and J. R. Delaney (1987), Two-phase separation and fracturing in mid-ocean ridge gabbros at temperatures greater than 700°C, *Earth Planet. Sci. Lett.*, *83*, 53–66.

Kennish, M. J., and R. A. Lutz (1998), Morphology and distribution of lava flows on mid-ocean ridges: A review, *Earth Sci. Rev.*, *43*, 63–90.

Kishima, N. (1989), A thermodynamic study on the pyrite-pyrrhotite-magnetite-water system at 300°–500°C with relevance to the fugacity/concentration quotient of aqueous H_2S, *Geochim. Cosmochim. Acta*, *53*, 2143–2155.

Le Bris, N., P. M. Sarradin, D. Birot, and A. M. Alayse-Danet (2000), A new chemical analyzer for in-situ measurement of nitrate and total sulfide over hydrothermal vent biological communities, *Mar. Chem.*, *72*, 1–15.

Le Bris, N., P. M. Sarradin, and S. Pennec (2001), A new deep-sea probe for in-situ pH measurement in the environment of hydrothermal vent biological communities, *Deep-Sea Res. I*, *48*, 1941–1951.

Le Bris, N., B. Govenar, C. Le Gall, and C. R. Fisher (2006), Variability of physico-chemical conditions in 9°50′N EPR diffuse flow vent habitats, *Mar. Chem.*, *98*, 167–182.

Lilley, M. D., J. A. Baross, and L. I. Gordon (1983), Reduced gases and bacteria in hydrothermal fluids; the Galapagos Spreading Center and 21°N East Pacific Rise, in *Hydrothermal Processes at Seafloor Spreading Centers*, edited by P. A. Rona, K. Bostroem, L. Laubier, and K. L. Smith Jr., pp. 411–449, Plenum Press, New York.

Lilley, M. D., J. E. Lupton, D. A. Butterfield, and E. Olson (2003), Magmatic events can produce rapid changes in hydrothermal vent chemistry, *Nature*, *422*, 878–881.

Lowell, R. P., B. W. Crowell, S. R. Gosnell, K. C. Lewis, L. Liu, and Y. Yang, this volume,

Lowell, R. P., S. Gosnell, and Y. Yang (2007), Numerical simulations of single-pass hydrothermal convection at mid-ocean ridges: Effects of the extrusive layer and temperature-dependent permeability, *Geochem. Geophys. Geosyst.*, *8*, Q10011, doi: 10.1029/2007GC001653.

Lowell, R. P., Y. Yao, and L. N. Germanovich (2003), Anhydrite precipitation and the relationship between focused and diffuse flow in seafloor hydrothermal systems, *J. Geophys. Res.*, *108*(89), 2424, doi: 10.1029/2002JB002371.

Luther, G. W., III, T. F. Rozan, M. Taillefert, D. B. Nuzzio, C. Di Meo, T. Shank, R. Lutz, and C. Cary (2001), Chemical speciation drives hydrothermal vent ecology, *Nature*, *410*, 813–816.

Lutz, R. A., T. M. Shank, D. J. Fornari, R. M. Haymon, M. D. Lilley, K. L. Von Damm, and D. Desbruyeres (1994), Rapid growth at deep-sea vents, *Nature*, *371*, 663–664.

Macdonald, K. C. (1982), Mid-ocean ridges: Fine scale tectonic, volcanic and hydrothermal processes within the plate boundary zone, *Annu. Rev. Earth Planet. Sci.*, *10*, 155–190.

Macdonald, K. C. (1998), Linkages between faulting, volcanism, hydrothermal activity and segmentation on fast spreading centers, in *Faulting and Magmatism at Mid-Ocean Ridges*, *Geophys. Monogr. Ser.*, vol. 106, edited by W. R. Buck, et al., pp. 27–58, AGU, Washington, D. C.

Martin, J. T., and R. P. Lowell (2000), Precipitation of quartz during high-temperature, fracture controlled hydrothermal upflow at ocean ridges: Equilibrium versus linear kinetics, *J. Geophys. Res.*, *105*, 869–882.

Massoth, G. J., H. B. Milburn, S. R. Hammond, D. A. Butterfield, R. E. McDuff, and J. E. Lupton (1989), The geochemistry of submarine venting fluids at Axial Volcano, Juan de Fuca Ridge: New sampling methods and a VENTS program rationale, in *Global Venting, Midwater, and Benthic Ecological Processes, National Undersea Research Program Res. Rep.*, 88-4, edited by M. P. De Luca and I. Babb, pp. 29–59, NOAA.

Morgan, J. P., and Y. J. Chen (1993a), Dependence of ridge-axis morphology on magma supply and spreading rate, *Nature*, *364*, 706–708.

Morgan, J. P., and Y. J. Chen (1993b), The genesis of oceanic crust: Magma injection, hydrothermal circulation, and crustal flow, *J. Geophys. Res.*, *98*, 6283–6297.

Mottl, M. J. (1983), Metabasalts, axial hot springs, and the structure of hydrothermal systems at mid-ocean ridges, *Geol. Soc. Am. Bull.*, *94*, 161–180.

Mottl, M. J., H. D. Holland, and R. F. Corr (1979), Chemical exchange during hydrothermal alteration of basalt by seawater II: Experimental results for Fe, Mn, and sulfur species, *Geochim. Cosmochim. Acta*, *43*, 869–884.

Mullineaux, L. S., C. R. Fisher, C. H. Peterson, and S. W. Schaeffer (2000), Tubeworm succession at hydrothermal vents: Use of biogenic cues to reduce habitat selection error?, *Oecologia*, *123*, 275–284.

Ohmoto, H., and A. C. Lasaga (1982), Kinetics of reactions between aqueous sulfates and sulfides in hydrothermal systems, *Geochim. Cosmochim. Acta*, *46*, 1727–1745.

Perram, L. J., and K. C. Macdonald (1994), An overlapping propagating spreading center at 87°30′W on the Galapagos Spreading Center, *Earth Planet. Sci. Lett.*, *121*, 195–211.

Saccocia, P. J., and K. M. Gillis (1995), Hydrothermal upflow zones in the oceanic crust, *Earth Planet. Sci. Lett.*, *136*, 1–16.

Seewald, J. S., and W. E. Seyfried Jr. (1990), The effect of temperature on metal mobility in subseafloor hydrothermal systems: Constraints from basalt alteration experiments, *Earth Planet. Sci. Lett.*, *101*, 388–403.

Seewald, J., A. Cruse, and P. Saccocia (2003), Aqueous volatiles in hydrothermal fluids from the Main Endeavour Field, northern Juan de Fuca Ridge: Temporal variability following earthquake activity, *Earth Planet. Sci. Lett.*, *216*, 575–590.

Seewald, J. S., M. Y. Zolotov, and T. McCollom (2006), Experimental investigation of single carbon compounds under hydrothermal conditions, *Geochim. Cosmochim. Acta*, *70*, 446–460.

Seyfried, W. E., Jr. (1987), Experimental and theoretical constraints on hydrothermal alteration processes at mid-ocean ridges, *Annu. Rev. Earth Planet. Sci.*, *15*, 317–335.

Seyfried, W. E., Jr., and K. Ding (1995), Phase equilibria in subseafloor hydrothermal systems: A review of the role of redox, temperature, pH and dissolved Cl on the chemistry of hot spring fluids at mid-ocean ridges, in *Seafloor Hydrothermal Systems: Physical, Chemical, Biological and Geological Interactions*, *Geophys. Monogr. Ser.*, vol. 91, edited by S. E. Humphris, R. A. Zierenberg, L. S. Mullineaux and R. E. Thompson, pp. 248–273, AGU, Washington, D. C.

Seyfried, W. E., Jr., D. R. Janecky, and M. J. Mottl (1984), Alteration of the oceanic crust: Implications for geochemical cycles of lithium and boron, *Geochim. Cosmochim. Acta*, *48*, 557–569.

Seyfried, W. E., Jr., J. S. Seewald, M. E. Berndt, K. Ding, and D. I. Foustoukos (2003), Chemistry of hydrothermal vent fluids from the Main Endeavour Field, Northern Juan de Fuca Ridge: Geochemical controls in the aftermath of June 1999 seismic events, *J. Geophys. Res.*, *108*(B9), 2429, doi: 10.1029/2002JB001957.

Shank, T., D. Fornari, K. L. Von Damm, M. Lilley, R. Haymon, and R. A. Lutz (1998), Temporal and spatial patterns of biological community development at nascent deep-sea hydrothermal vents (9°50′N, East Pacific Rise), *Deep-Sea Res. II*, *45*, 465–515.

Shank, T., et al. (2003), Deep submergence synergy: Alvin and ABE explore the Galapagos Rift at 86°W, *Eos Trans. AGU*, *84*, 425, doi: 10.1029/2003EO410001.

Shock, E. L., and C. M. Koretsky (1993), Metal-organic complexes in geochemical processes; calculation of standard partial molal thermodynamic properties of aqueous acetate complexes at high pressures and temperatures, *Geochim. Cosmochim. Acta*, *57*, 4899–4922.

Shock, E. L., D. C. Sassani, M. Willis, and D. A. Sverjensky (1997), Inorganic species in geologic fluids: Correlations among standard molal thermodynamic properties of aqueous ions and hydroxide complexes, *Geochim. Cosmochim. Acta*, *61*, 907–950.

Sinton, J. M., and R. S. Detrick (1992), Mid-ocean ridge magma chambers, *J. Geophys. Res.*, *97*, 197–216.

Sleep, N. H. (1975), Formation of oceanic crust: Some thermal constraints, *J. Geophys. Res.*, *80*, 4037–4042.

Sohn, R. A., D. J. Fornari, K. L. Von Damm, J. A. Hildebrand, and S. C. Webb (1998), Seismic and hydrothermal evidence for a cracking event on the East Pacific Rise crest at 9°50′N, *Nature*, *396*, 159–161.

Sohn, R. A., J. A. Hildebrand, and S. C. Webb (1999), A microearthquake survey of the high-temperature vent fields on the volcanically active East Pacific Rise (9°50′N), *J. Geophys. Res.*, *104*, 25,367–25,377.

Spiess, F. N., K. C. Macdonald, T. Atwater, R. Ballard, A. Carranza, D. Cordoba, C. Cox, V. M. Diazgarcia, J. Francheteau, J. Guerrero, J. Hawkins, R. Haymon, R. Hessler, T. Juteau, M. Kastner, R. Larson, B. Luyendyk, J. D. Macdougall, S. Miller, W. Normark, J. Orcutt, and C. Rangin (1980), East Pacific Rise: Hot springs and geophysical experiments, *Science*, *207*, 1421–1433.

Sverjensky, D. A., E. L. Shock, and H. C. Helgeson (1997), Prediction of the thermodynamic properties of aqueous metal complexes to 1000°C and 5 kb, *Geochim. Cosmochim. Acta*, *61*, 1359–1412.

Tivey, M. K., and J. R. Delaney (1986), Growth of large structures on the Endeavour segment of the Juan de Fuca Ridge, *Earth Planet. Sci. Lett.*, *77*, 303–317.

Tivey, M. A., and H. P. Johnson (2002), Crustal magnetization reveals subsurface structure of Juan de Fuca Ridge hydrothermal vent fields, *Geology*, *30*, 979–982.

Tsurumi, M., and V. Tunnicliffe (2001), Characteristics of a hydrothermal vent assemblage on a volcanically active segment of Juan de Fuca Ridge, northeast Pacific, *Can. J. Fish Aquat. Sci.*, *58*, 530–542.

Tunnicliffe, V., R. W. Embley, J. F. Holden, D. A. Butterfield, G. J. Massoth, and S. K. Juniper (1997), Biological colonization of new hydrothermal vents following an eruption on Juan de Fuca Ridge, *Deep-Sea Res. I*, *44*, 1627–1644.

Van Ark, E. M., R. S. Detrick, J. P. Canales, S. M. Carbotte, A. J. Harding, G. M. Kent, M. R. Nedimovic, W. S. D. Wilcock, J. B. Diebold, and J. M. Babcock (2007), Seismic structure of the Endeavour segment, Juan de Fuca Ridge: Correlations with seismicity and hydrothermal activity, *J. Geophys. Res.*, *112*, B02401, doi: 10.1029/2005JB004210.

Von Damm, K. L. (1988), Systematics of and postulated controls on submarine hydrothermal solution chemistry, *J. Geophys. Res.*, *93*, 4551–4562.

Von Damm, K. L. (1995), Controls on the chemistry and temporal variability of seafloor hydrothermal fluids, in *Seafloor Hydrothermal Systems: Physical, Chemical, Biological and Geological Interactions*, *Geophys. Monogr. Ser.*, vol. 91, edited by S. E. Humphris, et al., pp. 222–248, AGU, Washington, D. C.

Von Damm, K. L. (2000), Chemistry of hydrothermal vent fluids from 9°–10°N, East Pacific Rise: "Time zero," the immediate posteruptive period, *J. Geophys. Res.*, *105*, 11,203–11,222.

Von Damm, K. L. (2004), Evolution of the hydrothermal system at East Pacific Rise 9°50′N: Geochemical evidence for changes in the upper oceanic crust, in *Mid-Ocean Ridges: Hydrothermal Interactions Between the Lithosphere and Oceans*, *Geophys. Monogr. Ser.*, vol. 148, edited by C. R. German, J. Lin, and L. M. Parson, pp. 285–305, AGU, Washington, D. C.

Von Damm, K. L., J. L. Edmond, B. Grant, C. I. Measures, B. Walden, and R. F. Weiss (1985), Chemistry of submarine hydrothermal solutions at 21°N, East Pacific Rise, *Geochim. Cosmochim. Acta*, *49*, 2221–2237.

Von Damm, K. L., J. L. Bischoff, and R. J. Rosenbauer (1991), Quartz solubility in hydrothermal seawater; an experimental study and equation describing quartz solubility for up to 0.5 M NaCl solutions, *Am. J. Sci.*, *291*, 977–1007.

Von Damm, K. L., M. Lilley, W. C. Shanks, M. Brockington, A. Bray, K. M. O'Grady, E. Olson, A. Graham, and G. Proskurowski (2003), Extraordinary phase separation and segregation in vent fluids from the southern East Pacific Rise, *Earth Planet. Sci. Lett.*, *206*, 365–378.

Von Damm, K. L., and M. D. Lilley (2004), Diffuse flow hydrothermal fluids from 9°50′N East Pacific Rise: Origin, evolution and biogeochemical controls, in *The Subseafloor Biosphere at Mid-Ocean Ridges*, *Geophys. Monogr. Ser.*, vol. 144, edited by W. S. Wilcock, et al., pp. 245–268, AGU, Washington, D. C.

Welhan, J. A., and H. Craig (1979), Methane and hydrogen in East Pacific Rise hydrothermal fluids, *Geophys. Res. Lett.*, *6*, 829–831.

Wetzel, L. R., and E. L. Shock (2000), Distinguishing ultramafic from basalt-hosted submarine hydrothermal systems by comparing calculated vent fluid compositions, *J. Geophys. Res.*, *105*, 8319–8340.

Wilcock, W. S. D., and J. R. Delaney (1996), Mid-ocean ridge sulfide deposits: Evidence for heat extraction from magma chambers or cracking fronts?, *Earth Planet. Sci. Lett.*, *145*, 49–64.

N. J. Pester, Department of Geology and Geophysics, University of Minnesota, Minneapolis, Minnesota 55455, USA. (peste005@umn.edu)

Hydrothermal Fluid Composition at Middle Valley, Northern Juan de Fuca Ridge: Temporal and Spatial Variability

Anna M. Cruse and Jeffrey S. Seewald

Department of Marine Chemistry and Geochemistry, Woods Hole Oceanographic Institution, Woods Hole, Massachusetts, USA

Peter J. Saccocia

Department of Earth Sciences, Bridgewater State College, Bridgewater, Massachusetts, USA

Robert Zierenberg

Geology Department, University of California, Davis, California, USA

Hydrothermal fluids were collected in July 2000 from the Dead Dog and Ore Drilling Program (ODP) Mound vent fields at Middle Valley, a sediment-covered spreading center on the northern Juan de Fuca Ridge. Measured fluid temperatures varied from 187° to 281°C in focused flow vents and 40°C in ODP Hole 1035F. Cl concentrations indicate that ODP Mound fluids undergo phase separation in the subsurface, whereas Dead Dog fluids do not. The lack of phase separation at Dead Dog is consistent with other geochemical indicators of lower subsurface temperatures. Cooling and equilibration with quartz after phase separation at the ODP Mound results in exit temperatures and silica concentrations that are indistinguishable from those at Dead Dog. The sulfur isotopic composition of aqueous ΣH_2S indicates extensive reduction of seawater SO_4 and leaching of basaltic sulfur at both areas.

A new area of venting, which resulted from drilling operations during ODP Leg 169, was discovered on the eastern side of the ODP Mound. The fluids in the new area have compositions that are similar to those of Hole 1035H and Shiner Bock, except for lower H_2 and higher H_2S concentrations. These differences reflect the conversion of pyrite to pyrrhotite in the ODP Mound as fluids react with sulfide minerals during upflow.

Fluid temperatures and compositions have remained constant between 1990 and 2000 indicating that subsurface reaction zone conditions did not change over this period. Near constant concentrations of sediment-derived mobile trace elements suggest that the residence time of fluids in a high-temperature reservoir exceeds 10 years.

Magma to Microbe: Modeling Hydrothermal Processes at Ocean Spreading Centers
Geophysical Monograph Series 178

10.1029/178GM08

1. INTRODUCTION

Middle Valley, northern Juan de Fuca Ridge, is a sedimented mid-ocean ridge spreading center with two regions of hydrothermal venting: the Dead Dog and Ore Drilling Program (ODP) Mound fields (Figure 1). Within the relatively limited number of sedimented vent systems studied to date, Middle Valley is unique due to the formation of the large (>8.8 × 10^6 tons) Bent Hill massive sulfide deposit [*Zierenberg et al.*, 1998]. Numerous investigations, including two drilling expeditions (ODP Legs 139 and 169), have demonstrated that the presence of up to 1.5 km of hemipelagic and turbiditic sediments overlying the ridge axis plays an important role in regulating the chemical and physical evolution of hydrothermal fluids, the trapping of ore-forming metals, flow patterns of circulating fluids, and heat and mass fluxes between the lithosphere and water column. The composition of Middle Valley vent fluids sampled in 1990 and 1991 shows clear evidence of reaction with both sediment and basalt [*Butterfield et al.*, 1994a]. Maximum measured temperatures of 276°C, however, were substantially lower than the 350° to 400°C estimated for the formation of the Bent Hill deposit [*Goodfellow and Franklin*, 1993; *Peter et al.*, 1994], suggesting that the temperature and composition of vent fluids have changed over time.

The characterization of vent fluid composition and temporal variability is key to the assessment of massive sulfide formation at ridge crests and the impact of hydrothermal activity on global biogeochemical cycles. Studies on vent fluids from sedimented and unsedimented ridge crest hydrothermal systems have demonstrated that fluid composition can be remarkably stable over time scales of several years [*Butterfield et al.*, 1994b; *Campbell et al.*, 1988; *Von Damm et al.*, 2005], whereas other studies indicate that vent fluid temperatures and chemical compositions may fluctuate on time scales of weeks in response to magmatic activity [*Butterfield and Massoth*, 1994; *Von Damm et al.*, 1995; *Lilley et al.*, 2003; *Seewald et al.*, 2003; *Seyfried et al.*, 2003].

In addition to the characterization of temporal changes in hydrothermal activity, studies on spatial variability can greatly increase our understanding of chemical and physical processes that regulate vent fluid chemistry. Middle Valley represents an unprecedented opportunity in this regard due to extensive drilling in 1991 and 1996 during ODP Legs 139 and 169, respectively. Drilling at Middle Valley appears to have influenced the subsurface hydrologic regime. For example, venting of hydrothermal fluid was immediately observed in Holes 1035F and 1035H after drilling in 1996 [*Shipboard Scientific Party*, 1998a]. Moreover, a new area of venting was initiated on the eastern side of the ODP Mound in response to the penetration of an impermeable seal during drilling of Hole 1035H [*Cruse and Seewald*, 2006]. In contrast to drilling-induced venting, an underpressured reservoir was penetrated during drilling of Hole 858F in 1991, allowing seawater to be drawn into the formation at a rate of thousands of gallons per minute [*Shipboard Scientific Party*, 1992b].

We report the results of a vent fluid sampling program conducted at Middle Valley in 2000. The data are compared with samples collected in 1990 and 1991 [*Butterfield et al.*, 1994a] to provide a decade-long record of fluid chemistry and are used to constrain geochemical models describing processes that occur during the interaction of a hydrothermal fluid with a disequilibrium mineral assemblage in the upflow zone. Data for the chemical and isotopic composition of low molecular weight aqueous hydrocarbons and other volatile species (H_2, ΣCO_2, and ΣNH_3) in the 2000 samples are discussed by *Cruse and Seewald* [2006].

2. GEOLOGIC SETTING

Middle Valley is an axial rift valley located at 48°27′N (latitude) on the northern edge of the Juan de Fuca Ridge, an intermediate-spreading ridge with a half-spreading rate of 30 mm/yr [*Goodfellow and Blaise*, 1988; Figure 1]. Active spreading jumped westward from Middle Valley to West Valley during the last 200,000 years, and perhaps as late as 10,000 years ago [*Karsten et al.*, 1986]. Middle Valley was completely filled with terrigenous and hemipelagic sediments derived from the continental shelf during Pleistocene sea-level lowstands. The sediment cover ranges in thickness from a few hundred meters in the south to 1.5 km in the north [*Goodfellow and Blaise*, 1988]. Sedimentation was contemporaneous with active extension and volcanism, so that the uppermost basement consists of a basaltic-sill/sediment complex [*Davis and Villinger*, 1992]. Past hydrothermal activity has resulted in the creation of two massive sulfide deposits that formed from the direct precipitation of sulfide minerals on the seafloor: the Bent Hill massive sulfide deposit, which is estimated to contain over 8.8 million tons of ore, and the slightly younger ODP Mound massive sulfide deposit, which was sampled during the drilling of ODP Hole

Figure 1. (Opposite) Location maps of (a) Middle Valley, Juan de Fuca Ridge, (b) Dead Dog Vent Field, and (c) ODP Mound Vent Field. Sampled vents are indicated with triangles, and ODP drillholes are indicated with circles. Contours in panel (b) are drawn at 2-m intervals. Figures are modified from *Zierenberg and Miller* [2000].

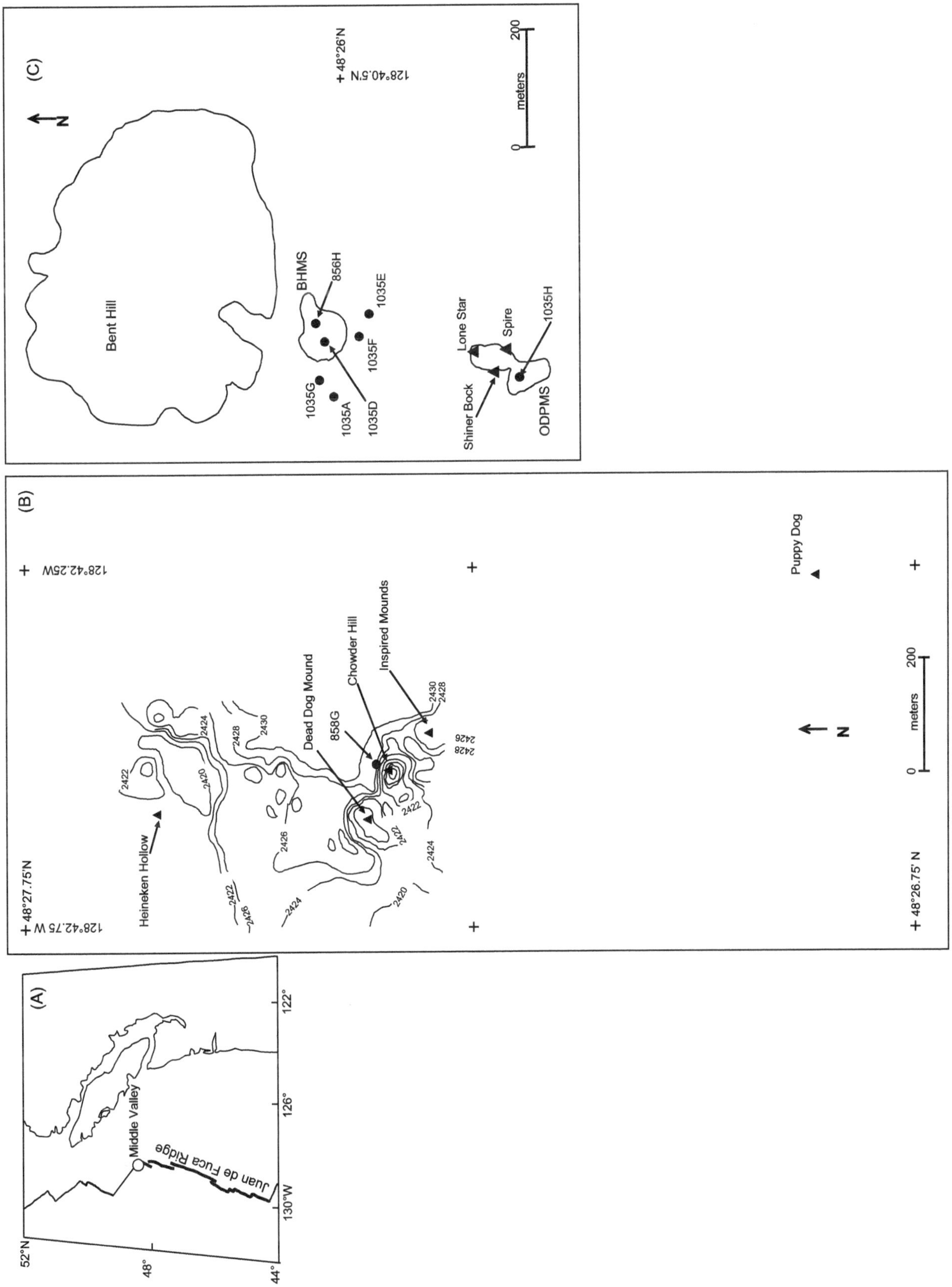
(A)
52°N
48°
44°
130°W
126°
122°
Middle Valley
Juan de Fuca Ridge
(B)
+ 48°27.75'N
128°42.75 W
128°42.25W
+ 48°26.75' N
Heineken Hollow
Dead Dog Mound
858G
Chowder Hill
Inspired Mounds
Puppy Dog
2420
2422
2424
2426
2428
2430
N
0
200
meters
(C)
+ 48°26'N
128°40.5'N
Bent Hill
BHMS
856H
1035E
1035F
1035G
1035A
1035D
Lone Star
Spire
1035H
Shiner Bock
ODPMS
N
0
200
meters

1035H [*Shipboard Scientific Party,* 1998a; *Zierenberg et al.*, 1998]. Mineralogical and fluid inclusion data from sulfides in the Bent Hill deposit indicate precipitation of the sulfide minerals from hydrothermal fluids that reached temperatures of 350° to 400°C [*Goodfellow and Franklin,* 1993; *Peter et al.*, 1994]. Such fluids were apparently derived from a hot reaction zone located in the basaltic basement with upflow focused along faults and fissures that were subsequently sealed by mineral precipitation [*Zierenberg et al.*, 1998].

Multiple episodes of hydrothermal venting are apparent at the ODP Mound, which consists of three stacked sequences of massive sulfide, each of which are underlain by sediment-hosted feeder-zone mineralization [*Shipboard Scientific Party*, 1998a; *Zierenberg et al.*, 1998]. Within the lowermost feeder zone is a narrow band rich in Cu-Fe sulfides that trends northward to the Bent Hill massive sulfide deposits, and is referred to as the Deep Copper Zone (DCZ) [*Shipboard Scientific Party*, 1998a; *Zierenberg et al.*, 1998]. The top of the DCZ is marked by an area of intensely silicified sediments that forms an impermeable caprock preventing hydrothermal fluids from reaching the seafloor, except when it is penetrated by fracturing or drilling (see below; *Zierenberg et al.*, 1998). Within the feeder zones, sulfide mineralization largely consists of veins and bands of pyrrhotite, chalcopyrite/isocubanite, pyrite, and minor sphalerite contained within hydrothermally altered mudstone and sandstones [*Shipboard Scientific Party*, 1998a]. Late anhydrite veins that crosscut the sulfide minerals are present within some sections of the feeder zones. Overlying the feeder zones are massive to semimassive sulfide assemblages that consist predominantly of sphalerite-pyrrhotite-pyrite-magnetite that is highly variable in terms of the relative mineral abundances [*Shipboard Scientific Party*, 1998a]. The topmost interval of Hole 1035H consists of sulfide breccias of pyrite, marcasite, Fe-rich sphalerite, minor Cu sulfides, and pyrrhotite that formed largely in situ with little or no sedimentary transport of the sulfides [*Shipboard Scientific Party*, 1998a]. Both pyrite and marcasite have been observed to replace primary hexagonal pyrrhotite in this facies and could be related to postdepositional weathering of the sulfide by seawater-like fluids [*Shipboard Scientific Party*, 1998a].

Current hydrothermal activity is localized in two areas at a water depth of 2400 m: the ODP Mound vent field and the Dead Dog Vent Field (Figure 1). Fluids sampled in July 2000 at the ODP Mound were collected from Shiner Bock vent, ODP Leg 169 drillhole 1035H, and Spire vent (Table 1). The Spire vent is located on the eastern side of the ODP Mound in an area characterized by small spire and beehive structures. Active venting was not observed during *Alvin* operations in 1998 (Zierenberg, unpublished data), and therefore was less than 2 years old when it was sampled as part of this study in 2000. Reactivation of fluid flow through the mound along preexisting conduits likely exposed the constituent sulfide mineral assemblage to fluids that are compositionally distinct from those responsible for their formation during earlier phases of venting. Fluids venting from ODP Leg 169 drillhole 1035F, located near the Bent Hill massive sulfide deposit, were also sampled. Measured exit temperatures were 272°, 263°, and 267°C for Shiner Bock, Spire, and 1035H, respectively, and only 40°C for the fluids venting from 1035F (Table 2). Within the Dead Dog Vent Field, fluids were sampled from five regions of focused venting at Heineken Hollow, Dead Dog Mound, Inspired Mounds, Chowder Hill, and Puppy Dog, where maximum exit temperatures were 187°, 274°, 281°, 261°, and 202°C, respectively (Table 2).

3. METHODS

Vent fluid samples were collected using isobaric gastight fluid samplers [*Seewald et al.*, 2002b] deployed from the submersible *Alvin.* These samplers prevent degassing of fluids before processing by maintaining seafloor pressures onboard ship while subsamples are withdrawn through a micrometering valve. With the exception of Dead Dog, at least two different gastight samples were collected at each vent. Additional samples were also taken at some vents using the 755-mL titanium syringe-style "major" samplers, which are not gastight [*Edmond et al.*, 1992; *Von Damm et al.*, 1985]. Temperatures were measured using the *Alvin* high-temperature probe and zero-point corrected by 2°C to account for the cold-junction temperature.

Fluids were subsampled into acid-cleaned polyethylene bottles and stored for shore based analyses. Aliquots for trace metal analysis were acidified with Optima® HCl to a pH <1. For selected samples, metal precipitates that had formed in the sampler were collected by filtration. An aliquot of the acidified trace metal sample was diluted 100-fold at sea for shore-based aqueous SiO_2 analysis. Dissolved concentrations of Na, Ca, K, Cl, Br, and SO_4 were determined using ion chromatography with conductivity detection. Because precautions were not taken to prevent oxidation of reduced sulfur species during storage, measured SO_4 concentrations represent maximum values. ΣNH_3 ($NH_{3(aq)}$ + NH_4^+) concentrations were measured spectrophotometrically by flow injection analysis using the Berthelot reaction [*Koroleff*, 1976]. Analytical uncertainties for the major ions and ΣNH_3 are estimated to be ±3 and ±5%, respectively. Dissolved Si concentrations were determined by inductively coupled plasma optical emission spectroscopy (ICP-OES), with estimated analytical uncertainties of ±5%. Dissolved Sr, Li, Rb, Cs, Fe, Mn, Zn, Cu, and Pb concentrations were determined by inductively coupled plasma mass spectrometry (ICP-MS). The estimated analytical uncertainties are ±2% for all species except Pb, which has an uncertainty of ±5%. Precipitates collected from the fluid sampler were dissolved in *aqua regia* before analysis and the measured amounts were added to concentrations measured in the acidified fluid samples to yield a final concentration, which is reported in Table 1. For vents that had fluid samples with and without sampler particulates included in the reported concentrations, only the former concentrations were used for regression of end-member values.

Aqueous ΣCO_2 ($CO_{2(aq)}$ + HCO_3^- + CO_3^{2-}), H_2, and CH_4 concentrations were measured using gas chromatography, with estimated analytical uncertainties of <10% for ΣCO_2 and <4% for H_2 and CH_4. A Ross® combination electrode was used to measure pH (25°C). Aqueous ΣH_2S ($H_2S_{(aq)}$ + HS^- + S^{2-}) concentrations were determined gravimetrically after precipitation as Ag_2S from a 5 wt % $AgNO_3$ solution. The analytical uncertainty is estimated to be <10%.

Because fluid samplers have a finite dead volume that is filled with bottom seawater before deployment, and seawater entrainment occurs to varying degrees during sampling, the fluid that is collected represents a mixture between bottom seawater and hydrothermal vent fluid. Laboratory experiments have demonstrated near-quantitative removal of Mg from seawater during hydrothermal seawater-sediment and seawater-basalt interaction at temperatures, pressures, and water/rock ratios that exist in ridge-crest hydrothermal systems [*Bischoff and Dickson*, 1975; *Mottl and Holland*, 1978; *Seyfried and Bischoff*, 1981b; *Thornton and Seyfried*, 1987; *Seewald et al.*, 1990]. Therefore, the composition of end-member hydrothermal vent fluids is calculated by extrapolating the concentrations of individual species to zero Mg using linear least squares regression of vent fluid and seawater compositions (Table 2, Figure 2).

The stable sulfur isotopic composition of aqueous ΣH_2S was measured at the University of California at Davis using an Isoprime® continuous flow gas ratio mass spectrometer interfaced with an automated Carlo Erba® elemental analyzer (EA). Ag_2S precipitates were combusted to SO_2 in a stream of high purity oxygen in the EA. The SO_2 was carried by a He stream through a chromatography column to separate it from other gases before being introduced into the mass spectrometer. Isotopic compositions are reported relative to the Canyon Diablo Troilite (CDT) standard (R_{CDT}; $^{34}S/^{32}S$ = 0.0450045; *Ohmoto and Rye*, 1979) using International Atomic Energy Agency reference materials S-1, S-2, and S-3 with accepted values of –0.30, 22.67, and –32.55‰, respectively [*Ding et al.*, 2001b]. Isotope ratios are reported using standard delta notation, where $\delta^{34}S$ is expressed in per mil (‰) units, and is calculated as the relative difference in the $^{34}S/^{32}S$ ratio of the sample, R_s, and CDT standard, R_{CDT}:

$$\delta^{34}S(‰) = \left[\frac{R_s - R_{CDT}}{R_{CDT}}\right] \times 1000 \quad (1)$$

The precision of the $\delta^{34}S$ analyses was 0.6‰ (1σ). One or two aliquots of Ag_2S were analyzed for each fluid sample collected. For a given vent, all measured Ag_2S values were averaged to give a single sulfur isotopic composition of aqueous ΣH_2S (Table 2).

Samples from the base of the chimney that formed over Hole 1035H were sampled on Alvin Dive 3239 in July 1998 after the chimney collapsed after insertion of the Alvin T probe into the vent. The chimney wall was composed of coarse bladed anhydrite, with only minor amounts of fine-gained hexagonal pyrrhotite grains. Clear anhydrite grains that were free of sulfide inclusions were handpicked under a binocular microscope and analyzed for sulfur isotopes at the University of California at Davis as described above for the AgS_2 precipitates.

4. RESULTS

A total of 33 fluid samples were collected during the 2000 cruise. For most vents, at least one of the sampled fluids had a measured Mg concentration of <20% of the seawater concentration, greatly reducing the uncertainty associated with extrapolating to zero Mg end-member values (Table 2, Figure 2). For ODP Hole 1035F, however, the lowest Mg concentration in the three samples collected was 43.2 mmol/kg (Table 1). Accordingly, significantly greater uncertainty is associated with extrapolated end-member concentrations for these fluids. Measured pH (25°C) in low Mg samples (<10 mmol/kg) varied from 5.2 to 6.3. There is no clear difference between measured pH (25°C) at the Dead Dog Vent Field and the ODP Mound. Aqueous silica concentrations range between 5.8 and 12 mmol/kg fluid, with no difference between the two vent fields.

Calculated end-member compositions in 2000 are listed in Table 2, along with end-member compositions for 1990 that were calculated from the data of *Butterfield et al.* [1994a]. Generally speaking, the data for each vent were characterized by high correlation coefficients, with the notable exceptions of some of the trace metals (Figure 2, Table 3). The end-member compositions of fluids from the Puppy Dog vent are almost identical to those of fluids from within the Dead Dog Vent Field, which is located ~800 m to the north (Figure 1b). This similarity indicates that the fluids are derived from the same source. Similarly, the end-member composition of

Table 1. Concentrations of Aqueous Species in Vent Fluids From Middle Valley, Northern Juan de Fuca Ridge.

Vent	Sample	Mg, m	Na,[a] m	K, m	Ca, m	Si, m	Li, μ	Rb, μ	Cs, μ	Sr, μ	Fe, μ	Mn, μ
Dead Dog												
Heineken	BGT-3597-1	51.0	458	10.6	12.9	0.32	40	1.8	0.04	93	3	2
Hollow	BGT-3597-2	2.37	415	19.1	78.4	8.72	486	23	1.5	245	12	65
	M-3596-12C	3.13	416	19.4	72.5		479	23	1.4	245	22	62
	M-3596-1C	2.25	407	19.8	74.2		490	23	1.4	250	19	64
	M-3597-12	45.9	455	11.9	23.2		101	4.8	0.23	120	11	10
	M3597-15	27.9	422	15.0	45.7		245	12	0.73	176	5.1	32
Dead Dog	M-3597-1C	1.51	416	19.7	73.9		483	23	1.5	275	19	72
	M-3597-12C	2.03	417	19.6	73.2		473	23	1.4	276	22[b]	72
Chowder Hill	BGT-3596-3	1.90	425	19.7	63.8		506	24	1.4	274	260[b]	70
	BGT-3596-4	1.12	430	19.7	63.2	11.6	504	24	1.4	273	32[b]	67
Inspired	BGT-3597-3	34.4	437	13.6	34.3	3.25	189	9.3	0.52	157	29[b]	28
Mounds	BGT-3597-4	41.0	437	12.4	26.7	2.49	134	6.2	0.34	133	25[b]	17
	M-3597-6C	30.3	418	14.3	40.8		223	11	0.65	171	49	40
	M-3597-11C	43.0	454	12.3	25.9		118	5.8	0.31	131	16	16
Puppy Dog	BGT-3599-3	6.16	423	18.5	67.4	7.15	417	21	1.3	254	74[b]	49
	BGT-3599-4	44.6	445	11.8	24.2	1.18	92	4.6	0.21	125	10	11
Bent Hill												
1035F	BGT-3598-3	43.2	449	10.7	14.1	1.23	63	2.7	0.10	99	14	50
	M-3598-6C	46.7	454	10.9	15.4		66	2.7	0.10	99	16	51
	M-3598-11C	46.7	441	10.6	15.7		65	2.8	0.10	100	19	54
ODP Mound												
Shiner Bock	BGT-3595-1	1.61	354	13.7	34.0	10.3	345	16	0.94	147	55[b]	136
	BGT-3595-2	3.40	357	13.5	33.8	9.85	334	15	0.89	148	62[b]	133
	M-3595-11C	5.99	356	13.3	36.0		315	14	0.82	147	54	129
	M-3595-6C	12.7	379	13.1	32.7		298	13	0.76	143	48	119
	M-3595-1C	28.7	406	11.8	27.8		180	8.1	0.44	129	93	66
	M-3595-12	3.10	361	14.2	36.2		339	15	0.93	152	35	161
Spire	BGT-3595-3	1.64	355	13.3	39.3	10.5	341	15	0.68	168	29	169
	BGT-3595-4	22.5	395	12.2	26.9	5.68	212	9.5	0.49	125	19	107
	M-3595-15	4.12	354	13.6	36.0		324	15	0.86	146	35	132
1035H	BGT-3599-1	17.0	394	12.2	28.8	6.58	221	10	0.58	132	125[b]	146
	BGT-3599-2	20.9	396	12.1	27.1	5.82	198	9.1	0.52	126	150[b]	131
	M-3599-12	29.3	421	11.7	24.4		154	7.0	0.41	119	36	98
	M-3599-15	30.3	419	11.5	23.7		152	7.0	0.39	117	31	93
BSW	M-3598-6C	53.0	459	10.4	11.3	0.18	28	2.0	0.003	87	0	0

Vent	Sample	Cu, μ	Zn, μ	Pb, n	Cl, m	Br, μ	SO_4^{2-}, m	H_2, m	H_2S, m	NH_4, μ	pH, n
Dead Dog											
Heineken	BGT-3597-1	0.3	0.13		544	815	25.6	0.03	0.2	64	7.2
Hollow	BGT-3597-2		0.54	16	589	1018	2.53	1.5	3.2	2792	5.8
	M-3596-12C	0.3	1.1	108	581	1042	2.04			2855	
	M-3596-1C	4.1	1.2	116	576	1020	1.26			2996	
	M-3597-12	0.1	0.12	51	557	841	23.8			282	5.7
	M3597-15	0.9	0.25	0.8	554	914	14.6			1350	6.0
Dead Dog	M-3597-1C	0.5	1.1	104	584	1016	0.83		3.0	3076	5.8
	M-3597-12C	0.4[b]	1.4[b]	99[b]	585	1037	1.11		2.9	2933	6.4
Chowder Hill	BGT-3596-3	6. 4[b]	7.2[b]	523[b]	572	1003	1.41	2.5	3.6	3045	5.9
	BGT-3596-4	33[b]	1.2[b]	161[b]	576	1026	0.68	2.6	3.7	3103	5.8

Table 1. (continued)

Vent	Sample	Cu, μ	Zn, μ	Pb, n	Cl, m	Br, μ	SO_4^{2-}, m	H_2, m	H_2S, m	NH_4, μ	pH, n
Dead Dog											
Inspired	BGT-3597-3	7.7[b]	6.8[b]	937[b]	551		18.4	0.81	0.8	993	6.1
Mounds	BGT-3597-4	4.8[b]	6.4[b]	1832[b]	542	871	21.2	0.53		524	6.2
	M-3597-6C	7.7	20	3244	541	912	16.4			1251	
	M-3597-11C	4.6	3.9	880	557		23.4			551	
Puppy Dog	BGT-3599-3	2.0[b]	2.7[b]	299[b]	582	975	3.29	1.8	3.4	2646	5.7
	BGT-3599-4	1.6	1.5	231	544		25.1	0.3	0.7	351	6.5
Bent Hill											
1035F	BGT-3598-3		1.1	11	527	806	23.2	0.39	0.3	232	6.4
	M-3598-6C				542	811	23.4		0.3	160	6.3
	M-3598-11C				528	821	23.9		0.3	121	
ODP Mound											
Shiner Bock	BGT-3595-1	2.4[b]	2.2[b]	75[b]	435	730	1.74	8.0	4.3	2262	5.3
	BGT-3595-2	2.7[b]	3.1[b]	85[b]	439	745	2.89	7.6	3.9	2099	5.4
	M-3595-11C	2.6	2.2	44	444	785	4.34			2130	
	M-3595-6C	1.5	1.9	36	467	778	7.35			1800	
	M-3595-1C	8.1	3.9	49	495	785	17.7			1070	
	M-3595-12	2.3	1.6	158	443	752	2.11			2240	
Spire	BGT-3595-3	0.3	0.4	12	434	754	7.81	2.1	7.8	2107	6.0
	BGT-3595-4	3.6	0.2	2.0	479	774	13.2	1.2	4.5	1264	5.7
	M-3595-15	1.7	11	61	442	798	2.64			2215	
1035H	BGT-3599-1	1.4[b]	3.3[b]	52[b]	479	786	8.84	4.2	2.7	1446	5.5
	BGT-3599-2	5.0[b]	4.3[b]	92[b]	481	773	11.1	3.7	2.4	1262	5.6
	M-3599-12	0.4	1.2	6	508	792	15.8			938	
	M-3599-15	0.5	1.6	53	506	796	15.6			855	
BSW	M-3598-6C	0	0	0	543	820	27.2	0	0	0.8	

[a]Charge balance.

[b]Measured concentrations contain redissolved precipitate from sampler; see text for discussion.

BSW, bottom seawater; m, mmol/kg; μ, μmol/kg; n, nmol/kg, except H_2 and CH_4, which are mmol/L. [Mg] value for BGT-3597-2 is incorrectly reported as 11.3 in Table 1 of *Cruse and Seewald* [2006].

fluids from ODP Hole 1035F, located near the BHMS (Figure 1c) are indistinguishable from the ODP Mound fluids located 300 m to the south, consistent with a shared source fluid. Comparison of fluids from the Dead Dog Vent Field and the ODP Mound reveals significant differences indicating they do not share a common source.

4.1. Anions

End-member chloride concentrations are quite distinct for the two vent fields. Concentrations in fluids from the ODP Mound vent field are depleted relative to seawater, ranging from 432 to 449 mmol/kg fluid, whereas at the Dead Dog field concentrations are slightly enriched relative to seawater, ranging between 546 and 587 mmol/kg fluid (Table 2). Bromide concentrations are likewise depleted relative to seawater in the ODP Mound fluids, ranging from 742 to 769 μmol/kg fluid, and enriched in the Dead Dog fluids, ranging from 995 to 1037 μmol/kg fluid. Measured SO_4 concentrations extrapolate to values very close to zero in both fields, consistent with the composition of vent fluids from virtually all sedimented and unsedimented ridge-crest hydrothermal systems, where quantitative removal of seawater SO_4 in end-member vent fluids is observed.

4.2. Cations

The major cations (Na, Ca, K) are also depleted in ODP Mound fluids relative to Dead Dog fluids. For example, Na concentrations range between 349 and 388 mmol/kg fluid in ODP Mound fluids and between 378 and 418 mmol/kg fluid in Dead Dog fluids. Sodium concentrations are depleted relative to seawater in all the vent fluids, whereas Ca and K concentrations are enriched. Strontium concentrations are

Table 2. Time Series (1990–2000) of Measured Exit Temperatures and Calculated Endmember Vent Fluid Compositions From Middle Valley, Northern Juan de Fuca Ridge.

Vent	Year	Temp, °C	Na,[a] m	K, m	Ca, m	Si, m	Li, μ	Rb, μ	Cs, μ	Sr, μ	Fe, μ	Mn, μ	Cu, μ
Dead Dog													
Heineken	1990	184	411	19.3	74.4	7.7	580	33		260	53	73	1.0
Hollow	2000	187	408	19.9	78.8	9.1	506	24	1.5	256	18	67	2.3
Dead Dog[b]	1990	268	401	18.5	80.0	9.4	487	31	1.4	259	16	66	0.97
	2000	274	416	20.0	75.7		493	24	1.5	282	20	74	0.37
Chowder Hill[b]	1990	276	399	18.1	82.1	10	291	31		257	19		
	2000	281	426	20.0	65.1	12	519	25	1.5	279	33	70	20
Inspired	1990	254	402	18.8	79.7	10	418	33	1.4	257	5.4	59	1.4
Mounds[b]	2000	261	378	19.5	79.5	9.3	486	23	1.5	287	91	85	22
Puppy Dog[b]	2000	202	418	19.5	75.3	8.1	467	23	1.5	277	84	56	2.3
Bent Hill													
1035F	2000	40	388	12.5	34.6	5.8	268	6.7	0.65	165	104	337	
ODP Mound													
Lone Star	1990	265	319	13.5	40.4	10	371	22	1.0	159	16	79	0.8
Shiner Bock[b]	2000	272	350	13.9	37.5	11	359	16	0.96	155	61	150	2.8
Spire	2000	263	349	13.6	39.1	11	349	15	0.82	160	34	163	1.8
1035H[b]	2000	267	363	13.1	38.1	9.6	311	14	0.87	153	212	216	4.8

Vent	Year	Zn, μ	Pb, n	Cl, m	Br, μ	SO_4^{2-}, μ	H_2, m	H_2S, m	CO_2, m	CH_4, m	NH_4, μ	$\delta^{34}S$ ‰	$\delta^{34}S$ 1σ	$\delta^{34}S$ *n*
Dead Dog														
Heineken	1990	1.5		563	1087	13.2								
Hollow	2000	0.96	79	583	1036	0.65	1.9	3.4	8.2	22.6	3180	+7.2	0.7	3
Dead Dog[b]	1990	1.6	125	577	1058	1.95		1.2						
	2000	1.2	101	586	1034	0.07		3.1			3108			
Chowder Hill[b]	1990			579	1087	1.87								
	2000	4.3	350	575	1020	0.28	2.6	3.7	8.4	20.3	3164	+5.7		1
Inspired	1990	0.98		575	1055	2.21		4.3						
Mounds[b]	2000	21	4133	546	1037	2.26	2.3	2.4	12	21.6	2866	+6.8		1
Puppy Dog[b]	2000	3.0	339	586	995	0.49	2.0	3.8	8.4	18.5	2974			
Bent Hill														
1035F	2000	6.1	61	448	761	2.36	2.1	2.0	8.2	3.0	1263	+10.7	0.2	2
ODP Mound														
Lone Star	1990	0.69	49	412	768	1.05		2.4						
Shiner Bock[b]	2000	2.8	83	434	750	1.26	8.2	4.3	12	7.1	2340	+9.7	0.6	6
Spire	2000	5.1	33	432	768	3.88	2.2	7.9	13	6.7	2270	+3.2	0.4	2
1035H[b]	2000	5.9	109	449	759	0.53	6.1	4.0	11	5.9	2096	+10.4	0.4	3

[a]Charge balance.

[b]Calculated concentrations of Fe, Cu, Zn, and Pb based only on combined fluid + redissolved precipitate samples. See text for discussion. m, mmol/kg; μ, μmol/kg; n, nmol/kg; except H_2 and CH_4, which are in mmol/L; $\delta^{34}S$ is calculated relative to the CDT standard. Endmember compositions in 1990 calculated from data of *Butterfield et al.* [1994a].

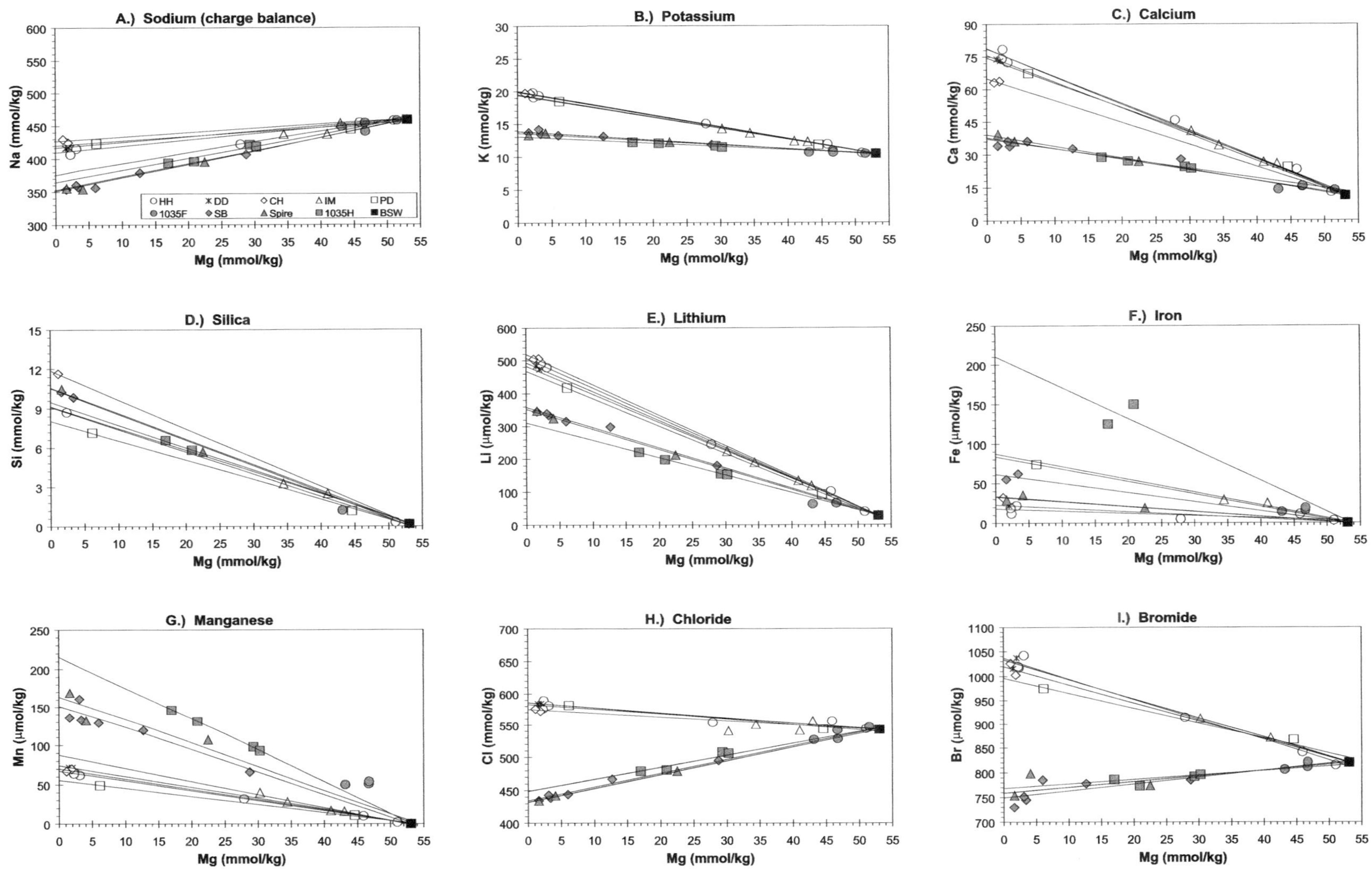

Figure 2. Graphs of selected aqueous species versus Mg in sampled vent fluids. (a) Sodium. (b) Potassium. (c) Calcium. (d) Silica. (e) Lithium. (f) Iron. (g) Manganese. (h) Chloride. (i) Bromide.

Table 3. Correlation Coefficients and Uncertainties for Endmember Compositions From July, 2000 Fluid Sampling.

Vent		Na,[†] m	K, m	Ca, m	Si, m	Li, μ	Rb, μ	Cs, μ	Sr, μ	Fe, μ	Mn, μ	Cu, μ	Zn, μ
Dead Dog													
Heineken	r^2	0.955	0.999	0.996	0.999	0.999	0.999	0.999	0.997	0.799	0.999	0.495	0.848
Hollow	[a]	12	0.6	2.4	0.5	10	0.5	0.03	5	0.4	1	0.7	0.1
Dead	r^2	0.999	0.999	1		0.999	0.999	0.999	0.999	0.999	0.999	0.774	0.984
Dog[‡]	[a]	13	0.6	2.3		10	0.5	0.03	6	0.4	2	0.1	0.1
Chowder	r^2	0.991	0.999	0.999	1	0.999	0.999	0.999	0.999	1	0.999	0.619	0.580
Hill[‡]	[a]	13	0.60	2.0	0.6	10	0.5	0.03	6	0.66	1	6.0	1.4
Inspired	r^2	0.894	0.998	0.997	0.993	0.999	0.998	0.998	0.996	0.975	0.988	0.999	0.955
Mounds[‡]	[a]	11	0.6	2.4	0.5	10	0.5	0.03	6	6	3	0.4	2.0
Puppy	r^2	0.953	0.999	0.998	0.999	0.999	0.998	0.999	0.998	1	0.998	1	1
Dog[‡]	[a]	13	0.6	2.3	0.4	9	0.5	0.03	6	2	1	0.1	0.1
Bent Hill													
1035F	r^2	0.636	0.689	0.768	1	0.899	0.891	0.909	0.902	0.823	0.912		1
	[a]	20	0.5	4.9	0.3	31	0.6	0.08	10	19	41		0.1
ODP Mound													
Shiner	r^2	0.997	0.985	0.975	0.999	0.998	0.998	0.998	0.981	0.991	0.988	0.991	0.991
Bock[‡]	[a]	11	0.4	1.1	0.6	7	0.3	0.02	3	2	4	0.1	0.1
Spire	r^2	0.999	0.993	0.998	0.997	0.999	0.999	0.974	0.973	0.981	0.983	0.288	0.288
	[a]	10	0.4	1.2	0.6	7	0.3	0.04	4	2	7	0.8	0.8
1035H[‡]	r^2	0.992	0.998	0.995	0.999	0.999	0.999	0.998	0.997	0.971	0.999	0.633	0.633
	[a]	11	0.4	1.1	0.5	6	0.3	0.02	3	14	4	1.4	1.4

Vent		Pb, n	Cl, m	Br, μ	SO_4^{2-}, m	H_2, m	H_2S, m	CO_2, m	CH_4, m	NH_4, μ
Dead Dog										
Heineken	r^2	0.526	0.931	0.992	0.999	0.999	0.999	0.997	0.999	0.998
Hollow	[a]	19	17	31	0.2	0.08	0.3	0.8	0.9	159
Dead	r^2	0.995	0.999	0.996	1		0.999			0.999
Dog[‡]	[a]	5	18	31	0.1		0.3			155
Chowder	r^2	0.710	0.997	0.996	0.999	0.999	0.999	0.983	0.999	0.999
Hill[‡]	[a]	85	17	31	0.1	0.1	0.4	0.8	0.8	158
Inspired	r^2	0.668	0.059	0.999	0.993	0.999		0.996	0.999	0.992
Mounds[‡]	[a]	1192	16	31	0.7	0.1	0.2	1.2	0.9	143
Puppy	r^2	1	0.974	1	0.991	0.999	0.999	0.999	0.999	0.998
Dog[‡]	[a]	17	18	30	0.1	0.1	0.4	0.8	0.7	149
Bent Hill										
1035F	r^2	1	0.716	0.652	0.957	0.974	0.880	1	1	0.987
	[a]	3	19	23	2.0	0.1	0.3	0.8	0.1	63
ODP Mound										
Shiner	r^2	0.991	0.997	0.876	0.996	0.999	0.998	0.999	0.999	0.999
Bock[‡]	[a]	4	13	23	0.4	0.3	0.4	1.2	0.3	117
Spire	r^2	0.492	0.999	0.752	0.974	0.999	0.999	0.999	0.999	0.999
	[a]	11	13	23	1.3	0.1	0.8	1.3	0.3	114
1035H[‡]	r^2	0.859	0.986	0.942	0.999	0.999	0.999	0.998	0.999	0.999
	[a]	17	13	23	0.2	0.2	0.4	1.1	0.2	105

[a]Uncertainty in reported values is estimated to be the larger of the analytical uncertainty or the error associated with the extrapolation to a zero Mg concentration.

r^2 values of 1 indicate that only one hydrothermal sample was available for the regression calculation.

†: charge balance.

‡: calculated concentrations of Fe, Cu, Zn and Pb based on combined fluid and redissolved precipitate samples. See text for discussion.

depleted in the ODP Mound fluids relative to the Dead Dog fluids, ranging between 153–165 and 256–287 μmol/kg fluid in the two fields, respectively. Middle Valley vent fluids are greatly enriched in the alkali trace metals Li, Rb, and Cs. For example, measured concentrations of Li and Rb at the Dead Dog Vent Field varied from 467 to 519 and 23 to 25, respectively, whereas Cs concentrations were 1.5 μmol/kg in all vents. In general, concentrations of Li, Rb, and Cs were lower in the ODP Mound fluids, where they varied from 268 to 359, 6.7 to 16, and 0.65 to 0.96 μmol/kg, respectively.

Measured concentrations of aqueous Mn are characterized by a strong linear correlation with aqueous Mg and yield end-member concentrations of 150 to 337 μmol/kg at the ODP Mound and 56 to 85 at the Dead Dog Vent Field. Incontrast, measured concentrations of Fe, Zn, Cu, and Pb show a high degree of scatter when plotted against aqueous Mg (Figure 2) resulting in relatively large uncertainties for the reported end-member concentrations (Table 3). Hence, these values (especially for Cu and Pb) should be used with caution. The poor correlations likely reflect precipitation of sulfides and/or oxides during cooling in the fluid samplers. Sample BGT-3596-3, collected at Chowder Hill, was characterized by Fe concentrations that are almost an order-of-magnitude higher than other samplers from the same vent (Table 1). We interpret this concentration to reflect entrainment of chimney material during sampling. This sample was not used for end-member regressions. Unlike aqueous Mn abundances that were higher at the ODP Mound relative to the Dead Dog Vent Field, end-member concentrations of other transition metals showed no systematic variation between the two areas. Aqueous Fe concentrations varied from 18 to 212 mol/kg, whereas Zn, Cu, and Pb were substantially lower varying from 0.96 to 21 μmol/kg, 0.37 to 22 μmol/kg, and 33 to 4133 nmol/kg, respectively.

4.3. Dissolved Gases

End-member concentrations of aqueous H_2 and H_2S are very similar for all vents in the Dead Dog field, ranging between 1.9 and 2.6 mmol/L and 2.4 and 3.8 mmol/kg, respectively. In contrast, at the ODP Mound field, a wide range in concentrations is observed between the three vents. End-member H_2 concentrations varied from 2.2 to 8.2 mmol/L, respectively, whereas end-member H_2S concentrations varied from 1.7 to 7.9 mmol/L (Table 2).

4.4. Sulfur Isotopes

The sulfur isotopic composition of aqueous H_2S in all sampled fluids is enriched in ^{34}S relative to basaltic sulfide (+0.1‰; *Sakai et al.* [1984]; Figure 3). At the Dead Dog

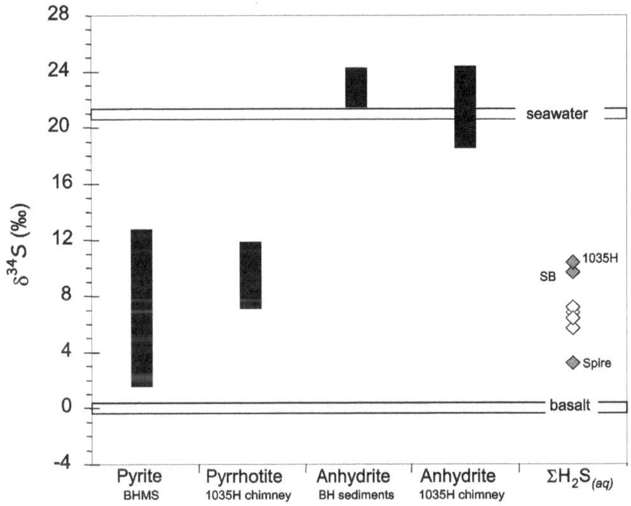

Figure 3. $\delta^{34}S$ of sulfate and sulfide minerals recovered from Bent Hill, the Bent Hill massive sulfide deposit, and the chimney at ODP Hole 1035H. Also shown in the stippled bars is the isotopic composition of seawater sulfate [*Rees et al.*, 1978] and basaltic sulfide [*Sakai et al.*, 1984]. Data for Bent Hill minerals are from *Zierenberg* [1994].

field, $\delta^{34}S$ ranged between +5.7 and +7.2‰. There is a striking difference between the stable sulfur isotopic compositions between the three vents located on the ODP Mound. At Spire vent, where aqueous H_2S was the most abundant, the $\delta^{34}S$ value of +3.2‰ is the most depleted in ^{34}S. H_2S from Shiner Bock and Hole 1035H fluids are the most enriched of all sampled fluids, having values of +10.4 and +9.7‰, respectively (Figure 3; Table 2). The sulfur isotopic composition of anhydrite from the collapsed chimney sampled at 1035H in 1998 varied from +11 to +20‰ (Figure 3).

5. DISCUSSION

The chemistry of vent fluids at Middle Valley reflects the integrated effects of reaction with basaltic rocks and sediments containing organic matter [*Butterfield et al.*, 1994a; *Cruse and Seewald*, 2006]. However, assessing the relative influence of basalt and sediment on vent fluid composition is not straightforward, because the direction and magnitude of chemical exchange for many chemical species are similar for reactions occurring at elevated temperatures and pressures. For example, high-temperature reaction of seawater-derived fluids with either sediment or basalt at fluid/rock (sediment) mass ratios less than 50 results in near-quantitative removal of aqueous Mg, and SO_4 from seawater and addition of Ca, K, Si, H_2, and H_2S [*Seyfried*, 1987; *Thornton and Seyfried*, 1987; *Seewald et al.*, 1990; *Von Damm*; 1991; *Cruse and

Seewald, 2001]. The strongest evidence for reaction with basalt or sediment comes from species that are substantially enriched and/or isotopically distinct in either substrate. Thus, the absolute abundance of Rb and Cs in Middle Valley vent fluids, which are substantially enriched in sediments relative to basalt, provides compelling evidence for sediment interaction (Table 3; *Butterfield et al.* [1994a]). Moreover, *Butterfield et al.* [1994a] have shown that Rb/Cs and Li/Rb ratios in Middle Valley fluids reflect the relative abundance of these elements in pelagic sediment, whereas their relative abundance in fluids from unsedimented systems reflects ratios in basalt. High concentrations of Br along with elevated Br/Cl molal ratios in these fluids (0.0017–0.001.9) relative to seawater (0.0015) are also indicative of fluid-sediment interaction [*You et al.*, 1994]. Finally, high abundances of aqueous species derived from sedimentary organic matter, such as CH_4, C_{2+} hydrocarbons, and ΣNH_3, also reflect a contribution of sediment alteration to fluid composition at Middle Valley [*Butterfield et al.*, 1994a; *Cruse and Seewald*, 2006].

The abundant evidence for extensive fluid-sediment interaction, notwithstanding, $^{87}Sr/^{86}Sr$ isotope ratios of 0.7043 for aqueous Sr are lower than seawater values of 0.70918 and values for sedimentary sources that range from 0.7066 for calcite to ≥0.708 for the bulk sediment [*Butterfield et al.*, 1994a; *Goodfellow and Blaise*, 1988; *Goodfellow et al.*, 1994]. The isotopic composition of Sr, however, is similar to values for basaltic Sr (0.703), providing compelling evidence for chemical interaction with basaltic rocks. Because the chemistry of Sr is intimately linked to Ca under hydrothermal conditions [*Berndt et al.*, 1988], derivation of vent fluid Sr from basaltic sources suggests similar processes may regulate aqueous Ca concentrations. Lead isotope ratios in massive sulfides from Middle Valley also indicate a predominantly basaltic source for the lead [*Goodfellow et al.*, 1993; *Bjerkgard et al.*, 2000; *Cousens et al.*, 2002].

5.1. Subsurface Conditions

In addition to chemical interaction with sediment and basalt, the composition of fluids at the Dead Dog Vent Field and ODP Mound are variably influenced by the effects of phase separation. Measured Cl concentrations at the ODP Mound are depleted by ~100 mmol/kg relative to ambient seawater (Table 1), suggesting that these fluids have experienced phase separation in subsurface environments [*Butterfield et al.*, 1994a; *Cruse and Seewald*, 2006]. For a seafloor depth of 2400 m, which corresponds to a seafloor pressure of 240 bars, minimum temperatures for phase separation are 385°C, and increase with subsurface depth [*Bischoff and Rosenbauer*, 1985]. The occurrence of phase separated fluids in conjunction with measured exit temperatures of ≤272°C indicates that ODP Mound vent fluids have cooled by at least 100°C. The possibility exists that phase separation was not actively occurring in 1991 and 2000, and the fluids exiting the ODP Mound vents represent a mixture of seawater and a low chlorinity vapor stored at depth. It is hard to envision, however, a mechanism by which highly buoyant vapor-rich fluids would have remained trapped in the subsurface. Regardless of the timing of phase separation, the hydrothermal fluids currently venting at the ODP Mound field have cooled by at least 100°C from their maximum temperatures.

In contrast to fluids from the ODP Mound, fluids from the Dead Dog Vent Field are slightly enriched (<10%) relative to ambient seawater. Although the slight enrichment could reflect venting of a brine phase formed by phase separation, it could also be produced by hydration of sediments during alteration [*You et al.*, 1994]. There are numerous lines of evidence supporting lower peak subsurface temperatures and the apparent absence of phase separation at the Dead Dog Vent Field. In particular, lower values for the ratios of C_1/C_{2+} alkanes and branched versus straight alkanes, and temperatures estimated from alkane-alkene thermodynamic equilibria and isotopic equilibration of CO_2 and CH_4 all indicate subsurface temperatures are lower at the Dead Dog Vent Field relative to the ODP Mound [*Cruse and Seewald*, 2006]. In contrast to the Bent Hill/ODP Mound area, massive sulfide deposits have not formed at the Dead Dog Vent Field, consistent with the absence of relatively high-temperature, metal-rich fluids during past and present hydrothermal activity. Higher Ca concentrations at the Dead Dog Vent Field also suggest lower temperatures because it is released to solution during albitization of plagioclase, a process that is favored at lower temperatures [*Berndt et al.*, 1989; *Seyfried et al.*, 1992; *Butterfield et al.*, 1994a].

The ability of hydrothermal fluids to transport metals from depth to near-seafloor environments where they may form massive sulfide deposits is critically dependent on the extent of fluid mineral equilibrium at the prevailing temperatures and pressures of subsurface reaction zones. In general, transition metal concentrations decrease with lower temperature and higher pressure during fluid-sediment or fluid-basalt reactions [*Seyfried*, 1987; *Seewald and Seyfried*, 1991; *Cruse and Seewald*, 2001]. Thus, depending on the extent of subsequent reequilibration, cooling of a high-temperature fluid can greatly influence its chemistry. For many basalt-hosted systems, cooling processes are sufficiently rapid that reequilibration does not occur to a significant degree and the composition of vent fluids at the seafloor reflects high-temperature processes [*Seewald and Seyfried*, 1991; *Seyfried et al.*, 2003]. Laboratory experiments have demonstrated that hydrothermal alteration of sediment or basalt at temperatures required for phase separation at Middle Valley (≥385°C) produces aqueous transition metal concentrations that are orders of magnitude higher than abundances observed in ODP Mound fluids [*Seewald et al.*, 1990, 1994; *Seewald and Seyfried*, 1990]. The relatively low Mn, Fe, Zn, Cu, and Pb concentrations in these fluids, despite having undergone phase separation, are indicative of extensive reequilibration after cooling. Although extensive decreases in aqueous Mn have occurred at both vent fields, slightly higher concentrations at the ODP Mound relative to Dead Dog may record higher abundances before cooling, which is consistent with the occurrence of phase separated fluids at the ODP Mound.

The concentrations of aqueous silica in vent fluids can be used to constrain temperature and pressure conditions during upflow if it is assumed that the fluids equilibrate with quartz in the subsurface. At Middle Valley, this is a reasonable assumption considering the abundance of quartz in unaltered sediment and veins and altered rocks formed in the hydrothermal upflow zones [*Leybourne and Goodfellow*, 1994]. Owing to the significant temperature and pressure dependence of quartz solubility at conditions typical of submarine hydrothermal systems, either of these variables can be estimated if the other is known. Using the quartz solubility model of *Von Damm et al.* [1991], dissolved silica concentrations in Middle Valley vent fluids characterized by temperatures >250°C are consistent with quartz saturation at 240 to 1200 bars with the majority <800 bars (Figure 4). These pressures place the zone of equilibration within the sediment-sill complex to a depth of no more than 1.2 km below the seafloor. Equilibration of fluids at temperatures significantly higher than the measured values is not possible due to the increase in quartz solubility with increasing temperature, even in the low pressure, near-critical region where the temperature dependence of quartz solubility undergoes a reversal [*Fournier*, 1983; *Von Damm and Bischoff*, 1991; *Foustoukos and Seyfried*, 2007].

Equilibration of aqueous silica with quartz during discharge is supported by reactive transport modeling of hydrothermal circulation at Middle Valley [*Alt-Epping and Smith*, 2001]. These models have shown that for flow rates, permeability structure, and temperature gradients that may be representative of the Middle Valley system, aqueous silica precipitation is sufficiently rapid to maintain aqueous concentrations that are near or at quartz solubility [*Alt-Epping and Smith*, 2001]. Unlike aqueous silica concentrations in Dead Dog and ODP Mound vent fluids that attain nearly identical values due to the temperature- and pressure-dependent homogenous precipitation of quartz from solution, similar concentrations of Ca, K, and Na are not expected because the abundance of these species is controlled by heterogeneous equilibration of vent fluid with minerals present in both basalt and sediment. In addition to temperature and pressure, vent fluid composition during heterogeneous fluid-mineral interaction is strongly influenced by the alteration mineralogy of basalt and/or sediment, which in turn is a sensitive function of water/rock ratio [*Alt-Epping and Smith*, 2001]. Under equilibrium conditions at low water/rock ratio, the composition of the mineral assemblage will buffer fluid chemistry at a given temperature and pressure. In contrast, under fluid-dominated conditions (high water/rock ratio), the composition of the fluid may overwhelm the chemistry of the solid phases and dictate the stable mineral assemblage without experiencing large changes in the abundance of major cations. Accordingly, cooling of Middle Valley fluids in hydrothermal upflow zones characterized by high water/rock ratios allows large differences in the abundance of Ca, K, and Na to be preserved in Dead Dog and ODP Mound fluids despite cooling and chemical equilibrium at similar temperature and pressure conditions.

Cooling of hydrothermal fluids at Middle Valley may occur in response to mixing with lower temperature fluids or conductive heat loss to the surrounding rocks and sediments. On the basis of large differences in subsurface temperatures estimated using variety of organic and inorganic geothermometers, *Cruse and Seewald* [2006] concluded that observed Cl concentrations and the abundance and isotopic composition of aqueous organic compounds at the ODP Mound were the result of subsurface mixing processes. The uniformity of temperatures between the Dead Dog and ODP Mound vent fields, however, suggests that some conductive heat loss may also be involved. A likely scenario may involve local mixing of high-temperature fluids into a regional hydrothermal reservoir such as the isothermal aquifer identified at ODP Hole 857 [*Goodfellow and Peter*, 1994]. Continued flow after mixing would likely result in loss of excess heat to surrounding rocks before upflow and venting.

5.2. Sulfur Isotopes

The isotopic composition of aqueous ΣH_2S in Middle Valley vent fluids varies from 3.2 to 10.4‰ and is comparable to the range of values observed in vent fluids from other sedimented and unsedimented ridge-crest hydrothermal systems [*Shanks*, 2001]. Sources of aqueous ΣH_2S at Middle Valley include leaching of basaltic sulfur ($\delta^{34}S$ = 0.1‰; *Sakai et al.* [1984]), leaching of bacteriogenic pyrite in unaltered sediment ($\delta^{34}S$ = –40.7 to –5.4‰; *Duckworth et al.* [1994]), and reduction of seawater SO_4 ($\delta^{34}S$ = +21‰; *Rees et al.* [1978]). Dissolved H_2S may also be derived from the decomposition of sedimentary organic matter ($\delta^{34}S$ = –25 to –5‰; *Werne et al.* [2003]), but this is unlikely to represent a major contributor to Middle Valley vent fluids. Middle Valley sediments have relatively low organic carbon content (≤0.5 wt % total organic carbon [TOC]; *Shipboard Scientific Party* [1992b]) and Recent to Upper Pleistocene marine sediments have relatively low sulfur/TOC ratios [*Anderson and Pratt*, 1995]. That the vent fluid ^{34}S values are bracketed by seawater values on one side and basalt values on the

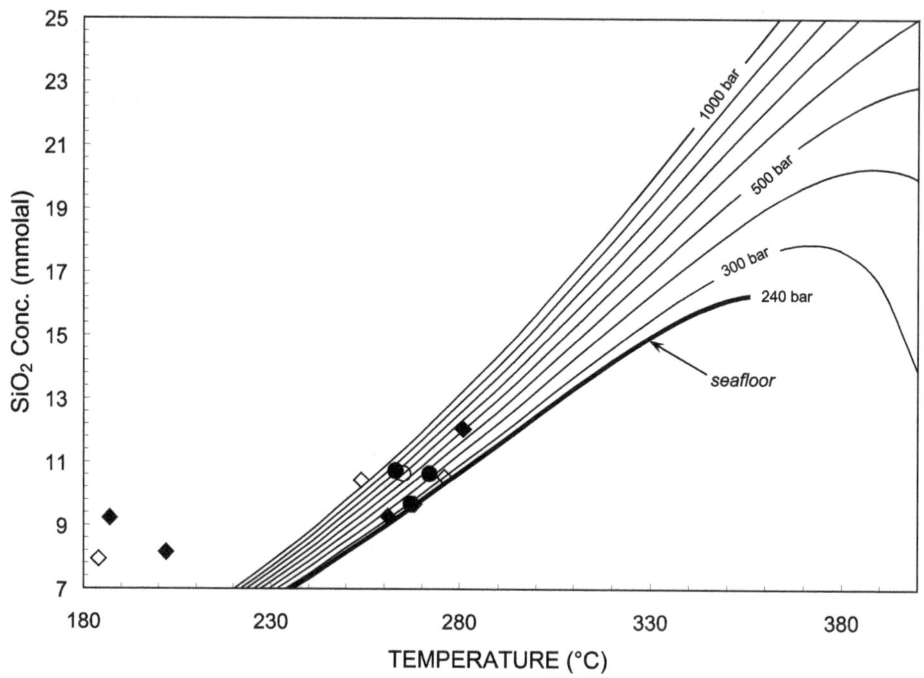

Figure 4. Isobars of aqueous silica concentration in seawater predicted from quartz saturation as a function of temperature. Vent fluids from the Dead Dog (diamonds) and ODP Mound (circles) vent fields are also plotted at their measured exit temperatures. Open symbols represent the 1991 data of *Butterfield et al.* [1994a], whereas the solid symbols represent the 2000 data presented in this study. Quartz solubility data requisite to the construction of this diagram are from *Von Damm et al.* [1991a].

other provides compelling evidence for extensive seawater SO_4 reduction and leaching of basalt as the major sources of aqueous reduced sulfur.

Extensive reduction of seawater SO_4 is consistent with the highly reducing nature of Middle Valley vent fluids. Aqueous H_2 concentrations are some of the highest values measured in either sediment-covered and basalt-hosted systems [*Lilley et al.*, 2003; *Seewald et al.*, 2003; *Von Damm et al.*, 2007], and even approach values observed in peridotite-hosted systems where serpentinization reactions produce extremely reducing conditions [*Charlou et al.*, 2002; *Schmidt et al.*, 2007]. High aqueous H_2 concentrations at Middle Valley have been attributed to the oxidation of aqueous CH_4 and longer chain hydrocarbons to CO_2 according to the general reaction:

$$C_nH_{2n+2(aq)} + 2nH_2O_{(l)} = nCO_{2(aq)} + (3n+1)H_{2(aq)} \quad (2)$$

where $C_nH_{2n+2(aq)}$ represents an aqueous hydrocarbon of chain length n [*Cruse and Seewald*, 2006]. Requisite hydrocarbons for reaction (2) are derived from the hydrothermal alteration of sedimentary organic matter [*Cruse and Seewald*, 2006].

Owing to the extremely low solubility of anhydrite at temperatures and pressures experienced by Middle Valley fluids, subsurface reaction zones will be characterized by very low aqueous SO_4 activity. Anhydrite is found throughout Middle Valley sediments, in alteration zones that surround the areas of active discharge [*Leybourne and Goodfellow*, 1994; *Shipboard Scientific Party*, 1998a, 1998b]. The anhydrite forms in response to the mixing of upward- and outward-flowing hydrothermal fluids with recharging heated seawater and conductively heated sedimentary porefluids [*Goodfellow and Peter*, 1994]. This anhydrite, then, represents the reservoir of seawater SO_4 available for reduction by Middle Valley vent fluids ascending from subsurface reaction zones. The stability of anhydrite in the presence of these fluids can be assessed using phase relations in the system Fe-S-O-H-Ca expressed in terms of aqueous H_2 and H_2S activities and the measured vent fluid composition (Figure 5). The anhydrite stability line in Figure 5 is shown for a range of in situ pH values likely to exist during fluid-sediment interaction under hydrothermal conditions [*Cruse and Seewald*, 2001; *Seyfried et al.*, 1991] and is further constrained by the activity of Ca^{2+} calculated for Middle Valley vent fluids using the

EQ3NR computer code [*Wolery*, 1992] at the measured exit temperatures. The measured concentrations of aqueous H_2 and H_2S at the Dead Dog Vent Field and ODP Mound plot in the pyrrhotite stability field, consistent with the mineralogy of active chimneys, indicate that anhydrite is unstable in the hydrothermal reaction zone. Thus, the presence of isotopically heavy ΣH_2S indicative of seawater SO_4 reduction is consistent with the thermodynamic drive for the reduction of anhydrite to ΣH_2S according to the reaction:

$$CaSO_{4(s)} + 2H^+ + 4H_{2(aq)} = Ca^{2+} + H_2S_{(aq)} + 4H_2O_{(l)} \quad (3)$$

It should be noted that reaction (3) requires a source of acidity to proceed to the right. Likely candidates include Ca fixation as epidote or calcite during fluid-plagioclase equilibria [*Berndt et al.*, 1989; *Seyfried et al.*, 1991; *Seewald et al.*, 1994; *Cruse and Seewald*, 2001] and reaction of illite to form albite and K-feldspar [*Cruse and Seewald*, 2001]. All of these minerals are present in altered sediments at Middle Valley.

The sulfur isotope composition of ΣH_2S in Middle Valley vent fluids is in good agreement with the composition of sulfide minerals analyzed from the Dead Dog Vent Field and the Bent Hill Massive sulfide deposits (Figure 3; *Goodfellow and Blaise* [1988]; *Duckworth et al.* [1994]; *Zierenberg* [1994]). In the latter case, the similarity of isotopic values in the fluids and sulfide minerals deposited at higher temperatures suggests that the extent of seawater SO_4 reduction occurring during the formation of the massive sulfide deposit was not significantly different than is currently observed in 270°C fluids.

5.3. Activation of Hydrothermal Venting at the ODP Mound

In contrast to the similar concentrations of ionic species in fluids sampled from the three vents at the ODP Mound field, aqueous H_2 and H_2S concentrations are strikingly different, despite their very close (<~40 m) spatial proximity. Hydrothermal activity was observed in the area of the Shiner Bock vent in 1996 during a camera survey that preceded drilling of Hole 1035H during ODP Leg 169. Vigorous hydrothermal venting was immediately initiated in ODP Hole 1035H when the subsurface hydrologic seal at 150 mbsf (meters below seafloor) was punctured during drilling in 1996 [*Shipboard Scientific Party*, 1992]. Hydrothermal fluids had been flowing up the drill hole for at least 4 years before this study. Activation of fluid flow through the ODP Mound also caused the development of a new field of small spires and beehives on the eastern side of ODP Mound that were not observed during an Alvin dive to this site in 1998 (Zierenberg, unpublished data). Fluids sampled from Shiner Bock vent and ODP Hole 1035H contain end-member aqueous H_2 concentrations of 8.2 and 6.1 mmol/L fluid, respectively, whereas the concentration in fluids from Spire vent located in the newly formed field was only 2.2 mmol/L fluid. End-member H_2S concentrations were higher in the Spire vent fluids at 7.9 mmol/L relative to 4.0 and 4.3 mmol/L in 1035H and Shiner Bock fluids, respectively (Table 2).

The close spatial proximity of Spire vent to the 1035H and Shiner Bock vents, similar exit temperatures, and nearly identical abundances of ionic species suggest that the three vents derive fluids from the same source region, with the lower H_2 and higher H_2S concentrations at Spire vent being the result of late-stage reactions that occur during upflow. Reduction of SO_4 according to reaction (3) represents a potential process that could consume H_2 and produce H_2S within upflow zones. However, if reduction of sulfate present as anhydrite was affecting vent fluid composition during upflow at Spire vent, the H_2S produced would be enriched in the heavy isotope of sulfur relative to that in Hole 1035H and Shiner Bock vent fluids (Figure 3). The measured value for $\delta^{34}S$ of aqueous H_2S from Spire vent is depleted by ~7‰ relative to Shiner Bock and Hole 1035H fluids, indicating that reaction (3) is not responsible for the observed differences in vent fluid composition. Moreover, the equimolar H_2 decrease and H_2S increase in Spire fluids relative to Shiner Bock and Hole 1035H fluids is not consistent with the stoichiometry of reaction (3) that consumes 4 mol of H_2 for each mole of H_2S generated.

Another possible reaction that can influence the concentrations of H_2 and H_2S during upflow through the ODP Mound is the reduction of pyrite to pyrrhotite:

$$\underset{\text{pyrite}}{FeS_{2(s)}} + H_{2(aq)} \rightarrow \underset{\text{pyrrhotite}}{FeS_{(s)}} + H_2S_{(aq)} \quad (4)$$

This reaction consumes H_2 and produces H_2S in equimolar amounts, which is consistent with the observed differences in aqueous H_2 and H_2S concentrations in ODP Mound fluids. Measured H_2 and ΣH_2S concentrations in the Shiner Bock and 1035H fluids that plot deep inside the pyrrhotite stability field for the Fe-S-O-H-Ca system at seafloor conditions of 270°C, 240 bars (Figure 5) indicate that pyrite is a thermodynamically unstable phase in these fluids and, in the absence of kinetic barriers, will react to form pyrrhotite. The location of Middle Valley fluids within the pyrrhotite stability field is consistent with the occurrence of pyrrhotite as the predominant iron sulfide mineral actively precipitating at the Dead Dog Vent Field and ODP Mound [*Ames et al.*, 1993].

To account for the observed compositional differences in fluids currently venting from ODP Mound, we invoke a

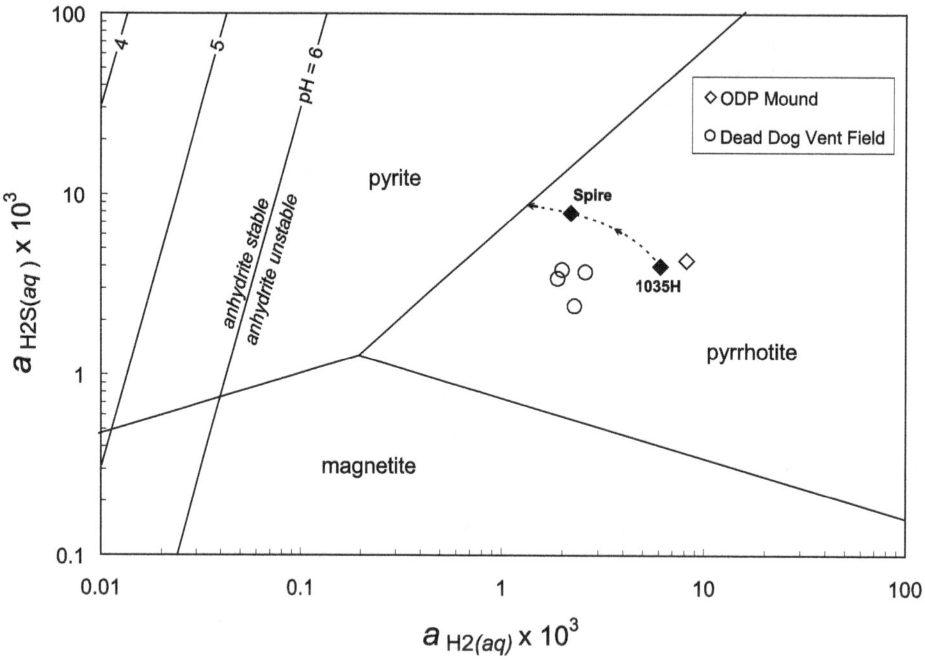

Figure 5. Phase diagram for the system Fe-S-O-H-Ca at 270°C, 240 bars, showing the stability fields for pyrite, pyrrhotite and magnetite. The solubility of anhydrite as a function of pH and calculated Ca^{2+} activity at measured exit temperatures (see text) is also indicated. Measured vent fluid compositions for Dead Dog Vent Field are indicated with white circles. Vent fluid compositions for ODP Mound Vent Field fluids are indicated with white and black diamonds. The reaction path calculated using EQ6 [*Wolery and Daveler*, 1992] for the reaction of 1035H fluids with pyrite is shown in the curved line. Note that this line passes through the measured concentrations for Spire vent fluids. Requisite thermodynamic data for the construction of this figure are from the SUPCRT92 database [*Johnson et al.*, 1992] and *Toulmin and Barton* [1964].

model involving the infiltration of 1035H fluids through the massive sulfide on the eastern side of the ODP Mound that began after the drillhole punctured the silicified layer that acted as a subsurface hydrologic seal (Figure 6). Preexisting pyrite in the mound is converted to pyrrhotite according to reaction (4) during the flow of fluids along tortuous paths to the Spire vent, resulting in equimolar consumption of H_2 and production of H_2S. Fluids venting at 1035H retain their composition from depth due to high fluid/pyrite ratios during flow in the zero-tortuosity drill hole. The effect of flow path geometry and fluid/mineral ratios are discussed further in *Alt-Epping and Dymond* [this volume]. Similarly, long-term venting at the Shiner Bock vent will have converted any pyrite that may have been present due to high integrated fluid/rock ratios resulting in fluid composition that is nearly identical to that at 1035H.

The above model was tested by calculating the reaction path for 1035H fluids during reaction with pyrite using the computer program EQ6 [*Wolery and Daveler*, 1992] and thermodynamic database generated from SUPCRT92 [*Johnson et al.*, 1992] at 250 bars, with data for $MgSO_4^0$ and $NaSO_4^-$ added from *McCollom and Shock* [1997]. Calculations were performed at 270°C, which corresponds to the exit temperatures measured at Spire and 1035H chimneys. Results of the reaction path modeling indicate that 1035H fluids evolve to the composition of Spire vent fluids when 3.9 mmol of pyrite have reacted with 1 kg of solution before eventually reaching saturation with pyrite (Figure 5). The slight displacement of the Spire fluids from the pyrite-pyrrhotite phase boundary in Figure 5 could be an artifact of uncertainties in the thermodynamic data used in the calculations. A second possibility is that these fluids have not yet fully equilibrated with pyrite during upflow. The lack of equilibration may reflect kinetic barriers or insufficient time for complete reaction along fluid flow paths in the mound.

The differences in the measured $\delta^{34}S$ of aqueous H_2S between the ODP Mound vents is also consistent with pyrite reduction in the mound at Spire vent. We modeled the evolution of $\delta^{34}S$ during the reaction of 1035H fluids using a simple mass balance approach that assumes pyrite in the mound is quantitatively dissolved without isotopic fractionation. At the same time, pyrrhotite precipitates from solution in iso-

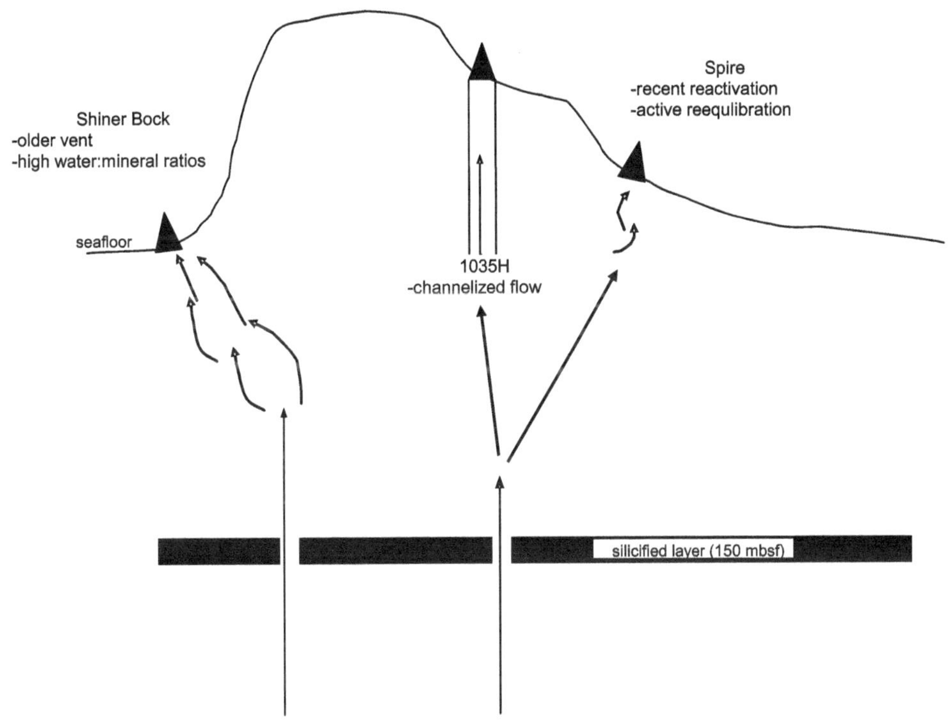

Figure 6. Schematic diagram of fluid circulation pathways in the ODP Mound vent field. Shiner Bock vent is several years old, leading to high integrated water/rock ratios. ODP Hole 1035H is a low-tortuosity channelized flow path, creating conditions of high water/rock ratios. Spire vent is from a new vent field that had formed within 2 years before sampling due to reactivation of flow through the mound during drilling of ODP Hole 1035H. Vent fluid composition at this location is strongly influenced by the reduction of pyrite to pyrrhotite.

topic equilibrium with the fluid at 270°C ($\alpha_{Po\text{-}H_2S} = 1.00034$; *Ohmoto and Rye* [1979]) and is then removed from the chemical system (i.e., no further reaction or isotopic reequilibration occurs). Changes in the ^{34}S content of the fluid were integrated into the model during the reaction of 3.9 mmol of pyrite with 1 kg of fluid to produce the observed changes in H_2 and H_2S between the 1035H and Spire vents. As might be expected, the isotopic composition of the fluid is strongly dependent on the isotopic composition of the reactant pyrite. The isotopic composition of pyrite necessary to reproduce the observed $\delta^{34}S$ value of the Spire vent fluids is 0.9‰. The isotopic composition of pyrite in the ODP Mound is presently unknown, but sulfides in the nearby Bent Hill massive sulfide deposit ranges between 1.9 and 12.4‰ [*Zierenberg*, 1994]. Thus, although the estimated isotopic composition of pyrite in the ODP is generally consistent with the range of pyrite values at Bent Hill, the small relative depletion may reflect the presence of more ^{34}S-depleted pyrite at the ODP Mound due to a greater proportion of basalt-derived sulfide, or potentially the involvement of isotopically depleted bacteriogenic pyrite. Unlike the Bent Hill massive sulfide deposit, which contains no sediment within the mound, the ODP Mound massive sulfide consists of three discrete lenses of massive sulfides interlayered with sediments that may represent the source of isotopically depleted pyrite [*Shipboard Scientific Party*, 1998a; *Zierenberg et al.*, 1998].

5.4. Temporal Variability

Previous work has shown that vent fluid compositions from both sediment-covered and bare-rock hydrothermal vent sites can remain stable over periods of years in the absence of tectonic or volcanic activity [*Butterfield et al.*, 1994b; *Campbell et al.*, 1988; *Von Damm et al.*, 2005]. *Butterfield et al.* [1994a] used data from all areas of venting to calculate an average end-member concentrations for the entire Dead Dog Vent Field in 1990. To allow for a comparison on the scale of individual vents, the data of *Butterfield et al.* [1994a] were used to recalculate end-member compositions for Heineken Hollow, Dead Dog mound, Inspired Mounds, and Chowder Hill vents from the Dead Dog field, and Lone Star vent from the ODP Mound field (Table 2).

Examination of Table 2 reveals that between 1990 and 2000, the composition of individual vents at the Dead Dog Vent Field remained stable. This is an interesting result in light of the fact that OPD Hole 858F, located at the Dead Dog Vent Field near Chowder Hill vent (Figure 1b), was underpressured and allowed thousands of gallons of seawater per minute to flow into the formation when drilled in 1991 [*Shipboard Scientific Party*, 1992b]. Three fluid samples collected from Chowder Hill vent 1991 were characterized by high Mg concentrations [*Butterfield et al.*, 1994a]. Rather than reflecting the difficulties of collecting vent fluid samples, these concentrations may have actually been "real," and caused by this influx of seawater through ODP Hole 858F. Comparison between the 1990 and 2000 data for Chowder Hill indicates that any impacts from the influx of seawater into the formation were not long-lived.

Direct comparison of individual vents at the ODP Mound in 1990 and 2000 is not possible because the Lone Star vent sampled in 1990 was not sampled in 2000. The similarity of Lone Star fluid chemistry in 1990 with the vents sampled in 2000, however, suggests that fluid composition did not change during this period. In the context of a model involving buffering of the fluid chemistry by fluid-mineral equilibria, this constancy indicates that the chemistry, mineralogy, temperature, and pressure in subsurface reaction zones did not change over a 10-year period. Nearly identical measured temperatures and aqueous silica concentrations likely controlled by pressure-dependent quartz solubility provide additional evidence for constant subsurface conditions. Measured temperatures of 184° and 187°C at Heineken Hollow in 1990 and 2000, respectively, are particularly intriguing because they were approximately 80°C lower than most of the other vents at Middle Valley, which clustered around 270°C. The reproducibility of this temperature on both dates suggests that the difference is real and not the result of poor thermocouple positioning in the vent orifice during measurement. Despite the substantial difference in temperature, fluid composition is indistinguishable from the other vents at the Dead Dog Vent Field. Owing to the strong temperature dependence of fluid mineral equilibria in hydrothermal systems, these data suggest that the Heineken Hollow fluids equilibrated at temperatures near 270°C, but have experienced extensive cooling during ascent in the upflow zone. Dissolved volatile and metal concentrations in Heineken Hollow fluids are the same as other vents at the Dead Dog field, indicating that cooling during upflow is sufficiently rapid so as to preclude reequilibration to 180°C. Cooling due to mixing with seawater is not indicated due to the low Mg concentrations of these fluids (Table 1). Thus, in addition to constant conditions in relatively deep-seated reaction zones, flow rates and pathways in the upflow zone have remained remarkably stable resulting in virtually identical extents of conductive heat loss between 1990 and 2000.

Carbon dioxide concentrations are approximately twice as high in 2000 relative to 1990 for the Dead Dog field, but are identical in the ODP Mound fluids. However, as the non-gastight titanium "majors" samplers were used in 1990, it is unclear if this difference reflects changes in geochemical conditions or limitations of the samplers.

Variations in concentrations of ΣNH_3 cannot be assessed at the scale of individual vents, because only an average value for the entire Dead Dog field was reported by *Butterfield et al.* [1994a]. The average concentration was 2.9 mmol/kg fluid in 1990, which is nearly identical to the average value of 3.1 mmol/kg fluid calculated from the 2000 results. Laboratory experiments have demonstrated that hydrothermal alteration of organic-bearing sediments results in rapid and near quantitative release of ΣNH_3 to solution [*Seewald et al.*, 1990, 1994]. Accordingly, measured ΣNH_3 concentrations can be used as indicator of fluid/sediment mass ratios in subsurface environments if ΣNH_3 concentrations in the sediment are known. For Middle Valley vent fluids, measured ΣNH_3 concentrations indicate fluid/sediment mass ratios of ~14 at the Dead Dog Vent Field and slightly higher values of 18 to 20 at the ODP Mound [*Cruse and Seewald*, 2006]. Near-constant ΣNH_3 ratios over the 10-year period from 1990 to 2000 indicates that there has been no change in the amount of fresh sediment with which the fluids venting at the Dead Dog field interact. Constant fluid/sediment mass ratios are also indicated by measured concentrations of sediment Li and Cs, whereas a 25% decrease in Rb concentrations suggests fluid/sediment ratios may have increased slightly. Fluid sediment/mass ratios estimated from Rb concentrations assuming an average Rb content of 65 ppm for unaltered sediments at Middle Valley [*Goodfellow and Peter*, 1994] range from 23 to 56. These values are slightly higher, but generally consistent with values estimated from ΣNH_3 abundances.

The observation that the concentrations of "soluble" or "mobile" species remain constant over a 10-year period has important implications for understanding the residence time and flow paths of fluids in subsurface environments. Time-invariant abundances of Li, B, Rb, and Cs have been used in basalt-hosted systems as evidence that fresh rock is continually exposed to circulating fluids [*Seyfried et al.*, 1991]. The model introduced by *Lister* [1983] involving a "cracking front" that develops in response to thermal stresses induced during cooling from magmatic temperatures is often invoked as a mechanism to create a continuous supply of fresh basalt. The occurrence of a "cracking front" type mechanism in basaltic basement at Middle Valley, however, cannot be used to explain the constant supply of mobile species that are clearly derived from sediments. Tectonic activity and associated faulting and fracturing represents an additional mechanism that could create new flow paths that would expose fresh sediment to circulating fluids. However, this cannot account for the stable fluid compositions at Middle Valley, because episodic exposure of new sediment during tectonic processes would result in step changes in the concentrations of mobile trace elements and would also likely be accompanied by variations in major-element concentrations due to sudden changes in fluid/sediment ratios and temperature and pressure conditions.

An alternative explanation to account for time-invariant concentrations of mobile trace elements at Middle Valley is that the reservoir of high-temperature fluids feeding seafloor vents is sufficiently large to maintain constant fluid chemistry on a time scale of tens of years. In other words, the residence time of vent fluids in this subsurface reservoir is longer than the duration of time-series datasets for mid-ocean ridge vent fluids and sampled fluids represent the first pass through fresh sediment. *Goodfellow and Peter* [1994] postulated that the hydrothermal aquifer intersected at ODP Site 857 located 3 km to the south of the Dead Dog field is representative of the source region for venting fluids, based on similarities in the composition and mineralogy of alteration assemblages. The igneous basement below the impermeable sediment cover at Middle Valley is approximately isothermal at temperatures similar to those measured in the hydrothermal vents [*Davis and Villinger*, 1992], and has sufficient permeability to allow for subsurface convection [*Stein and Fisher*, 2001]. This region of the crust may contribute to a fluid reservoir that is sufficiently large to maintain venting of fluids of constant temperature and composition over a 10-year period.

The large differences in absolute concentrations notwithstanding, mobile trace elements show similar temporal stability in both sedimented and unsedimented systems. Although the results presented here cannot be used to preclude the existence of a "cracking front" in bare-rock vent systems, they clearly document the occurrence of an alternative process that can maintain time-invariant abundances of mobile trace elements. The possibility that such a process may be regulating the abundance of mobile trace elements in vent fluids from unsedimented systems warrants serious consideration.

6. SUMMARY

The concentrations of dissolved ionic and volatile species were analyzed in vent fluids collected at two areas of hydrothermal venting at sediment-covered Middle Valley in July 2000. The composition of fluids at the Dead Dog Vent Field and ODP Mound field reflect extensive chemical interaction of seawater-derived hydrothermal fluids with both sediment and basalt. Aqueous Cl concentrations that are depleted relative to seawater indicate the occurrence of phase separation at the ODP Mound. In contrast, near-seawater Cl concentrations at the Dead Dog field suggest phase separation is not occurring. The absence of phase separation in conjunction with numerous other geochemical indicators suggests that subsurface temperatures are lower at the Dead Dog Vent Field than at the ODP Mound field, despite similar measured vent temperatures. Vent temperatures that are substantially lower than conditions necessary for phase separation require that fluids at the ODP Mound have cooled extensively before venting. Nearly identical measured temperatures and aqueous silica and transition metal concentrations (Fe, Zn, Cu, and Pb) at both vent fields suggest residence times after cooling are of sufficient duration to allow chemical and thermal reequilibration of the ODP Mound vent fluids before venting.

The sulfur isotope composition of aqueous ΣH_2S indicates derivation from reduction of seawater sulfate and leaching of basalt. Oxidation of sediment-derived aqueous organic compounds is responsible for the extremely reducing conditions necessary for SO_4 reduction. At the ODP Mound, aqueous concentrations of H_2 and ΣH_2S and the sulfur isotopic composition of ΣH_2S in Spire vent fluids exhibit a large variation relative to fluids from Shiner Bock and 1035H despite their close spatial proximity. These variations are consistent with the conversion of pyrite to pyrrhotite in the ODP mound after the penetration of an impermeable seal during the drilling of ODP Hole 1035H and the initiation of hydrothermal flow through permeable conduits. The large and rapid changes demonstrate that the redox-dependent equilibration of sulfide minerals in upflow zones is rapid and may influence the chemistry of hydrothermal fluids during the ascent from deep-seated reaction zones to the seafloor.

Comparison of fluid composition measured in 2000 with fluid compositions observed in 1990 and 1991 indicates that the concentrations of major and minor inorganic ions at the Dead Dog Vent Field and ODP Mound remained constant over a 10-year period. Time-invariant fluid compositions and vent temperatures indicate that the composition, mineralogy, temperature, and pressure in subsurface reaction zones did not change. A key observation is that the abundance of sediment-derived mobile trace elements (Li, Cs, ΣNH_3) did not vary, suggesting that the volume of the hydrothermal reservoir is sufficient to produce residence times for vent fluids that are greater than 10 years.

Acknowledgments. The outcome of this study was highly dependent on the skills, professionalism, and able assistance of the captain and crew of the *R/V Atlantis* and *DSV Alvin* team. We thank them for a highly successful and enjoyable cruise. Tom McCollom, Karen Hurley, Hallie Marbet, and Dane Percy provided invaluable laboratory assistance aboard ship. We thank Marv Lilley and an anoymous reviewer for comments that helped improve this manuscript. This work was funded by NSF OCE-9906752.

REFERENCES

Anderson, T. F., and L. M. Pratt (1995), Isotopic evidence for the origin of organic sulfur and elemental sulfur and marine sediments, in *Geochemical Transformations of Sedimentary Sulfur*, edited by M. A. Vairavamurthy and M. A. A. Schoonen, pp. 378–396, American Chemical Society, Washington, D. C.

Alt-Epping, P., and L. Smith (2001), Computing geochemical mass transfer and water/rock ratios in submarine hydrothermal systems: Implications for estimating the vigour of convection. *Geofluids* 1, 163–181.

Berndt, M. E., W. E. Seyfried Jr., and J. W. Beck (1988), Hydrothermal alteration processes at midocean ridges: Experimental and theoretical constraints from Ca and Sr exchange reactions and Sr isotopic ratios, *J. Geophys. Res*, *93*, 4573–4583.

Berndt, M. E., W. E. Seyfried, and D. R. Janecky (1989), Plagioclase and epidote buffering of cation ratios in mid-ocean ridge hydrothermal fluids: Experimental results in and near the supercritical region, *Geochim. Cosmochim. Acta*, *53*, 2283–2300.

Berner R. A., A. C. Lasaga, and R. M. Garrels (1983), The carbonate-silicate geochemical cycle and its effect on atmospheric carbon dioxide over the past 100 million years. *Am. J. Sci.*, *283*, 641–683.

Bischoff, J. L., and F. W. Dickson (1975), Seawater-basalt interaction at 200°C and 500 bars: Implications for the origin of seafloor heavy-metal deposits and the regulation of seawater chemistry, *Earth Planet. Sci. Lett.*, *25*, 385–397.

Bischoff, J. L., and R. J. Rosenbauer (1985), An empirical equation of state for hydrothermal seawater (3.2 percent NaCl), *Am. J. Sci.*, *285*, 725–763.

Bjerkgard, T., B. Cousens, and J. M. Franklin (2000), The Middle Valley sulfide deposits, northern Juan de Fuca Ridge; radiogenic isotope systematics, *Econ. Geol.*, *95*, 1473–1488.

Butterfield, D. A., and G. J. Massoth (1994), Geochemistry of north Cleft segment vent fluids: Temporal changes in chlorinity and their possible relation to recent volcanism, *J. Geophys. Res.*, *99*, 4951–4968.

Butterfield, D. A., R. E. McDuff, J. Franklin, and C. G. Wheat (1994a), Geochemistry of hydrothermal vent fluids from Middle Valley, Juan de Fuca Ridge, in *Proc. Ocean Drill. Program, Sci. Results*, vol. 139, edited by M. J. Mottl, E. E. Davis, A. T. Fisher, and J. F. Slack, pp. 395–410.

Butterfield, D. A., R. E. McDuff, M. J. Mottl, M. D. Lilley, J. E. Lupton, and G. J. Massoth (1994b), Gradients in the composition of hydrothermal fluids from the Endeavour segment vent field: Phase separation and brine loss, *J. Geophys. Res.*, *99*, 9561–9583.

Campbell, A. C., T. S. Bowers, C. I. Measures, K. K. Falkner, M. Khadem and J. M. Edmond (1988), A time series of vent fluid compositions from 21°N, East Pacific Rise (1979, 1981, 1985) and the Guaymas Basin, Gulf of California (1982, 1985), *J. Geophys. Res.*, *93*, 4537–4549.

Canfield, D. E. (2001), Biogeochemistry of sulfur isotopes, in *Stable Isotope Geochemistry*, edited by J. W. Valley and D. R. Cole, pp. 607–636, Mineralogical Society of America, Washington, D. C.

Charlou, J. L., J. P. Donval, Y. Fouquet, P. Jean-Baptiste, and N. Holm (2002), Geochemistry of high H_2 and CH_4 vent fluids issuing from ultramafic rocks at the Rainbow hydrothermal field (36°14′N, MAR), *Chem. Geol.*, *191*, 345–359.

Chen, A. C.-T., C.-M. Lin, B.-T. Huang, and L.-F. Chang (1996), Stoichiometry of carbon, hydrogen, nitrogen, sulfur and oxygen in particulate matter of the western Pacific marginal seas, *Mar. Chem.*, *54*, 179–190.

Cousens, B. L., J. Blenkinsop, and J. M. Franklin (2002), Lead isotope systematics of sulfide minerals in the Middle Valley hydrothermal system, northern Juan de Fuca Ridge, *Geochem. Geophys. Geosyst*, *3*(5), 1029, doi: 10.1029/2001GC000257.

Cruse, A. M., and J. S. Seewald (2001), Metal mobility in sediment-covered ridge-crest hydrothermal systems: Experimental and theoretical constraints. *Geochim. Cosmochim. Acta*, *65*(19), 3233–3247.

Cruse, A. M. and J. S. Seewald (2006), Geochemistry of low-molecular weight hydrocarbons in hydrothermal fluids from Middle Valley, northern Juan de Fuca Ridge, *Geochim. Cosmochim. Acta*, *70*, 2073–2092.

Davis, E. E., and H. Villinger (1992), Tectonic and thermal structure of the Middle Valley sedimented rift, northern Juan de Fuca Ridge, in *Proc. Ocean Drill. Program, Init. Rep.*, vol. 139, edited by E. E. Davis, M. J. Mottl, A. T. Fisher, et al., pp. 9–41.

Ding, K., and W. E. Seyfried Jr. (1992), Determination of Fe-Cl complexing in the low pressure supercritical region (NaCl fluid)—Iron solubility constraints on pH of subseafloor hydrothermal fluids, *Geochim. Cosmochim. Acta*, *56*, 3681–3692.

Ding, K., and W. E. Seyfried Jr. (1994), Effect of conductive cooling on chemistry of mid-ocean ridge hydrothermal fluids: Experimental and theoretical constraints, *Mineral. Mag.*, *58A*, 231–232.

Ding, K., W. E. Seyfried Jr., M. K. Tivey, and A. M. Bradley (2001), In situ measurement of dissolved H_2 and H_2S in high-temperature hydrothermal fluids at the Main Endeavour Field, Juan de Fuca Ridge, *Earth Planet. Sci. Lett.*, *186*, 417–425.

Ding, T., S. Valkiers, H. Kipphardt, P. De Bièvre, P. D. T. Taylor, R. Gonfiantini, and R. Krouse (2001), Calibrated sulfur isotope abundance ratios of three IAEA sulfur isotope reference materials and V-CDT with a reassessment of the atomic weight of sulfur, *Geochim. Cosmochim. Acta*, *65*, 2433–2437.

Duckworth, R. C., A. E. Fallick, and D. Rickard (1994), Mineralogy and sulfur isotopic composition of the Middle Valley massive sulfide deposit, northern Juan de Fuca Ridge, in *Proc. Ocean Drill. Program, Sci. Results*, vol. 139, edited by M. J. Mottl, E. E. Davis, A. T. Fisher, and J. F. Slack, pp. 373–385.

Edmond, J. M., G. Massoth, and M. D. Lilley (1992), Submersible-deployed samplers for axial vent waters, *RIDGE Events*, *3*, 23–24.

Fournier, R. O. (1983), A method of calculating quartz solubilities in aqueous sodium chloride solutions, *Geochim. Cosmochim. Acta*, *47*, 579–586.

Foustoukos, D. I., and W. E. Seyfried Jr. (2007), Quartz solubility in the two-phase and critical region of the NaCl–KCl–H_2O system: Implications for submarine hydrothermal vent systems at 9°50′N East Pacific Rise, *Geochim. Cosmochim. Acta*, *71*, 186–201.

Goodfellow, W. D., and B. Blaise (1988), Sulfide formation and hydrothermal alteration of hemipelagic sediment in Middle Valley, northern Juan de Fuca Ridge, *Can. Mineral.*, *26*, 675–696.

Goodfellow, W. D., and J. M. Franklin (1993), Geology, mineralogy, and chemistry of sediment-hosted clastic massive sulfides in shallow cores, Middle Valley, Northern Juan de Fuca Ridge, *Econ. Geol.*, *88*, 2037–2068.

Goodfellow, W. D. and J. M. Peter (1994), Geochemistry of hydrothermal altered sediment, Middle Valley, northern Juan de Fuca Ridge, in *Proc. Ocean Drill. Program, Sci. Res.*, vol. 139, edited by M. J. Mottl, E. E. Davis, A. T. Fisher, and J. F. Slack, pp. 207–289.

Johnson, J. W., E. H. Oelkers, and H. C. Helgeson (1992), SUPCRT92: A software package for calculating the standard molal thermodynamic properties of minerals, gases, aqueous species, and reactions from 1 to 5000 bar and 0 to 1000°C, *Comput. Geosci.*, *18*, 899–947.

Karsten, J. L., S. R. Hammond, E. E. Davis, and R. Currie (1986), Detailed morphology and neotectonics of the Endeavour segment, Northern Juan de Fuca Ridge: New results from Seabeam swath mapping, *Geol. Soc. Am. Bull.*, *97*, 213–221.

Koroleff, F. (1976), Determination of nutrients, in *Methods of Seawater Analysis*, edited by K. Grasshoff, pp. 125–157, Verlag-Chemie, Weinheim, Germany.

Leybourne, M. I., and W. D. Goodfellow (1994), Mineralogy and mineral chemistry of hydrothermally altered sediment, Middle Valley, Juan de Fuca Ridge, in *Proc. Ocean Drill. Program, Sci. Results*, vol. 139, edited by M. J. Mottl, E. E. Davis, A. T. Fisher, and J. F. Slack, pp. 155–206.

Lister, C. R. B. (1983), The basic physics of water penetration into hot rock, in *Hydrothermal Processes at Seafloor Spreading Centers*, edited by P. A. Rona, K. Boström, and K. L. Smith Jr., pp. 141–168, Plenum, New York.

McCollom, T. M., and E. Shock (1997), Geochemical constraints on chemolithoautotrophic metabolism by microorganisms in seafloor hydrothermal systems, *Geochim. Cosmochim. Acta*, *61*, 4375–4391.

Mottl, M. J., and H. D. Holland (1978), Chemical exchange during hydrothermal alteration of basalt by seawater—I. Experimental results for major and minor components of seawater, *Geochim. Cosmochim. Acta*, *42*, 1103–1115.

Ohmoto, H., and R. O. Rye (1979), Isotopes of carbon and sulfur, in *Geochemistry of Hydrothermal Ore Deposits*, edited by H. L. Barnes, pp. 798, John Wiley, New York.

Peter, J. M., W. D. Goodfellow, and M. I. Leybourne (1994), Fluid inclusion petrography and microthermometry of the Middle Valley hydrothermal system, northern Juan de Fuca Ridge, in *Proc. Ocean Drill. Program, Sci. Results*, vol. 139, edited by M. J. Mottl, E. E. Davis, A. T. Fisher, and J. F. Slack, pp. 411–428.

Rees, C. E., W. J. Jenkins, and J. Monster (1978), The sulphur isotopic composition of ocean water sulphate, *Geochim. Cosmochim. Acta*, *42*, 377–381.

Richter, F. M., D. B. Rowley, and D. J. DePaolo (1992), Sr isotope evolution of seawater: The role of tectonics, *Earth Planet. Sci. Lett.*, *109*, 11–23.

Sakai, H., D. J. Des Marais, A. Ueda, and J. G. Moore (1984), Concentrations and isotope ratios of carbon, nitrogen and sulfur in ocean-floor basalts, *Geochim. Cosmochim. Acta*, *48*, 2433–2441.

Schmidt, K., A. Koschinsky, D. Garbe-Schönberg, L. M. de Carvalho, R. Seifert (2007), Geochemistry of hydrothermal fluids from the ultramafic-hosted Logatchev hydrothermal field 15°N on the Mid-Atlantic Ridge: Temporal and spatial investigation, *Chem. Geol.*, *242*, 1–21.

Seewald, J. S., and W. E. Seyfried Jr. (1990), The effect of temperature on heavy and base metal mobility in subseafloor hydrothermal systems: Constraints from basalt alteration experiments and field studies, *Earth Planet. Sci. Lett.*, *101*, 388–403.

Seewald, J. S., W. E. Seyfried Jr., and E. C. Thornton (1990), Organic-rich sediment alteration: An experiment and theoretical study at elevated temperatures and pressures, *Appl. Geochem.*, *5*, 193–209.

Seewald, J. S., A. M. Cruse, and P. Saccocia (2002a), Aqueous volatiles in hydrothermal fluids from the Main Endeavour Field, northern Juan de Fuca Ridge: Temporal variability following earthquake activity, *Earth Planet. Sci. Lett.*, *216*, 575–590.

Seewald, J. S., K. W. Doherty, T. R. Hammar, and S. P. Liberatore (2002b), A new gas-tight isobaric sampler for hydrothermal fluids, *Deep-Sea Res. I*, *49*, 189–196.

Seyfried, W. E., Jr. (1987), Experimental and theoretical constraints on hydrothermal alteration processes at mid-ocean ridges, *Annu. Rev. Earth Planet. Sci.*, *15*, 317–335.

Seyfried, W. E., Jr., and J. L. Bischoff (1979), Low-temperature basalt alteration by seawater: an experimental study at 70°C and 150°C, *Geochim. Cosmochim. Acta*, *43*, 1937–1947.

Seyfried, W. E., Jr., and J. L. Bischoff (1981a), Experimental seawater-basalt interaction at 300°C, 500 bars, chemical exchange, secondary mineral formation and implications for the transport of heavy metals, *Geochim. Cosmochim. Acta*, *45*, 135–147.

Seyfried, W. E., Jr., and J. L. Bischoff (1981b), Experimental seawater-basalt interaction at 300°C, 500 bars: Chemical exchange, secondary mineral formation and implications for the transport of heavy metals, *Geochim. Cosmochim. Acta*, *45*, 135–149.

Seyfried, W. E., Jr., and K. Ding (1993), The effect of redox on the relative mobilities of Cu and Fe in Cl bearing aqueous fluids at elevated temperatures and pressures: An experimental study with application to subseafloor hydrothermal systems, *Geochim. Cosmochim. Acta*, *57*, 1905–1917.

Seyfried, W. E., Jr., K. Ding, and M. E. Berndt (1991), Phase equilibria constraints on the chemistry of hot spring fluids at mid-ocean ridges, *Geochim. Cosmochim. Acta*, *55*, 3559–3580.

Seyfried, W. E., K. Ding, and B. Rao (2002), Experimental calibration of metastable plagioclase-epidote-fluid equilibria at elevated temperatures and pressures: Applications to the chemistry of hydrothermal fluids at mid-ocean ridges, in *Water-Rock Interactions, Ore Deposits, and Environmental Geochemistry: A Tribute to David A. Crerar*, edited by R. Hellmann and S. A.

Wood, *Geochem. Soc., Spec. Publ.*, *7*, 257–277, The Geochemical Society, St. Louis, Missouri.

Seyfried, W. E., Jr., J. S. Seewald, M. E. Berndt, K. Ding, and D. I. Foustoukos (2003), Chemistry of hydrothermal vent fluids from the Main Endeavour Field, northern Juan de Fuca Ridge: Geochemical controls in the aftermath of June 1999 seismic events, *J Geophys. Res.*, *108*(B9), 2429, doi: 10.1029/2002JB001957.

Shipboard Scientific Party (1992a), Site 855, in *Proc. Ocean Drill. Program, Init. Rep.*, vol. 139, edited by E. E. Davis, M. J. Mottl, A. T. Fisher et al., pp. 101–160.

Shipboard Scientific Party (1992b), Site 858, in *Proc. Ocean Drill. Program, Init. Rep.*, vol. 139, edited by E. E. Davis, M. J. Mottl, A. T. Fisher et al., pp. 431–569.

Shipboard Scientific Party (1998a), Middle Valley: Bent Hill Area (Site 1035), in *Proc. Ocean Drill. Program, Init. Rep.*, vol. 169, edited by Y. Fouquet, R. A. Zierenberg, D. J. Miller et al., pp. 35–152.

Shipboard Scientific Party (1998b), Middle Valley: Dead Dog Area (Site 1036), in *Proc. Ocean Drill. Program, Init. Rep.*, vol. 169, edited by Y. Fouquet, R. A. Zierenberg, D. J. Miller et al., pp. 153–203.

Stein, J. S., and A. T. Fisher (2001), Multiple scale hydrothermal circulation in Middle Valley, northern Juan de Fuca Ridge: Physical constraints and geologic models, *J. Geophys. Res.*, *106*, 8563–8580.

Toulmin, P., III, and P. B. Barton Jr. (1964), A thermodynamic study of pyrite and pyrrhotite, *Geochim. Cosmochim. Acta*, *28*, 641–671.

Von Damm, K. L. (1991), A comparison of Guaymas Basin hydrothermal solutions with other sedimented systems and experimental results, in *The Peninsular Province of the Californias*, edited by J. P. Dauphin and B. R. T. Simoneit, *AAPG Mem.*, *47*, 743–751.

Von Damm, K. L., J. M. Edmond, B. Grant, C. I. Measures, B. Walden, and R. F. Weiss (1985), Chemistry of submarine hydrothermal solutions at 21°N, East Pacific Rise, *Geochim. Cosmochim. Acta*, *49*, 2197–2220.

Von Damm, K. L., J. L.Bischoff, and R. J. Rosenbauer (1991), Quartz solubility in hydrothermal seawater: an experimental study and equation describing quartz solubility for up to 0.5M NaCl solutions, *Am. J. Sci.*, *291*, 977–1007.

Von Damm, K. L., S. E. Oosting, R. Kozlowski, L. G. Buttermore, D. C. Colodner, H. N. Edmonds, J. M. Edmond, and J. M. Grebmeier (1995), Evolution of East Pacific Rise hydrothermal vent fluids following a volcanic eruption, *Nature*, *375*, 47–50.

Von Damm, K. L., C. M. Parker, R. A. Zierenberg, M. D. Lilley, E. J. Olson, D. A. Clague, and J. S. McClain (2005), The Escanaba Trough, Gorda Ridge hydrothermal system: Temporal stability and subseafloor complexity, *Geochim. Cosmochim. Acta*, *68*, 4971–4984.

Werne, J. P., D. J. Hollander, T. W. Lyons, and J. S. Sinninghe Damsté (2003), Organic sulfur biogeochemistry: Recent advances and future research directions, in *Sulfur Biogeochemistry*, edited by J. P. Amend, K. J. Edwards, and T. W. Lyons, pp. 135–150, Geological Society of America, Boulder, Colorado.

Wolery, T. J. (1992), EQ3NR, *A Computer Program for Geochemical Aqueous Speciation-Solubility Calculations: Theoretical Manual, User's Guide, and Related Documentation* (Version 7.0), Lawrence Livermore National Lab.

Wolery, T. J., and S. A. Daveler (1992), *EQ6, A Computer Program for Reaction Path Modeling of Aqueous Geochemical Systems: Theoretical Manual, User's Guide, and Related Documents*, Lawrence Livermore National Lab.

You, C.-F., D. A. Butterfield, A. J. Spivak, J. M. Gieskes, T. Gamo, and A. J. Campbell (1994), Boron and halide systematics in submarine hydrothermal systems: Effects of phase separation and sedimentary contributions, *Earth Planet. Sci. Lett.*, *123*, 227–238.

Zierenberg, R. A. (1994), Data report: Sulfur content of sediment and sulfur isotope values of sulfide and sulfate minerals from Middle Valley, in *Proc. Ocean Drill. Program, Sci. Results*, vol. 139, edited by M. J. Mottl, E. E. Davis, A. T. Fisher, and J. F. Slack, pp. 739–748.

Zierenberg, R. A., and D. J. Miller (2000), Overview of Ocean Drilling Program Leg 169: Sedimented Ridges II, in *Proc. Ocean Drill. Program, Sci. Results*, vol. 169, edited by W. R. Normark, pp. 1–39.

Zierenberg, R. A., et al. (1998), The deep structure of a sea-floor hydrothermal deposit, *Nature* 392, 485–488.

A. M. Cruse, School of Geology, Oklahoma State University, 105 Noble Research Center, Stillwater, Oklahoma 74078-3031, USA. (anna.cruse@okstate.edu)

Reactive Transport and Numerical Modeling of Seafloor Hydrothermal Systems: A Review

Peter Alt-Epping and Larryn W. Diamond

Institute of Geological Sciences, University of Bern, Bern, Switzerland

Sites of oceanic hydrothermal activity constitute complex systems in which numerous physical and chemical processes interact at various rates. Most numerical studies of these systems that have been carried out to date have focused on either the physical or the chemical aspects of the hydrothermal activity, while greatly simplifying the other aspects or neglecting them altogether. Because transport of solutes through geological systems is what drives fluid/rock interaction, it is evident that in order to design realistic models, coupling of physical and chemical processes is essential. This implies that fluid flow, heat flow, solute transport, and chemical reaction will have to be integrated into a single model. This single model can then be used to describe how these processes are linked, interact, and evolve through space and time. With the recent advance of reactive transport models in combination with the enhanced performance of computers, there is now an opportunity to apply coupled models to oceanic hydrothermal systems and explore the interaction between the physical and chemical processes that take place in those settings. Current reactive transport models are not capable of simulating all facets of oceanic hydrothermal systems, but existing reactive transport models may provide the basis for developing a coupled model that is specifically designed for this problem. In this work we provide an overview of recent attempts to model oceanic hydrothermal systems and illustrate the salient features and possibilities of such models using new results from two simple, generic simulations carried out with existing reactive transport codes.

1. INTRODUCTION

Reactive transport models have become an important tool for the analysis of coupled physical, chemical, and biological processes in geological systems. These models have been used in a variety of disciplines, such as chemical engineering, petroleum engineering, and soil science, and in different environments such as groundwater systems, contaminant plumes, marine settings, and metamorphic and magmatic systems. The most active areas of research that spurred the development of reactive transport models are the treatment of subsurface contamination and the investigation of potential sites for the disposal of nuclear waste.

Reactive transport models are typically designed to address a specific type of problem and there is currently no model that considers the full range of chemical and physical processes that occur in geological systems. A review of the current state of reactive transport models can be found in the work of *MacQuarrie and Mayer* [2005], including a list of selected reactive transport codes and their capabilities, and *Steefel et al.* [2005]. *Raffensperger* [1996], *Lichtner* [1996], and *Steefel and MacQuarrie* [1996] provide an introduction to the theory of reactive transport modeling and its application.

Magma to Microbe: Modeling Hydrothermal Processes at Ocean Spreading Centers
Geophysical Monograph Series 178

10.1029/178GM09

Reactive transport models integrate natural processes, field observations, and laboratory experiments into one theoretical framework [*Steefel et al.,* 2005]. Geochemical processes, for instance, can be viewed in the context of the behavior of the entire system, such as in relation to fluid flow, solute transport, heat transport, and microbiological processes. These processes are often coupled and show a complex interplay at different spatial and temporal scales. For example, the dissolution and precipitation of minerals affect the solute concentrations and mineral volume fractions, which alter the porosity and permeability of the medium, which in turn feed back to influence the flow velocity. The temperature distribution, which depends on the fluid flow pattern, affects the local thermodynamic conditions and/or the rates of mineral precipitation and dissolution. The chemical reactions may in turn have a feedback impact on the thermal regime. Thus, the processes that are incorporated into a model framework influence the evolution and the distribution of properties of the system to varying degrees. Reactive transport simulation is a tool to unravel the role and relative importance of such fundamental processes in natural systems [*Steefel et al.,* 2005].

2. EARLY MODEL DEVELOPMENTS

The development of reactive transport models was initiated by merging two modeling approaches. The first emphasized the physical aspects of the geological environment, that is, the transport of fluids, gases, solids, and heat while keeping the formulation of fluid/rock interaction simple, or ignoring it entirely. The second line of models focused on the development and application of quantitative models for irreversible chemical reactions. We want to focus in this paper on the evolution, formulation, and application of these chemical models to oceanic hydrothermal systems. In the first section of this paper, we review the historical development of chemical models and their theoretical basis. In the second part, we provide an overview of previous efforts to model oceanic systems. Finally, we present two new reactive transport models to illustrate the insight that can be gained by coupling transport and chemistry.

Brinkley [1947] published the first algorithm that solved numerically for the equilibrium state of a multicomponent system based on the law of mass action. His approach required the equilibrium constants for the reactions occurring in the system. *White et al.* [1958] published a method for solving for the equilibrium state by minimizing the Gibbs free energy. Both methods, the equilibrium constant and the Gibbs minimization approach, differ in the way that a chemical simulation problem is set up and solved, but are computationally and conceptually equivalent [*Zeleznik and Gordon,* 1960, 1968; *Brinkley,* 1960]. This is because the equilibrium constant is itself a mathematical expression of minimized free energy [*Bethke,* 1996]. However, discussions on which approach is simpler or shows better convergence are still ongoing. Early equilibrium speciation computer programs such as SOLMNEQ [*Kharaka and Barnes,* 1973] and WATEQ [*Truesdell and Jones,* 1974] were designed to calculate the distribution of aqueous species, mineral saturation states, and gas fugacities from the bulk composition of a closed chemical system. These speciation models served as the basis for reaction path and reactive transport models.

Garrels and Mackenzie [1967] simulated the evaporation of spring water by running a sequence of speciation calculations each followed by the extraction of a certain amount of water. This "reaction path" approach extended the closed system equilibrium speciation calculations to an open-system process simulation [*Bethke,* 1996]. Reaction path models represent the simplest and most commonly used form of an open system. These models describe a sequence of equilibrium states in a beaker or a batch reactor as a result of a stepwise mass transfer between phases in the system through the incremental addition or removal of a reactant, and/or a stepwise temperature change. Equilibrium constraints, such as equilibrium with a gas phase in the atmosphere, can be imposed on the system during each step (Figure 1). If a mineral is added in a stepwise fashion, the added reactant dissolves irreversibly and during that process other minerals may become saturated and precipitate or existing minerals may dissolve. The process continues until the reactant is in equilibrium with the fluid or becomes exhausted [*Bethke,* 1996].

Helgeson and coworkers [*Helgeson,* 1968; *Helgeson et al.,* 1969, 1970] developed the reaction path program PATH1 and computerized the approach of *Garrels and Mackenzie* [1967]. After this early work, conceptual improvements

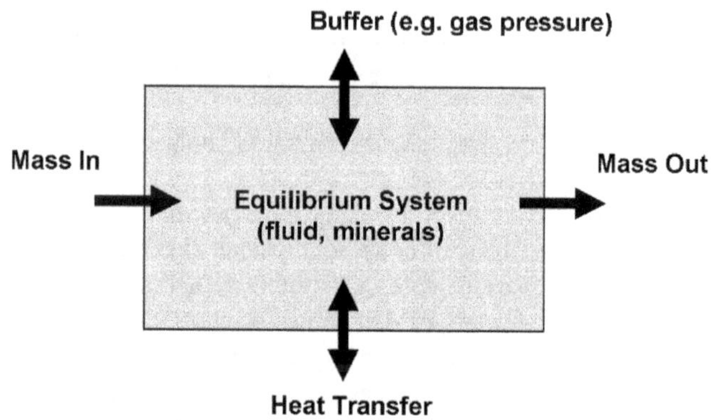

Figure 1. The principle of an equilibrium batch reactor in a reaction path model. Mass may be added or removed from the system, the temperature may be increased or decreased and the system may be buffered by an external gas phase in a series of equilibrium states along the reaction path [after *Bethke,*1996].

were made as well as improvements regarding numerical efficiency, such as simpler formulations of the governing equations. *Wolery* [1978] extended the mass transfer program by Helgeson and coworkers and calculated chemical equilibria between fluid and mineral phases in a basalt-seawater system for temperatures up to 350°C. *Wolery* [1979, 1983] developed the reaction path code EQ3/6, which was a derivative from the early work of Helgeson and coworkers. The EQ3/6 code was subsequently used in a number of studies of oceanic hydrothermal systems [e.g., *Bowers et al.,* 1985, 1988; *Bowers and Taylor,* 1985; *Janecky,* 1982; *Janecky and Seyfried,* 1984; *Janecky and Shanks,* 1988; *McCollom and Shock,* 1998], which will be discussed in greater detail in the second part of this paper.

Several scenarios can be simulated with reaction path models. For instance, in flow-through models the evolution of the composition of a fluid volume is monitored as it moves through an aquifer. This is simulated by removing reaction products from the system so that the fluid reacts only with fresh aquifer minerals. Alternatively, the evolution of the mineralogical composition of the aquifer can be computed if unreacted fluid is added to the system. Chemical and thermal mixing of fluids can also be accommodated in reaction path models. For instance, the mixing between an endmember hydrothermal solution and seawater can be approximated by titrating increasing amounts of seawater into the hydrothermal solution. The temperature of the mixed system can be calculated as a function of the mixing ratio. For example, *Janecky and Seyfried* [1984] used a linear relationship between mixing ratio and temperature while neglecting changes in heat capacity and conductive heat transfer.

Kinetic effects on mineral precipitation or dissolution can be implemented in reaction path models by suppressing the formation of certain minerals or by reducing their primary amounts so that the minerals will be completely dissolved during reaction progress. Other reaction path codes such as PHREEQC [*Parkhurst and Appelo,* 1999] include an optional kinetic formulation.

Common to all reaction path models is that transport (of solutes and heat) is not calculated explicitly but approximated by a sequence of equilibrium states with specified bulk composition, temperature, pressure, and water/rock (w/r) ratio. The w/r ratio is the mass ratio of fluid to rock in the reactor. It constitutes a crude link between the chemistry and the hydrology. Care has to be taken that the size of the reaction step (i.e., the change in temperature or w/r ratio) carried out along a reaction path is small enough such that no critical process is "overstepped."

Because of their simplified representation of the physical characteristics of a system, reaction path models are of limited use for problems where transport is the driving force of chemical reactions, that is, where chemical species are moved through the system by advection, dispersion, and diffusion [*Steefel et al.,* 2005]. Consequently, the next step in the development of quantitative chemical models was to couple (i.e., to solve together) the equilibrium or kinetic chemical models with fluid flow and mass transport models that had been used in the hydrology and engineering communities [e.g., *Person et al.,* 1996].

The multiple reaction path formulation provides a description of moving fronts by treating the system as a sequence of stationary states [*Lichtner,* 1988]. A stationary state reaction path is attained if the properties of the rock and minerals change slowly such that the local mineral reaction rates and fluid composition remain constant in time. From the perspective of an aliquot of fluid moving through the rock, the reaction progress variable can be thought of as travel time along the flowpath. An advantage of this approach is that unlike usual finite difference techniques, in which requirements of stability and accuracy restrict the size of the time step, this approach allows simulation of fluid/rock interaction over geological time scales.

Coupled multicomponent reactive-transport models based on the continuum approach originated in the mid-1980s [e.g., *Lichtner,* 1985, 1996; *Yeh and Tripathi,* 1989]. The models predict chemical processes and the position of reaction fronts along a groundwater flowpath as a function of space and time (Figure 2). In the continuum representation of porous media, the properties of the fluid and the solid are defined locally and are averaged and assumed to be uniform

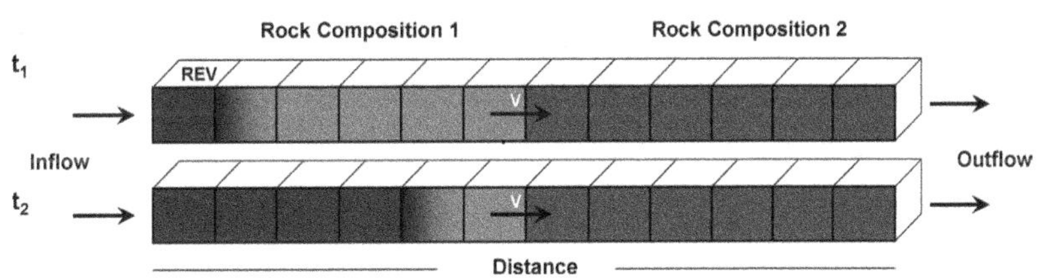

Figure 2. The principle of reactive transport along a 1-D flowpath. Fluid of a certain composition moves at a flowrate *v* along the flowpath while reacting with a rock composition at a certain temperature specified for each grid cell (REV). Fluid and rock compositions become a function of space and time t ($t_2 > t_1$).

over a representative elementary volume (REV) (Figure 2). These properties include, for instance, the porosity, mineralogy, fluid composition, fluid velocity, and the reactions rates. Similar to the beaker concept in reaction path models, the REV is a well-mixed reactor without internal gradients. This approach ignores the details of processes at a scale smaller than the REV (e.g., the pore scale). The scale of a REV may not always represent the scale at which field data are available or may not reflect the distribution of heterogeneities in the real medium. Consequently, the input of data or the characterization of the material in a reactive transport model typically involves some form of up-scaling or down-scaling. This implies that it is virtually impossible to reproduce or predict the real state of a system at any particular point in space and time [*Steefel et al.,* 2005].

3. MODEL PARAMETERS

3.1. Governing Equations

The building blocks for a reactive transport model of a non-isothermal system consist of equations for heat transport, fluid movement, transport of multiple species, and a set of (usually nonlinear) equations describing chemical reactions within the fluid phase (homogeneous reactions) and reactions between the minerals and the fluid or between a gas phase and the fluid (heterogeneous reactions) [e.g., *Rubin,* 1983]. Also needed are constitutive laws for important properties such as the density and viscosity of the fluid or the permeability evolution of the solid. The equations for heat transport and fluid flow are linked through the advection of heat and the temperature dependence of the fluid density and viscosity.

A convection-conduction equation without heat production or consumption due to chemical reactions or radioactive decay can be written as [e.g., *Steefel and Lasaga,* 1994]:

$$\rho_m c_m \frac{\partial T}{\partial t} = \nabla \cdot \left(\lambda_m \nabla T - \rho_f c_f \mathbf{q} T\right), \qquad (1)$$

where λ_m is the thermal conductivity of the bulk medium, T is the temperature, ρ_m and ρ_f are the densities of the bulk medium and the fluid, respectively, c_f and c_m are the specific heat of the fluid and the medium, respectively, and $\mathbf{q}$ is the Darcy flux vector according to:

$$\mathbf{q} = -\frac{\mathbf{k}}{\mu}\left(\nabla P - \rho_f \mathbf{g}\right) \qquad (2)$$

where $\mathbf{k}$ is the permeability tensor, μ is the dynamic viscosity, P is the fluid pressure, and $\mathbf{g}$ is the gravity vector [*Steefel and Lasaga,* 1994].

The fluid continuity equation can be written as:

$$\frac{\partial\left(\phi\rho_f\right)}{\partial t} = -\nabla \cdot \left(\rho_f \mathbf{q}\right) \qquad (3)$$

which simplifies for an incompressible fluid and matrix to

$$\nabla \cdot \mathbf{q} = 0 \qquad (4)$$

where ϕ is the porosity [*Steefel and Lasaga,* 1994]. To solve coupled fluid flow and heat transport problems, appropriate equations of state relating fluid density and viscosity to temperature are also required.

The chemical system is defined by a set of chemical components, secondary aqueous species, minerals, and gases. Chemical components are defined to be linearly independent chemical entities. Every (secondary) aqueous species, mineral, or gas that is not part of the component set can be uniquely expressed in terms of a combination of these components, yet no one component is a combination of other components [e.g., *Raffensperger and Garven,* 1995].

The governing equation for N_c dissolved species describing the conservation of solute mass in the aqueous phase of a system that includes both transport and reaction in a saturated porous medium can be written as [e.g., *Mayer et al.,* 2002]:

$$\frac{\partial}{\partial t}(\phi C_j) - \nabla \cdot \mathbf{q} C_j + \nabla \cdot D \nabla C_j = R_j \qquad j = 1, N_c \quad (5)$$

where C_j is the total aqueous component concentration (mol/L), D is the tensor of hydrodynamic dispersion, which incorporates diffusive and dispersive transport of all species in the system, and R_j (mol L^{-1} s^{-1}) is the rate of addition of species j to the fluid by all heterogeneous (mineral dissolution/precipitation), sorption, and homogeneous (aqueous) reactions. R_j is negative if species j is removed from solution via the precipitation of a mineral or via its adsorption on a mineral surface.

If a total component concentration is defined as [*Kirkner and Reeves,* 1988; *Yeh and Tripathi,* 1989; *Steefel and Lasaga,* 1994; *Lichtner,* 1996; *MacQuarrie and Mayer,* 2005]

$$C_j^{tot} = C_j^c + \sum_{i=1}^{N_s} \nu_{ij}^s C_i^s \quad j = 1, N_c \qquad (6)$$

where C_j^c is the concentration of a component j in solution, C_i^s is the concentration of a secondary species i, ν_{ij}^s is the stoichiometric coefficient of the component in the dissociation or association reaction of the secondary species, and N_s is the number of secondary species in solution, then the governing differential equation (equation (5)) can be written in terms of total concentration where only aqueous species are involved [*Kirkner and Reeves,* 1988]. The total concentration of a component in solution is the sum of the concentration of the component and the amount of that component present in all the secondary species. The total concentrations of all components account for all of the solute mass in the system. If total concentrations rather than species concentrations are substituted into the transport equation the number of transport equations that need to be solved can be reduced considerably [*Yeh and Tripathi,* 1989].

The state of a geochemical system can be calculated by assuming either equilibrium between its reactants and reaction products, or a kinetic relationship, or both. A set of governing equations fully describing the state of the system consists of two parts: mass balance equations that require mass to be conserved and either mass action equations or kinetic rate laws that describe equilibrium or a time-dependent relationship between the activities of reactants and reaction products, respectively. For any given chemical reaction at equilibrium, the distribution between reactants and reaction products is defined by an equilibrium constant according to [e.g., *MacQuarrie and Mayer,* 2005]:

$$C_i^s = (K_i^s \gamma_i^s)^{-1} \prod_{j=1}^{N_c} (\gamma_j^c C_j^c)^{v_{ij}^s} \qquad i = 1, N_c \tag{7}$$

where C_j^c is the concentration of a primary species in solution, C_i^s is the concentration of a secondary species, K_i^s is the equilibrium constant for the dissociation reaction into components of secondary species i, and γ_i^s and γ_j^c are the activity coefficients for the secondary species and the primary species, respectively. Activity coefficients for solutions with low ionic strength ($I < 0.1$–0.7 M) are commonly calculated from the extended Debye-Hückel or the Davis equation. For high ionic-strength solutions such as brines, the Pitzer approach for activity corrections may be required [e.g., *Pitzer,* 1979, 1987]. The law of mass action (equation (7)) provides an algebraic link between the components and secondary species for each reversible reaction. The final form of the governing equations describing a chemical system at equilibrium is given by substituting the mass action equations into the mass balance equations and expressing the concentration of secondary species in terms of the components.

It is commonly assumed that aqueous reactions are more rapid than the normal rates of fluid flow and hence that the equilibrium assumption for aqueous reactions is valid. This assumption has a computational advantage because through the law of mass action the number of unknowns is reduced from the total number of aqueous species by the independent chemical reactions between the components [*Steefel and Lasaga,* 1994].

Describing mineral dissolution/precipitation reactions at equilibrium and finding a stable mineral assemblage is somewhat more complex than finding the equilibrium fluid composition. Some models employ a procedure of elimination where undersaturated minerals are removed from the assemblage and oversaturated minerals added to it. The sequence of removal or addition is given by the degree of over- or undersaturation. This process proceeds until the calculation predicts a system at equilibrium [*Bethke,* 1996].

If fluid/rock interaction is described kinetically, the reaction rate of a mineral can be formulated as a function of a rate constant, an equilibrium constant, the reactive surface area of the mineral, the ionic strength, and the composition of the fluid. A rate law for mineral precipitation and dissolution can be written as [e.g., *Lasaga,* 1981a, 1981b, 1984; *Aagaard and Helgeson,* 1982; *Steefel and Van Capellen,* 1990; *Steefel and Lasaga,* 1994]:

$$r_m = \operatorname{sgn}\left(\log\left(\frac{Q_m}{K_m}\right)\right) A_m k_m \left(\prod_{i=1}^{N_i} a_i^p\right) \times \left|\left(\frac{Q_m}{K_m}\right)^M - 1\right|^n, \tag{8}$$

where k_m is the dissolution rate constant of mineral m at 25°C, a_i is the activity of an inhibiting or catalyzing species i in solution raised to an empirically determined power p, N_i is the total number of species, M and n are empirically determined exponents to account for nonlinear rate laws, Q_m is the saturation product, K_m is the equilibrium constant of mineral m, and $\operatorname{sgn}\left(\log\left(\frac{Q_m}{K_m}\right)\right)$ denotes the sign of the affinity term $\left|\left(\frac{Q_m}{K_m}\right)^M - 1\right|$, which is negative when the fluid is undersaturated and positive when the fluid is oversaturated with respect to the mineral. The affinity term ensures that there is no net dissolution of a mineral phase once the reaction attains equilibrium.

The reactive surface area of a mineral is probably the least constrained parameter in the rate expression. It is virtually impossible to measure the reactive surface within a system at a scale larger than a petrographic thin section. In addition, the reactive surface area can be expected to vary spatially and with time [*Steefel and VanCapellen,* 1990]. One approach to calculate changes in the reactive surface area after mineral dissolution and precipitation is to relate the surface area to changes in the porosity, according to

$$A_m = A_m^0 \left(\frac{\phi_m}{\phi_m^0}\right)^{\frac{2}{3}}, \quad (9)$$

where A_m^0 and ϕ_m^0 are the initial reactive surface area and the initial porosity, respectively [*Lichtner,* 1996].

The temperature dependence of the rate constant is commonly expressed through the Arrhenius equation.

$$k = k_{25} \exp\left[\frac{-E_a}{R}\left(\frac{1}{T} - \frac{1}{298.15}\right)\right], \quad (10)$$

where E_a is the activation energy, k_{25} is the rate constant at 25°C, R is the gas constant, and T is the temperature in Kelvin.

3.2. Equilibrium Versus Kinetic Approach

Reactive transport models can be classified according to whether reactions are "sufficiently fast" or "insufficiently fast" relative to the local flow rate to justify local equilibrium [*Rubin,* 1983]. Local equilibrium is a concept used for open systems and implies that equilibrium has been attained over some portion of the system (i.e., in terms of a continuum formulation over some REV). Thus, unlike in a closed system where at equilibrium forward and backward reactions rates are equal and have opposite signs, mass transfer may take place under a local equilibrium condition and is determined by transport across the boundary of that volume. At local equilibrium a fluid is in equilibrium with the rock over the scale of the REV but may be in disequilibrium with its neighboring REVs [*Lichtner,* 1996].

A kinetic formulation of mineral dissolution/precipitation reactions is the more general approach and allows for greater flexibility in describing individual reactions. Equilibrium can be approximated in such a case by assigning fast reaction rates. It is only the equilibrium approach that involves assumptions with regard to the system's behavior [*MacQuarrie and Mayer,* 2005]. However, the user has to be aware that choosing a fast rate for a certain mineral reaction will make this reaction the dominant one in the system. The relative magnitude of reaction rates therefore requires information about the relative importance of chemical processes in the natural system.

The relation between the time scales of advective transport and of chemical reaction and an expression for the closeness of a system to local equilibrium can be written in terms of the advective Damköhler number [*Damköhler,* 1936]:

$$Da = \frac{l_a k}{v} \quad (11)$$

where l_a is the advection length of interest, v is the linear advective flow velocity, and k is the effective rate constant for a simple first-order kinetic reaction. If $Da \gg 1$, the time scales of reactions are shorter than those of advective transport, such that mass transfer between the rock and the fluid is transport-limited and the reactions can be expressed as equilibrium reactions. Conversely, if $Da \ll 1$, advective transport is faster than chemical reactions and a kinetic formulation is required to describe the reactions. In this case, an equilibrium formulation would overestimate the mass transfer. The Damköhler number implies uniform and constant flow velocities and reaction rates over the length scale of interest. Therefore, in hydrothermal systems the length scale must be chosen carefully to take account of the dependence of reaction rates on temperature gradients.

In heterogeneous, transient systems with multiple reacting minerals, the equilibrium assumption may be a reasonable approximation in certain regions of the system and/or at certain times and/or for certain minerals. However, in other regions, or at other times, or for other minerals in the system, the approximation may not be reasonable [*MacQuarrie and Mayer,* 2005]. It is advantageous to have a reaction formulation implemented that allows for the selection of either an equilibrium or kinetic description of the reaction progress.

3.3. Solution Methods for Coupling of Transport and Reaction Processes

Multicomponent reactive transport simulations can be computationally extremely intensive, because the numerical solution of a chemical system requires, for each time-step, the solution of a minimum of $N_j \times M$ equations, where N_j is the number of component species and M the number of grid cells in the system [*Steefel and Lasaga,* 1994]. Thus, in addition to appropriate computational hardware, efficient numerical methods are required to make reactive transport problems tractable.

There are two distinct formulations for solving reactive transport processes. The global implicit or one-step and the sequential or operator splitting approach [e.g., *Engesgaard and Christensen,* 1988; *Yeh and Tripathi,* 1989, 1991; *Mangold and Tsang,* 1991; *Saaltink et al.,* 2001]. The tightest coupling between chemistry and transport is achieved in the global implicit method, where the equations for equilibrium or kinetic chemical reactions are substituted directly into the transport equation. The resulting set of nonlinear partial differential equations can then be solved simultaneously with a Newton-Raphson-type solution method. In the operator splitting method, transport and reaction equations are more loosely coupled and solved sequentially. Only the governing partial differential equations for transport show spatial derivatives, whereas the nonlinear algebraic equations describing chemistry are solved as an isolated set at each point within the flow domain, using the total aqueous concentrations from the transport equations from the previous iteration or time step. Fluid/rock reactions are calculated and the mass transfer between the fluid and the rock is treated as a source or sink in the transport equation [i.e., the R term in equation (5)].

An accurate solution of the global set of equations in the sequential approach requires iteration between the reaction and transport terms [*Yeh and Tripathi,* 1991]. Without such iteration, the sequential method may introduce an operator splitting error arising from the assumption that physical transport is sufficiently rapid so that the reactions only begin after the transport is complete. However, provided that the coupling between reaction and transport is weak, or that the time-step taken is small, the reaction and transport terms may be solved separately [*Steefel and Lasaga,* 1994]. This error may be reduced by using a multistep method such as embedding the reaction step between two transport steps.

One advantage of the operator splitting method over the global implicit method is a lower computational memory requirement, because there is no need to invert very large matrices. However, with the recent advances in computational memory, this advantage is becoming less significant. Because reaction and transport can be treated separately, the operator splitting method is more flexible in terms of the time discretization of transport and reaction [e.g., *Prommer et al.,* 2003] or in the selection and coupling of already existing routines for solving flow, transport, and reaction (e.g., the code PHAST; *Parkhurst et al.,* 2002). On the other hand, because of limitations with regard to the size of the time-step, excessive computer time may be required to track the system over geologic time scales. In contrast, the simultaneous solution of transport and reaction may show better and faster global convergence than the decoupled methods [*Steefel and Lasaga,* 1994]. In addition, the global implicit approach may allow larger time steps to be taken than the two-step approach, making it more appropriate for simulating geological time scales [*MacQuarrie and Mayer,* 2005].

3.4. Reactive Transport Environments

Chemical reactions that accompany transport processes occur where the fluid changes its thermodynamic state as it moves through the system. These changes can be induced by gradients in the composition of the solid phases, spatial and temporal gradients in temperature and pressure, or spatial and temporal gradients in the composition of the aqueous phase. Thus, according to *Phillips* [1991], three main reactive transport environments can be distinguished. Fluid/rock reactions can be initiated through changes in the rock composition when a fluid that is buffered by a mineral assemblage comes in contact with a different mineral assemblage (Figure 3, first panel). Alternatively, fluid/rock reactions occur if the rock composition is uniform but if the injected fluid is in disequilibrium with the rock. Both processes lead

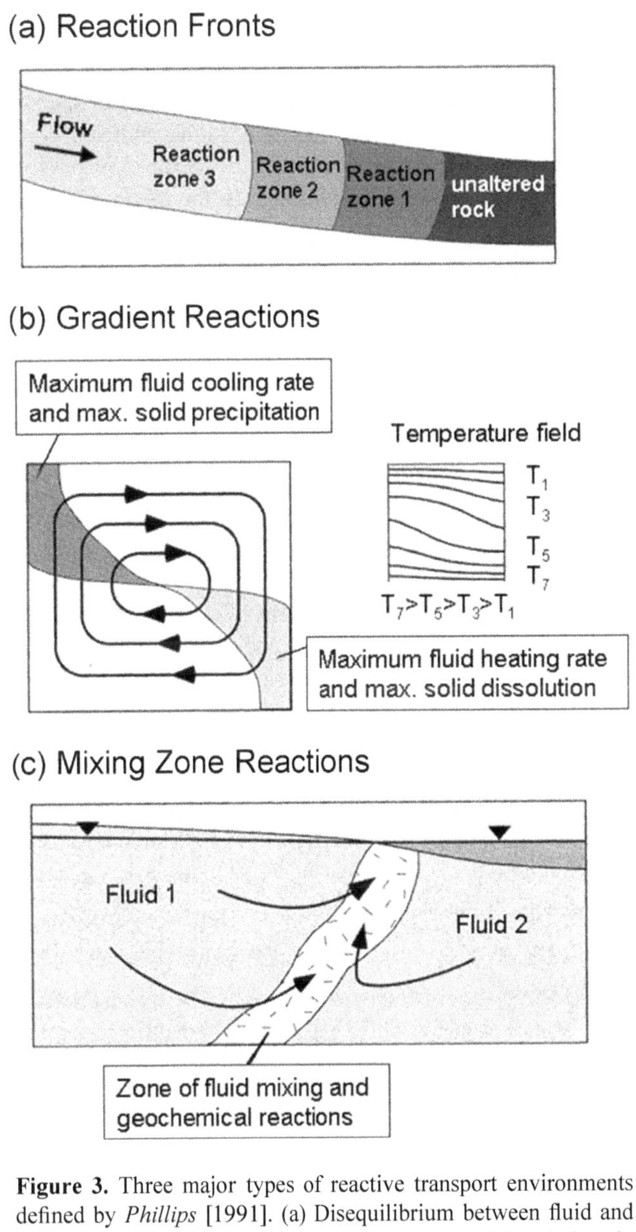

Figure 3. Three major types of reactive transport environments defined by *Phillips* [1991]. (a) Disequilibrium between fluid and rock composition, (b) gradient reactions due to spatial variations in temperature and pressure, (c) mixing of fluids [after *Person et al.*, 1996; modified after *Ingebritsen et al.*, 2006].

to reaction zones in the rock that move through space and evolve through time.

If the rock composition is uniform, reactions are also initiated by fluids carrying the solutes across temperature and/or pressure gradients (Figure 3, second panel). The result is the dissolution or precipitation of mineral phases toward a new steady state or local equilibrium state. *Phillips* [1991] defined for hydrothermal systems the potential for rock alteration as "rock alteration index":

$$\mathrm{RAI} = \mathbf{q}\nabla T \tag{12}$$

where $\mathbf{q}$ is the specific discharge and ∇T is the temperature gradient. Positive values of RAI indicate areas of convection and heating, whereas negative values indicate convection and cooling. A value of zero indicates no-flow or isothermal conditions. The RAI is based on the assumption that fluid flow across temperature gradients is the dominant process driving chemical reactions, and that pressure and compositional gradients are negligible.

Finally, reactions may also occur when two or more fluids mix. Mixing zone reactions tend to produces more localized alteration or mineral precipitation (Figure 3, third panel).

In oceanic hydrothermal systems, all of these reactive transport environments occur and often the processes associated with these environments occur simultaneously. In the recharge zone, for instance, reactions due to fluid/rock disequilibrium are combined with gradient reactions, whereas in the discharge zone fluid mixing may become an important process in addition to gradient reactions.

Unlike in reaction path models, where fluid mixing is implemented either by titrating one solution into another solution or by specifying a mixing ratio between two solutions at one reaction step, in reactive transport models fluid mixing occurs due to the physical process of mechanical dispersion and molecular diffusion. The degree of mixing is therefore determined by the coefficient of hydrodynamic dispersion (equation (5)), which can be written as [e.g., *Freeze and Cherry,* 1979]:

$$D_l = \alpha_l \bar{v} + D_m^*, \tag{13}$$

where D_l is the coefficient of hydrodynamic dispersion in the longitudinal direction, $\boldsymbol{\alpha}_l$ is the dispersivity of the medium, $\bar{v}$ is the average linear flow velocity, and D_m^* is the coefficient of molecular diffusion. Mixing ratios may change in space and time. One approach to track mixing ratios in reactive transport models is to trace each solution involved in the mixing process with a nonreactive tracer such that the concentration ratios of the tracers correspond to the mixing ratios.

3.5. Modeling Porosity/Permeability Changes Due to Chemical Reactions

Several studies have addressed the effects of coupling between reactions and fluid flow and reaction-induced porosity and permeability changes [e.g., *Wood and Hewett,* 1982; *Ortoleva et al.,* 1987; *Hoefner and Fogler,* 1988; *Sanford and Konikow,* 1989; *Lowell et al.,* 1993; *Steefel and Lasaga,* 1990, 1994; *Bolton et al.,* 1996, 1997; *Martin and Lowell,* 1997, 2000]. In particular, over long time scales permeability changes due to alteration will impact on the evolution of hydrological, thermal, and chemical conditions [*Bolton et al.,* 1999]. Assessment of the relationship between chemically induced porosity changes and corresponding changes in permeability requires that fluid flow, heat and solute transport, and chemistry be fully coupled. This means that the flow field has to be recalculated during the simulation with an updated permeability field that incorporates porosity/permeability changes due to chemical reactions.

The permeability of a rock is probably the most important parameter for determining flow rates and, unfortunately, like the reactive surface area, it is difficult to measure. The permeability of a porous medium describes the interconnectedness of the pores [*Freedman et al.,* 2003] and its magnitude depends not only on the porosity, but also on the pore-size distribution, the geometry of the pores, and their connectivity [*Islam et al.,* 2001]. The permeability of a rock and thus the flow rates or flow pattern will be affected as precipitating minerals fill up pore and fracture space, clog a pore throat and/or change the geometry of pores or fractures, or as dissolution of minerals creates pore space and opens new pathways for fluids [*MacQuarrie and Mayer,* 2005]. A review of processes leading to permeability changes can be found in the report of *Saripalli et al.* [2001].

The most commonly used relationship between porosity and permeability is the Carman-Kozeny model [e.g., *Bolton et al.,* 1996, 1997], which is based on the assumption that the porous medium is granular, well-sorted, and unconsolidated. The porosity/permeability relationship can be written as [*Carman,* 1956]:

$$k = \frac{\phi^3}{5A^2(1-\phi)^2} \tag{14}$$

where k is the permeability, ϕ is the porosity, and A is the solid surface area exposed to the fluid. The Carman-Kozeny relationship does not account for the effect of pore geometry and tends to underestimate permeability reduction [*Saripalli et al.,* 2001]. Modified versions of the Carman-Kozeny or other relationships that do account for the pore geometry then require additional parameters such as particle size, cementation factor, shape factors, and surface roughness [e.g., *Panda and Lake,* 1994, 1995].

Whereas much of the research on processes affecting porosity/permeability changes has focused on porous media, permeability changes in fractured, crystalline rocks have received relatively little attention [*Steefel and Lasaga,* 1990, 1994; *Oelkers,* 1996; *Le Gallo et al.,* 1998; *Saripalli et al.,* 2001]. In most reactive transport models, porosity changes are calculated from the net volume change due to mineral precipitation/dissolution reactions according to [e.g., *Steefel and Lasaga,* 1994]:

$$\phi = 1 - \sum_{i=1}^{N_m} \phi_i^m \quad (15)$$

where ϕ_i^m is the volume fraction of mineral *i*. Correspondingly, temporal changes in fracture aperture caused by mineral precipitation/dissolution can be formulated as [*Steefel and Lichtner,* 1998]:

$$\frac{\partial \delta}{\partial t} = -\partial \sum_{i=1}^{N_m} \frac{\partial \phi_i^m}{\partial t} \quad (16)$$

where δ is the fracture aperture. Models addressing porosity and permeability changes in fractured media are commonly based on simple fracture geometries such as parallel fractures with smooth walls [e.g., *Steefel and Lasaga,* 1994; *Martin and Lowell,* 2000]. For a medium with N parallel fractures per unit distance ($1/L$) and fracture aperture δ (L), the fracture porosity can be written as [e.g., *Snow,* 1968]

$$\phi_f = N\delta \quad (17)$$

The relationship between the permeability and the geometry of fractures in a rock can then be written as [*Snow,* 1968]:

$$k = \frac{N\delta^3}{12} \quad (18)$$

where k is the permeability. This relationship holds where Darcy's law is valid and it can be applied to a volume of rock of sufficient size to act as a Darcian continuum [*Freeze and Cherry,* 1979]. *Steefel and Lasaga* [1994] found that significant departure from equilibrium with silicate phases can occur in thermal boundary layers such as in discharge zones at the seafloor. For flow rates sufficiently slow that the temperature gradient is linear, local equilibrium may be a reasonable assumption unless fractures are widely spaced. For a two-dimensional, convective system, the authors showed that reaction-induced permeability change are significant enough to prevent the convection cell from attaining a hydrodynamic or thermal steady state. This is because the permeability tends to be reduced in the upflow zone and enhanced in the downflow zone of the convection cell due to cooling and mineral precipitation and heating and mineral dissolution, respectively. This leads to more diffuse flow in the upflow zone and more channeled flow in the discharge zone. *Bolton et al.* [1996] found that because of the unsteady nature of a convective system, reversals of the quartz saturation state occur in some regions of the convective domain and these may induce zonation in precipitated quartz crystals. Also, if the fluid is far from equilibrium with quartz, it can remain under- or oversaturated even if it moves down or up a temperature gradient, respectively.

Martin and Lowell [1997] investigated the sealing of cracks accounting for both thermoelastic stresses and silica precipitation as fluid moves down a temperature gradient. The results from their numerical analysis were in good agreement with laboratory experiments on granite core samples. *Dobson et al.* [2003] used a 1-D plug-flow numerical model to simulate the precipitation of amorphous silica and the corresponding porosity and permeability reduction in a single fracture under nonisothermal conditions. Their calculated rate of permeability reduction was consistent with laboratory experiments. A kinetic control of reactions approximated the distribution of amorphous silica along the fracture more closely than equilibrium control. The simulation was sensitive to a number of parameters, such as the initial mineral assemblage, thermodynamic parameters used, and, most importantly, the reactive surface area.

Most studies assume a positive correlation between porosity and permeability, but exceptions occur when the morphology of minerals, such as clay minerals, reduces the permeability while the porosity increases [e.g., *Le Gallo at al.,* 1998]. Clearly, it is essential to implement a representative porosity/permeability relationship in fully coupled models.

3.6. Spatial and Temporal Scales in Reactive Transport

The processes that are incorporated in reactive transport simulations may operate and be effective on different spatial and temporal scales. Addressing fractured media, for instance, involves multiple length- and time scales that are characterized by distinct properties for the fracture network, on the one hand, and for the rock matrix, on the other. Infiltration of a fluid through widely spaced fractures occurs faster (owing to higher permeabilities) and the ratios of surface area to fluid volume (i.e., the potential for water/rock exchange) are lower than in the matrix that contains pore spaces, microcracks, and grain boundaries. Thus, mass transfer in a fractured system will be greatly reduced and even though the rock may see strong alteration along the fracture margins, the fluid composition will undergo only minor changes along the flowpath [e.g., *Cruse et al.*, this volume]. Porosity, permeability, and reactive surface areas may be either enhanced or reduced through chemical alteration processes over time in both a fracture and the matrix. As a consequence, the composition of water and the type of mineral reactions in a fracture and in the matrix may be different, a situation that requires a dual continuum model to describe each length scale.

Depending on the spatial scale of interest, different formulations can be used to describe reactive transport in porous media: (1) continuum models, (2) pore-scale models (pore network, lattice Boltzmann models), and (3) multiple continua, including hybrid models (combined pore scale and continuum scale models). *Steefel et al.* [2005] and *MacQuarrie and Mayer* [2005] provide an overview of these different approaches in greater detail.

3.7. Thermodynamic Databases

Numerical models that use a thermodynamic formulation to calculate the chemical state of a system require a thermodynamic database. Codes that use the Gibbs minimization approach read in standard thermodynamic properties and calculate the equilibrium state from the system's bulk composition at the pressure and temperature condition of interest. In the equilibrium constant approach, the database contains molar properties of minerals, aqueous species, and gases, and the equilibrium constants (log K) of the dissociation or association reactions of all minerals and aqueous species, as well as the constants for dissolution of gases. Usually, the equilibrium constants are tabulated at different temperature points so that the code can interpolate to intermediate temperatures. The pressure dependence of the equilibrium constants is considered only implicitly through the pressure that is associated with the equilibrium constant in the database. Tailored log K-based databases can be constructed over a certain temperature and pressure range from databases containing standard thermodynamic properties, such as the SUPCRT92 database [*Johnson et al.,* 1992].

All thermodynamic data are subject to uncertainties. These may arise from difficulties in characterizing experimental reactants and products, from random and systematic errors in the property measurements, and from the inevitable leeway in extrapolating and interpreting the results. However, few databases report these uncertainties and no codes attempt to propagate them through reactive transport calculations. Moreover, databases are not always internally consistent, because they may represent a compilation of values from various sources without having been subjected to global optimization. Thermodynamic databases may also be incomplete, that is, they may lack important phases and solid solutions, and they may contain trivial (e.g., typographic) errors. Comparison of the databases in chemical simulation packages from different evolutionary lines (e.g., EQ3/6, PHREEQC, MINTEQA2) is often hindered by differences in their format and standard states. Databases such as FreeGs [*Bastrakov et al.,* 2004] or the CTDP project [*Van der Lee and Lomenech,* 2004] are an attempt to compile available thermodynamic data and include references to the origin and, in most cases, a consistency and/or quality assessment of the data. Critically reviewed databases are typically limited to a selected subset of thermodynamic data, such as the Nuclear Energy Agency database for the uranium system. *Nordstrom and Munoz* [1994], *Haas and Fisher* [1976], *Helgeson et al.* [1978], and *Johnson et al.* [1992] provide overviews of this issue.

4. CHEMICAL MODELS OF OCEANIC HYDROTHERMAL SYSTEMS

Oceanic hydrothermal systems involve movement of fluid from the seafloor into the oceanic crust and back to the seafloor (Figure 4). This induces fluid/rock reactions that are driven by initial disequilibria between the fluid and the rock as well as fluid movement through temperature gradients. The fluid's pathway determines both alteration patterns in the rock and fluid composition at any point along the flowpath. Chemical reactions will alter the porosity and the permeability of the rock through time, affecting the flow pattern, heat transport, and mass transport. Ideally, a model designed to simulate oceanic hydrothermal systems considers these processes simultaneously with appropriate formulations for the feedback between individual processes. However, simpler modeling approaches with reaction path models have provided valuable insights into the chemical aspects of oceanic hydrothermal systems.

Wolery [1978] carried out the first reaction path equilibrium and mass transfer calculations of seawater/basalt interaction at mid-ocean ridges and demonstrated that observations from experimental studies can be reproduced by theoretical models. He predicted the occurrence of minerals associated with greenschist facies alteration at elevated temperatures and demonstrated the general validity of the thermodynamic database. *Reed* [1983], in theoretical calculations, and *Janecky and Seyfried* [1984], with the computer code EQ3/6 [*Wolery,* 1983], explored the formation of massive sulfide deposits as a result of cooling and mixing of endmember fluids with seawater. Their model successfully reproduced mineral assemblages that are observed in and

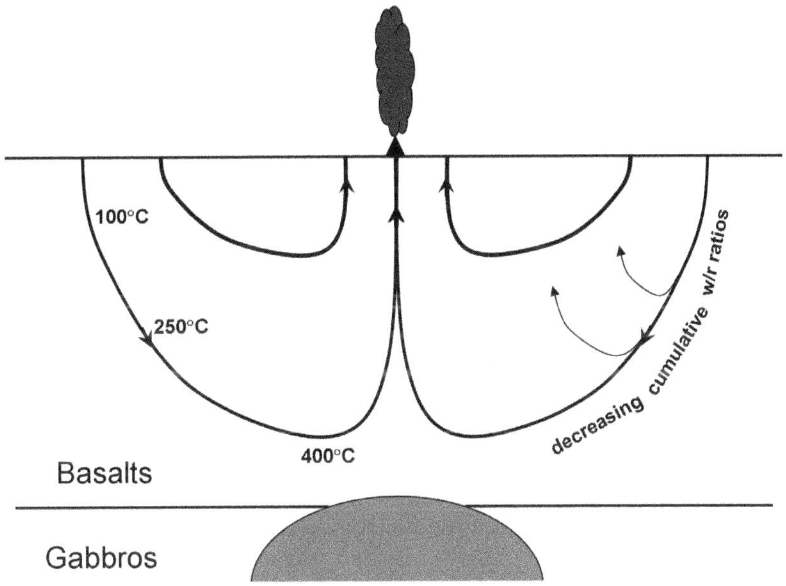

Figure 4. Schematic diagram showing the concept of a single-pass flow-through system at a ridge crest. Seawater recharges into the upper oceanic crust, reacts with a basaltic rock at high temperatures and discharges back into the ocean [after *Bowers and Taylor*, 1985].

around chimneys from 21°N on the East Pacific Rise (EPR). The formation of chalcopyrite, anhydrite, and pyrrhotite at high temperatures and at low mixing ratios is followed by the precipitation of sphalerite and barite at lower temperatures and at high mixing ratios. Subsequent replacement of early formed pyrrhotite with pyrite is apparently kinetically inhibited and the calculated bulk composition of the mixing product is very similar to the bulk composition of known massive sulfide deposits such as that on Cyprus. Anhydrite precipitation occurs through mixing and conductive heating of seawater on the chimney walls. However, the abundance of quartz and talc predicted by the mixing simulation was inconsistent with the absence of quartz and talc at 21°N. The lack of quartz in massive sulfide deposits, its abundance in feeder zones, and the fact that the solutions at 21°N are not in overall equilibrium with the rock led *Bowers et al.* [1985] to conclude that kinetic effects play an important role. The authors suggest that the silicification of the feeder zone may occur during the latest, cooling stage of the system when fluid residence times are high enough to allow for quartz precipitation. Evidence for potential disequilibrium conditions with quartz also comes from experimental studies. The experiments by *Berndt et al.* [1989] demonstrated that a fluid attains silica equilibrium with a diabase only under high w/r ratios, and that quartz equilibrium is also more likely if the rock has a glassy texture.

Bowers et al. [1985] used EQ3/6 to look in greater detail at the importance of conductive cooling versus fluid mixing. They showed that conductive cooling of a hydrothermal fluid from 21°N (EPR) without interaction with seawater produces sulfide mineral assemblages consistent with those found in chimneys and conduits plus quartz. If mixing with seawater is included in the model, anhydrite appears in the mineral assemblage. Similar calculations were applied to the Guayamas Basin in the Gulf of California, which is part of an active spreading center but overlain by several hundred meters of terrigenous sediments. These calculations demonstrated that the characteristics of the fluid and of the mineral alteration profile found in the sediments can be explained by the interaction between EPR-type fluids and the sediments. Thus, the model results were consistent with the idea that the Guayamas Basin may represent a system that is actively forming.

Bowers and Taylor [1985] used EQ3/6 to track chemical and isotopic changes due to seawater/basalt interaction in a ridge crest hydrothermal system. A pathway of a fluid migrating from the seafloor into the reaction zone in the oceanic crust was simulated in a sequence of steps of specified temperature and w/r ratio representing conditions at certain depths in the crust (Figure 4). The exchange of O^{16}/O^{18} was included in the simulation to provide an additional constraint on the reaction evolution pathway. The authors state that the two most important factors controlling the outcome of this reaction path simulation were the effective w/r ratios, that is, the amount of rock that chemically interacts with a fluid, and temperature. In oceanic hydrothermal systems, w/r ratios decrease with depth because (1) not all of the recharging fluid descends into the high-temperature region of the crust and (2) reaction rates increase with depth thereby allowing more rocks to react with the fluid [*Bowers and Taylor,* 1985]. Decreasing w/r ratios were modeled by increasing the amount of rock that was titrated into solution. Even though some discrepancies between observations and model results exist, in particular with regard to the pH and mineral saturation states, the model by *Bowers and Taylor* [1985] produced mineral assemblages that agree very well with samples from the oceanic crust and with observations from ophiolite complexes. Also, the computed composition of the fluid agreed with observations. The small discrepancies mentioned above could be attributed to uncertainties in or to the incompleteness of the thermodynamic dataset, to the simplifying assumptions underlying this simulation, such as the use of mineral endmembers rather than solid solutions, or to neglecting pressure effects [*Bowers and Taylor*, 1985]. In addition, the size of the temperature and w/r ratio steps taken along the reaction path may have been too large and a pH-affecting process at intermediate temperatures may have been overlooked.

Janecky and Shanks [1988] used a combination of chemical reactions and sulfur isotope exchange to demonstrate that fluid/rock interaction and fluid mixing are required in seafloor vent systems to produce the observed isotopic characteristics and metal zonation. Their reaction pathway simulations included different temperature scenarios, adiabatic versus conductive temperature changes, and equilibrium versus disequilibrium reactions.

One of the most important observations from individual vents at the seafloor is that, in the absence of seismic activity, the major-element composition of the vent fluid remains uniform through time [*Campbell et al.,* 1988; *Von Damm*, 1995]. In contrast, seismic activity may cause dramatic changes in composition, such as those encountered at the Main Endeavour Field [*Seyfried et al.*, 2003]. *Bowers et al.* [1988] addressed the observation of temporal uniformity with a numerical simulation involving EQ3/6 and a revised set of thermodynamic data. Their study demonstrated that the uniformity is controlled by equilibrium buffering of the fluid with respect to minerals of a greenschist assemblage at depth. For the vent fluids at 13°–11°N and 21°N (EPR), the mineral assemblage buffering the fluid appears to consist of various aluminosilicates, quartz, and an Fe sulfide phase. Variations between vents are due to temperature differences, whereas the local mineral assemblage is controlled by w/r ratios, permeability, and/or depth [*Bowers et al.,* 1988]. The inclusion of the exchange of sulfur isotopes in a following study [*Bowers,* 1989] indicated that, consistent with the major element chemistry, the $\delta^{34}S$ values in the vent fluid are controlled by rock buffering at high temperature and are less useful for analyzing the cumulative effect of water/rock interaction than oxygen and hydrogen isotopes or trace elements. Fluid buffering can imply local equilibrium between the fluid and the minerals of the rock at depth [*Reed and Spycher,* 1984; *Bowers et al.*, 1988] but, just as well, a non-equilibrium state where the ratio of dissolution and precipitation rates attain some steady state value.

Wells and Ghiorso [1991] explored the issue of thermodynamic equilibrium versus chemical steady state for a simple 1-D quartz + $SiO_2(aq)$ system with uniform Darcy flux and constant geothermal gradient and tested the applicability of the quartz geobarometer [*Von Damm et al.,* 1985; *Von Damm and Bischoff,* 1989] in hydrothermal upflow zones. The authors showed that the quartz geobarometer, which is based on the assumptions that the fluid at depth is at quartz equilibrium and that no silica mass transfer occurs during the ascent of the fluid, has to be used with caution. Conductive cooling during ascent, low w/r ratios (i.e., low flow velocities and/or narrow fracture spacing) may result in silica mass transfer that renders the quartz geobarometer useless. *Lowell et al.* [1993] modeled the sealing of a vertical fracture through silica precipitation in a hydrothermal upflow zone using an equilibrium model with temperature- and pressure-dependent silica solubility and a constant flow rate. The authors found that for high permeabilities and high temperatures in the reaction zone, sealing in the near-surface part of the upflow zone can occur on time scales of less than a decade. *Martin and Lowell* [2000] focused on quartz precipitation in a fractured hydrothermal discharge zone and found that, depending on whether an equilibrium or a kinetic formulation is used for silica mass transfer, in systems with high initial permeability the time scale for crack closure may differ by orders of magnitude (i.e., from months to several decades for the equilibrium and kinetic cases, respectively). The kinetic formulation also showed that thinner cracks will fill with silica more rapidly than wider cracks, which is consistent with seismic velocity data. *Fontaine et al.* [2001] investigated the precipitation of anhydrite and sulfide minerals in the discharge zone and found that the steep thermal gradients near the seafloor lead to anhydrite precipitation and the formation of a low-permeability layer that separates a high-temperature reaction zone from a lower-temperature top layer. The authors suggest that this process may be the cause of diffuse venting. High-temperature vents thus require the breaching of this low-permeability layer by faults or fractures through tectonic or volcanic events.

McCollom and Shock [1998] carried out a reaction path simulation of a fluid descending into the lower oceanic crust. The fluid is initially equilibrated with a basaltic rock at 300°C and 250 bars. It then traverses a linear *P/T* trajectory starting at 300°C and 500 bars and ending at 900°C and 2000 bars. The *P/T* path consists of temperature and pressure pairs in 100°C and 250-bar intervals, respectively. The pressure in the lower oceanic crust was overestimated to ensure single-phase conditions for water. One model scenario investigated the effect of w/r ratios on the alteration assemblage. *McCollom and Shock* [1998] found that, contrary to common belief, the amount of hydrous minerals (i.e., amphibole) is not an indicator of the extent of fluid flow. Instead, temperature and bulk composition are more important factors controlling the proportion of amphibole. The simulation demonstrated the feasibility of using thermodynamic models to examine quantitatively the evolution of fluid and mineral compositions during alteration at lower crustal conditions.

Wetzel et al. [2001] used a decoupled fluid flow, heat transport, and mass transfer simulation to assess conditions in young (0.2–1 Ma) oceanic crust. A series of reaction path simulations based on the earlier study by *McCollom and Shock* [1998] was carried out. Fluid flow and heat transport simulations provided constraints for the physical, spatial, and temporal context for the reaction path simulations. The study illustrated the need to develop and incorporate more realistic permeability models that better match field observations, and emphasized the need for fully coupled reactive transport models to capture the complete dynamics of the system.

Mottl [1983], *Sleep* [1991], and *Lowell and Yao* [2002] addressed the issue of anhydrite ($CaSO_4$) precipitation, the associated porosity/permeability reduction, and its effect on the flow rates in the recharge zone. Anhydrite precipitates when seawater is heated to temperatures in excess of ~150°C [*Blount and Dickson,* 1969; *Bischoff and Seyfried,* 1978]. The rate and distribution of anhydrite precipitation depends on the interplay between the flow rate, the thermal gradient, and the slope of the anhydrite solubility curve [*Lowell and Yao,* 2002]. The greatest permeability reduction through anhydrite precipitation occurs at the depth of the 150°C isotherm, that is, when anhydrite starts to become oversaturated. As the permeability becomes sealed, the flow rate slows and the isotherm and the region of anhydrite precipitation become shallower. Based on the calculations of *Lowell and Yao* [2002], sealing through anhydrite precipitation occurs rapidly if the recharge area is of the same size as the discharge area. Because anhydrite precipitation does not appear to impact high-temperature hydrothermal circulation, *Lowell and Yao* [2002] suggest that fluid recharge is diffuse rather than focused, occurring over a much larger area than fluid discharge.

Reactive transport simulations have been used to estimate flow rates through the sediments on ridge flanks matching calculated sediment alteration patterns and pore fluid composition to measured data. The simulation by *Fisher et al.* [2001] incorporated sediment burial, fluid advection, species-dependent diffusion, organic matter degradation, and mineral diagenesis to infer flow rates of bottom seawater through the sediments of 2–10 mm/yr at the east flank of the Alarcon Ridge. These low velocities are consistent with conductive seafloor heat flow. *Giambalvo et al.* [2002], in a similar study of sediments on the eastern flank of the Juan de Fuca Ridge, found upwelling velocities through the sediments on the order of 1.9 mm/yr (site 1030) and 3.2 mm/yr (site 1031).

Alt-Epping and Smith [2001] used a loosely coupled reactive transport simulation of conditions in the upper oceanic crust to explore the relationship between w/r ratios, chemical mass transfer, and flow velocities in a flow-through system. We discussed above that w/r ratios are used in reaction path models to represent, from the perspective of a volume of rock, the time-integrated fluid flux reacting with the rock. From the perspective of a volume of fluid, w/r ratio corresponds to the cumulative amount of rock with which the fluid has reacted. Considering a certain length of the flowpath in a flow-through system, the w/r ratio is a function of flow rates versus reaction rates and its meaning is equivalent to that of the Damköhler number (equation (11)).

If the mass transfer over a certain interval of the flowpath is at steady state, w/r ratios are directly related to the flow velocity. If W is the mass of water that has passed along the flowpath in a certain time and M is the total mass of rock that the fluid has reacted with, then the chemical w/r ratio can be written as [*Alt-Epping and Smith,* 2001]:

$$\frac{W}{M} = \frac{\Delta C_{i(r)}}{\Delta C_{i(w)}}, \tag{19}$$

where $\Delta C_{i(r)}$ is the loss/gain of species i per kilogram rock and $\Delta C_{i(w)}$ is the gain/loss of species i in the fluid phase over some interval of the flowpath. Increasing chemical w/r ratios is equivalent to increasing flow rates (W) or lowering $\Delta C_{i(w)}$ at constant reactivity $\Delta C_{i(r)}$, or increasing the reactivity and decreasing the amount of rock (M) at constant flow rates.

5. EXAMPLES OF REACTIVE TRANSPORT IN OCEANIC HYDROTHERMAL SYSTEMS

We illustrate the application of reactive transport models to oceanic hydrothermal systems with two relatively simple examples that build on studies discussed in the previous section.

These examples are generic and designed to elucidate the approaches and capabilities of coupled models.

5.1. Chemical Reaction and Porosity Changes (Discussion of First Example)

In the first example, we use the fully coupled reactive transport code FLOTRAN [*Lichtner,* 2007] to extend the study by *Lowell and Yao* [2002] and look for the first time at the effect of anhydrite precipitation in combination with fluid/rock interaction in the recharge zone. In addition to accounting for porosity changes due to anhydrite precipitation induced by the heating of seawater, we also include in this model the porosity changes and the release of additional Ca^{+2} (and thus the formation of additional anhydrite) induced by the alteration of the primary mineralogy. The model includes the chlorite endmembers clinochlore and daphnite as alteration phases. The primary igneous mineralogy consists of 58.73 vol % plagioclase ($an_{75}ab_{25}$) and 31.27 vol % clinopyroxene ($di_{80}he_{20}$). The initial porosity is 10%. We also include a suite of aqueous complexes that may form in the fluid. We assume the same rate constant for all alteration minerals (1×10^{-8} mol/m^2 s^{-1}) and the same rate constant and reactive surface area for all primary minerals (1×10^{-11} mol/m^2 s^{-1} and 10 m^2/m^3, respectively). The rates for clinopyroxene and plagioclase fall in the range of experimentally determined weathering rates reported by *Brantley and Chen* [1995] and *Blum and Stillings* [1995], respectively. High precipitation rates were assigned to the secondary minerals on the assumption that their formation would be limited by the recharge rate of seawater or by the supply rate of chemical constituents from the primary mineralogy. Differences in the reactivity among the primary and secondary minerals are a function of the affinity term and the reactive surface area of each mineral, which in turn depends on the volume fraction of each mineral.

The domain is set up in 2-D but the flowfield is essentially a 1-D flowpath embedded between two thermally conductive, nonreactive layers (Figure 5). The flow velocity is controlled by a fixed pressure gradient and by the permeability. The permeability is allowed to change as a result of mineral precipitation/dissolution reactions. Porosity and permeability are related through a simple cubic relationship:

$$k = k_0 \left(\frac{\phi}{\phi_0} \right)^3, \tag{20}$$

where k and ϕ are the permeability and porosity, and k_0 (1×10^{-12} m^2) and ϕ_0 are the initial permeability and initial porosity, respectively. Equation (20) states that if we assume a lower initial porosity for the system, the same porosity reduction would lead to a lower permeability. For instance, for an initial porosity of 1% (instead of 10%), a porosity reduction of 0.5% would lead to a permeability that is lower by a factor of about 6.9. However, this factor is not necessarily equivalent to the rate of sealing via mineral reactions because the porosity affects the reactive surface area of minerals (equation (9)) and thus the mineral reaction rate through equation (8).

A constant heat flux of 1 W/m^2 is applied to the bottom boundary, representing a geological setting similar to that at Middle Valley, Juan de Fuca Ridge. This heat flow value is somewhat higher than the basal heat flux of 750 mW/m^2 used by *Bessler et al.* [1994] in a numerical study on fluid flow and heat transport at this site. The temperatures at the top boundary, the left (inflow), and the right (outflow) side of the domain are fixed at 5°C (the lowest permissible temperature in this version of FLOTRAN). The simulation consists of two steps. In the first step, we run the flow and temperature fields to steady state and then in the second step we "switch on" the chemistry, that is, we inject seawater at the left-hand boundary and compute fluid/mineral reactions as the fluid moves along the high permeability layer.

The steady state velocity (i.e., specific discharge) and temperature profiles (time = 0, with respect to chemistry) are shown in Figures 6 and 7, respectively. The specific discharge

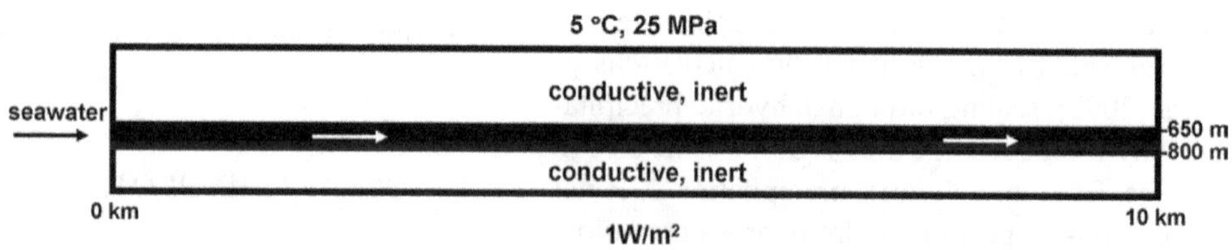

Figure 5. Model domain of a fully coupled reactive transport simulation (not drawn to scale). The seawater (arrow) is injected into a high permeability (black) layer which is heated from below. The flowpath is situated between two conductive layers. (Top layer (650 m): density, 2.7×10^3 kg/m^3; thermal conductivity, 2.0 J/s/m_K; specific heat; 800 J/kg_K. Permeable (50 m) and bottom layer (100 m): density, 2.9×10^3 kg/m^3; thermal conductivity, 2.0 J/s/m_K; specific heat; 837 J/kg_K.)

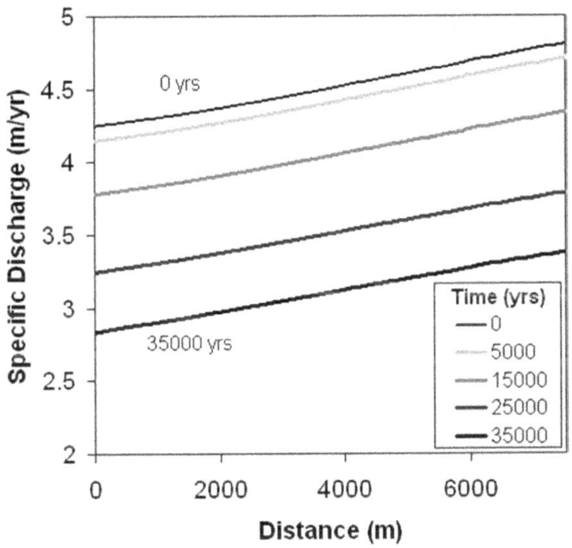

Figure 6. Evolution of the velocity profile. The velocity decreases with time due to the clogging of pore space by anhydrite precipitation.

increases along the flowpath because the rise in temperature expands the fluid. This effect of thermal expansion would go unnoticed if our fluxes were based on fluid mass because mass is conserved in the system. In Figure 8, the mineral alteration patterns are shown in terms of volume changes over 1000-year periods for two examples, one early in the modeled history (between 2000 and 3000 years) and one later in the history (between 34,000 and 35,000 years). Magnesium-rich chlorite starts to precipitate as soon as seawater enters the system. The amount of chlorite and the amount of dissolved primary minerals increase with temperature, consistent with the temperature dependence of the reaction rate. The similar shape of the chlorite and plagioclase profiles suggests that the rate of chlorite precipitation is limited by the supply rate of chemical constituents from plagioclase in the rock, rather than by the supply of Mg^{+2} derived from seawater or from the dissolution of clinopyroxene. Because $SiO_2(aq)$ increases and Al^{+3} decreases throughout this interval (Figure 9), the rate of chlorite formation is likely to be limited by the release of Al^{+3} from plagioclase. As the Fe^{+2} concentration in solution increases due to clinopyroxene dissolution, the Fe-rich chlorite precipitates at about the same position as the peak in clinopyroxene dissolution (Figure 8). Anhydrite starts to precipitate at elevated temperatures, removing SO_4^{-2} and Ca^{+2} from seawater and consuming the Ca^{+2} released from dissolving primary minerals. As time progresses, the zone affected by this alteration pattern shifts upstream and the intensity of alteration decreases (Figure 8). This is due to the combination of increasing temperatures (Figure 7), faster reaction rates, decreasing flow velocities, and reduced recharge of seawater (Figure 6). Note that Figure 8 shows alteration patterns over a fixed time interval during the simulation and not the cumulative pattern since time = 0. The cumulative pattern for anhydrite since time = 0 is displayed in Figure 10. This figure shows that the zone of anhydrite precipitation extends upstream but the location of maximum anhydrite precipitation is nearly stationary. This is due to the cumulative effect of anhydrite precipitation and the relatively large amounts of anhydrite formation during the early state of the system when recharge rates are highest. In other words, the instantaneous maximum rate of anhydrite precipitation moves upstream faster than the cumulative maximum anhydrite precipitation. This upstream migration of the region of anhydrite precipitation is consistent with the results reported by *Lowell and Yao* [2002]. Figure 10 includes the anhydrite profile at 20,000 years from a simulation with a lower recharge rate. The influx of cold seawater was reduced by a factor of 1.5. This reduced recharge leads to an upstream shift of the temperature profile and of the anhydrite peak. The zone of anhydrite precipitation is narrower than for higher fluxes as a result of the reduced supply of seawater-derived Ca^{+2} and SO_4^{-2} in combination with a higher ratio of precipitation rate to flow rate [i.e., a higher Damköhler number (equation (11)), and conditions closer to equilibrium). However, the maximum amount of anhydrite precipitation is about the same in both scenarios (Figure 10).

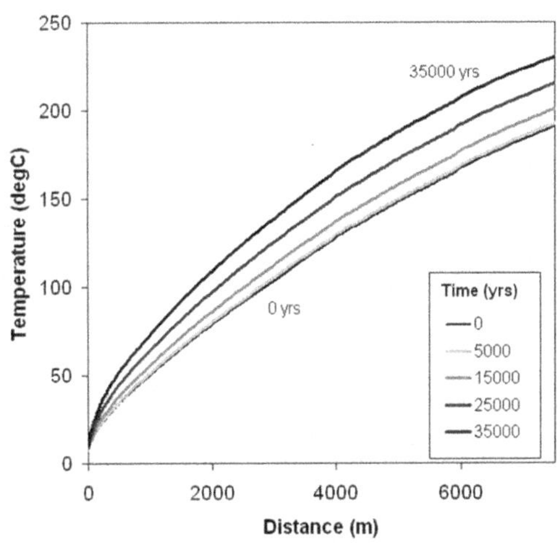

Figure 7. Evolution of the temperature profile. The temperature increase along the profile at time = 0 reflects the steady-state balance between fluid flow rate and heat supply. The temperature increase with time is due to the reduced recharge of cold seawater caused by the clogging of pore space by anhydrite precipitation (Figure 6).

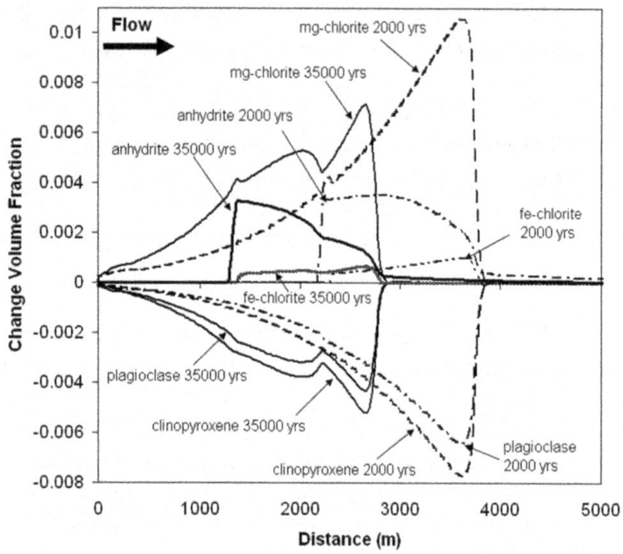

Figure 8. A generic model exploring feedback mechanisms between chemical reactions, hydrology, and thermal conditions in a recharge zone. Shown are changes of mineral volume fractions at two different times, integrated respectively over a period of 1000 years. Positive values are minerals precipitated, negative values are minerals dissolved.

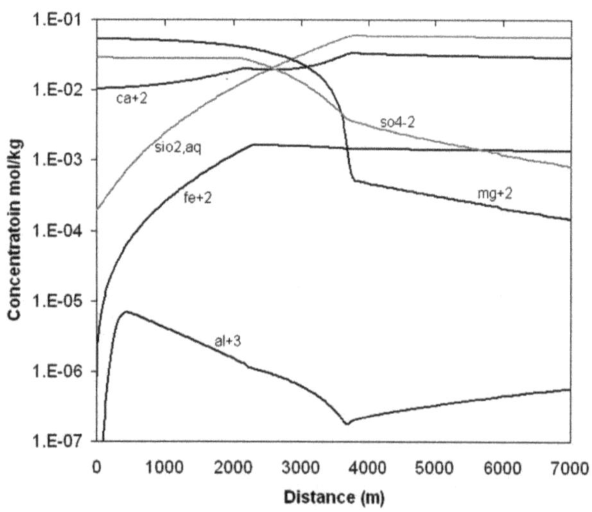

Figure 9. Steady-state fluid composition profile (total concentrations of components) along the flowpath at 2000 years. The concentrations of Mg^{+2} and Al^{+3} decrease as chlorite precipitates, whereas $SiO_2(aq)$, Fe^{+2}, and Ca^{+2} increase. The release of Ca^{+2} from the primary mineralogy exceeds the removal through anhydrite precipitation. Anhydrite precipitation is reflected in the depletion of SO_4^{-2}.

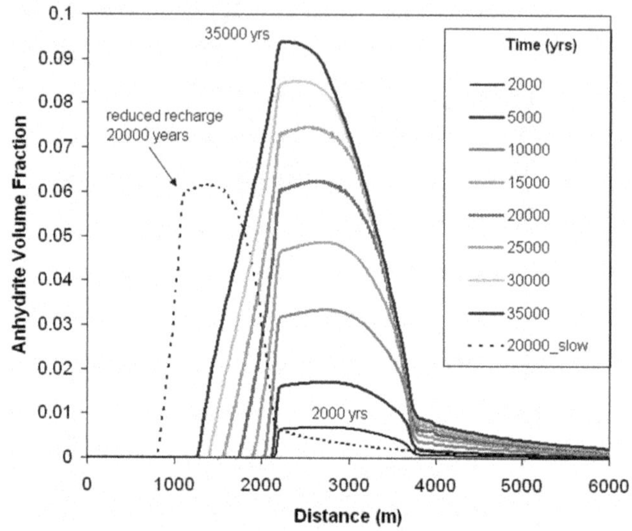

Figure 10. Cumulative amounts of anhydrite precipitation. The zone of anhydrite precipitation moves upstream faster than the anhydrite precipitation maximum. Reducing the recharge rate by a factor of 1.5 shifts the zone of anhydrite precipitation upstream and the zone becomes narrower. The maximum volume fraction of anhydrite is nearly unchanged.

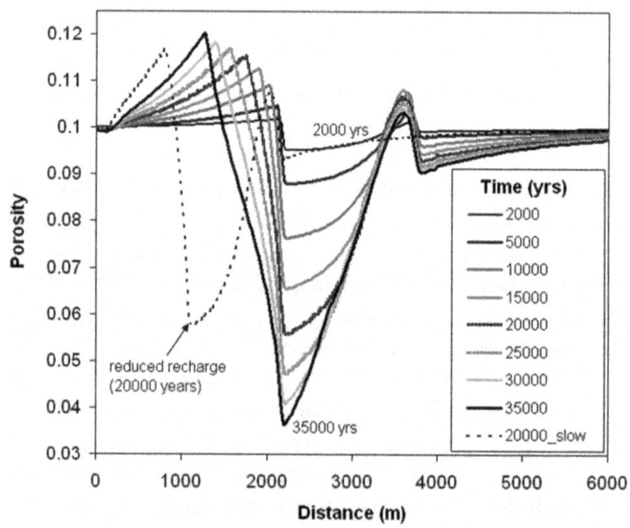

Figure 11. Cumulative porosity change representing the net volume change due to mineral dissolution/precipitation reactions. The maximum porosity reduction correlated with the maximum in the anhydrite profile (Figure 10). The chloritization of the primary mineral assemblage leads to a porosity increase and is counteracted by anhydrite precipitation. Note the similar degree of porosity reduction for high and low recharge rates at 20,000 years.

The evolution of the porosity profile represents the net volume change of all fluid/rock reactions. The large porosity reduction at 2100 m (Figure 11) correlates with the region of anhydrite precipitation (Figure 10), demonstrating that anhydrite precipitation is the dominant process of porosity reduction. The similar magnitude of porosity decrease for high and low recharge rates suggests that the hydrological consequences of anhydrite precipitation are not dependent on the recharge rates but rather on the lifespan of the recharge zone. Figure 11 also shows that the dissolution of primary minerals leads to a net porosity increase, because the volume loss from the dissolution of these minerals is only partly compensated by chlorite precipitation. Thus, water/rock interaction counteracts the porosity reduction caused by anhydrite precipitation. In regions where anhydrite precipitation is not effective enough, the porosity actually increases. This effect may be one reason why the sealing of the recharge zone due to anhydrite precipitation is much slower than that calculated by *Lowell and Yao* [2002], who used the same initial porosity but did not simulate wall-rock alteration. On the other hand, in contrast to *Lowell and Yao* [2002], we account for the additional anhydrite precipitation caused by the release of Ca^{+2} from the rock, which should make sealing more effective. One other possible explanation for the lower effectiveness of anhydrite precipitation in reducing porosity could be our choice of the reaction rate constant for anhydrite, which may have been too low to ensure transport-limited precipitation. More work is needed to explore these relationships in detail. Future simulations should also include chlorite solid solutions and other low-temperature alteration phases, such as other clay minerals or zeolites, which may all affect the distribution and the evolution of the porosity profile in the recharge zone.

5.2. Water/Rock Ratios and Alteration Assemblages (Discussion of Second Example)

In the second example, we calculate a continuous alteration profile along the fluid's pathway as it moves from the seafloor through the upper oceanic crust and back to the seafloor again (Figure 4). In this study, we examine alteration assemblages in the high-temperature reaction zone (400°C) as a function of w/r ratios. As discussed above, w/r ratios for open, flow-through systems are a function of the flow rate and the reaction rates. By keeping the flow rates fixed, we can modify w/r ratios by changing the dissolution rates of the primary minerals. Thus, increasing the rates corresponds to a lower w/r ratio and vice versa.

The model is based on the reaction path study by *McCollom and Shock* [1998], which we discussed above. In that simulation, the *P/T* path consisted of a sequence of relatively few temperature and pressure points at which the equilibrium chemical state of the system was calculated. The intervals between these temperature and pressure points were rather large, amounting to 100°C and 25 MPa, respectively. We simulate a geological problem similar to that of *McCollom and Shock* 1998], but we incorporate mass transport, kinetics, and a finer grid of *PT* points. By comparing our results with those of *McCollom and Shock* [1998], we see the effects that our modifications have on the calculated alteration mineralogy.

However, the main purpose of this reactive transport simulation is to compute the alteration processes as a function of space and time. The fluid travels along a 5.25-km-long flowpath, traversing the same temperature and pressure points as in the work of *McCollom and Shock* [1998] up to the maximum temperature of 400°C. Once the fluid is heated to 400°C, it remains at that temperature (i.e., we assume the fluid moves laterally) over a distance that is long enough to ensure that the fluid will be buffered by the rock. At the end of the flowpath, the fluid ascends back to the seafloor along the same, but reversed, *P/T* path. We calculate the equilibrium constants for all phases in the system at the same temperature and pressure points and let the code (OS3D; *Steefel and Yabusaki,* 1996) interpolate between these points to obtain the equilibrium constants for intermediate temperatures and, implicitly, for intermediate pressures. In this simulation, we couple flow, heat transport, and chemical reactions only loosely and do not solve the flow and heat transport equation. Rather, we keep the flow rate fixed at 2.5 m^3/m^2_yr and assume an arbitrary, constant exponential temperature path (Figure 12).

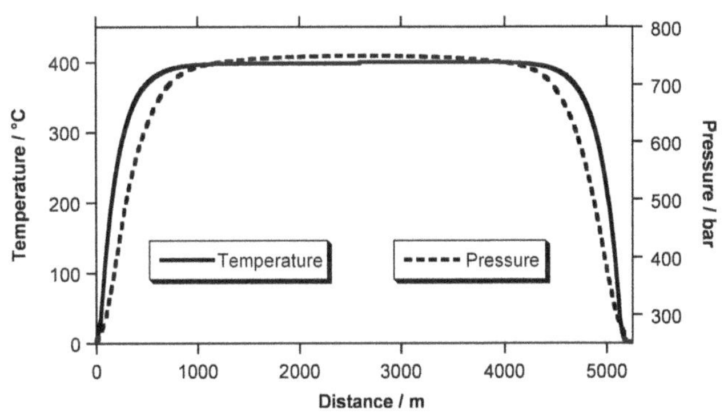

Figure 12. Temperature and pressure path of the fluid traversing the upper oceanic crust. The temperature path is given explicitly, whereas the pressure path is given implicitly by the equilibrium constants at discrete pressure/temperature points. Only the temperature profile is used to interpolate between the equilibrium constants at these points.

We use a similar primary mineral assemblage to that in *McCollom and Shock* [1998]. The primary rock, which we assume to be compositionally homogeneous over the entire length of the flowpath, is an olivine-bearing basalt composed of plagioclase ($an_{60}ab_{40}$), clinopyroxene ($di_{80}he_{20}$), and olivine ($fo_{75}fa_{25}$). Unlike in the study by *McCollom and Shock* [1998], we also add pyrrhotite to the system. We set these minerals to be present in excess to avoid complete leaching of minerals before one pore volume of fluid has moved through the system. This is necessary because we are interested in the stable alteration assemblage of the primary mineralogy in the high temperature region. This constraint has no additional affects on the outcome of the simulation. We use relative rates among the primary minerals to approximate their proportions in the rock. Flow velocities are not updated in response to porosity changes from mineral dissolution or precipitation. Thus the porosity remains fixed at 10%. For illustration, we nevertheless calculate porosity from the sum of the mineral volume fractions. The rate constants for plagioclase and clinopyroxene are equal, as are the rate constants for olivine and pyrrhotite. All secondary minerals have precipitation rates several orders-of-magnitude higher than the dissolution rates of primary minerals so that, if oversaturated, the precipitation of minerals is limited by transport or by the dissolution rate of primary minerals. Relative reaction rates among the primary minerals and among the secondary minerals are controlled by the affinity term and by the reactive surface area. The initial reactive surface area of all minerals is specified to be 10 m^2/m^3 rock. Few published data are available to constrain this parameter, so the value we have adopted is best viewed as an educated guess. Solid solutions assuming ideal mixing are included for epidote, plagioclase, clino- and orthopyroxene, chlorite, and amphibole. The code OS3D does not have the capability to treat solid solution compositions as a flexible, free parameter that is determined by the reaction progress. Instead, a selection of intermediate members of the solid solutions are included as discrete minerals with fixed composition and thermodynamic properties and modeled in the same fashion as pure mineral phases. Because we start the simulation at the seafloor, we add the lower-temperature phases anhydrite, calcite, zeolites, and clay minerals to the system. The thermodynamic database was produced with SUPCRT92 (*Johnson et al.*, 1992).

McCollom and Shock [1998] calculated alteration assemblages at temperatures between 300° and 450°C that include the following: albitic plagioclase, actinolitic amphibole, ±diopsidic clinopyroxene, ±epidote, ±chlorite, ±prehnite, ±analcite, and ±quartz. This is the assemblage with which we want to compare our high temperature mineralogy produced by considering transport processes.

In the first run, we use a rate constant of 1×10^{-13} mol/ m^2 s^{-1} for primary plagioclase and clinopyroxene. This rate is lower than that in the previous example and lower than the rates reported from dissolution experiments involving anorthite and diopside [*Brantley and Chen*, 1995; *Blum and Stillings*, 1995]. Considering the uncertainty of the experimentally measured rates and the fact that they tend to overestimate actual rates in natural systems, often by several orders of magnitude [*Brantley*, 1992, and references therein], the rate we have assumed seems reasonable. The results of this simulation are summarized in Figure 13 and presented as profiles along the flowpath as mineral volume changes over a time interval of 150 years. A negative volume change is equivalent to dissolution. The profiles show that a stable alteration assemblage consisting of epidote, tremolite, albite, diopside, ferropargasite, and minor quartz evolves in the highest temperature region. The epidote in this region (labeled epidote25/75) is of intermediate composition with a relatively low Fe content ($Fe_{0.25}Al_{2.75}$). Except for the lack of analcite and the presence of ferropargasite instead of actinolitic amphibole (which can easily be attributed to small differences in thermodynamic properties of the solid solutions), we produce a mineral assemblage at 400°C that is very similar to that calculated by *McCollom and Shock* [1998]. This confirms that the choice of reaction steps made by *McCollom and Shock* [1998] successfully captured the major alteration processes along the flowpath.

The stable high-temperature assemblage occurs between about 770 and 4750 m (Figure 13). Before 770 m the alteration is more heterogeneous because the fluid still bears characteristics of seawater and/or moves across steep temperature gradients. In the cooling section of the flowpath, which starts at 4000 m, prehnite appears in the assemblage at about 300°C.

Figure 14 shows in greater detail the alteration assemblages in the first 770 m of the flowpath where seawater evolves into a rock-buffered hydrothermal fluid. Figures 15 and 16 show the corresponding fluid composition profiles in terms of total component concentrations and the pH, respectively. Three alteration intervals can be distinguished:

(1) Between 0 and 350 m, there is an interval dominated by seawater/rock reaction and low to medium temperatures. Characteristic minerals are anhydrite, chlorite, talc, paragonite, and some hematite. The fluid becomes depleted in Mg^{+2} and SO_4^{-2} due to the precipitation of chlorite, talc, and anhydrite, respectively. The uptake of Mg^{+2} into secondary mineral phases, such as smectites at lower temperatures (which are not included in the model) and chlorite at elevated temperatures, is typical for the recharge zone [*Alt,* 1995]. Experimental results suggest that this uptake is accompanied by a release of Ca^{+2} from dissolution of primary minerals in

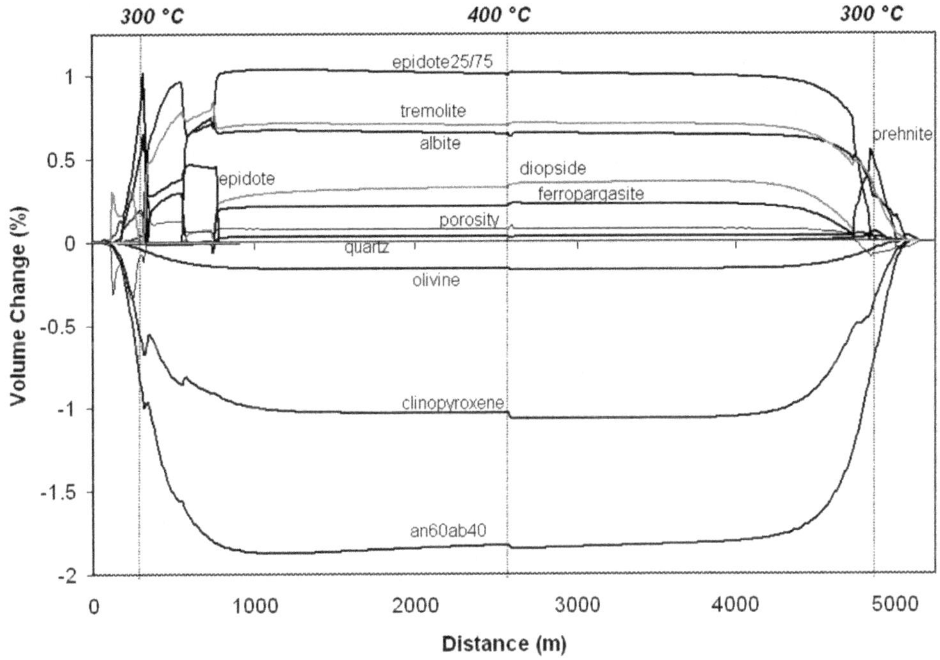

Figure 13. Profiles of mineral volume changes (%) over a time interval of 150 years. The simulation produced a stable mineral assemblage at high temperatures similar to that calculated in a reaction path simulation by *McCollom and Shock* [1998]. The mineral epidote constitutes the Fe-rich endmember of the epidote solid solution; epidote25/75 is of intermediate composition with a lower Fe content ($Fe_{0.25}Al_{2.75}$).

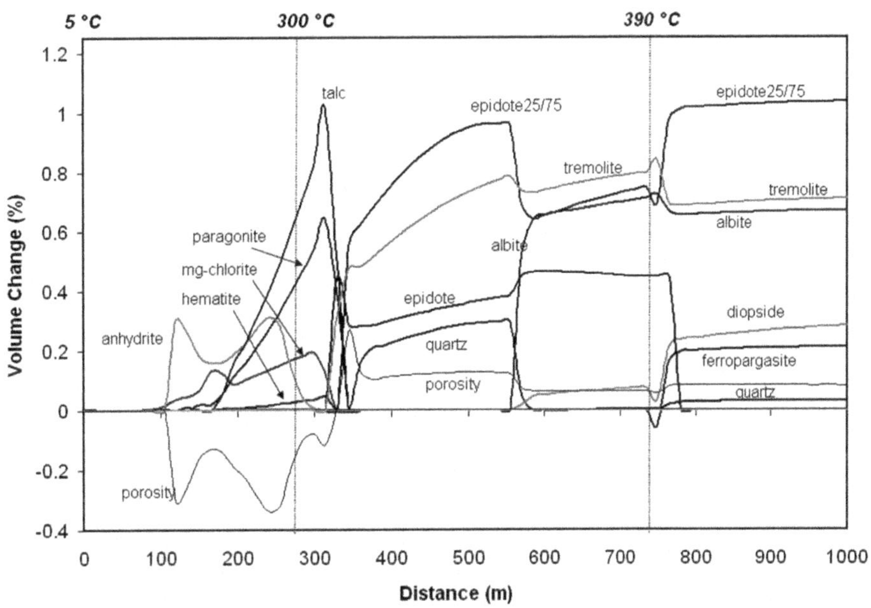

Figure 14. Mineral profiles of the up-temperature section of the flowpath showing the transition from reactions due to seawater/basalt disequilibrium and heating towards a stable mineral alteration assemblage. The mineral epidote constitutes the Fe-rich endmember of the epidote solid solution; epidote25/75 is of intermediate composition with a lower Fe content ($Fe_{0.25}Al_{2.75}$).

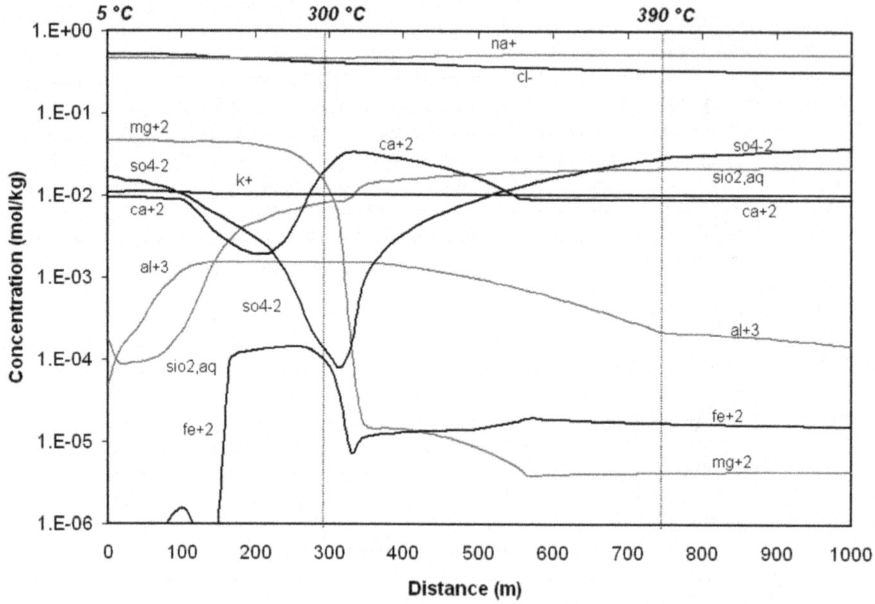

Figure 15. The profiles of total aqueous concentrations of the components in the system. Mg^{+2} is strongly depleted due to alteration reactions. Total sulfur (labelled SO_4^{-2}) recovers because of dissolving pyrrhotite after an initial drop due to anhydrite precipitation. Anhydrite precipitation causes the Ca^{+2} concentration to drop before it recovers and approaches uniformity. The Fe^{+2} and $SiO_2(aq)$ concentrations increase substantially compared to seawater levels, and Fe^{+2} decreases again at high temperature. These trends show good agreement with observations from oceanic systems [e.g., *Von Damm,* 1995].

the rock [*Mottl*, 1983; *Seyfried*, 1987; *Seyfried et al.*, 1988], which is consistent with our model results (Figure 15). This source of Ca^{+2} somewhat compensates for the removal of Ca^{+2} from the fluid via anhydrite precipitation. In contrast, sulfate removal is much more pronounced. The Ca^{+2} concentration increases after about 200 m, and Fe^{+2} shows a maximum at 250 m (Figure 15). The pH attains its lowest value (4.11) at about 250 m (Figure 16).

(2) Between 350 and 550 m, an assemblage consisting of epidote (both the Fe-rich endmember and the low-Fe intermediate epidote), tremolite, and quartz becomes stable. The appearance of this assemblage coincides with a drop in Mg^{+2} and a recovery of total S in solution and of the pH (Figures 14, 15, and 16).

(3) Between 550 and 770 m, a higher-temperature assemblage composed of tremolite, the low-Fe epidote, albite, and diopside is formed.

Figure 14 also includes the porosity profile, showing the porosity changes due to mineral precipitation and dissolution reactions. The strongest porosity reduction occurs in the first alteration interval and the profile mirrors the anhydrite precipitation pattern. A slight porosity increase occurs throughout the high temperature interval, but these changes are small and not likely to have major impact on the flow system.

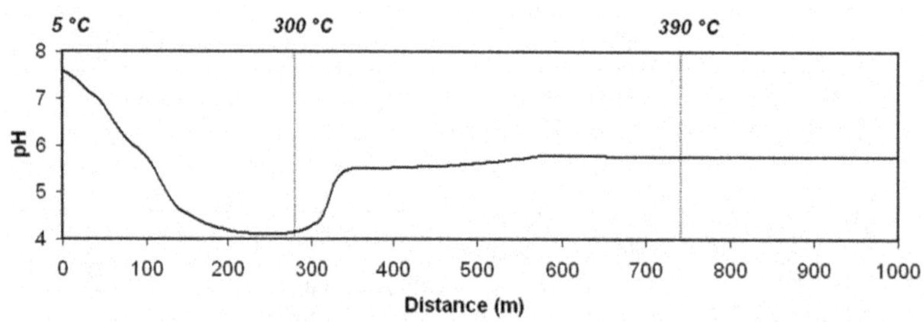

Figure 16. Profile of pH corresponding to the concentration profile in Figure 15. The transition to a stable alteration assemblage at high temperatures is marked by a strong decrease of the pH due to the formation of OH-bearing minerals such as chlorite. The uniformity of the pH at high temperatures (5.74) is consistent with a rock-buffered fluid composition.

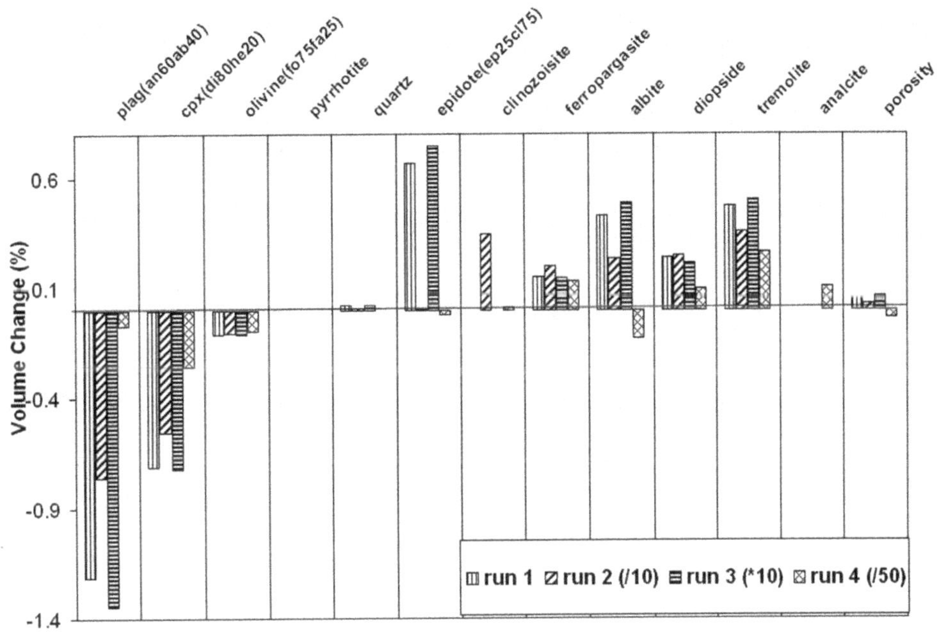

Figure 17. Mineral alteration assemblage at 400°C as a function of the dissolution rates of primary minerals. The amounts of minerals are normalized according to the factor used to increase or decrease the reaction rates relative to run 1. The alteration assemblage at high temperatures changes (albite and epidote redissolve, analcite forms) once primary plagioclase is completely altered (run 4).

To assess the effect of w/r ratios on the alteration assemblage in the highest temperature region, we ran three additional simulations and multiplied the dissolution rates of the primary minerals (1×10^{-13} mol/m^2 s^{-1}) by factors of 0.1, 10, and 50. Figure 17 shows the alteration assemblage of a model rock taken from within the highest temperature region at 3500 m. The plotted mineral fractions are normalized according to the factor used to increase or decrease the dissolution rate of the primary minerals. The alteration assemblage is quite robust and differences occur mostly in terms of the intensity of alteration, that is, alteration is stronger for higher rates. The replacement of the intermediate, low-Fe epidote by iron-free clinozoisite is one obvious compositional difference for lower w/r ratios. If we increase the reaction rate by a factor of 50, the primary mineral assemblage or parts thereof become completely altered over the time interval considered (run 4). We include this scenario here to illustrate the transients that can be generated from a perturbation of fluid/rock buffering anywhere along the flowpath. In this scenario, primary plagioclase is altered to a degree that leads to redissolution of secondary albite and epidote. Analcite appears in the assemblage as a new alteration phase, probably replacing albite (Figure 17). These mineralogical changes also affect the porosity, which now decreases. Thus, if alteration of the rock has proceeded such that a mineral phase is no longer exposed and in contact with the fluid, the system adjusts to a new steady state or equilibrium between the fluid and the modified assemblage. This is similar to the situation when the primary mineralogy is no longer exposed to the fluid because alteration phases along fracture walls shield the fresh rock from contact with the fluid.

6. DISCUSSION

We have presented an overview of approaches and concepts of modeling chemical processes in oceanic hydrothermal systems. Most numerical models have, so far, treated physical and chemical processes in the oceanic crust either independently or have linked these processes by strongly simplifying either one or the other. Physical and chemical models of oceanic hydrothermal systems have been successful in reproducing key observations. Results from speciation and reaction path models suggest that, despite some discrepancies, the thermodynamic database developed initially by Helgeson and coworkers, extended by Wolery and others and now compiled in SUPCRT92 [*Johnson et al.* 1992], produces results that agree quite well with observations from various sites and settings over a large range of temperatures. Similarly, coupled hydrological and heat transport models are capable of reproducing heat flow patterns and thermal constraints from oceanic ridge crests or ridge flanks [e.g., *Fisher and Becker,* 2002; *Stein and Fisher,* 2003]. The significance of biological processes has been recognized over the past years as has the need to relate these processes to hydrological, thermal, and chemical models. Numerical models are needed that integrate and explore the combined effect of these different processes.

Reactive transport models have become an important tool to link and integrate the physical and chemical and, in some cases, biological aspects of the evolution of geological systems. The development of reactive transport codes has been driven mainly by research related to groundwater contamination. However, several reactive transport studies have successfully reproduced observations from hydrothermal systems related to the formation of ore deposits [e.g., *Raffensperger and Garven,* 1995; *Garven,* 1995; *White,* 1995; *Ague and Brimhall,* 1989; *Xu et al.*, 2001] or have explored generic issues related to coupling of fluid flow, heat transport, and chemistry in hydrothermal systems [e.g., *Steefel and Lasaga,* 1994; *Bolton et al.,* 1996, 1997; *Alt-Epping and Smith,* 2001]. Existing reactive transport codes can be used to study coupled processes in the upper oceanic crust, as our examples demonstrate.

There are limits to the applicability of these models, of course. Indeed, the complete characterization and exact modeling of processes in oceanic hydrothermal systems may never be possible. The greatest limitation is the present restriction to single-fluid-phase systems. The development of models capable of handling high-temperature, multiphase, variable-salinity systems for simulating high-temperature multiphase flow and reactions in a ridge-crest setting is still in its infancy. Recent progress was made by the studies of *Geiger* [2006a, 2006b], *Lewis and Lowell* [2004, 2006], and *Lowell et al.* [this volume]. Other difficulties may include modeling conditions in high permeability or fracture zones, mechanical deformation, and the extension into three dimensions. On the other hand, other aspects that are relevant to oceanic systems, for instance, the effect of salinity on fluid flow, biological processes, or temperature changes due to chemical reactions, could be implemented without rewriting existing codes.

Incorporating different processes into a single model compounds the complexities of each individual process with those of their coupling [*Steefel et al.,* 2005]. As a first step towards developing reactive transport codes specifically designed to address questions related to oceanic hydrothermal systems, existing fluid flow and heat transport codes for oceanic hydrothermal systems could be loosely coupled with chemistry or reactive transport modules. This could be done either by an operator splitting approach through a common interface or in sequence, where flow and temperature fields are produced first and then read into a reactive transport code.

In addition to the development of numerical tools, more field and experimental data are needed to constrain the input to these models. Critical input parameters include the composition of solid and fluid phases, porosity and permeability, chemical and hydrological heterogeneity, kinetic rates, and reactive surface areas. Furthermore, reactive transport models require realistic boundary and initial conditions, that is, they have to be put into a realistic spatial and temporal framework.

Coupling different processes requires explicit relationships between processes and properties. At present, these relationships are often based on empirical or semiempirical data and are not generally representative. In oceanic hydrothermal systems, the most critical relationship is probably that between the porosity and the permeability. Most of the porosity/permeability relationships in the literature have been developed for porous media. Relationships between fracture geometry and distribution, porosity, and permeability are typically based on simple, unrealistic fracture geometries.

In their current state, reactive transport models are not unique and hence they are mainly useful for testing hypotheses and "what-if" scenarios. Simple model designs such as 1-D, single-pass, flow-through models, or a loose coupling of existing codes, along with new field and laboratory data, would be useful steps towards building up the insight necessary to understand the various processes in oceanic hydrothermal systems and towards developing more realistic, fully coupled models of these systems in the future.

Acknowledgments. The authors thank Peter Lichtner for his help during the implementation of the FLOTRAN reactive transport code and for his review. The paper benefited as well from thoughtful reviews by Bob Lowell and one anonymous reviewer.

REFERENCES

Aagaard, P., and H. C. Helgeson (1982), Thermodynamic and kinetic constraints on reaction rates among minerals and aqueous solutions, 1. Theoretical considerations, *Am. J. Sci.*, *282*, 237–285.

Ague, J. J., and G. H. Brimhall (1989), Geochemical modeling of steady state fluid flow and chemical reaction during supergene enrichment of porphyry copper deposits, *Econ. Geol.*, *84*, 506–528.

Alt, J. C. (1995), Subseafloor processes in mid-ocean ridge hydrothermal systems, in *Seafloor Hydrothermal Systems: Physical, Chemical, Biological, and Geological Interactions, Geophys. Monogr. Ser.*, vol. 91, edited by S. E. Humphris, et al., pp. 85–114, AGU, Washington, D. C.

Alt-Epping, P., and L. Smith (2001), Computing geochemical mass transfer and water/rock ratios in submarine hydrothermal systems: Implications for estimating the vigour of convection, *Geofluids*, *1*, 163–181.

Bastrakov, E., Y. Shvarov, S. Girvan, J. Cleverley, and L. Wyborn (2004), FreeGs: Web-enabled thermodynamic database for modelling of geochemical processes, in *Dynamic Earth: Past, Present and Future*, edited by J. McPhie and P. McGoldrick, p. 52, Abstracts of the 17th Australian Geological Convention, Hobart, Tasmania, Australia, February 8–13, No. 73.

Berndt, M. E., W. E Seyfried Jr., and D. R Janecky (1989), Plagioclase and epidote buffering of cation ratios in mid-ocean ridge hydrothermal fluids; experimental results in and near the supercritical region, *Geochim. Cosmochim. Acta*, *53*(9), 2283–2300.

Bessler, J. U., L. Smith, and E. E. Davis (1994), Hydrologic and thermal conditions at a sediment/basalt interface: Implications for interpretation of field measurements at Middle Valley, *Proc. Ocean Drill. Program, Sci. Results*, 139, 667–675.

Bethke, C. M. (1996), *Geochemical Reaction Modeling*, 397 pp., Oxford Univ. Press, New York.

Bischoff, J. L., and W. E. Seyfried (1978), Hydrothermal chemistry of seawater from 25° to 350°C, *Am. J. Sci.*, *278*, 838–860.

Blount, C. F., and F. W. Dickerson (1969), The solubility of anhydrite ($CaSO_4$) in NaCL-H_2O from 100°C to 450 °C and 1 to 1000 bars, *Geochim. Cosmochim. Acta*, *33*, 227–245.

Blum, A. E., and L. L. Stillings (1995), Feldspar dissolution kinetics, in *Chemical Weathering Rates of Silicate Minerals*, edited by A. F. White and S. L. Brantley, *Mineral. Soc. Am.*, *31*, 291–351.

Bolton, E. W., A. C. Lasaga, and D. M. Rye (1996), A model for the kinetic control of quartz dissolution and precipitation in porous media flow with spatially variable permeability: Formulation and examples of thermal convection, *J. Geophys. Res.*, *101*, 22,157–22,187.

Bolton, E. W., A. C. Lasaga, and D. M. Rye (1997), Dissolution and precipitation via forced flux injection in porous medium with spatially variable permeability: Kinetic control in two dimensions, *J. Geophys. Res.*, *102*, 12,159–12,173.

Bolton, E. W., A. C. Lasaga, and D. M. Rye (1999), Long-term flow/chemistry feedback in porous medium with heterogeneous permeability: Kinetic control of dissolution and precipitation, *Am. J. Sci.*, *299*, 1–68.

Bowers, T. S. (1989), Stable isotope signatures of water-rock interaction in mid-ocean ridge hydrothermal systems: Sulfur, oxygen, and hydrogen, *J. Geophys. Res.*, *94*(B5), 5775–5786.

Bowers, T. S., and H. P. Taylor Jr. (1985), An integrated chemical and stable isotope model of the origin of midocean ridge hot spring systems, *J. Geophys. Res.*, *90*, 12,583–12,606.

Bowers, T. S., K. L. Von Damm, J. M. Edmond (1985), Chemical evolution of mid-ocean ridge hot springs, *Geochim. Cosmochim. Acta*, *49*, 2239–2252.

Bowers, T. S., A. C. Campbell, C. I. Measures, A. J. Spivack, M. Khadem, and J. Edmond (1988), Chemical controls on the composition of vent fluids 13°–11°N and 21°N, East Pacific Rise, *J. Geophys. Res.*, *93*, 4522–4536.

Brantley, S. L. (1992), Kinetics of dissolution and precipitation—experimental and field results, in *Water-Rock Interaction 13–18 July, 1992*, edited by Y. K. Kharaka and A. S. Maest, Balkema, Rotterdam, pp. 3–6.

Brantley, S. L., and Y. Chen (1995), Chemical weathering rates of pyroxenes and amphiboles, in *Chemical Weathering Rates of Silicate Minerals*, edited by A. F. White and S. L. Brantley, *Mineral. Soc. Am.*, *31*, 119–172.

Brinkley, S. R., Jr. (1947), Calculation of the equilibrium composition of systems of many components, *J. Chem. Phys.*, *15*, 107–110.

Brinkley, S. R., Jr. (1960), Discussion of "A brief survey of past and current methods of solution for equilibrium composition" by H. E. Brandmaier and J. J. Harnett, in *Kinetics, Equilibria and Performance of High Temperature Systems*, edited by G. S. Bahn and E. E. Zukovski, p. 73, Butterworths, Washington, D. C.

Campbell, A. C., T. S. Bowers, C. I. Measures, K. K. Falkner, M. Khadem, and J. M. Edmond (1988), A time series of vent fluid compositions from 21°N, East Pacific Rise (1979, 1981, 1985) and the Guyamas Basin, Gulf of California (1982, 1985), *J. Geophys. Res.*, *93*(B5), 4537–4549.

Carman, P. C. (1956), *The Flow of Gases Through Porous Media*, Academic Press, New York.

Cruse, A., J. S. Seewald, P. J. Saccocia, and R. Zierenberg, Hydrothermal fluid composition at Middle Valley, northern Juan de Fuca Ridge: Temporal and spatial variability, this volume.

Damköhler, G. (1936), Einflüsse der Strömung, Diffusion und des Wärmeübergangs auf die Leistung von Reaktoröfen, *Z. Elektrochem.*, *42*(12), 846–862.

Dobson, P. F., T. J. Kneafsey, E. L. Sonnenthal, N. Spycher, and J. A. Apps (2003), Experimental and numerical simulation of dissolution and precipitation: Implications for fracture sealing at Yucca Mountain, Nevada, *J. Contam. Hydrol.*, *62–63*, 459–476.

Engesgaard, C., and T. H. Christensen (1988), A review of chemical solute transport models, *Nord. Hydrol.*, *19*, 183–216.

Fisher, A. T., and K. Becker (2002), Channelized fluid flow in oceanic crust reconciles heat-flow and permeability data, *Nature*, *403*, 71–74.

Fisher, A. T., E. Giambalvo, J. Sclater, M. Kastner, B. Ransom, Y. Weinstein, and P. Lonsdale (2001), Heat flow, sediment and pore fluid chemistry, and hydrothermal circulation on the east flank of Alarcon Ridge, Gulf of California, *Earth Planet. Sci. Lett.*, *188*, 521–534.

Fontaine, F. J., M. Rabinowicz, and J. Boulègue (2001), Permeability changes due to mineral diagenesis in fractured crust: Implications for hydrothermal circulation at mid-ocean ridges, *Earth Planet. Sci. Lett.*, *184*, 407–425.

Freedman, V. L., K. P. Saripalli, and P. D. Meyer (2003), Influence of mineral precipitation and dissolution on hydrologic properties of porous media in static and dynamic systems, *Appl. Geochem.*, *18*, 589–606.

Freeze, R. A., and J. A. Cherry (1979), *Groundwater*, Prentice-Hall, Englewood Cliffs, NJ.

Garrels, R. M., and F. T. Mackenzie (1967), Origin of the chemical compositions of some springs and lakes, in *Equilibrium Concepts in Natural Waters*, *Adv. Chem. Ser.*, vol. 67, pp. 222–242, American Chemical Society, Washington, D. C.

Garven, G. (1995), Continental-scale groundwater flow and geologic processes, *Annu. Rev. Earth Planet. Sci.*, *24*, 89–117.

Geiger, S., T. Driesner, C. A. Heinrich, and S. K. Matthai (2006a), Multiphase thermohaline convection in the Earth's crust: I. A new finite element-finite volume solution technique combined with a new equation of state for NaCl-H_2O, *Transp. Porous Media, 63*, 399–434.

Geiger, S., T. Driesner, C. A. Heinrich, and S. K. Matthai (2006b), Multiphase thermohaline convection in the Earth's crust: II. Benchmarking and application of a finite element-finite volume solution technique combined with a new equation of state for NaCl-H_2O, *Transp. Porous Media, 63*, 435–461.

Giambalvo, E. R., C. I. Steefel, A. T. Fisher, N. D. Rosenberg, and C. G Wheat (2002), Effect of fluid-sediment reaction on hydrothermal fluxes of major elements, eastern flank of the Juan de Fuca Ridge, *Geochim. Cosmochim. Acta, 66*(10), 1739–1757.

Haas, J. L., Jr., and J. R. Fisher (1976), Simultaneous evaluation and correlation of thermodynamic data, *Am. J. Sci., 276*, 525–545.

Helgeson, H. C. (1968), Evaluation of irreversible reactions in geochemical processes involving minerals and aqueous solutions: I. Thermodynamic relations, *Geochim. Cosmochim. Acta, 32*, 853–877.

Helgeson, H. C., R. M. Garrels, F. T. MacKenzie (1969), Evaluation of irreversible reactions in geochemical processes involving minerals and aqueous solutions: II. Applications, *Geochim. Cosmochim. Acta, 33*, 455–482.

Helgeson, H. C., T. H. Brown, A. Nigrini, and T. A. Jones (1970), Calculation of mass transfer in geochemical processes involving aqueous solutions, *Geochim. Cosmochim. Acta, 34*, 569–592.

Hoefner, M. L., and H. S. Fogler (1988), Pore evolution and channel formation during flow and reaction in porous media, *AIChE J., 34*, 45–54.

Ingebritsen, S., W. Sanford, and C. Neuzil (2006), *Groundwater in Geologic Processes*, 2nd ed., Cambridge Univ. Press, New York.

Islam, J., N. Singhal, and M. O'Sullivan (2001), Modeling biogeochemical processes in leachate contaminated soils: A review, *Transp. Porous Media., 43*, 407–440.

Janecky, D. R. (1982), Serpentinization of peridotite within the oceanic crust: Experimental and theoretical investigations of seawater-peridotite interaction at 200°C and 300°C, 500 bars, PhD thesis, University of Minnesota, Minneapolis, Minn.

Janecky, D. R., and W. E. Seyfried Jr. (1984), Formation of massive sulfide deposits on oceanic ridge crests, incremental reaction models for mixing between hydrothermal solution and seawater, *Geochim. Cosmochim. Acta, 48*, 2723–2738.

Janecky, D. R., and W. C. Shanks III (1988), Computational modeling of chemical and sulfur isotopic reaction processes in seafloor hydrothermal systems, chimneys, massive sulfides, and subjacent alteration zones, *Can. Mineral., 26*, 805–825.

Johnson, J. W., E. H. Oelkers, and H. C. Helgeson (1992), SUPCRT92: A software package for calculating the standard molal thermodynamic properties of minerals, gases, aqueous species and reactions from 1 to 5000 bar and 0 to 1000°C, *Comput. Geosci., 18*, 899–947.

Kharaka, Y. K., and I. Barnes (1973), SOLMNEQ: Solution-mineral equilibrium computations, *U. S. Geol. Surv. Comput. Contrib. Rep. PB*, 215-899.

Kirkner, D. J., and H. W. Reeves (1988), Multicomponent mass transport with homogeneous and heterogeneous chemical reactions: The effect of the chemistry on the choice of numerical algorithms: part 1. Theory, *Water Resour. Res., 24*(10), 1719–1728.

Lasaga, A. C. (1981a), Rate laws of chemical reactions, in *Kinetics of Geochemical Processes*, edited by A. C. Lasaga and R. J. Kirkpatrick, pp. 1–68, Mineralogical Society of America, Washington, D. C.

Lasaga, A. C. (1981b), Transition state theory, in *Kinetics of Geochemical Processes*, edited by A. C. Lasaga and R. J. Kirkpatrick, pp. 1–68, Mineralogical Society of America, Washington, D. C.

Lasaga, A. C. (1984), Chemical kinetics of water-rock interactions, *J. Geophys. Res., 89*, 4009–4025.

LeGallo, Y., O. Bildstein, E. Brosse (1998), Coupled reaction-flow modeling of diagenetic changes in reservoir permeability, porosity and mineral compositions, *J. Hydrol., 209*, 366–388.

Lewis, K. C., and R. P. Lowell (2004), Mathematical modeling of phase separation of seawater near an igneous dike, *Geofluids, 4*, 197–209.

Lewis, K. C., R. P. Lowell, and W. Xu (2006), FISHES—a numerical code for modeling NaCl-H_2O hydrothermal systems with applications to mid-ocean ridges, Ridge Theoretical Institute, Mammoth Lakes, CA, June 25–30.

Lichtner, P. C. (1985), Continuum model for simultaneous chemical reactions and mass transport in hydrothermal systems, *Geochim. Cosmochim. Acta, 49*, 779–800.

Lichtner, P. C. (1988), The quasi-stationary state approximation to coupled mass transport and fluid-rock interaction in a porous medium, *Geochim. Cosmochim. Acta, 52*, 143–165.

Lichtner, P. C. (1996), Continuum formulation of multicomponent-multiphase reactive transport, in *Reactive Transport in Porous Media, Rev. Mineral.*, vol. 34, edited by P. C. Lichtner, C. I. Steefel, and E. H. Oelkers, pp. 1–81, Mineralogical Society of America, Washington, D.C.

Lichtner, P. C. (2007), *FLOTRAN Users Manual: Two-Phase Non-Isothermal Coupled Thermal-Hydrologic-Chemical (THC) Reactive Flow and Transport Code, Version 2*, Los Alamos National Laboratory, Los Alamos, N.M.

Lowell, R. P., and Y. Yao (2002), Anhydrite precipitation and the extent of hydrothermal recharge zones at ocean ridge crests, *J. Geophys. Res., 107*(B9), 2183, doi:10.1029/2001JB001289.

Lowell, R. P., P. Van Capellen, and L. N. Germanovich (1993), Silica precipitation in fractures and the evolution of permeability in hydrothermal upflow zones, *Science, 260*(5105), 192–194.

Lowell, R. P., B. W. Crowell, S. R. Gosnell, K. C. Lewis, L. Liu, and Y. Yang, Modeling multiphase, multi-component processes at oceanic spreading centers: Magma to microbe, this volume.

MacQuarrie, K. T. B., and K. U. Mayer (2005), Reactive transport modeling in fractured rock: A state-of-the-science review, *Earth-Sci. Rev., 72*, 189–227.

Mangold, D. C., and C. H. F. Tsang (1991), A summary of subsurface hydrological and hydrochemical models, *Rev. Geophys., 29*, 51–79.

Martin, J. T., and R. P. Lowell (1997), On thermoelasticity and silica precipitation in hydrothermal systems: Numerical modeling of laboratory experiments, *J. Geophys. Res.*, *102*(B6), 12,095–12,107.

Martin, J. T., and R. P. Lowell (2000), Precipitation of quartz during hydrothermal circulation at mid-ocean ridges, *J. Geophys. Res.*, *105*, 869–882.

Mayer, K. U., E. O. Frind, and D. W. Blowes (2002), A numerical model for the investigation of reactive transport in variably saturated media using a generalized formulation for kinetically controlled reactions, *Water Resour. Res.*, *38*(9), 1174, 1301–1321.

McCollom, T. M., and E. L. Shock (1998), Fluid-rock interactions in the lower oceanic crust: Thermodynamic models of hydrothermal alteration, *J. Geophys. Res.*, *103*(B1), 547–575.

Mottl, M. J. (1983), Metabasalts, axial hot springs and the structure of hydrothermal systems at mid-ocean ridges, *Geol. Soc. Am. Bull.*, *94*, 161–180.

Nordstrom, D. K., and J. L. Munoz (1994), *Geochemical Thermodynamics*, 2nd ed., Blackwell, Boston.

Oelkers, E. H. (1996), Physical and chemical properties of rocks and fluids for chemical mass transport calculations, in *Reactive Transport in Porous Media*, edited by P. C. Lichtner, C. I. Steefel and E. H. Oelkers, *Rev. Mineral.*, 34.

Ortoleva, P., J. Chadam, E. Merino, and A. Sen (1987), Geochemical self-organization II: The reactive flow instability, *Am. J. Sci.*, *287*, 1008–1040.

Panda, M. N., and L. W. Lake (1994), Estimation of single-phase permeability from parameters of particle-size distribution, *AAPG Bull.*, *78*, 1028–1039.

Panda, M. N., and L. W. Lake (1995), A physical model of cementation and its effects on single-phase permeability, *AAPG Bull.*, *79*, 431–443.

Parkhurst, D. L., and C. A. J. Appelo (1999), User's guide to PHREEQC (Version 2)—A computer program for speciation, batch reaction, one-dimensional transport, and inverse geochemical calculations, *USGS Water Resour. Invest. Rep.* 99-4259.

Parkhurst, D. L., K. L. Kipp, P. Engesgaard, and S. R. Charlton (2002), PHAST—a program for simulating ground-water flow and multicomponent geochemical reactions, (http://www.brr.cr.usgs.gov/projects/GWC_coupled/phast/index.html).

Person, M., J. P. Raffensperger, S. Ge, and G. Garven (1996), Basin-scale hydrogeologic modeling, *Rev. Geophys.*, *34*(1), 61–87.

Phillips, O. M. (1991), *Flow and Reactions in Permeable Rocks*, 285 pp., Cambridge Univ. Press, New York.

Pitzer, K. S. (1979), Theory: Ion interaction approach, in *Activity Coefficients in Electrolyte Solutions*, vol. 1, edited by R. M. Pytkowitz, pp. 157–208, CRC Press, Boca Raton.

Pitzer, K. S. (1987), A thermodynamic model for aqueous solutions of liquid-like density, in *Thermodynamic Modeling of Geological Materials: Minerals, Fluids and Melts*, edited I. S. E. Carmichael and H. P. Eugster, *Rev. Mineral.*, *17*, 97–142.

Prommer, H., D. A. Barry, and C. Zheng (2003), MODFLOW/MT-3DMS based reactive multi-component transport model, *Ground Water*, *42*, 247–257.

Raffensperger, J. P. (1996), Numerical simulation of sedimentary basin-scale hydrochemical processes, in *Advances in Porous Media*, edited by Y. C. Corapcioglu, 440 pp., Elsevier Science, Amsterdam, The Netherlands.

Raffensperger, J. P., and G. Garven (1995), The formation of unconformity-type uranium ore deposits—2. Coupled hydrochemical modeling, *Am. J. Sci.*, *295*, 639–696.

Reed, M. H. (1983), Seawater-basalt reaction and the origin of greenstones and related ore deposits, *Econ. Geol.*, *78*, 466–485.

Reed, M. H., and N. F. Spycher (1984), Calculation of pH and mineral equilibria in hydrothermal waters with application to geothermometry and studies of boiling and dilution. *Geochim. Cosmochim. Acta*, *48*, 1479–1492.

Rubin, J. (1983), Transport of reacting solutes in porous media: Relation between mathematical nature of problem formulation and chemical nature of reactions, *Water Resour. Res.*, *19*, 1231–1252.

Saaltink, M. W., J. Carrera, and C. Ayora (2001), On the behavior of approaches to simulate reactive transport, *J. Contam. Hydrol.*, *48*, 213–235.

Saripalli, K. P., P. D. Meyer, D. H. Bacon, and V. L. Freedman (2001), Changes in hydrologic properties of aquifer media due to chemical reactions: A review, *Crit. Rev. Environ. Sci. Technol.*, *31*, 311–349.

Sanford, W. E., and L. F. Konikow (1989), Simulation of calcite dissolution and porosity changes in saltwater mixing zones in coastal aquifers, *Water Resour. Res.*, *25*, 655–667.

Seyfried, W. E. (1987), Experimental and theoretical constraints on hydrothermal alteration processes at mid-ocean ridges, *Annu. Rev. Earth Planet. Sci.*, *15*, 317–335.

Seyfried, W. E., M. E. Berndt, and J. S. Seewald (1988), Hydrothermal alteration processes at mid-ocean ridges: Constraints from diabase alteration experiments, hot-spring fluids and composition of the oceanic crust, *Can. Mineral.*, *26*, 787–804.

Seyfried, W. E., J. S. Seewald, M. E. Berndt, K. Ding, and D. I. Foustoukos (2003), Chemistry of hydrothermal vent fluids from the Main Endeavour Field, northern Juan de Fuca Ridge; geochemical controls in the aftermath of June 1999 seismic events, *J. Geophys. Res.*, *108*(B9), 1–23.

Sleep, N. H. (1991), Hydrothermal circulation, anhydrite precipitation and thermal structure at ridge axes, *J. Geophys. Res.*, *96*, 2375–2387.

Snow, D. T. (1968), Rock fracture spacings, openings and porosities, *J. Soil Mech. Found. Div., Proc. Am. Soc. Civil Eng.*, *94*, 73–91.

Steefel, C. I., and A. C. Lasaga (1990), Evolution of dissolution patterns: Permeability change due to coupled flow and reaction, in *Chemical Modeling of Aqueous Systems II, ACS Symp. Ser.*, vol. 416, edited by D. C. Melchoir and R. L. Bassett, pp. 221–225, American Chemical Society.

Steefel, C. I., and A. C. Lasaga (1994), A coupled model for transport of multiple chemical species and kinetic precipitation/dissolution reactions with application to reactive flow in single phase hydrothermal systems, *Am. J. Sci.*, *294*, 529–592.

Steefel, C. I., and P .C. Lichtner (1998), Multicomponent reactive transport in discrete fractures: I. Controls on reaction front geometry, *J. Hydrol.*, *209*, 186–199.

Steefel, C. I., and K. T. B. MacQuarrie (1996), Approaches to modeling reactive transport in porous media, in *Reactive Transport in Porous Media*, edited by P. C. Lichtner, C. I. Steefel, and E. H. Oelkers, *Rev. Mineral.*, *34*, 83–125.

Steefel, C. I., and P. Van Capellen, 1990, A new kinetic approach to modeling water-rock interaction, the role of nucleation, precursors, and Ostwald ripening, *Geochim. Cosmochim. Acta*, *54*, 2657–2677.

Steefel, C. I., and S. B. Yabusaki (1996), *OS3D/GIMRT: Software for Modeling Multicomponent-Multidimensional Reactive Transport, User Manual and Programmer's Guide*, PNL-11166, Pacific Northwest National Laboratory, Richland, WA.

Steefel, C. I., D. J. dePaolo, and P. C. Lichtner (2005), Reactive transport modeling: An essential tool and a new research approach for Earth sciences, *Earth Planet. Sci. Lett.*, *240*, 539–558.

Stein, J. S., and A. T. Fisher (2003), Observations and models of lateral hydrothermal circulation on a young ridge flank: Numerical evaluation of thermal and chemical constraints, *Geochem., Geophys., Geosyst.*, *4*(3), 1026, doi:10.1029/2002GC000415.

Truesdell, A. H., and B. F. Jones (1974), WATEQ, a computer program for calculating chemical equilibria of natural waters, *U. S. Geol. Surv. J. Res.*, *2*, 233–248.

Van der Lee, J., C. Lomenech (2004), Towards a common thermodynamic database for speciation models, *Radiochim. Acta*, *92*(9), 811–818.

Von Damm, K. L. (1995), Controls on the chemistry and temporal variability of seafloor hydrothermal fluids, in *Seafloor Hydrothermal Systems: Physical, Chemical, Biological, and Geological Interactions*, *Geophys. Monogr. Ser.*, vol. 91, edited by S. E. Humphris, et al., pp. 222–247, AGU. Washington, D. C.

Von Damm, K. L., and J. L. Bischoff (1987), Chemistry of hydrothermal solutions from the southern Juan de Fuca Ridge, *J. Geophys. Res.*, *92*, 11,334–11,346.

Von Damm, K. L., J. M. Edmond, B. Grant, C. I. Measures, B. Walden, and R. F. Weiss (1985), Chemistry of submarine hydrothermal solutions at 21°N, East Pacific Rise, *Geochim. Cosmochim. Acta*, *49*, 2197–2220.

Wells, J. T., and M. S. Ghiorso (1991), Coupled fluid flow and reaction in mid-ocean ridge hydrothermal systems: The behavior of silica, *Geochim. Cosmochim. Acta*, *55*, 2467–2481.

Wetzel, L. R., J. P. Raffensperger, and E. L. Shock (2001), Predictions of hydrothermal alteration within near-ridge oceanic crust from coordinated geochemical and fluid flow models, *J. Volcanol. Geotherm. Res.*, *110*, 319–342.

White, S. P. (1995), Multiphase non-isothermal transport of systems of reacting chemicals, *Water Resour. Res.*, *31*, 1761–1772.

White, W. B., S. M. Johnson, and G. B. Dantzig (1958), Chemical equilibrium in complex mixtures, *J. Chem. Phys.*, *28*, 751–755.

Walder, J., and A. Nur (1984), Porosity reduction and crustal pore pressure development, *J. Geophys. Res.*, *89*, 11,539–11,548.

Wolery, T. J. (1978), Some chemical aspects of hydrothermal processes at mid-ocean ridges, a theoretical study, 1. Basalt-seawater reaction and chemical cycling between the oceanic crust and the oceans, 2. Calculation of chemical equilibrium between aqueous solutions and minerals, Ph.D. dissertation, Northwestern University, Evanston, IL.

Wolery, T. J. (1979), *Calculation of Chemical Equilibrium Between Aqueous Solution and Minerals: The EQ3/6 Software Package*, Lawrence Livermore National Laboratory, UCRL-52658.

Wolery, T. J. (1983), *EQ3NR: A Computer Program for Geochemical Aqueous Speciation-Solubility Calculations: User's Guide and Documentation*, Lawrence Livermore National Laboratory Report UCRL-53414.

Wood, J. R., and T. A. Hewett (1982), Fluid convection and mass transfer in porous sandstones—A theoretical model, *Geochim. Cosmochim. Acta*, *46*, 1707–1733.

Xu, T., E. Sonnenthal, N. Spycher, K. Pruess, G. Brimhall, and J. Apps (2001), Modeling multiphase non-isothermal fluid flow and reactive geochemical transport in variably saturated fractured rocks: 2. Applications to supergene copper enrichment and hydrothermal flows, *Am. J. Sci.*, *301*, pp. 34–59.

Yeh, G. T., and V. S. Tripathi (1989), A critical evaluation of recent developments in hydrogeochemical transport models of reactive multichemical components, *Water Resour. Res.*, *25*, 93–108.

Yeh, G. T., and V. S. Tripathi (1991), A model for simulating transport of reactive multispecies components: Model development and demonstration, *Water Resour. Res.*, *27*, 3075–3094.

Zeleznik, F. J., and S. Gordon (1960), An analytical investigation of three general methods of calculating chemical equilibrium composition, NASA Technical Note D-473, Washington, D. C.

Zeleznik, F. J., and S. Gordon (1968), Calculation of complex chemical equilibria, *Ind. Eng. Chem.*, *60*, 27–57.

Observational, Experimental, and Theoretical Constraints on Carbon Cycling in Mid-Ocean Ridge Hydrothermal Systems

Thomas M. McCollom

CU Center for Astrobiology and Laboratory for Atmospheric and Space Physics, University of Colorado, Boulder, Colorado, USA

Carbon occurs in many different forms in mid-ocean ridge hydrothermal systems, and there are many processes that convert one form of carbon to another. Because most transformations of carbon compounds in these environments are closely linked to biological and geological processes, a broad understanding of the contributors to carbon cycling can provide valuable insights into microbial activity and geochemical processes, particularly in the subsurface. Interpretations of carbon cycling in mid-ocean ridge hydrothermal systems currently rely heavily on analyses of carbon compounds in vent fluids discharged at the seafloor. The compositions of these fluids indicate the carbon pool receives inputs from multiple sources, including magmatic volatiles, carbon compounds in seawater, in situ biological metabolism, thermal degradation of organic matter, and abiotic organic synthesis. However, defining the magnitude of the relative contributions of these different sources is problematic for most compounds, and is hampered by limited data. Development of comprehensive, quantitative models of carbon cycling at mid-ocean ridges will require more extensive analyses of organic compounds and biomass in fluid and rock samples, evaluation of the scope of biological activity throughout the system, and improved knowledge of the rates of organic chemical reactions under hydrothermal conditions.

1. INTRODUCTION

Carbon is present in a variety of forms in mid-ocean ridge (MOR) hydrothermal systems, including carbonate minerals, inorganic and organic compounds dissolved in circulating fluids, gases trapped in fluid inclusions or dissolved in basaltic glass, insoluble organic matter, and biomass. Because carbon compounds are key substrates in biological metabolism and are also principal components of biomass, the ability to trace the fate of carbon is critical for interpretation of biological processes in MOR systems. Hydrothermal circulation also mediates exchange of carbon between the crust and the overlying ocean, connecting the ocean crust with the broader oceanic and global carbon cycles. Hydrothermal effluents also provide organic matter to the overlying water column that can serve as energy and fixed carbon sources for autotrophic and heterotrophic organisms, stimulating biological activity in a part of the ocean that is otherwise depleted in these nutrients. For these and many other reasons, carbon cycling is a crucial element in understanding the links between biological and geological processes within MOR hydrothermal systems, and the links between ridges and the overlying ocean.

There are a number of potential sources and sinks for both inorganic and organic carbon compounds in MOR hydrothermal environments as well as numerous processes that transfer carbon from one reservoir to another throughout the

Magma to Microbe: Modeling Hydrothermal Processes at Ocean Spreading Centers
Geophysical Monograph Series 178

10.1029/178GM10

system (Figure 1). However, although it is relatively straightforward to conceptually identify the major contributors to carbon cycling at mid-ocean ridges, currently available data do not allow quantitative constraints to be placed on the sizes of many of the carbon sources and sinks, and on the processes that transfer carbon between them. As a consequence, comprehensive, quantitative models of carbon cycling for the mid-ocean ridge environment have not been forthcoming. However, considerable progress has been made in understanding some aspects of the overall system, particularly with regard to the major carbon components carbon dioxide (CO_2) and methane (CH_4).

One critical factor contributing to the continuing uncertainty on this topic is the difficulty involved in obtaining samples directly from subsurface environments near the ridge axis for analysis of carbon compounds. The restricted access to rock and fluid samples from the subsurface means that interpretation of the carbon cycling in MOR hydrothermal systems must rely heavily on information inferred from analysis of carbon compounds in hydrothermal fluids discharged at the seafloor. Given that there are many potential sources for both organic and inorganic carbon compounds as fluids circulate through the system (Figure 1), the compositions of fluids discharged at the seafloor are likely to reflect the accumulated contributions of multiple sources and processes, complicating interpretation of the data. Disentangling the relative contributions of the various sources and processes to carbon cycling requires consideration of the chemical and isotopic composition of carbon compounds in vent fluids as well as in rock, sediment, and biological samples, combined with constraints provided by laboratory experiments and theoretical models.

This article provides a brief overview of the major factors contributing to carbon cycling in MOR hydrothermal systems as they are presently understood. For the sake of brevity, only carbon cycling in hydrothermal regions along the mid-ocean ridge axis is considered in this review, although the composition of carbon compounds in the ocean

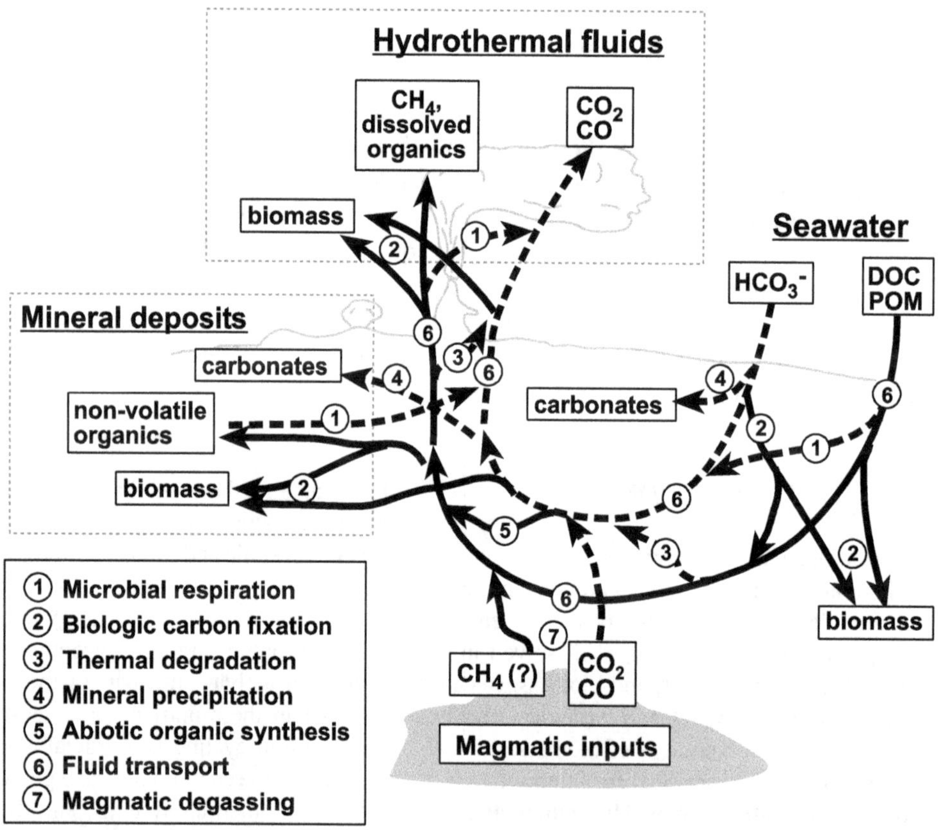

Figure 1. Schematic drawing illustrating the major sources and sinks of carbon compounds in deep-sea hydrothermal systems, and the processes effecting carbon transfer among these reservoirs. The inorganic and organic carbon cycles are represented separately (dashed lines, inorganic; solid lines, organic), although they are strongly linked by a number of different processes in different parts of the system. DOC, dissolved organic carbon; POM, particulate organic carbon (includes biomass).

crust continues to evolve as the crust cools and ages [e.g., *Alt and Teagle,* 1999]. Because the distribution of carbon compounds in sediment-covered MOR hydrothermal systems such as Guaymas Basin and Middle Valley is largely dominated by thermal processing of sedimentary organic matter, this review will primarily concentrate on carbon cycling in unsedimented systems in order to focus on contributions indigenous to the hydrothermal environment. Readers who are interested in more detailed discussion of the thermal alteration of organic matter in sediment-covered deep-sea hydrothermal systems are directed toward reviews of this topic by *Simoneit* [1988, 1993, 1994]. In addition, this article largely focuses on CO_2 and CH_4 because these are the most abundant carbon compounds in MOR hydrothermal systems and they are the compounds for which the most data are available to interpret their origin. Lastly, owing to limited analyses of carbon compounds in other types of samples, this overview primarily discusses constraints on carbon cycling provided by analyses of high-temperature, axial hydrothermal vent fluids.

The review begins with a summary of the occurrence of carbon compounds in MOR environments, followed by a discussion of potential sources for these compounds and the processes that transfer carbon between reservoirs. In the final section, the potential for development of more comprehensive models of MOR carbon cycling is considered, along with an outline of the types of additional data that will be needed to construct such models.

2. CARBON COMPOUNDS IN MID-OCEAN RIDGE HYDROTHERMAL SYSTEMS

Carbon dioxide is by far the most abundant carbon compound present in hydrothermal fluids discharged at mid-ocean ridges. Measured CO_2 concentrations in most unsedimented MOR hydrothermal fluids range between 3 and 30 m*M* (Figure 2a; Table 1). However, concentrations as high as 200 m*M* have been reported in some vents on the East Pacific Rise where CO_2 abundances are enriched by phase separation in the subsurface or by release of magmatic volatiles after eruptions [e.g., *Von Damm and Lilley*, 2004], and transient elevated CO_2 concentrations up to 106 m*M* have been recorded after tectonic activity at Endeavour on the Juan de Fuca Ridge [*Lilley et al.*, 2003; *Seewald et al.*, 2003]. High concentrations are also observed in fluids that have interacted with sediments, where decomposition of organic matter can contribute CO_2 to the fluids.

Methane is also present in substantial concentrations in most MOR hydrothermal fluids, and is by far the most abundant organic compound observed (Figure 2b; Table 1). Measured concentrations of CH_4 in unsedimented systems range up to 2.5 m*M*, with concentrations especially elevated in

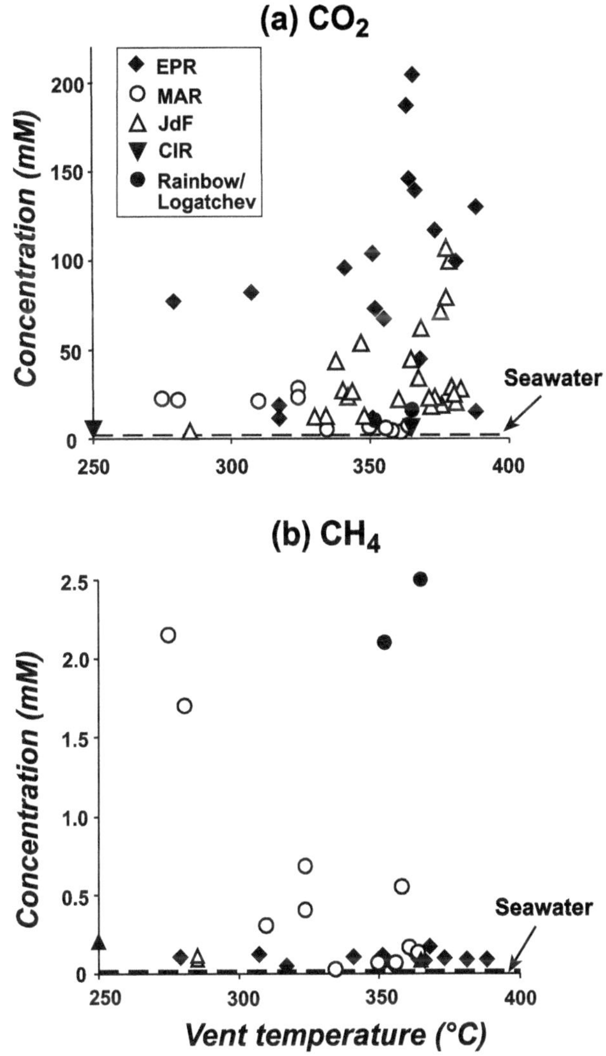

Figure 2. Measured concentrations of dissolved CO_2 (a) and CH_4 (b) in unsedimented mid-ocean ridge hydrothermal vent fluids. EPR, East Pacific Rise; MAR, Mid-Atlantic Ridge; JdF, Juan de Fuca Ridge; CIR, Central Indian Ridge. Hydrothermal fluids from Rainbow and Logatchev on the MAR are plotted separately because their compositions appear to be influenced by interaction with ultramafic and gabbroic rocks in the subsurface. Methane concentrations at Endeavour on the JdF are not shown because CH_4 in vent fluids at this location appears to be derived largely from interaction with buried sediments [*Lilley et al.,* 1993]. Data are from *Craig et al.* [1980], *Lilley et al.* [1983], *Welhan and Craig* [1983], *Merlivat et al.* [1987], *Evans et al.* [1987], *Campbell et al.* [1988], *Jean-Baptiste et al.* [1991], *Charlou et al.* [1996a, 1996b, 1998, 2000, 2002], *Lein et al.* [2000], *Lilley et al.* [2003], *Seewald et al.* [2003], *Von Damm and Lilley* [2004], and *Takai et al.* [2004].

Table 1. End-member concentrations and isotopic compositions of dissolved carbon compounds for representative MOR hydrothermal vent fluids.

Location	Seawater	Lucky Strike[a] MAR	Menez Gwen[b] MAR	Broken Spur[c] MAR	TAG[d] MAR	Plume Vent[e] JdF	18°26′S[f] EPR	21°N[g] EPR	Kairei[h] CIR	Rainbow[i] MAR	Logatchev[i] MAR	Guaymas[j]	Middle Valley[k] JdF
Host rock		basalt	basalt	basalt	basalt	basalt	basalt	basalt	basalt	ultramafic/gabbro	ultramafic/gabbro	basalt	basalt
Sediment covered?		no	no	no	no	no	no	no	no	no	no	Yes	yes
T, °C	2	324	275	364	321	246	310		365	365	347	298	281
pH	7.8	3.7	4.3	–	3.1	–	3.0		–	2.8	3.3	6.7	
H_2, mM	0.0004	0.12	0.044	1.03	0.15	0.35	1.30	0.65	2.5	16	12	3	2.6
CO_2, mM	2.3	23.0	22.6	7.1	2.9	3.92	22	5.7	7.8	16	10.1	40	8.4
$\delta^{13}C_{CO_2}$, ‰	~0	−10.6	−9.1	−9.0	−8.4	−4.4	−5.8	−7	−6.1	−3.15	−4.3	−8.3	−23.8
CH_4, mM	0.0003	0.68	2.15	0.13	0.12	0.084	0.032	0.06	0.084	2.5	2.1	54	20.3
$\delta^{13}C_{CH_4}$, ‰	–	−12.7	−19.6	−19	−8.0	−20.8	−23.5	−15	−13.8	−15.8	−13.6	−43.3	−54.3
$\delta^{2}H_{CH_4}$, ‰	–	–	–	–	–	–	–	−102	–	–	−109	−101	–
CO, µM	0.0003	–	–	–	–	–	–	0.110	0.5	5	–	–	–
C_2H_6, nM	–	159	808	–	–	180	156	35	–	1097	–	444,000	232,000
C_3H_8, nM	–	9	18	–	–	46	36	35	–	48	–	31,000	56,600
n-C_4H_{10}, nM	–	–	–	–	–	5.7	–	47	–	–	–	1400	6200
$T_{equilibrium}$, °C[l]	–	>1200	690	710	>1200	470	440	845	870	590	755		

Values in table represent analyses for a single vent within the designate vent field. MAR, Mid-Atlantic Ridge; JdF, Juan de Fuca Ridge; EPR, East Pacific Rise; CIR, Central Indian Ridge.

Data sources: [a]Eiffel vent, *Charlou et al.* [2000]; [b]Fontaine vent, *Charlou et al.* [2000]; [c]*Lein et al.* [2000]; [d]*Charlou et al.* [1996a, 2002]; [e]*Evans et al.* [1988]; [f]Fromveur vent, *Charlou et al.* [1996b]; [g]*Craig et al.* [1980], *Lilley et al.* [1983], *Welhan and Craig* [1983], *Welhan and Lupton* [1987]; [h]Kali chimney, *Takai et al.* [2004]; [i]*Charlou et al.* [2002]; [j]*J. Seewald* (personal communication); [k]Chowder Hill, Dead Dog Field, *Cruse and Seewald* [2006].

[l]Temperature at which the measured carbon isotope compositions of CO_2 and CH_4 would be in equilibrium, calculated using the CO_2-CH_4 isotope fractionation curve of *Horita* [2001].

axial hydrothermal systems where the fluids have interacted at depth with ultramafic and gabbroic rocks, such as those at Rainbow and Logatchev (Figure 2b) [e.g., *Charlou et al.*, 2002]. Comparably high CH_4 concentrations (1–2 m*M*) have also been observed in warm hydrothermal fluids at the off-axis, ultramafic-hosted Lost City system [*Kelley et al.*, 2005; *Proskurowski et al.*, 2006]. Even higher concentrations (20 m*M* and higher) are observed in sediment-covered systems, where methane is generated from decomposing sedimentary organic matter (Table 1) [e.g., *Cruse and Seewald,* 2006].

Other than CH_4, measurements of organic compounds in hydrothermal vent fluids from unsedimented systems have been extremely limited. To date, analyses of dissolved light hydrocarbons (C_2-C_4 compounds) have been reported for only a few unsedimented hydrothermal fluids (Table 1). These compounds are present in levels much lower than CH_4, with C_1/C_{2+} (molar ratio of CH_4 to the total of C_2H_6 and other light hydrocarbons) ranging from ~100 to >4000. Abundances of higher hydrocarbons in vent fluids from unsedimented systems have been reported in only a few instances. *Brault et al.* [1988] extracted nonvolatile hydrocarbons from a >250°C vent fluid at 13°N on the East Pacific Rise, and found that total hydrocarbons in dissolved and particulate fractions were 57 and 100 times more abundant, respectively, than in bottom seawater. Even higher levels (400 times seawater) were observed in warm (15°C) fluids discharging from an adjacent fissure. Saturated linear hydrocarbons (*n*-alkanes) to >C_{35} and trace amounts of biomarker compounds (e.g., hopanes) were identified in the samples, but the total extracts were dominated by an unresolved complex mixture of unidentified hydrocarbons. Fatty acids were also observed in the particulate fraction at this site [*Brault et al.*, 1984]. *Holm and Charlou* [2001] also reported trace amounts of C_{16} to C_{29} *n*-alkanes in hydrothermal fluids at Rainbow.

Of course, carbon compounds are not limited to fluids, but also occur in rocks and sediments in MOR hydrothermal systems. Freshly erupted mid-ocean ridge basalt contains both carbon dissolved in glass (~60–120 ppm C) and magmatic CO_2 gas trapped in vesicles (~100–600 ppm CO_2) [*Gerlach,* 1989; *Bottinga and Javoy,* 1989; *Blank et al.*, 1993; *Kelley et al.*, 2004]. Although traces of CH_4 and light hydrocarbons have also been reported in MOR basalts, the abundance and composition of organic compounds (and biomass) in fresh MOR basalt remains undefined.

Nonvolatile organic compounds at levels up to several 10s of μg/g have been extracted from bulk samples of sulfide chimneys and metaliferous sediments from unsedimented MOR systems on the East Pacific Rise and Mid-Atlantic Ridge [*Brault et al.*, 1984, 1988, 1989; *Simoneit et al.*, 1990, 2004; *Glynn et al.*, 2006]. Compounds identified in the organic extracts include branched and linear alkanes, fatty acids, polycyclic aromatic hydrocarbons (PAH), and various biomarkers (e.g., pristine/phytane, terpenoids), but an unresolved complex mixture (UCM) of unidentified compounds frequently constituted a large fraction of the extracts. Bioorganic compounds in the form of microbial biomass and biofilms are also present in sulfide chimneys and sediments [*Takai et al.*, 2001; *Schrenk et al.*, 2003; *Kormas et al.*, 2006; *Glynn et al.*, 2006], but the composition, abundance, and spatial distribution of these compounds has not been reported.

Much more abundant and diverse organic compounds have been reported in vent fluids and mineral samples from sedimented MOR hydrothermal systems. Measured levels of light hydrocarbons are substantially higher in sediment-covered than unsedimented systems (Table 1), resulting in much lower C_1/C_{2+} ratios (e.g., 70–245 at Middle Valley; *Cruse and Seewald* [2006]). Semivolatile aromatic compounds including benzene, toluene, and xylene are also observed in these fluids [*Simoneit et al.*, 1988; *Cruse and Seewald,* 2006]. In many sediment-covered deep-sea hydrothermal systems, a separate organic phase (hydrothermal petroleum) is present and discharges at the seafloor [e.g., *Simoneit,* 1988, 1993, 1994]. The compositions of these petroleums reflect a high degree of thermal maturity, including high abundances of PAH. Sulfide chimneys in these organic-rich sediment-covered systems can contain many milligrams of organic matter per gram of sample [*Kawka and Simoneit,* 1987; *Simoneit et al.*, 1992].

3. SOURCES AND SINKS FOR CARBON COMPOUNDS IN MID-OCEAN RIDGE HYDROTHERMAL SYSTEMS

In this section, the origins of carbon compounds in hydrothermal fluids and mineral samples are discussed, along with some of the processes that transform one type of carbon compound to another. Because they largely arise from different sources, CO_2, CH_4 plus other single-carbon compounds, and more complex (C_{2+}) organic compounds are discussed separately.

3.1. Carbon Dioxide

Concentrations of CO_2 in vent fluids are substantially higher than the amount of dissolved inorganic carbon (DIC; = $CO_{2(aq)}$ + HCO_3^- + CO_3^{2-}) in seawater, which is primarily present as bicarbonate (HCO_3^-) at seawater pH (Figure 2). Consequently, hydrothermal circulation results in a net gain of DIC in circulating fluids and, because this is the predominant form of carbon, circulation also therefore results in a significant increase in total dissolved carbon. Carbon isotope compositions of CO_2 ($\delta^{13}C_{CO_2}$) in hydrothermal fluids from unsedimented MOR systems fall within a narrow range of

–3 to –11‰ (Figure 3). These values overlap the isotopic composition of magmatic CO_2 (≈–4 to –9‰) [*Blank et al.*, 1993] and are significantly depleted in ^{13}C relative to seawater bicarbonate (~0‰) (Figure 3), indicating that magmatic volatiles are the primary contributor of CO_2 to the fluids. In contrast to unsedimented systems, measured $\delta^{13}C_{CO_2}$ values for some sediment-covered hydrothermal systems (i.e., Middle Valley and Guaymas) fall outside the magmatic range (Figure 3). Isotopic compositions in these systems reflect significant contributions of inorganic carbon from thermal and biological degradation of organic matter present within the sediments.

There are two principal ways that magmatic volatiles can be incorporated into circulating aqueous fluids at mid-ocean ridges. First, CO_2 and other volatiles exsolve from crystallizing magmas at depth, and these gases migrate toward the seafloor. Along the way, these gases may redissolve into hydrothermal fluids circulating through the shallower parts of the crust. Second, magmatic volatiles trapped in vesicles in crystallized basalt or dissolved in basaltic glass are extracted during fluid circulation through the crust. Because fresh MOR basalts are estimated to contain ~100–600 ppm magmatic CO_2 [*Gerlach*, 1989; *Bottinga and Javoy*, 1989; *Blank et al.*, 1993; *Kelley et al.*, 2004], complete extraction of this gas at a 1:1 water/rock ratio would increase the dissolved CO_2 concentration in circulating fluids by between ~1.5 and 9 m*M*, comparable to the concentrations observed in many axial hydrothermal fluids (Table 1).

Although magmatic and sedimentary sources appear to be the predominant contributors to CO_2 in MOR hydrothermal fluids, the abundance and isotopic compositions of CO_2 allows for the possibility that seawater bicarbonate might persist throughout circulation and remain as a secondary contributor to measured concentrations. Indeed, some isotopic studies have assumed that seawater DIC persists in the fluid as it circulates through the crust and have subtracted out this presumed contribution to arrive at an estimate for the abundance and isotopic composition of a pure end-member "magmatic" CO_2 component [e.g., *Welhan and Lupton*, 1987; *Evans et al.*, 1988]. However, combined ^{13}C and ^{14}C isotopic analyses for CO_2 in hydrothermal fluids from the Endeavour system indicate that most, if not all, of seawater bicarbonate is removed from circulating fluids before discharge

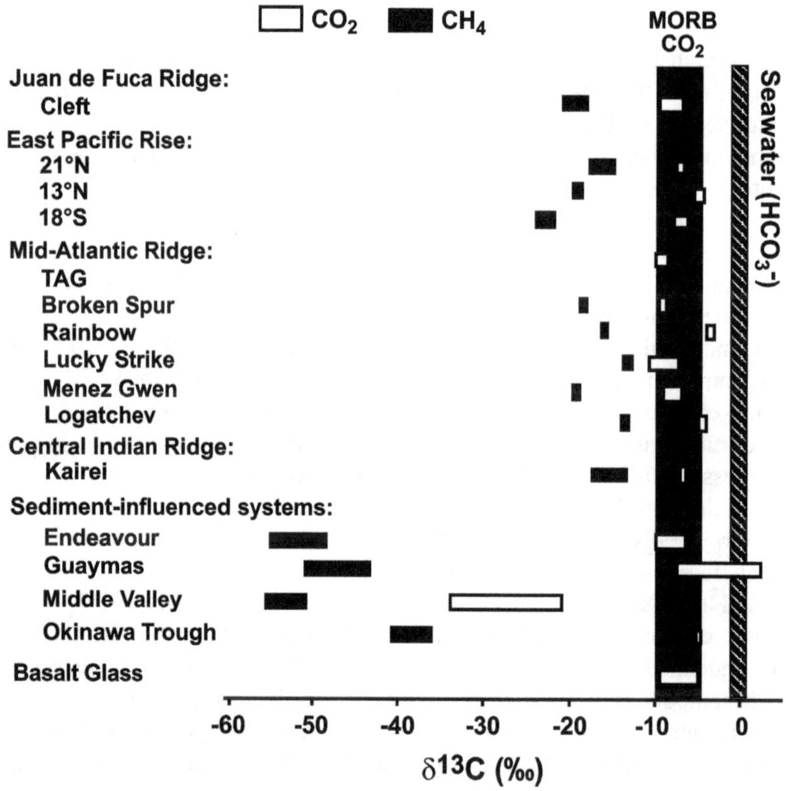

Figure 3. Range of carbon isotopic compositions for dissolved CO_2 and CH_4 in mid-ocean ridge hydrothermal systems. Data sources are as listed for Figure 2, with additional data from *Welhan and Lupton* [1987], *Lilley et al.* [1993], *Ishibashi et al.* [1995], and *Proskurowski et al.* [2004]. MORB, mid-ocean ridge basalt.

at hydrothermal vents on the seafloor [*Proskurowski et al.,* 2004]. If this observation holds true for other systems, it would appear that seawater DIC makes little or no contribution to the CO_2 in hydrothermal vent fluids, and is instead largely or entirely removed during circulation through the crust. Thermal decomposition of organic compounds transported into the crust by downwelling seawater in the recharge limbs of hydrothermal circulation cells might also contribute some CO_2 to high-temperature fluids discharging at the seafloor, but as concentrations of dissolved organic carbon (DOC) in deep seawater are only ~30–40 μM [*Lang et al.,* 2006], this can only be a minor contributor.

It remains to be determined what happens to the seawater DIC that is removed during hydrothermal circulation. One possible sink is precipitation of calcite or other carbonate minerals. Examinations of drill cores indicate that the upper sections of basaltic crust on the flanks of mid-ocean ridges contain, on average, around 0.14–0.21 wt % CO_2, which is largely present as calcite and other carbonate minerals [*Staudigel et al.,* 1989; *Alt and Teagle,* 1999; *Bach et al.,* 2003]. However, because some unknown fraction of these minerals may have precipitated during low-temperature circulation off-axis, the amount of carbonates formed from downwelling seawater in the recharge zone of axial hydrothermal systems remains uncertain. Other potential sinks for seawater DIC include biological carbon fixation by autotrophic microorganisms and conversion to organic compounds through abiotic synthesis. At present, the relative contributions of these processes in the removal of seawater DIC are unknown.

3.2. Methane and Other C_1 Compounds

Like CO_2, CH_4 is present in MOR hydrothermal fluids at abundances that are significantly enriched relative to seawater, indicating that hydrothermal circulation is a net source for this compound as well (Figure 2b; Table 1). However, although interpretation of the primary source of CO_2 in MOR hydrothermal fluids appears to be fairly straightforward, determining the source of CH_4 is more problematic. There are a variety of possible contributors to the CH_4 in hydrothermal vent fluids, including biological metabolic activities ("biogenic methane"), thermal degradation of organic matter ("thermogenic methane"), direct injection of magmatic gases, extraction of magmatic gases trapped within basalt, and synthesis from reduction of CO_2 during hydrothermal circulation ("abiotic methane") [*Welhan,* 1988; *McCollom and Seewald,* 2007]. Furthermore, the CH_4 that is discharged in hydrothermal fluids at the seafloor is likely to reflect the integrated contributions of several of these sources, as well as losses of CH_4 from biological and abiotic processes. Unraveling the potential contributions of these different processes requires consideration of the isotopic composition and abundance of CH_4 in hydrothermal fluids together with constraints provided by experimental studies and thermodynamic considerations.

One criterion that has been commonly used to evaluate the source of CH_4 in geologic systems is a comparison of its C and H isotopic compositions [*Schoell,* 1988]. Based on empirical observations of gases from a variety of environments, it has been suggested that CH_4 from thermogenic and biogenic sources can be differentiated by their coupled C and H isotopic signatures (Figure 4) [*Schoell,* 1988]. The biogenic field is further divided into CH_4 derived from fermentation of organic matter and that derived from reduction of CO_2 by methanogenic organisms. Coupled hydrogen and carbon isotope values for CH_4 have only been reported for a few unsedimented MOR hydrothermal systems (21°N on the East Pacific Rise and Logatchev), but the available data fall at one extreme end of the thermogenic trend (Figure 4). In contrast, coupled C and H isotopic values for CH_4 from the sediment-covered Guaymas Basin system fall near the thermogenic and biogenic fields in Figure 4, consistent with derivation primarily from a combination of thermogenic and fermentative biogenic sources.

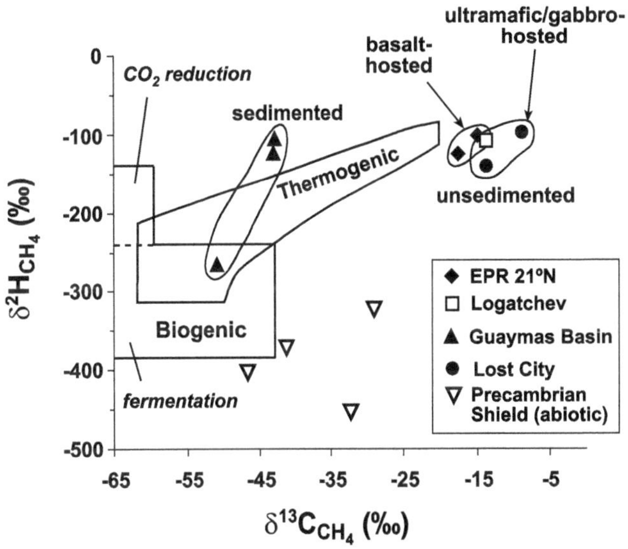

Figure 4. Carbon and hydrogen isotopic compositions of CH_4 in vent fluids from unsedimented and sedimented deep-sea hydrothermal systems. The unsedimented measurements include both basalt-hosted (EPR) and ultramafic/gabbro-hosted (Logatchev, Lost City) systems. Also shown for reference are fields encompassing typical isotopic compositions of thermogenic and biogenic CH_4, with the biogenic field separated into CH_4 formed by CO_2 reduction and by fermentation [after *Schoell,* 1988], and that proposed to have an abiotic origin in crystalline Precambrian Shield rocks [*Sherwood Lollar et al.,* 2006]. Lost City data from *Kelley et al.* [2005] and *Proskurowski et al.* [2006].

Methane sources can be further differentiated by comparison of carbon isotopes with the ratio of CH_4 to higher hydrocarbons (C_1/C_{2+} ratio). In this respect, biogenic CH_4 typically has lighter ^{13}C compositions and higher C_1/C_{2+} than CH_4 from thermogenic sources (Figure 5). On a plot of C_1/C_{2+} versus $\delta^{13}C_{CH_4}$, light hydrocarbons in unsedimented deep-sea hydrothermal systems are clearly distinct from both typical biogenic and thermogenic sources (Figure 5). Hydrocarbons in fluids from sediment-covered ridges are intermediate between the thermogenic and biogenic fields, consistent with dominant inputs from microbial methanogenesis and thermal alteration of sedimentary organic matter. The criteria shown in Figures 4 and 5 suggest that the CH_4 from most unsedimented, axial MOR hydrothermal systems is compositionally and isotopically distinct from typical biogenic and thermogenic sources.

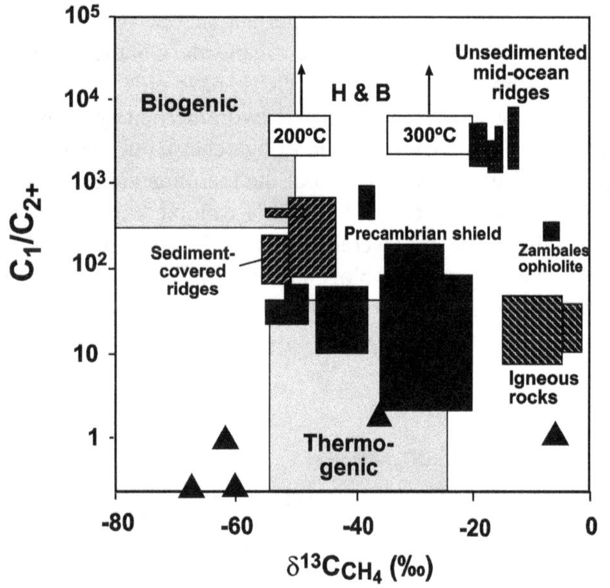

Figure 5. Carbon isotopic composition of methane versus C_1/C_{2+} ratios for volatile hydrocarbons in high-temperature fluids from mid-ocean ridge hydrothermal systems. Fields in light gray are ranges of values conventionally attributed to biogenic and thermogenic hydrocarbon gases. Shown for comparison are values for gases proposed to have an abiotic origin in Precambrian shield rocks, igneous rocks, and gases venting from a continental serpentinite (Zambales ophiolite) (see *Horita and Berndt* [1999, and references therein]). The fields marked "200°C" and "300°C" show the range of values for abiotic CH_4 generated in hydrothermal experiments of *Horita and Berndt* [1999]; because C_{2+} compounds were below detection limits, C_1/C_{2+} ratios for these experiments are minimum values. Triangles (▲) represent hydrocarbons formed in Fischer-Tropsch synthesis experiments [*Lancet and Anders*, 1970; *Hu et al.*, 1998; *McCollom and Seewald*, 2006]. Fields for unsedimented ridges include data for EPR 21°N, Rainbow, Menez Gwen, and Lucky Strike, while sediment-covered systems include Middle Valley, Guaymas, and Endeavour. Although Endeavour is currently unsedimented, it is included with the sediment-covered systems in this plot since hydrocarbons at this site appear to derive from buried sediments [*Lilley et al.*, 1993]. Data sources for vent fluids as in Figures 1 and 2, with additional data from *Cruse and Seewald* [2006].

Additional constraints on the origin of CH_4 are provided by comparison of the relative $\delta^{13}C$ compositions of CO_2 and CH_4 in vent fluids with the expected carbon isotopic fractionation between these compounds at isotopic equilibrium. Because the equilibrium fractionation between these compounds varies as a function of temperature [*Horita,* 2000], the measured isotopic compositions of CO_2 and CH_4 in vent fluids can potentially be used as an indicator of the temperature at which these compounds may have equilibrated. The calculated temperature at which the measured isotopic compositions of CO_2 and CH_4 would be in equilibrium is given for several vent fluids in Table 1. Figure 6 shows the measured isotopic compositions of coexisting CO_2 and CH_4 for a more comprehensive set of MOR hydrothermal fluids, with the superimposed lines representing equilibrium isotopic fractionation between these compounds for a range of temperatures from 100° to 1000°C. As shown by the data in the figure and table, the measured isotopic compositions correspond to equilibrium temperatures ranging from ~450°C to well over 1000°C, which in most cases far exceed the measured venting temperature of the fluids at the seafloor.

These data have commonly been interpreted to indicate that the CH_4 equilibrated with CO_2 during cooling of magmatic gases at higher temperatures deep within the hydrothermal system, with the highest values perhaps indicating equilibration under magmatic or mantle conditions [e.g., *Welhan and Craig,* 1983; *Evans et al.,* 1988; *Shanks et al.,* 1995; *Charlou et al.,* 1996b, 2000]. However, although the relative isotopic compositions of CO_2 and CH_4 might appear to indicate equilibration at high temperature, this interpretation is inconsistent with thermodynamic and observational constraints on the relative abundance of CH_4. At temperatures >~ 400°C, thermodynamic considerations strongly favor CO_2 relative to CH_4. This can be illustrated by examining the equilibrium constant (K) for the reaction:

$$CO_2 + 4H_2 \leftrightarrow CH_4 + 2H_2O \quad (1)$$

As shown in Figure 7a, K for this reaction strongly favors the compounds on the right side of the reaction at low temperatures, but increasingly favors the compounds on the left side at higher temperature. Equilibrium CO_2/CH_4 ratios calculated for H_2 levels typical of MOR systems yield val-

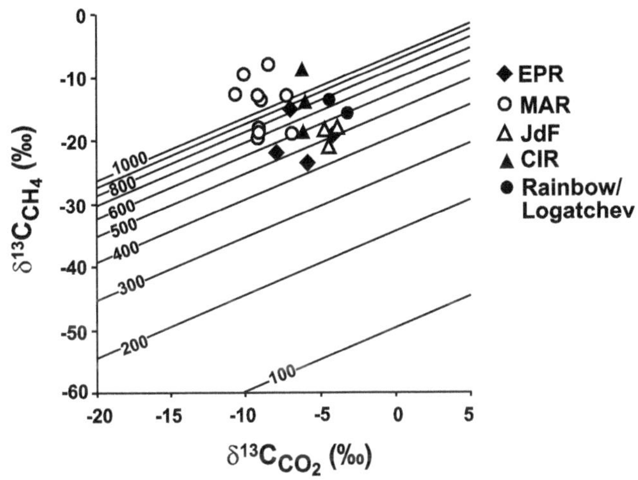

Figure 6. Carbon isotope compositions of coexisting CO_2 and CH_4 in MOR hydrothermal fluids. Lines represent equilibrium isotopic fractionation between these compounds as a function of temperature, based on data from *Horita* [2001]. Abbreviations and data sources as in Figures 1 and 2.

ues >>1 at temperatures above 500°C (Figure 7b), indicating that CO_2 should be the predominant form of carbon at equilibrium, with CH_4 present at only trace levels. As a result, the abundance of CH_4 in mid-ocean ridge magmatic gases should be vanishingly small [e.g., *Mathez,* 1984; *Kelley,* 1996]. Chemical analyses of magmatic volatiles trapped in vesicles within seafloor basalts confirm that magmatic gases are characterized by very high CO_2/CH_4 ratios [e.g., *Pineau and Javoy,* 1983] that are much higher than ratios observed in hydrothermal vent fluids. It is unlikely, therefore, that direct magmatic degassing is a significant source of CH_4 in MOR hydrothermal systems and, although some CH_4 may be generated in CO_2-rich magmatic gases as they cool, thermodynamic considerations indicate that only small amounts will be generated by this process at temperatures above ~400–500°C. Thus, it appears more likely that the apparent isotopic equilibrium temperatures are coincidental, and the carbon isotopic values of CH_4 probably reflect contributions of lower temperature processes that are in isotopic disequilibrium with magmatic CO_2 [*McCollom and Seewald,* 2007].

The thermodynamic constraints shown in Figure 7 also indicate that decreasing temperatures and higher H_2 concentrations favor CH_4 relative to CO_2. As a consequence, cooling of CO_2- and H_2-enriched fluids from high temperatures or the production of H_2 during fluid-rock interactions may make the reduction of CO_2 to CH_4 thermodynamically favorable. There are a variety of environments within hydrothermal systems where these conditions might arise [*Shock,* 1990, 1992]. For example, reaction of seawater with basalt in the recharge limb of hydrothermal circulation zones will produce H_2, favoring reduction of dissolved CO_2 (or HCO_3^-). During this process, dissolution of magmatic gases trapped within the rocks can supply additional CO_2 and H_2, providing further substrates and thermodynamic impetus for synthesis reactions. As fluids approach high-temperature reaction zones in the deeper parts of the system, increased reaction rates may supply additional H_2 from reactions with basalt or other rocks, whereas injection of CO_2-dominated magmatic gases into the system will provide additional carbon and increase the thermodynamic drive for abiotic synthesis (provided, of course, temperatures are not too high; Figure 7).

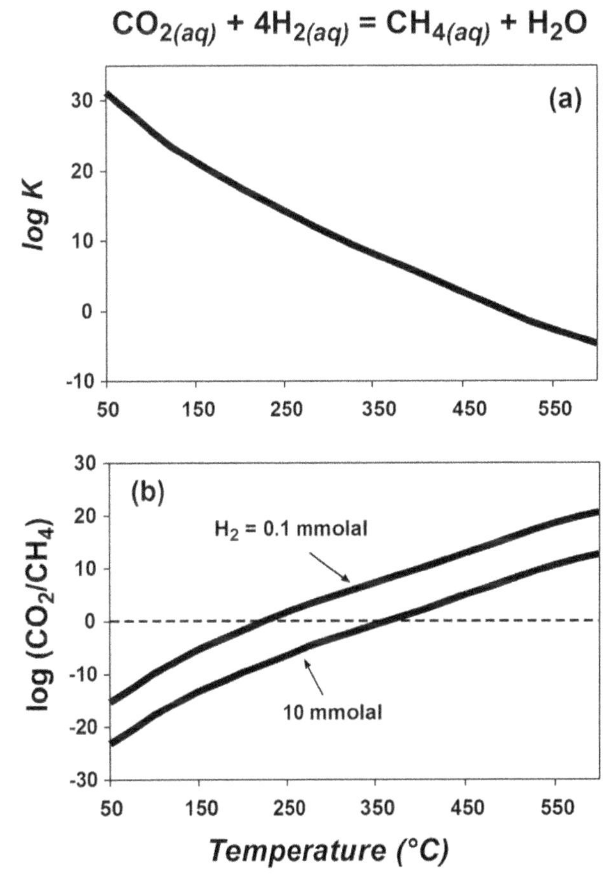

Figure 7. Thermodynamic relationships between dissolved CO_2 and CH_4. (a) Log K for reduction of CO_2 to CH_4 (reaction (1)). (b) Calculated equilibrium (CO_2/CH_4) ratios as a function of temperature at two values of H_2 concentration that bracket measured concentrations for most MOR systems (activity coefficients assumed to be equal to 1 for this calculation). Values for log K calculated using thermodynamic data from SUPCRT92 [*Johnson et al.,* 1992] and *Shock et al.* [1989].

The thermodynamic favorability of methane formation will also increase if CO_2- and H_2-charged high-temperature fluids cool by conduction and/or mixing with seawater during ascent to the seafloor [*Shock,* 1992].

Although CO_2 is dissolved in hydrothermal fluids at the elevated pressures of the subseafloor (>25 MPa), a number of recent experimental studies have demonstrated that dissolved CO_2 can be reduced to CH_4 under such conditions [e.g., *Horita and Berndt,* 1999; *McCollom and Seewald,* 2001, 2003, 2007; *Foustoukos and Seyfried,* 2004; *Seewald et al.,* 2006]. Although several of these studies suggest that minerals such as magnetite or chromite may catalyze the formation of CH_4, the role of mineral catalysis in CO_2 reduction remains uncertain and CH_4 formation has also been observed in experiments without any minerals present (*Seewald et al.* [2006]; see *McCollom and Seewald* [2007] for an extended discussion of this topic). In any case, these experimental studies demonstrate that abiotic CH_4 can be produced in MOR hydrothermal systems provided thermodynamically favorable conditions occur.

Although the above experiments show that CO_2 reduction to methane can proceed under hydrothermal conditions, it should be noted that, for the most part, reaction rates for reduction of CO_2 to CH_4 observed in experiments are sluggish, reflecting a strong kinetic inhibition to the reaction at temperatures below about 450°C. For example, most of the experiments cited above were performed at H_2 concentrations (~40 to >300 mmol kg^{-1}) that are elevated well above those observed in natural hydrothermal vent fluids, and under conditions where complete conversion of CO_2 to CH_4 was strongly thermodynamically favored. Nevertheless, only a small fraction of the available CO_2 was converted to CH_4 despite extended reaction times and the presence of potentially catalytic minerals including magnetite and chromite (e.g., ~0.5% conversion during 44 days of heating at 390°C [*Foustoukos and Seyfried,* 2004], ~0.5% conversion during 87 days at 260°C [*McCollom and Seewald,* 2003]). Even slower reaction rates would be expected at lower temperatures, and at the lower H_2 concentrations present in natural systems. Of course, the sluggish reaction rates can be overcome by other minerals that may be more effective catalysts, and *Horita and Berndt* [1999] have shown that reduction of dissolved CO_2 to CH_4 proceeds rapidly when catalyzed by Ni-Fe alloy (awaruite) at 200°–400°C and 50 MPa. In this case, essentially complete conversion of 12 m*m* ΣCO_2 to methane occurred in less than 350 h at 300°C. Ni-Fe alloys form only under very strongly reducing conditions, but such conditions are apparently sometimes attained in subseafloor environments during hydrothermal alteration of ultramafic rocks, where olivine provides a source of Ni and serpentinization produces strongly reducing environments [e.g., *Frost,* 1985; *Alt and Shanks,* 1998, 2003; *Früh-Green et al.,* 2004]. However, catalysis of abiotic CH_4 formation by Ni-Fe alloy is highly unlikely in basalt-hosted systems because redox conditions are not sufficiently reducing for native metal alloys such as awaruite to be present.

Current estimates indicate that the residence time of fluids in the high temperature zones of MOR hydrothermal systems are less than a couple of years [e.g., *Kadko and Butterfield,* 1998; *Fisher,* 2003, 2004]. The available experimental data suggest that slow reaction rates may prohibit reduction of CO_2 to CH_4 from reaching equilibrium levels on these time scales unless effective catalysts are present. However, a rigorous evaluation of this issue would have to consider the time-temperature-composition paths of the fluids and the effects of exposure of dissolved reactants to large surface areas of potential catalysts during fluid migration along fractures.

Abiotic formation of CH_4 in MOR hydrothermal systems has frequently been inferred to occur by Fischer-Tropsch synthesis, which is a surface-catalyzed reaction for the conversion of CO or CO_2 to organic compounds [e.g., *Charlou et al.,* 2000, 2002; *Kelley et al.,* 2005]. Fischer-Tropsch synthesis has been extensively studied as a gas-phase process for industrial applications, but surface-catalyzed, Fischer-Tropsch-type reactions have not yet been demonstrated to occur under the conditions of MOR hydrothermal systems (see review by *McCollom and Seewald* [2007]). Fischer-Tropsch-type synthesis usually produces hydrocarbons with a much lower C_1/C_{2+} ratio than is observed in MOR hydrothermal fluids and, although the available data are limited, the process also appears to produce CH_4 that is substantially more depleted in ^{13}C relative to the source CO_2 than the CH_4 reported in unsedimented MOR hydrothermal fluids (Figure 5). Thus, although Fischer-Tropsch synthesis may make some level of contribution to the CH_4 and other organic compounds observed in MOR hydrothermal systems, it appears unlikely to be the sole, or even predominant, source of CH_4 in most cases.

The Ni-Fe alloy catalyzed process studied by *Horita and Berndt* [1999] provides an alternative pathway for abiotic reduction of CO_2 to CH_4 in MOR hydrothermal systems, but its contribution is probably limited to systems that have interacted with ultramafic rocks in the subsurface such as Rainbow and Logatchev, or the lower-temperature off-axis system at Lost City [*Kelley et al.,* 2005] (although the abiotic CH_4 generation reported by *Horita and Berndt* [1999] has frequently been referred to as Fischer-Tropsch-type synthesis in the literature, it should be emphasized that the Fischer-Tropsch process refers explicitly to the catalyzed synthesis of complex organic compounds through sequential formation of C–C bonds; the Ni-catalyzed reduction of CO_2 to CH_4

without production of higher hydrocarbons has a distinctly different mechanism and is more properly referred to as the Sabatier process). Two results of their study are worthy of particular note with regard to interpretation of the source of CH_4 in MOR hydrothermal fluids. First, because higher hydrocarbons were below detection limits in the experiments of *Horita and Berndt* [1999], catalysis of CO_2 reduction by Ni-Fe alloy appears to be exclusive for CH_4 formation, and would therefore generate dissolved light hydrocarbons with a very high C_1/C_{2+} ratio. Second, the CH_4 generated in their experiments was depleted in ^{13}C by 50–66‰ relative to ΣCO_2 during reaction at 200°C and by 35–40‰ at 300°C. Thus, at high temperatures (300°C and higher), the Ni-Fe alloy catalyzed reaction generates CH_4 with an isotopic signature and high C_1/C_{2+} ratio that closely resembles light hydrocarbons observed in unsedimented hydrothermal systems, including Rainbow (Figure 5). At low temperatures (≤200°C), the process generates methane with a strongly depleted ^{13}C composition and a high C_1/C_{2+} ratio that very closely resembles CH_4 produced by biological organisms (Figure 5), suggesting that it may be very difficult to distinguish abiotic from biogenic methane in some lower temperature settings [*Horita and Berndt,* 1999].

In the absence of mineral catalysis, the reduction of dissolved CO_2 to CH_4 appears to take place through a series of incremental steps, with formic acid (HCOOH), carbon monoxide (CO), formaldehyde (H_2CO), and methanol (CH_3OH) occurring as intermediates (Figure 8) [*Seewald et al.,* 2006]. The first few steps in this sequence proceed rapidly in either direction under hydrothermal conditions, but reduction of H_2CO to CH_3OH proceeds more slowly. Further reduction of CH_3OH to CH_4 is even slower, as indicated by the small amounts of CH_4 that accumulate during experimental studies of CO_2 reduction [*McCollom and Seewald,* 2001, 2003; *Foustoukos and Seyfried,* 2004; *Seewald et al.,* 2006].

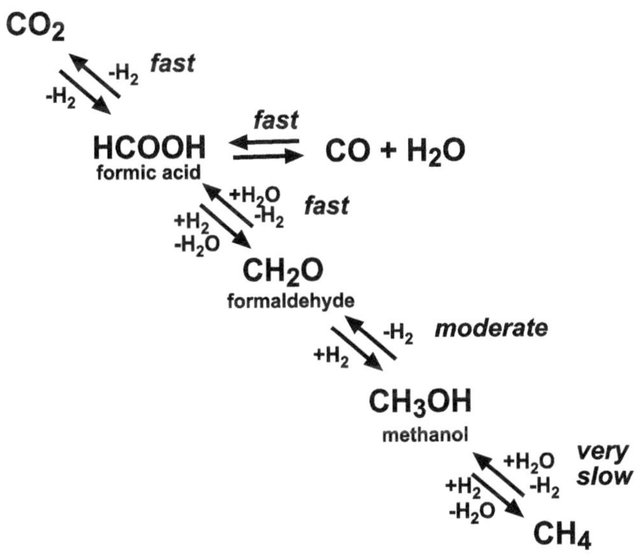

Figure 8. Reaction scheme for the sequential reduction of CO_2 to CH_4.

The rapid rates of reaction among dissolved CO_2, CO, and HCOOH should allow these compounds to reach thermodynamic equilibrium in MOR hydrothermal fluids [*McCollom and Seewald,* 2003; *Seewald et al.,* 2006]. However, thermodynamic constraints indicate that only trace amounts of these intermediates should accumulate in hydrothermal fluids at equilibrium, particularly at low temperatures. For instance, Figure 9a shows the distribution of single-carbon compounds that should be present at stable thermodynamic equilibrium for levels of total carbon and H_2 typical of deep-sea hydrothermal systems over a range of temperatures. At the high temperatures of most end-member hydrothermal vent fluids (>300°C), equilibrium concentrations of single-carbon compounds other than CO_2 are below micromolal levels. In the few instances where CO has been reported in hydrothermal vent fluids, its concentration appears to be consistent with the low levels expected for thermodynamic equilibrium with the measured CO_2 and H_2 concentrations (Table 1) [*Seewald et al.,* 2006]. Concentrations of other single-carbon compounds such as CH_3OH and HCOOH have not yet been reported for any high-temperature vent fluids from unsedimented MOR systems, but the very low concentrations of these compounds expected at equilibrium (Figure 9a) indicates that such measurements could be analytically challenging.

If complete equilibrium with CH_4 is maintained, the levels of other single-carbon compounds would be expected to drop off dramatically in the lower temperature parts of hydrothermal systems (Figure 9a). However, kinetic inhibitions to CH_4 formation might allow larger concentrations of the intermediates of CO_2 reduction to accumulate. Figure 9b shows the equilibrium distribution of single-carbon compounds that would be attained if CH_4 is excluded from the speciation model. In such circumstances, levels of CH_3OH and HCOOH approaching 100s of μM might accumulate in solution [*Shock,* 1990, 1992; *Seewald et al.,* 2006]

In lower temperature environments where CO_2 reduction may be thermodynamically favorable but kinetically inhibited, the conversion of CO_2 to CH_4 can be catalyzed by microorganisms to gain metabolic energy (microbial methanogenesis). Other organisms can generate CH_4 through degradation of more complex organic compounds (fermentation). Either mechanism could contribute to CH_4 observed in hydrothermal fluids. Of course, for biologically generated CH_4 to be observed in high-temperature vent fluids at the ridge

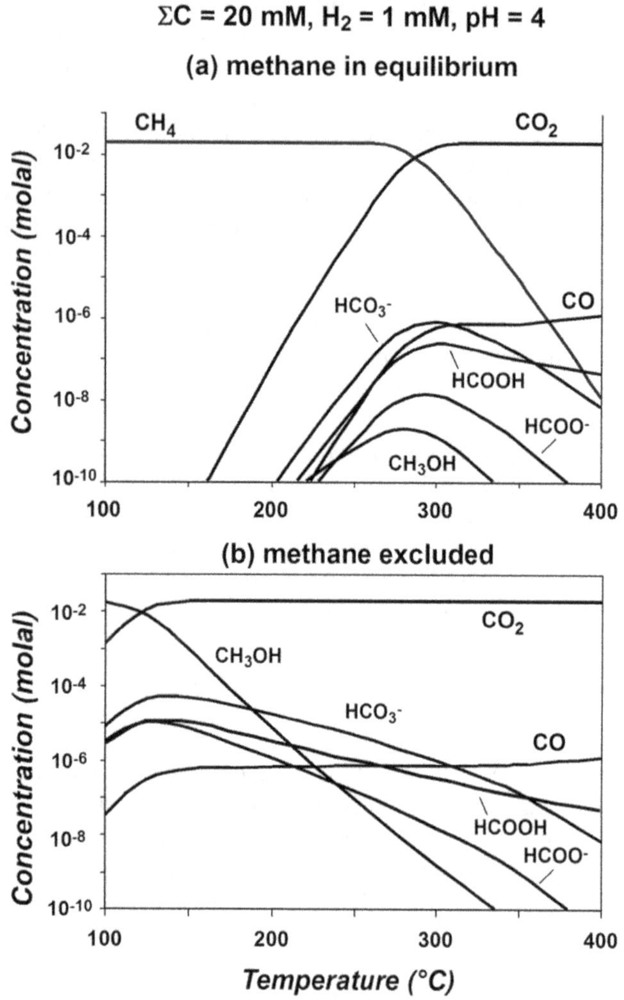

Figure 9. Speciation of dissolved single carbon compounds as a function of temperature for typical pH, H_2 concentration, and total dissolved carbon of unsedimented mid-ocean ridge hydrothermal fluids (20 m*m* total carbon, 1 m*m* H_2, pH = 4). (a) Species that are present at thermodynamic equilibrium for these conditions, whereas in (b) it is assumed that CH_4 formation is kinetically inhibited but all other species are in equilibrium. The bottom figure represents a metastable equilibrium speciation because most carbon would be present as CH_4 at stable equilibrium.

axis, it must be produced in the lower temperature parts of the system that allow life to exist and then be incorporated into circulating hydrothermal fluids. Microbial methanogenesis might occur, for instance, in downflowing fluids in the recharge limb of hydrothermal circulation cells with H_2 supplied from reaction of basalt with seawater [e.g., *Bach and Edwards,* 2003] or through fermentative decomposition of organic matter by heterotrophic organisms. Alternatively, methanogenesis might occur within the subsurface on the flanks of upflow zones and then be flushed into rising hydrothermal fluids.

The measured isotopic compositions of CH_4 in hydrothermal fluids from unsedimented MOR systems falls well outside the range of values typically attributed to microbial methanogenesis (Figures 4 and 5); however, this does not necessarily eliminate biogenic sources as a significant contributor to CH_4 in these environments. For example, if a large fraction of seawater HCO_3^- is converted to CH_4 by microorganisms in the cooler zones where seawater flows into the crust in the recharge limb of hydrothermal circulation cells, the CH_4 generated may be more enriched in ^{13}C than is typically observed for methanogenesis owing to isotopic reservoir effects [*McCollom and Seewald,* 2007]. That is, as a larger fraction of HCO_3^- is converted, the ^{13}C isotopic composition of the CH_4 produced will approach that of seawater HCO_3^- (~0‰) rather than the $<\sim -50$‰ typically observed for microbially generated CH_4 (Figure 4). There is sufficient HCO_3^- in seawater to account for the levels of CH_4 observed in hydrothermal vent fluids from unsedimented systems (Table 1), and its removal by microorganisms in recharge zones would be consistent with isotopic data indicating seawater HCO_3^- is quantitatively removed during hydrothermal circulation [*Proskurowski et al.,* 2004].

The presence of CH_4-rich fluid inclusions in rocks from deep within the ocean crust suggests that deeply circulating hydrothermal fluids are another potential contributor to CH_4 in axial vent fluids. Two families of CH_4-rich fluid inclusions have been observed within crystalline gabbroic rocks from the lower oceanic crust at the Southwest Indian Ridge [*Kelley,* 1996; *Kelley and Früh-Green,* 1999, 2001]. One type of fluid inclusion, formed at super-solidus to greenschist facies metamorphic conditions, contain CO_2-CH_4-H_2O-rich fluids with CH_4 concentrations up to 43 mol %, while a second set of CH_4-H_2O-H_2-bearing inclusions with CH_4 concentrations >40 mol % is inferred to have formed under a more limited temperature range near 400°C. The high CH_4 abundances in the inclusions are attributed to reduction of magmatic CO_2 under conditions of high H_2 activity created by precipitation of graphite during cooling and/or serpentinization of olivine-bearing ultramafic rocks elsewhere in the crust [*Kelley,* 1996; *Kelley and Früh-Green,* 1999, 2001]. The ^{13}C isotope compositions of CH_4 in the CH_4-H_2O-H_2 inclusions ranged from −9.1 to −33.8‰, which overlaps the range of carbon isotopic composition of CH_4 in vent fluids from unsedimented systems (Figure 3). If the fluids represented by these inclusions interact with the more shallowly circulating fluids that discharge in hydrothermal vents at the ridge axis, they may be responsible for some part of the CH_4 observed in the vent fluids. At present, however, the contribution of deeply circulating fluids to the fluids venting at the seafloor is unknown.

Although there has been considerable discussion in the literature regarding the potential contribution of abiotic CO_2 reduction to CH_4 in MOR hydrothermal fluids, it must be emphasized that most deep-sea hydrothermal fluids discharging at the seafloor actually have an *excess* of CH_4 relative to equilibrium at their venting temperature. This is evident, for instance, in Figure 10, which compares the CH_4 abundance in unsedimented MOR hydrothermal fluids with the amounts that would be in equilibrium with the measured concentrations of H_2 and CO_2 in the same fluid. With the exceptions of the H_2-rich fluids from the Logatchev and Rainbow fields and one vent from the southern East Pacific Rise, nearly all MOR vent fluids have an excess of CH_4 relative to equilibrium. In these fluids, CH_4 should decompose to CO_2 and H_2 (or other intermediate products; Figure 8) rather than forming from CO_2 reduction (i.e., reaction (1) should

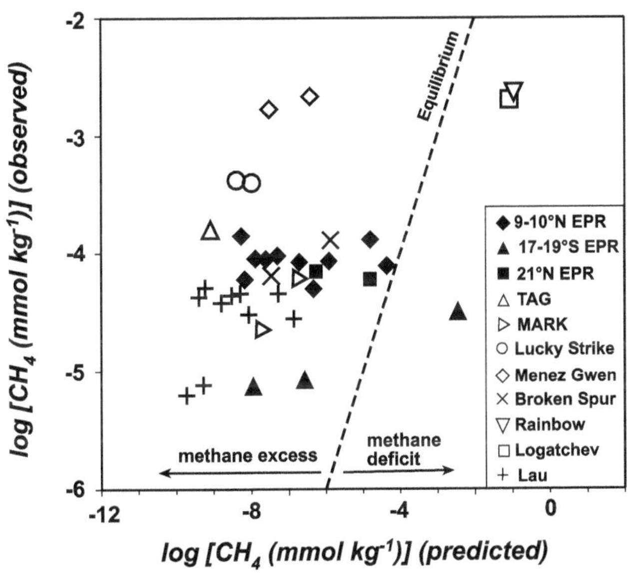

Figure 10. Comparison of measured aqueous CH_4 concentrations in high-temperature (>300°C) axial vent fluids from unsedimented MOR systems with values predicted for equilibrium according to reaction (1) at the measured concentrations of CO_2 and H_2 and temperatures of the same fluids. The solid line indicates a 1:1 relationship (i.e., equilibrium between CO_2 and CH_4). Measured values that are greater than the predicted equilibrium indicate that there is an excess of CH_4 in the fluids relative to equilibrium. In these fluids, there is no thermodynamic impetus for reduction of CO_2 to CH_4; instead, CH_4 should be converted to CO_2 to approach equilibrium. Data from sources listed in Figure 2, with additional data for Lau (J. Seewald et al., unpublished data).

proceed from right to left instead of from left to right). It is worth noting that although methane is in excess in most fluids at their venting temperatures, cooling of these fluids would make reduction of CO_2 to CH_4 thermodynamically favorable (Figure 7) [see, e.g., *Shock,* 1990, 1992].

Potentially, this excess could be explained by loss of CO_2 and/or H_2 as the fluids ascend to the seafloor after reaching equilibrium with CH_4 in the subsurface, but the amounts that would have to be lost are fairly large and appropriate sinks for these compounds are not immediately apparent. Consequently, if the excess CH_4 in these fluids is derived from CO_2 reduction, it must have formed at some point in the system where the reaction was more thermodynamically favorable than it is at the venting temperatures. One possible scenario to explain this excess is that the CH_4 is generated through biological or abiotic CO_2 reduction at lower temperatures in the recharge zone where the reaction is thermodynamically favorable, and it then remains dissolved in the fluid during circulation through the higher temperature part of the system. If this is the case, the presence of excess CH_4 in high-temperature vent fluids indicates that its decomposition is kinetically inhibited to some degree, since it must persist in excess relative to equilibrium through the high-temperature zone before venting at the seafloor. If the decomposition rates of CH_4 were known, this observation could be used to place some additional constraints on possible sources of CH_4 and how long the fluids remained at high temperature after the CH_4 was added to the fluid. However, at present there are insufficient experimental data available to assess the decomposition rate of CH_4 under hydrothermal conditions.

It is apparent from the foregoing discussion that there are a number of different sources that may contribute to CH_4 in MOR hydrothermal fluids. At present, however, it is unclear what the relative contributions of these various sources are, or whether one source is the predominant contributor to CH_4 in the fluids. Although the isotopic composition of CH_4 from several unsedimented MOR systems are clearly distinct from those conventionally attributed to biogenic and thermogenic CH_4 in other geologic settings (Figures 4 and 5), these compositions could plausibly arise from either biological or abiotic sources within the hydrothermal environment. Thus, the origin (or origins) of CH_4 remains somewhat uncertain. A possible exception is the CH_4 observed in fluids at Rainbow and Logatchev, where isotopic composition, thermodynamic constraints, and the possible existence of suitable catalysts in subsurface reaction zones all appear to be consistent with abiotic synthesis by CO_2 reduction. Even in this case, however, the possibility that microbial methanogenesis in low-temperature recharge zones may have contributed a significant fraction of the CH_4 in the fluids cannot be entirely dismissed.

3.3. Higher Hydrocarbons and Other Compounds

There are three primary sources of higher (C_{2+}) hydrocarbons and other organic compounds (including bioorganic compounds) in unsedimented deep-sea hydrothermal systems. First, organic matter and biomass is transported into the crust along with seawater where fluids enter the crust in hydrothermal recharge zones. Second, biological activities in the subsurface can generate organic compounds, including cellular biomass as well as soluble and insoluble metabolic by-products. Third, organic compounds can be formed from inorganic carbon by abiotic organic synthesis reactions. Sinks for organic compounds within the system include incorporation into biomass, metabolism to CO_2 or to other organic compounds by heterotrophic organisms, precipitation as insoluble organic matter, adsorption on mineral surfaces, and thermal degradation to form CO_2 and other thermogenic organic compounds. Furthermore, organic compounds can be transported by circulating fluids throughout the system, either in dissolved form, as particulates, or as biomass.

Organic compounds in seawater are transported into recharge zones in several different forms, including DOC, particulate organic matter (POM), and cellular biomass. *Lang et al.* [2006] reported that DOC in high-temperature hydrothermal vent fluids at Endeavour (15 ± 5 μ*M*) and Axial Volcano (17 ± 8 μ*M*) on the Juan de Fuca ridge are both significantly lower than bottom seawater (35–38 μ*M*). If these values are representative, they indicate an overall loss of DOC during hydrothermal circulation. It should be noted, however, that these measurements are made on depressurized samples and so do not include CH_4 and other volatile organic compounds that are present in the hydrothermal fluids, nor would they include any organic compounds that are soluble in the high-temperature fluids but precipitate during cooling of the fluids after sampling. In contrast, *Brault et al.* [1988] reported dissolved nonvolatile hydrocarbons were >50 times more abundant in hydrothermal fluids from the East Pacific Rise than in bottom seawater, whereas organic fractions associated with particulates were 100 times more abundant in the hydrothermal fluids. The fates of seawater DOC, POM, and biomass during fluid circulation through MOR hydrothermal systems has not been determined, but much of it may be degraded by heterotrophic microorganisms in low temperature environments or by thermal decomposition in higher temperature environments [*Lang et al.* 2006]. Some resistant compounds may also survive circulation through the crust, but it may be difficult to differentiate these compounds from those produced in situ within the crust.

Biological activities contribute to carbon cycling in MOR hydrothermal systems through the formation of biomass, the generation of organic metabolic by-products, and conversion of organic matter to CO_2 and CH_4. Biological activity is, of course, restricted to the parts of the system where temperatures allow organisms to function (<~130°C according to current knowledge of the upper temperature limits for life; *Kashefi and Lovley* [2003]). However, these conditions are present in recharge zones where seawater enters the crust in the downwelling limb of hydrothermal circulation cells, and in upflow zones where high-temperature hydrothermal fluids rise toward the surface and cool by conduction or mixing with cold seawater.

At present, very little is known about microbial populations and activity in hydrothermal recharge zones, but *Bach and Edwards* [2003] proposed that oxidants transported into the crust in recharge zones could support substantial amounts of carbon fixation by chemolithoautotrophic microorganisms, including sulfur- and iron oxidizers (chemolithoautotrophs are microorganisms that utilize inorganic chemical energy sources to obtain metabolic energy and use that energy to synthesize their own biomass from CO_2 and other inorganic compounds). Additional carbon fixation by sulfate reducers and methanogens could be supported by H_2 generated from the reaction of basalt with seawater. To estimate the amount of inorganic carbon that could be fixed by microorganisms within these environments, *Bach and Edwards* [2003] examined the extent of oxidation evident in altered oceanic crust as recorded in drill cores, and then performed thermodynamic calculations to estimate the amount of chemical energy liberated by these oxidation reactions that could potentially be used for metabolic energy by microorganisms. Their calculations yielded estimates that as much as ~10^9 kg of carbon could be fixed into biomass globally per year by autotrophic organisms living within the ocean crust, although some of this biomass production may not be associated with axial hydrothermal circulation. At an estimated total fluid flux of ~4×10^{15} kg yr^{-1} through MOR hydrothermal systems [*Elderfield and Schultz*, 1996], this would equate to around 20 μmol of C fixed per kg seawater, which would represent only a small fraction of the inorganic carbon transported into the crust as HCO_3^- in seawater. The biomass and organic by-products produced by these chemolithoautotrophs, as well as any organic matter transported into recharge zones by circulating seawater, could support additional populations of heterotrophic organisms (i.e., organisms that consume organic matter by respiration or fermentation). However, without more definite constraints on the actual size, spatial distribution, and metabolic activity of microbial populations in hydrothermal recharge zones, it is difficult to estimate their impact on carbon cycling.

Hydrothermal upflow zones at mid-ocean ridges have long been recognized to support substantial biological communities [*Karl,* 1995; *Kelley et al.,* 2002]. Microbial communities living in the interiors of sulfide chimneys [*Takai et al.*, 2001; *Schrenk et al.*, 2003; *Kormas et al.*, 2006] and diffuse subsurface mixing zones on the flanks of high-temperature vents [*Summit and Baross*, 2001; *Huber et al.*, 2002, 2003] contribute to the organic matter and biomass contained within the rocks and sediments of MOR environments. In many instances, dissolved organic compounds, particulate organic matter, and microbial cells produced by these communities may be flushed into hydrothermal fluids and discharged at the seafloor. The presence of particulate DNA and intact microbial cells in fluids discharging from both high-temperature and warm diffuse vents provide evidence that biomass produced in near-surface environments can be swept into hydrothermal fluids and transported to the seafloor and overlying ocean [e.g., *Straube et al.*, 1990; *Deming and Baross*, 1993; *Summit and Baross*, 2001; *Huber et al.*, 2002, 2003]. Because dissolved organic compounds produced by these communities may be present in both high- and low-temperature vent fluids [*Lang et al.*, 2006], it may be difficult to discern which organic compounds in the fluids are derived from processes deeper within the system and which are generated in near-surface environments.

Currently, there are no measurements of the amount of carbon fixed by the microbial communities living in hydrothermal upflow zones, and only a few reports of the amount of biomass present. For example, *Schrenk et al.* [2003] reported microbial populations were 10^8 cells/g in the outer parts of a sulfide chimney retrieved from the seafloor at the Juan de Fuca. In the absence of more detailed measurements, some constraints can be placed on the extent of microbial processes and populations by considering the amount of chemical energy provided to the microbial community for production of biomass [*McCollom and Shock*, 1997]. Numerical models have been developed to examine chemical disequilibria that develop during the mixing of hydrothermal fluids discharged at the seafloor with seawater, such as occurs within hydrothermal plumes, the walls of vent chimneys, and in diffuse mixing zones of the shallow subsurface adjacent to high-temperature vents [*McCollom and Shock*, 1997; *McCollom*, 2000, 2007; *Tivey*, 2004; *Shock and Holland*, 2004]. Because chemical disequilibria can provide energy for fixation of CO_2 into biomass, quantification of the amount of chemical energy available places a constraint on the extent of carbon fixation and biomass production. Results of the mixing models suggest that on the order of about 4–8 kJ of chemical energy is available from each kilogram of hydrothermal vent fluid, which is sufficient energy to support about 30–120 mg (dry weight) of biomass production, equivalent to ~1.1–4.5 mmol carbon fixed for each kilogram of vent fluid. These calculations, however, require assumptions about the efficiency of chemical energy utilization and conversion to biomass that are highly uncertain at present, so these values must be considered as rough estimates.

Abiotic synthesis is another possible source of organic matter in MOR hydrothermal systems, although organic compounds with an unambiguous abiotic origin have not yet been identified from these environments [*McCollom and Seewald*, 2007]. However, the same kinetic inhibition to CH_4 formation discussed above that allow CH_3OH and HCOOH to accumulate in hydrothermal fluids may also allow synthesis of detectable quantities of more complex organic compounds (i.e., compounds with C–C bonds) from reduction of inorganic carbon [*Shock*, 1992]. The potential for abiotic organic synthesis in the deep-sea hydrothermal environments has been explored through geochemical models by Shock and colleagues [*Shock* 1990, 1992; *Amend and Shock*, 1998; *Shock and Schulte*, 1998]. These authors use geochemical models to evaluate chemical environments during fluid circulation through the crust or during the mixing of hydrothermal fluids with seawater, and use these data to evaluate the energetic favorability of abiotic organic synthesis reactions such as:

$$2CO_2 + 8H_2 \rightarrow \underset{\text{acetic acid}}{H_3CCOOH} + 2H_2O \tag{2}$$

and

$$2CO_2 + NH_4^+ + 3H_2 \rightarrow \underset{\text{glycine}}{C_2H_5NO_2} + 2H_2O + H^+ \tag{3}$$

As is the case for CH_4, formation of more complex organic compounds tends to become more thermodynamically favorable with decreasing temperature. *Shock* [1992] and *Shock and Schulte* [1998] used equilibrium fluid speciation calculations like those shown in Figure 9, but expanded to include compounds with carbon–carbon bonds and performed at the in situ conditions defined by the geochemical models, to investigate the amounts of various organic compounds that should be created in subsurface if equilibrium conditions were achieved (CH_4 excluded). Their results indicate that nanomolal to millimolal concentrations of several organic compounds should be present at equilibrium, including carboxylic and amino acids. These results suggest that these organic compounds could be forming in measurable concentrations in MOR hydrothermal environments. Because measurements of these types of compounds have not yet reported for axial MOR hydrothermal fluids, it is not currently possible to determine whether the predicted products of abiotic organic synthesis are present in the natural system.

Of course, the potential for abiotic synthesis implied by the models can only proceed if suitable reaction pathways exist under hydrothermal conditions. In addition to its possible role in abiotic CH_4 formation, Fischer-Tropsch synthesis has been suggested as a source of higher hydrocarbons in MOR hydrothermal fluids, particularly in ultramafic-hosted systems [e.g., *Berndt et al.,* 1996; *Charlou et al.,* 2000, 2002; *Holm and Charlou,* 2001; *Foustoukos and Seyfried,* 2004]. The products of Fischer-Tropsch-type synthesis are predominantly straight-chain saturated hydrocarbons, alcohols, and carboxylic acids, that decrease in abundance with increasing number of carbons [e.g., *McCollom et al.,* 1999; *McCollom and Seewald,* 2006]. However, although Fischer-Tropsch synthesis proceeds readily in the gas phase with industrial catalysts, it has not yet been demonstrated to occur under the conditions of deep-sea hydrothermal systems where the reactants (CO_2, CO, H_2) are dissolved in water and the presence of suitable catalysts is uncertain [*McCollom and Seewald,* 2007]. *Berndt et al.* [1996] observed the generation of light hydrocarbons, including CH_4, C_2H_6 (ethane), and C_3H_8 (propane), during hydrothermal alteration of olivine (serpentinization) in laboratory experiments, and inferred that these compounds were formed by Fischer-Tropsch synthesis from reduction of dissolved HCO_3^- with magnetite as the catalyst. Subsequent experiments, however, showed that the C_2H_6 and C_3H_8 generated under the experimental conditions are from background sources and not abiotic synthesis products [*McCollom and Seewald,* 2001]. More recent experiments by *Foustoukos and Seyfried* [2004] performed at higher temperatures and with different minerals present demonstrated that C_2H_6 and C_3H_8 can be formed from reduction of dissolved CO_2 under hydrothermal conditions and, although the mechanism for the reaction is not certain, the results indicate that light hydrocarbons can be synthesized abiotically within MOR systems. Of course, there may be other unexplored pathways besides Fischer-Tropsch synthesis by which abiotic organic compounds may form in MOR hydrothermal environments [see *McCollom and Seewald,* 2007].

The presence of several potential sources of organic compounds and the lack of quantitative constraints for most of them makes it difficult to assess the relative contributions of the different sources to the organic composition of hydrothermal fluids in MOR systems. For example, the nonvolatile hydrocarbons observed in hydrothermal vent fluids from the East Pacific Rise [*Brault et al.,* 1988] might come from thermal decomposition of organic matter transported into the high-temperature zones of the system by circulating fluids, or they might be derived from flushing of organic compounds out of the walls of adjacent chimney structures just before venting. Additionally, *Holm and Charlou* [2001] suggested that trace amounts of hydrocarbons extracted from vent fluids at Rainbow may have formed abiotically by Fischer-Tropsch synthesis in the subsurface. However, organic extracts of sulfide chimneys at this locale are dominated by a similar suite of hydrocarbons, but in this case they are accompanied by biomarkers and membrane fatty acids [*Simoneit et al.,* 2001], suggesting that the compounds could have a biological origin.

4. DEVELOPING MODELS OF CARBON CYCLING IN MID-OCEAN RIDGE HYDROTHERMAL SYSTEMS

At present, there are no detailed, quantitative models of carbon cycling for the mid-ocean ridge hydrothermal environment. Indeed, even qualitative models encompassing the entire system have not yet been published. As discussed above, however, some aspects of the system have been addressed through modeling, including abiotic organic synthesis and biological carbon fixation. Even in these cases, however, there are sufficient uncertainties in the models in that they provide only limited constraints on overall carbon cycling and sources of carbon compounds in MOR environments.

It is possible, at least conceptually, to outline the basic components of a first-order model to encompass the major components of carbon cycling within the system. One useful approach for such a model would be to follow a packet of fluid as it circulates through the crust and to trace the evolution of carbon compounds in the packet as it goes along. Laying out the primary contributors to carbon cycling within this framework is relatively straightforward. As the seawater enters the recharge zone, it carries with it dissolved inorganic compounds and organic matter from the overlying ocean. Biological activity within the recharge zone may fix these compounds into biomass or transform them into other soluble and insoluble organic compounds, whereas interaction with rocks may cause inorganic carbon to precipitate as carbonate minerals. As the fluid migrates through the crust and is heated, water-rock interactions with surrounding rocks leaches magmatic gases and extracts reduced compounds, such as H_2, Fe^{2+}, and S^{2-}, that could stimulate further growth of microorganisms. As the fluid is heated to temperatures above the limits for life, organic matter and biomass transported by the fluid can undergo thermal degradation, and the fluid will incorporate exsolved magmatic gases from the underlying magma and crystallizing rocks. Reducing conditions created by water-rock interactions as the fluids are heated may induce abiotic synthesis of organic compounds from dissolved inorganic carbon of magmatic or seawater origin. As the hot fluids rise buoyantly to the seafloor, mixing with seawater and convective cooling allow conditions for conversion of inorganic carbon to biomass by microorganisms and, possibly, further abiotic synthesis of organic matter. Organic compounds transported by the fluids serve as additional substrates for microbial activities, and may also precipitate out within mineral deposits at the seafloor as the fluid cools. The outcome of all these processes will be the soluble organic compounds found in hydrothermal fluids, and insoluble organic matter left behind in the rocks.

Placing quantitative constraints on this framework will require acquisition of considerable additional data. One major barrier to construction of quantitative models of carbon cycling in MOR environments is the almost complete lack of information regarding the size and activity of microbial populations in subsurface environments. It is quite likely that microorganisms participate in the fixation of inorganic carbon and formation of bioorganic compounds in hydrothermal recharge zones [*Bach and Edwards,* 2003]. However, the extent of the microbial contribution in this part of the system is currently unknown. Similarly, there is increasing evidence that active microbial communities are fixing and transforming carbon in diffuse subsurface mixing environments surrounding high-temperature upflow zones [*Holden et al.,* 1998; *Summit and Baross,* 2001; *Huber et al.,* 2002, 2003; *Von Damm and Lilley,* 2004], but these environments have also been almost entirely unexplored to date.

Addressing this data gap requires collection and analysis of subsurface samples, both to define the size and diversity of the standing microbial community and to measure their metabolic activities. In addition, characterization of the types and amounts of organic compounds in subsurface rock and fluid samples is needed. Because this would involve drilling in difficult, bare rock environments where the drilling machinery may encounter very high temperatures even at shallow depths, obtaining these samples may necessitate development of new technologies.

Also needed to develop more detailed models of carbon cycling will be improved knowledge of the time-temperature relationships along flow paths as fluids circulate through the hydrothermal system, as well as better knowledge of the rates of reactions involving organic compounds under hydrothermal conditions. Coupling the rates of such reactions with the time-temperature history of the fluids will allow additional constraints to be placed on the origin of organic compounds. For instance, coupling rates of CH_4 oxidation with the temperature along the fluid path and the duration of heating will allow for evaluation of the plausibility that CH_4 generated at low temperatures within recharge zones persists through high-temperature reaction zones. Because thermal decomposition of organic compounds (including biomass) is dependent on both time and temperature, improved knowledge of these factors will also provide for better prediction of the contribution of thermal processing to the organic matter found in fluids and rocks. Similarly, tighter constraints on the rates of abiotic synthesis reactions under hydrothermal conditions and of the conditions that allow such reactions to occur (i.e., necessary catalysts, dependence on reactant concentrations) will allow clearer evaluation of the possible abiotic contribution to organic compounds in rocks and fluids [*McCollom and Seewald,* 2007].

Although there are considerable challenges to overcome in order to develop comprehensive, quantitative models of carbon cycling in the MOR environment, the benefits of such models make their pursuit worthwhile. Because biological, geological, and geochemical processes are all involved in carbon inputs and transformations, models of carbon cycling are likely to provide valuable insights into the connections between the different elements of the system. Indeed, given the limited availability of samples, building quantitative models of carbon cycling at mid-ocean ridges by combining data from field studies, laboratory experiments, and thermodynamic studies may provide the best means for interpreting the links between biological and geochemical processes in these environments. After all, one does not truly understand a system until one can build a model of it.

Acknowledgments. The author is grateful for the helpful comments of the reviewers that significantly improved the accuracy and readability of the text. Preparation of this article was supported by funds from the NSF Marine Geology and Geophysics and Chemical Oceanography programs through grants OCE-0241579 and OCE-0550800.

REFERENCES

Alt, J. C., and W. C. Shanks III (1998), Sulfur in serpentinized oceanic peridotites: Serpentinization processes and microbial sulfate reduction, *J. Geophys. Res., 103,* 9917–9929.

Alt, J. C., and W. C. Shanks III (2003), Serpentinization of abyssal peridotites from the MARK area, Mid-Atlantic Ridge: Sulfur geochemistry and reaction modeling, *Geochim. Cosmochim. Acta, 67,* 641–653.

Alt, J. C., and D. A. H. Teagle (1999), The uptake of carbon during alteration of oceanic crust, *Geochim. Cosmochim. Acta, 63,* 1527–1535.

Amend, J. P., and E. L. Shock (1998), Energetics of amino acid synthesis in hydrothermal systems, *Science, 281,* 1659–1662.

Bach, W., and K. J. Edwards (2003), Iron and sulfide oxidation within the basaltic ocean crust: Implications for chemolithoautotrophic microbial biomass production, *Geochim. Cosmochim. Acta, 67,* 3871–3887.

Bach, W., B. Peucker-Ehrenbrink, S. R. Hart, and J. S. Blusztajn (2003), Geochemistry of hydrothermally altered oceanic crust: DSDP/ODP Hole 504B—Implications for seawater-crust exchange budgets and Sr- and Pb-isotopic evolution of the mantle, *Geochem. Geophys. Geosyst., 4,* doi:10.1029/2002GC000419.

Berndt, M. E., D. E. Allen, and W. E. Seyfried Jr. (1996), Reduction of CO_2 during serpentinization of olivine at 300°C and 500 bar, *Geology, 24*, 351–354.

Blank, J. G., J. R. Delaney, and D. J. Des Marais (1993), The concentration and isotopic composition of carbon in basaltic glasses from the Juan de Fuca Ridge, Pacific Ocean, *Geochim. Cosmochim. Acta, 70*, 2073–2092.

Bottinga, Y., and M. Javoy (1989), MORB degassing: Evolution of CO_2, *Earth Planet. Sci. Lett., 95*, 215–225.

Brault, M., J. C. Marty, and A. Saliot (1984), Fatty acids from particulate matter and sediment in hydrothermal environments from the East Pacific Rise, near 13°N, *Org. Geochem., 6*, 217–222.

Brault, M., B. R. T. Simoneit, J. C. Marty, and A. Saliot (1988), Hydrocarbons in waters and particulate material from hydrothermal environments at the East Pacific Rise, 13°N, *Org. Geochem., 12*, 209–219.

Brault, M., B. R. T. Simoneit, and A. Saliot (1989), Trace petroliferous organic matter associated with massive hydrothermal sulfides from the East Pacific Rise at 13 and 21°N, *Oceanol. Acta, 12*, 405–415.

Campbell, A. C., M. R. Palmer, G. P. Klinkhammer, T. S. Bowers, J. M. Edmond, J. L. Lawrence, J. F. Casey, G. Thompson, S. Humphris, P. Rona, and J. A. Karson (1988), Chemistry of hot springs on the Mid-Atlantic Ridge, *Nature, 335*, 514–519.

Charlou, J. L., J. P. Donval, P. Jean-Baptiste, A. Dapoigny, and P. A. Rona (1996a), Gases and helium isotopes in high temperature solutions sampled before and after ODP Leg 158 drilling at TAG hydrothermal field (26°N MAR), *Geophys. Res. Lett., 23*, 3491–3494.

Charlou, J. L., Y. Fouquet, J. P. Donval, J. M. Auzende, P. Jean-Baptiste, and M. Stievenard (1996b), Mineral and gas chemistry of hydrothermal fluids on an ultrafast spreading ridge: East Pacific Rise, 17° to 19°S (Naudur cruise, 1993) phase separation processes controlled by volcanic and tectonic activity, *J. Geophys. Res., 101*, 15,899–15,919.

Charlou, J. L., J. P. Donval, E. Douville, P. Jean-Baptiste, J. Radford-Knoery, Y. Fouquet, A. Dapoigny, and M. Stievenard (2000), Compared geochemical signatures and the evolution of Menez Gwen (37°50′N) and Lucky Strike (37°17′N) hydrothermal fluids, south of the Azores Triple Junction on the Mid-Atlantic Ridge, *Chem. Geol., 171*, 49–75.

Charlou, J. L., J. P. Donval, Y. Fouquet, P. Jean-Baptiste and N. Holm (2002), Geochemistry of high H_2 and CH_4 vent fluids issuing from ultramafic rocks at the Rainbow hydrothermal field (36°14′N, MAR), *Chem. Geol., 191*, 345–359.

Craig, H., J. A. Welhan, K. Kim, R. Poreda, and J. E. Lupton (1980), Geochemical studies of the 21°N EPR hydrothermal fluids *EOS Trans. AGU, 61*, 992.

Cruse, A. M., and J. S. Seewald (2006), Geochemistry of low-molecular weight hydrocarbons in hydrothermal fluids from Middle Valley, northern Juan de Fuca Ridge, *Geochim. Cosmochim. Acta, 70*, 2073–2092.

Deming, J. W., and J. A. Baross (1993), Deep-sea smokers: Windows to a subsurface biosphere?, *Geochim. Cosmochim. Acta,* 57, 3219–3230.

Elderfield, H., and A. Schultz (1996), Mid-ocean ridge hydrothermal fluxes and the chemical compositions of the oceans, *Annu. Rev. Earth Planet. Sci., 24,* 191–224.

Evans, W. C., L. D. White, and J. B. Rapp (1988), Geochemistry of some gases in hydrothermal fluids from the Southern Juan de Fuca Ridge, *J. Geophys. Res., 93*, 15,305–15,313.

Fisher, A. T. (2003), Geophysical constraints on hydrothermal circulation, in *Energy and Mass Transfer in Marine Hydrothermal Systems*, edited by P. E. Halbach, V. Tunnicliffe, and J. R. Hein, pp. 29–52, Dahlem University Press, Berlin.

Fisher, A. T. (2004), Rates of flow and patterns of circulation, in *Hydrogeology of the Oceanic Lithosphere,* edited by E. E. Davis and H. Elderfield, pp. 339–377, Cambridge University Press, Cambridge.

Foustoukos, D. I., and W. E. Seyfried Jr. (2004), Hydrocarbons in hydrothermal vent fluids: The role of chromium-bearing catalysts, *Science, 304*, 1002–1005.

Frost, B. R. (1985), On the stability of sulfides, oxides, and native metals in serpentinite, *J. Petrol., 26*, 31–63.

Früh-Green, G. L., J. A. D. Connolly, A. Plas, D. S. Kelley, and B. Grobéty (2004), Serpentinization of ocean peridotites: Implications for geochemical cycles and biological activity, in *The Subseafloor Biosphere at Mid-Ocean Ridges*, edited by W. S. D. Wilcock, E. F. DeLong, D. S. Kelley, J. A. Baross, and S. C. Cary, pp. 119–136, American Geophysical Union, Washington D. C.

Gerlach, T. M. (1989), Degassing of carbon dioxide from basaltic magma at spreading centers: II. Mid-oceanic ridge basalts, *J. Volcanol. Geotherm. Res., 39*, 221–232.

Glynn, S., R. A. Mills, M. R. Palmer, R. D. Pancost, S. Severmann, and A. J. Boyce (2006), The role of prokaryotes in supergene alteration of submarine hydrothermal sulfides, *Earth Planet. Sci. Lett., 244*, 170–185.

Holden, J. F., M. Summit, and J. A. Baross (1998), Thermophilic and hyperthermophilic microorganisms in 3–30°C hydrothermal fluids following a deep-sea volcanic eruption, *FEMS Microbiol. Ecol., 25*, 33–41.

Holm, N. G., and J.-L. Charlou (2001), Initial indications of abiotic formation of hydrocarbons in the Rainbow ultramafic hydrothermal system, Mid-Atlantic Ridge, *Earth Planet. Sci. Lett., 191*, 1–8.

Horita, J. (2001), Carbon isotope exchange in the system CO_2-CH_4 at elevated temperatures, *Geochim. Cosmochim. Acta, 65*, 1907–1919.

Horita, J., and M. E. Berndt (1999), Abiogenic methane formation and isotopic fractionation under hydrothermal conditions, *Science, 285*, 1055–1057.

Hu, G., Z. Ouyang, X. Wang, and Q. Wen (1998), Carbon isotopic fractionation in the process of Fisher-Tropsch reaction in primitive solar nebula, *Sci. China Ser. D, 41*, 202–207.

Huber, J. A., D. A. Butterfield, and J. A. Baross (2002), Temporal changes in archaeal diversity and chemistry in a mid-ocean ridge subseafloor habitat, *Appl. Environ. Microbiol., 68*, 1585–1594.

Huber, J. A., D. A. Butterfield, and J. A. Baross (2003), Bacterial diversity in a subseafloor habitat following a deep-sea volcanic eruption, *FEMS Microbiol. Ecol., 43*, 393–409.

Ishibashi, J., Y. Sano, H. Wakita, T. Gamo, M. Tsutsumi, and H. Sakai (1995), Helium and carbon geochemistry of hydrothermal fluids from the Mid-Okinawa Trough Back Arc Basin, southeast Japan, *Chem. Geol., 123*, 1–15.

Jean-Baptiste, P., J. L. Charlou, M. Stievenard, J. P. Donval, H. Bougault, and C. Mevel (1991), Helium and methane measurements in hydrothermal fluids from the mid-Atlantic ridge: The Snake Pit site at 23°N, *Earth Planet. Sci. Lett., 106*, 17–28.

Johnson, J. W., E. H. Oelkers, and H. C. Helgeson (1992), SUPCRT92: A software package for calculating the standard molal thermodynamic properties of minerals, gases, aqueous species, and reactions from 1 to 5000 bar and 0 to 1000°C, *Comput. Geosci., 18*, 899–947.

Kadko, D., and D. A. Butterfield (1998), The relationship of hydrothermal fluid composition and crustal residence time to maturity of vent fields on the Juan de Fuca Ridge, *Geochim. Cosmochim. Acta, 62*, 1521–1533.

Karl, D. M. (1995), Ecology of free-living, hydrothermal vent microbial communities, in *The Microbiology of Deep-Sea Hydrothermal Vents,* edited by D. M. Karl, pp. 35–124, CRC Press, Boca Raton.

Kashefi, K., and D. Lovley (2003), Extending the upper temperature limit for life, *Science, 301*, 934.

Kawka, O. E., and B. R. T. Simoneit (1987), Survey of hydrothermally-generated the petroleums from the Guaymas Basin spreading center, *Org. Geochem., 11*, 311–328.

Kelley, D. S. (1996), Methane-rich fluids in the oceanic crust, *J. Geophys. Res., 101*, 2943–2962.

Kelley, D. S., and G. L. Früh-Green (1999), Abiogenic methane in deep-seated mid-ocean ridge environments: Insights from stable isotope analyses, *J. Geophys. Res., 104*, 10,439–10,460.

Kelley, D. S., and G. L. Früh-Green (2001), Volatile lines of descent in submarine plutonic environments: Insights from stable isotope and fluid inclusion analyses, *Geochim. Cosmochim. Acta, 65*, 3325–3346.

Kelley, D. S., J. A. Baross, and J. R. Delaney (2002), Volcanoes, fluids, and life at mid-ocean ridge spreading centers, *Annu. Rev. Earth Planet. Sci., 30*, 385–491.

Kelley, D. S., M. D. Lilley, and G. L. Früh-Green (2004), Volatiles in submarine environments: Food for life, in *The Subseafloor Biosphere at Mid-Ocean Ridges*, edited by W. S. D. Wilcock, E. F. DeLong, D. S. Kelley, J. A. Baross, and S. C. Cary, pp. 167–189, American Geophysical Union, Washington, D. C.

Kelley, D. S., J. A. Karston, G. L. Früh-Green, D. R. Yoerger, T. M. Shank, D. A. Butterfield, J. M. Hayes, M. O. Schrenk, E. J. Olson, G. Proskurowski, M. Jakuba, A. Bradley, B. Larson, K. Ludwig, D. Glickson, D. Buckman, A. S. Bradley, W. J. Brazelton, K. Roe, M. J. Elend, A. Delacour, S. M. Bernasconi, M. D. Lilley, J. A. Baross, R. E. Summons, and S. P. Sylva (2005), A serpentinite-hosted ecosystem: The Lost City hydrothermal field, *Science, 307*, 1428–1434.

Kormas, K. A., M. K. Tivey, K. Von Damm, and A. Teske (2006), Bacterial and archaeal phylotypes associated with distinct mineralogical layers of a white smoker spire from a deep-sea hydrothermal vent site (9°N, East Pacific Rise), *Environ. Microbiol., 8*, 909–920.

Lang, S. Q., D. A. Butterfield, M. D. Lilley, H. P. Johnson and J. I. Hedges (2006), Dissolved organic carbon in ridge-axis and ridge-flank hydrothermal systems, *Geochim. Cosmochim. Acta, 70*, 3830–3842.

Lancet, M. S., and E. Anders (1970), Carbon isotope fractionation in the Fischer-Tropsch synthesis and in meteorites, *Science, 170*, 980–982.

Lein, A. Y., D. V. Grichuk, E. G. Gurvich, and Y. A. Bogdanov (2000), A new type of hydrogen- and methane-rich hydrothermal solutions in the Rift zone of the Mid-Atlantic Ridge, *Dokl. Earth Sci., 375A*, 1304–1391.

Lilley, M. D., J. A. Baross, and L. I. Gordon (1983), Reduced gases and bacteria in hydrothermal fluids: The Galapagos Spreading Center and 21°N East Pacific Rise, in *Hydrothermal Processes at Seafloor Spreading Centers,* edited by P. A. Rona, et al., pp. 411–447, Plenum Press, New York.

Lilley, M. D., D. A. Butterfield, E. J. Olson, J. E. Lupton, S. A. Macko, and R. E. McDuff (1993), Anomalous CH_4 and NH_4^+ concentrations at an unsedimented mid-ocean-ridge hydrothermal system, *Nature, 364*, 45–47.

Lilley, M. D., D. A. Butterfield, J. E. Lupton, and E. J. Olson (2003), Magmatic events can produce rapid changes in hydrothermal vent chemistry, *Nature, 422*, 878–881.

Mathez, E. A. (1984), Influence of degassing on oxidation states of basaltic magma, *Nature, 310*, 371–375.

McCollom, T. M. (2000), Geochemical constraints on primary productivity in submarine hydrothermal vent plumes, *Deep-sea Res. I, 47*, 85–101.

McCollom, T. M. (2007), Geochemical constraints on sources of metabolic energy for chemolithoautotrophy in ultramafic-hosted deep-sea hydrothermal systems, *Astrobiology, 7*, 933–950.

McCollom, T. M., and E. L. Shock (1997), Geochemical constraints on chemolithoautotrophic metabolism by microorganisms in seafloor hydrothermal systems, *Geochim. Cosmochim. Acta, 61*, 4375–4391.

McCollom, T. M., and J. S. Seewald (2001), A reassessment of the potential for reduction of dissolved CO_2 to hydrocarbons during serpentinization of olivine, *Geochim. Cosmochim. Acta, 65*, 3769–3778.

McCollom, T. M., and J. S. Seewald (2003), Experimental constraints on the hydrothermal reactivity of organic acids and acid anions: I. Formic acid and formate, *Geochim. Cosmochim. Acta, 67*, 3625–3644.

McCollom, T. M., and J. S. Seewald (2006), Carbon isotope composition of organic compounds produced by abiotic synthesis under hydrothermal conditions, *Earth Planet. Sci. Lett., 243*, 74–84.

McCollom, T. M., and J. S. Seewald (2007), Abiotic synthesis of organic compounds in deep-sea hydrothermal environments, *Chem. Rev., 107*, 382–401.

McCollom, T. M., G. Ritter, and B. R. T. Simoneit (1999), Lipid synthesis under hydrothermal conditions by Fischer-Tropsch-type reactions, *Orig. Life Evol. Biosph., 29*, 153–166.

Merlivat, L., F. Pineau, and M. Javoy (1987), Hydrothermal vent waters at 13°N on the East Pacific Rise: Isotopic composition and gas concentration, *Earth Planet. Sci. Lett., 84*, 100–108.

Pineau, F., and M. Javoy (1983), Carbon isotopes and concentrations in mid-ocean ridge basalts, *Earth Planet. Sci. Lett., 62*, 239–257.

Proskurowksi, G., M. D. Lilley, and T. A. Brown (2004), Isotopic evidence of magmatism and seawater bicarbonate removal at the Endeavour hydrothermal system, *Earth Planet. Sci. Lett., 225*, 53–61.

Proskurowksi, G., M. D. Lilley, D. S. Kelley, and E. J. Olson (2006), Low temperature volatile production at the Lost City Hydrothermal Field, evidence from a hydrogen stable isotope geothermometer, *Chem. Geol., 229*, 331–343.

Schoell, M. (1988), Multiple origins of methane in the Earth, *Chem. Geol., 71*, 1–10.

Schrenk, M. O., D. S. Kelley, J. R. Delaney, and J. A. Baross (2003), Incidence and diversity of microorganisms within the walls of an active deep-sea sulfide chimney, *Appl. Environ. Microbiol., 69*, 3580–3592.

Seewald, J. S., M. Yu. Zolotov, and T. M. McCollom (2006), Experimental investigation of single carbon compounds under hydrothermal conditions, *Geochim. Cosmochim. Acta, 70*, 446–460.

Seewald, J. S., A. M. Cruse, and P. Saccocia (2003), Aqueous volatiles in hydrothermal fluids from the Main Endeavour Field, northern Juan de Fuca Ridge: Temporal variability following earthquake activity, *Earth Planet. Sci. Lett., 216*, 575–590.

Shanks, W. C., III, J. K. Böhlke, and R. R. Seal II (1995), Stable isotopes in mid-ocean ridge hydrothermal systems: Interactions between fluids, minerals, and organisms, in *Seafloor Hydrothermal Systems: Physical, Chemical, Biological, and Geological Interactions*, edited by S. E. Humphris, R. A. Zierenberg, L. S. Mullineaux, and R. E. Thomson, pp. 194–221, American Geophysical Union, Washington, D. C.

Sherwood Lollar, B., G. Lacrampe-Couloume, G. F. Slater, J. Ward, D. P. Moser, T. M. Gihring, L.-H. Lin, and T. C. Onstott (2006), Unraveling abiogenic and biogenic sources of methane in the Earth's deep subsurface, *Chem. Geol., 226*, 328–339.

Shock, E. L. (1990), Geochemical constraints on the origin of organic compounds in hydrothermal systems, *Orig. Life Evol. Biosph., 20*, 331–367.

Shock, E. L. (1992), Chemical environments of submarine hydrothermal systems, *Orig. Life Evol. Biosph., 22*, 67–107.

Shock, E. L., and M. D. Schulte (1998), Organic synthesis during fluid mixing in hydrothermal systems, *J. Geophys. Res., 103*, 28,513–28,517.

Shock, E. L., and M. E. Holland (2004), Geochemical energy sources that support the subsurface biosphere, in *The Subseafloor Biosphere at Mid-Ocean Ridges*, edited by W. S. D. Wilcock, E. F. DeLong, D. S. Kelley, J. A. Baross, and S. C. Cary, pp. 167–189, American Geophysical Union, Washington, D. C.

Shock, E. L., H. C. Helgeson, and D. A. Sverjensky (1989), Calculation of the thermodynamic and transport properties of aqueous species at high pressures and temperatures: Standard partial molal properties of inorganic neutral species, *Geochim. Cosmochim. Acta, 53*, 2157–2183.

Simoneit, B. R. T. (1988), Petroleum generation in submarine hydrothermal systems: An update, *Can. Mineral., 26*, 827–840.

Simoneit, B. R. T. (1993), Aqueous high-temperature and high-pressure organic geochemistry of hydrothermal environments, *Geochim. Cosmochim. Acta, 57*, 3231–3243.

Simoneit, B. R. T. (1994), Organic matter alteration and fluid migration in hydrothermal systems, in *Geofluids: Origin, Migration and Evolution of Fluids in Sedimentary Basins*, edited by J. Parnell, *Geol. Soc. Spec. Publ.*, 261–274.

Simoneit, B. R. T., O. E. Kawka, and M. Brault (1988), Origin of gases and condensates in the Guaymas Basin Hydrothermal system (Gulf of California), *Chem. Geol., 71*, 169–182.

Simoneit, B. R. T., M. Brault, and A. Saliot (1990), Hydrocarbons associated with hydrothermal minerals, vent waters and talus on the East Pacific Rise and Mid-Atlantic Ridge, *Appl. Geochem., 5*, 115–124.

Simoneit, B. R. T., W. D. Goodfellow, and J. M. Franklin (1992), Hydrothermal petroleum at the seafloor and organic matter alteration in sediments of Middle Valley, Northern Juan de Fuca Ridge, *Appl. Geochem., 7*, 257–254.

Simoneit, B. R. T., A. Yu. Lein, V. I. Peresypkin, and G. A. Osipov (2004), Composition and origin of hydrothermal petroleum and associated lipids in the sulfide deposits of the Rainbow Field (Mid-Atlantic Ridge at 36°N), *Geochim. Cosmochim. Acta, 68*, 2275–2294.

Staudigel, H., S. R. Hart, H. U. Schmincke, and B. M. Smith (1989), Cretaceous ocean crust at DSDP site 417 and site 418—Carbon uptake from weathering versus loss by magmatic outgassing, *Geochim. Cosmochim. Acta, 53*, 3091–3094.

Straube, W. L., J. W. Deming, C. C. Somerville, R. R. Colwell, and J. A. Baross (1990), Particulate DNA in smoker fluids: Evidence for existence of microbial populations in hot hydrothermal systems, *Appl. Environ. Microbiol., 56*, 1440–1447.

Summit, M., and J. A. Baross (2001), A novel microbial habitat in the mid-ocean ridge subseafloor, *Proc. Natl. Acad. Sci. U. S. A., 98*, 2158–2163.

Takai, K, T. Komatsu, F. Inagaki, and K. Horikoshi (2001), Distribution of Archaea in a black smoker chimney structure, *Appl. Environ. Microbiol., 67*, 3618–3629.

Takai, K, T. Gamo, U. Tsunogai, N. Nakayama, H. Hirayama, K. H. Nealson, and K. Horikoshi (2004), Geochemical and microbiological evidence for a hydrogen-based, hyperthermophilic subsurface lithoautotrophic microbial ecosystem (HyperSLIME) beneath an active deep-sea hydrothermal field, *Extremophiles, 8*, 269–282.

Tivey, M. K. (2004), Environmental conditions within active seafloor vent structures: Sensitivity to vent fluid composition and fluid flow, in *The Subseafloor Biosphere at Mid-Ocean Ridges*, edited by W. S. D. Wilcock, E. F. DeLong, D. S. Kelley, J. A. Baross, and S. C. Cary, pp. 167–189, American Geophysical Union, Washington, D. C.

Von Damm, K. L., and M. D. Lilley (2004), Diffuse flow hydrothermal fluids from 9°50′N East Pacific Rise: Origin, evolution and biogeochemical controls, in *The Subseafloor Biosphere at Mid-Ocean Ridges*, edited by W. S. D. Wilcock, E. F. DeLong, D. S. Kelley, J. A. Baross, and S. C. Cary, pp. 245–268, American Geophysical Union, Washington, D. C.

Welhan, J. A. (1988), Origins of methane in hydrothermal systems, *Chem. Geol., 71*, 183–198.

Welhan, J. A., and H. Craig (1983), Methane, hydrogen, and helium in hydrothermal fluids at 21°N on the East Pacific Rise, in *Hydrothermal Processes at Seafloor Spreading Centers,* edited by P. A. Rona, et al., pp. 391–409, Plenum Press, New York.

Welhan, J. A., and J. E. Lupton (1987), Light-hydrocarbon gases in Guaymas Basin hydrothermal fluids—Thermogenic versus abiogenic origin, *Bull. Am. Assoc. Petrol. Geol., 71*, 215–223.

T.M. McCollom, CU Center for Astrobiology and Laboratory for Atmospheric and Space Physics, Campus Box 392, University of Colorado, Colorado 80309-0392, USA. (mccollom@lasp.colorado.edu)

Modeling the Impact of Diffuse Vent Microorganisms Along Mid-Ocean Ridges and Flanks

Julie A. Huber

Josephine Bay Paul Center, Marine Biological Laboratory, Woods Hole, Massachusetts, USA

James F. Holden

Department of Microbiology, University of Massachusetts, Amherst, Massachusetts, USA

The subseafloor crustal environment is one of the largest and least accessible ocean environments, and our understanding of this vast natural resource is very limited. Low-temperature diffuse vent fluids, ubiquitous at hydrothermal systems, provide one of the few access points to the subseafloor environment. This review will focus on describing and modeling microorganisms from diffuse hydrothermal vent fluids and the subseafloor at basalt-hosted mid-ocean ridges and flanks. At the vent-field scale, microorganisms likely have a net impact on fluid chemistry, rock permeability, and vent ecology, but models of these processes are in early stages of development. Modeling the impact of basalt-hosted hydrothermal microorganisms on large spatial scales will require additional and improved diffuse fluid flux estimates, laboratory experiments with pure and mixed cultures under environmentally relevant conditions to determine growth yields and metabolic fluxes, and continued development of ocean drilling and real-time observatory technologies to better constrain and model the chemical and physical characteristics of the subseafloor environment. In this work we describe the diffuse flow and subseafloor environment, the metabolic diversity, distribution, life strategies, and biosignatures of microorganisms observed in diffuse fluids, and a prospectus of needs for modeling this complex and dynamic environment. Collectively, such modeling efforts should provide a clearer picture of life within hydrothermally influenced regions of the subseafloor.

1. INTRODUCTION

At active hydrothermal systems, reduced hydrothermal fluids mix with oxygen-saturated seawater in the crust, creating low-temperature diffuse vents on the seafloor that serve as a dynamic and energy-rich environment for both microbial and macrofaunal communities. Crustal mixing also creates subseafloor habitats that span gradients in temperature, energy availability, and physical parameters, thus enabling diverse subseafloor microbial communities to grow. Evidence is mounting that an active microbial community lives within the subseafloor at both on-axis hydrothermal systems [*Holden et al.*, 1998; *Holland et al.*, 2004; *Huber et al.*, 2002, 2003; *Summit and Baross*, 2001; *Takai et al.*, 2004a] and warm ridge flank environments [*Cowen et al.*, 2003; *Cowen*, 2004; *Ehrhardt et al.*, 2007; *Huber et al.*, 2006b]. Unlike most of the ocean, the subseafloor environment at unsedimented mid-ocean ridges receives very little, if any, organic matter derived from photosynthesis. Instead, energy and material for growth

Magma to Microbe: Modeling Hydrothermal Processes at Ocean Spreading Centers
Geophysical Monograph Series 178

10.1029/178GM11

comes from a combination of rock- and magmatically derived gases and metals and seawater-derived compounds [*Bach and Edwards*, 2003]. However, our knowledge of which microorganisms are present in the subseafloor, their role in biogeochemical transformations, and how organisms are distributed in this dynamic geochemical environment over time and space is fragmentary. The current needs for modeling the impact of microorganisms on diffuse fluids and the subseafloor are to predict community composition, microbial distribution, rates of biogeochemical reactions, and biomass production. These must be coupled with the development of first-order flow models with varying physical constraints (e.g., temperature, flow rates, permeability) and chemical constraints (e.g., nutrient limitation, pH, reduction potential) for the purpose of modeling the impact of microbial communities on fluid chemistry and rock properties in the crustal environment.

2. THE DIFFUSE FLOW AND SUBSEAFLOOR ENVIRONMENTS

Diffuse, low-temperature fluids are ubiquitous along mid-ocean ridges, and it is estimated that they are responsible for approximately half of the heat flux from the crust [*Rona and Trivett*, 1992; *Veirs et al.*, 2006]. Understanding how fluids circulate within the subseafloor and the chemistry of these fluids is essential for understanding the microbial ecology of this environment. Reviews on the structure of oceanic crust and hydrothermal circulation are available [*Alt and Bach*, 2003; *Fisher*, 2003]; therefore, only biologically relevant aspects will be discussed here. It is estimated that mixing zones within the upper crust contribute the largest amount of primary productivity relative to other environments at deep-sea vents by more than tenfold [*McCollom*, 2005]. A combination of high porosity, permeability, and fluid circulation with favorable temperatures (up to 120°C) make the upper extrusive layer the most likely habitat for subseafloor microbial populations (Figure 1). Newly formed basalt from magma extrusions typically has both high porosity and high permeability [*Fisher*, 1998; *Gouze and Coudrain-Ribstein*, 2000]. Porosity decreases with rock age, resulting in decreased permeability due to crack filling by mineral precipitation and subsequent tectonic and volcanic activity [*Jacobson*, 1992; *Johnson and Pruis*, 2003]. Estimates based on gravity studies suggest that the porosity of the upper volcanic layer of oceanic crust at mid-ocean ridges ranges from 9 to 34% [*Johnson and Pruis*, 2003]. This variability is due both to measurement uncertainty and structural differences of the crust such as voids and cracks in the pillow basalts rather than density differences. Overall, it is estimated that the average porosity of the upper oceanic crust is 12% and that this porous layer contains more than 2% of the global volume of seawater [*Johnson and Pruis*, 2003].

Precisely where seawater enters the upper crust and into the recharge zone for subsequent hydrothermal circulation is an

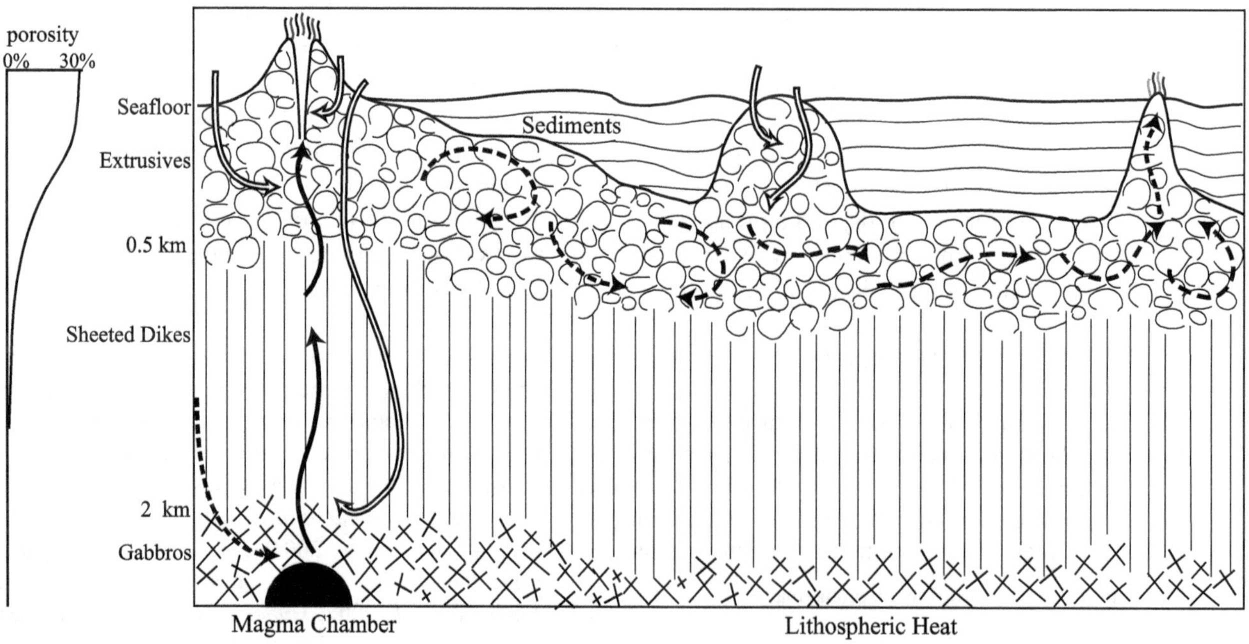

Figure 1. Schematic of oceanic crust, showing generalized porosity, heat sources, and fluid circulation patterns. Open arrows, seawater; solid arrows, hydrothermal fluid; dotted arrows, a mixture of the two.

Plate 1. Digital camera images of diffuse venting from the southeast rift zone of Axial Seamount, including (a) Cloud vent in 2003, (b) iron oxides found during transit from Bag City to Coquille vent in 2003, (c) Marker N3 vent in 2004, and (d) a "snowblower" vent found shortly after the 1998 eruption. Image Credits: NOAA Vents Program and NeMO Project.

ongoing topic of study, but it is assumed to cover large surface areas [*Alt and Bach*, 2003]. However, it is also likely that some degree of the recharge is focused locally along faults with the largest volume of fluids entering through high permeability volcanic regions [*Alt and Bach*, 2003]. Whether microorganisms within this zone are active is also a topic of ongoing studies. Even when magnesium concentrations of sampled fluids are extremely low, suggesting minimal input of seawater (seawater contains 52 mM Mg^{2+}, whereas pure high-temperature hydrothermal fluid lacks Mg^{2+}), marine pelagic microorganisms are routinely observed suggesting that they are present and active within the recharge zone [*Huber et al.*, 2006b].

Much of the observed chemistry of diffuse hydrothermal fluids can be explained by subseafloor mixing of hydrothermal fluids with seawater [*Butterfield and Massoth*, 1994; *Butterfield et al.*, 1997, 2004; *Edmonds and Edmond*, 1995; *James and Elderfield*, 1996; *Koschinsky et al.*, 2002]. However, other subseafloor processes can also modify the chemistry of low-temperature fluids such as mineral precipitation, conductive heating of crustal seawater, conductive cooling of hydrothermal vent fluid, and biologically mediated reactions [*Butterfield et al.*, 1997, 2004; *Cooper et al.*, 2000; *Edmond et al.*, 1979; *Jannasch and Mottl*, 1985; *Johnson et al.*, 1988; *Lilley et al.*, 1984; *Von Damm and Lilley*, 2004]. Spatial and temporal linkages between hydrothermal vent microorganisms, fluid discharge, and chemistry are beginning to emerge [*Huber et al.*, 2002, 2003, 2006a, 2007; *Nakagawa et al.*, 2005a]. Gas-enriched fluids from the Mid-Okinawa Trough hosted more abundant chemolithoautotrophic communities with gas-dependent energy metabolisms in comparison to gas-depleted fluids [*Nakagawa et al.*, 2005a]. In addition, studies on Axial Seamount show that variations in geochemical characteristics of the fluids due to dilution with seawater and changes in the end-member hydrothermal fluid over time have a large influence on microbial community diversity [*Huber et al.*, 2002, 2003]. However, because so few vents have been sampled and studied thoroughly, determining the underlying patterns and controls on the distribution of microorganisms in the subseafloor remains a challenge.

The study of diffuse fluids is useful as an indicator of subseafloor processes associated with magmatic eruptive events as well. Microbiological studies of eruption sites [e.g., 9°N East Pacific Rise (EPR), CoAxial segment, Gorda Ridge, Axial Seamount] provide compelling evidence of subseafloor microbial populations exemplified by frequently observed "snow blower" vents at new eruption sites (Plate 1), the presence of indicator organisms in event plumes, and subsequent new diffuse fluid venting following an eruption [*Holden et al.*, 1998; *Huber et al.*, 2002; *Juniper et al.*, 1995; *Juniper and Tebo*, 1995; *Summit and Baross*, 1998].

Finally, as mentioned previously, it is estimated that the upper oceanic crust contains more than 2% of the global volume of seawater [*Johnson and Pruis*, 2003]. In this regard, the crustal aquifer beyond the mid-ocean ridge is an important component of the deep sea from both biological and geochemical perspectives. The linkages between the deep sea, sediments, and crustal fluids are apparent, but there is considerable need to further define the relationships and fluxes between these systems. The only known source of recharge and discharge is on the flank of the Juan de Fuca Ridge, where recharge through a large volcanic outcrop was measured and discharge through a smaller outcrop located 50 km away was confirmed [*Fisher et al.*, 2003]. Microbial observations of ridge flank crustal fluids suggest a diverse community of bacteria and archaea distinct from those found on-axis [*Cowen et al., 2003; Cowen, 2004; Ehrhardt et al., 2007; Huber et al., 2006b; Nakagawa et al., 2006a*]. In addition, estimates suggest a potential microbial biomass production within ridge flanks up to ~1×10^{12} g C/yr [*Bach and Edwards*, 2003]. However, with only a few sites examined to date from a microbial perspective, there is little known about the ubiquitous and important ridge flank crustal environment.

3. MICROORGANISMS FROM DIFFUSE FLOW ENVIRONMENTS

3.1. Basics of Hydrothermal Vent Microbial Ecology

In environments where temperatures range from 1° to 2°C in background seawater to more than 120°C (the highest known growth temperature for an organism in culture), it is useful to classify organisms based on their optimum growth temperature (T_{opt}). In this way, microorganisms are classified as psychrophiles (T_{opt}, <20°C), mesophiles (T_{opt}, 20°–55°C), thermophiles (T_{opt}, 55°–80°C), and hyperthermophiles (T_{opt}, >80°C). Microorganisms are also classified on the basis of the carbon sources used for growth where autotrophs and heterotrophs refer to those organisms that use inorganic C_1 compounds (typically CO_2) and organic compounds, respectively, as carbon sources. However, these classifications can only be applied to those organisms that have been cultured or inferred through incubation studies. Because most microorganisms are not yet cultured, their taxonomic classification depends on the use of molecular biological techniques, the most common of which is comparing DNA nucleotide sequences (so-called phylogenetic comparisons) from the small-subunit ribosomal RNA gene. Sequence comparisons of this gene, which is highly conserved among all known life forms, allows for classification of all organisms into one of three domains: the Eukaryotes, which all contain a true nucleus and are often multicellular; the Bacteria; and the Archaea [*Woese et al.*, 1990]. These latter two groups are unicellular and lack a true nucleus. In addition to phylogenetic differences, bacteria and archaea differ in their cell wall compositions, transcriptional and translational machineries, and antibiotic resistances and are as different from each other as each is from eukaryotes.

Diffuse vents host a phylogenetically and physiologically diverse microbial community, representing a wide spectrum of thermal tolerances and metabolic strategies from both the subseafloor and the deep sea [*Corre et al.*, 2001; *Huber et al.*, 2002, 2003, 2007; *Reysenbach et al.*, 2000; *Takai and Horikoshi*, 1999]. Recently, it was shown that a relatively small number of different organisms dominate bacterial communities of diffuse flow hydrothermal vents, but thousands of low-abundance species account for most of the observed phylogenetic diversity [*Huber et al.*, 2007; *Sogin et al.*, 2006]. The presence of this "rare biosphere" has several important evolutionary and ecological implications, including the possibility that rare members may serve as a nearly inexhaustible source of genomic innovation or as keystone species, in particular subseafloor habitats responding when environmental parameters change and favor the growth of these low abundance organisms. However, although it is generally accepted that the composition of microbial populations changes with temperature and geochemistry, spatial and temporal linkages between hydrothermal vent microorganisms and fluid discharge and chemistry are only beginning to emerge [*Huber et al.*, 2002, 2003, 2006a, 2007; *Nakagawa et al.*, 2005a, 2006b].

3.2. Studied Microbes From Diffuse Vents

The primary types of microorganisms observed and studied in diffuse hydrothermal fluid discharge to date using molecular, microscopic, and culturing techniques are sulfide- and iron-oxidizing proteobacteria, epsilon-proteobacteria, Aquificales, methanogens, anaerobic methane oxidizers, sulfate reducers, and Thermococcales. One of the most visually obvious features associated with diffuse hydrothermal outflow is the presence of microbial mats along rock surfaces near the point of discharge, especially immediately after magma intrusions (Plate 1). Although these mats are primarily observed on the seafloor surface, they may also be present within the subseafloor. They are often pervasive in zones of diffuse fluid discharge and are indicative of fluid chemistry, and thus may prove useful for vent-scale modeling efforts to estimate the area of a field that is venting low-temperature fluids using camera tows. These mats are composed largely of mesophilic proteobacteria that grow in filaments contained within hollow tubular sheaths and oxidize either sulfide or iron. Within a given region, mats are either predominantly sulfide oxidizers or iron oxidizers (often white and brown in appearance, respectively, when newly grown), and this is tied directly to the chemistry of the fluid. If sulfide is present in diffuse hydrothermal fluids, with or without iron present, then sulfide oxidizers predominate because they readily outcompete iron oxidizers for resources. Iron oxidizers predominate when fluids are depleted in sulfide but still rich in iron.

After the 1993 volcanic eruption along the CoAxial segment of the Juan de Fuca Ridge, two zones of diffuse hydrothermal fluid discharge formed known as the "Floc" and "Flow" sites that were 30 km apart [*Baker et al.*, 1998]. Three months after the eruption, the Flow site, which was at the distal end of a lateral dike injection where new seafloor was formed on the surface, was colonized by a brown microbial mat up to 10 cm thick. Thin section electron microscopy of the mat showed microbial sheaths coated in metal particles believed to be ferric iron that are typical of iron-oxidizing, *Thiobacillus*-like bacteria [*Holden and Adams*, 2003; *Juniper et al.*, 1995]. Chemical analyses showed that the fluids were depleted in sulfide but high in iron [*Butterfield et al.*, 1997]. The Floc site had been heavily colonized by a white bacterial mat and was the source of white flocculent material ejected from the diffuse venting into the overlying seawater [*Juniper et al.*, 1995]. Electron micrographs of this mat material also showed the presence of wide-diameter bacterial sheaths coated with particles believed to be sulfur and were typical of sulfide-oxidizing, *Beggiatoa*-like bacteria [*Holden and Adams*, 2003; *Juniper et al.*, 1995]. The diffuse fluids from Floc were high in sulfide at the time [*Butterfield et al.*, 1997]. In addition, a morphologically and physiologically diverse group of iron oxidizers and bacteria that oxidize hydrogen sulfide to elemental sulfur have been identified from eruption-associated diffuse fluids and hydrothermal plumes [*Butterfield et al.*, 2004; *Cowen et al.*, 1999; *Huber et al.*, 2003; *Juniper et al.*, 1998; *Winn et al.*, 1986].

A common group of autotrophic bacteria found in diffuse hydrothermal fluids are the epsilon-proteobacteria. They were initially discovered by using molecular techniques rather than culture-dependent methods. These studies and others show that the epsilon-proteobacteria are a dominant group of organisms found within all hydrothermal vent biotopes including diffuse fluids [*Campbell et al.*, 2006; *Corre et al.*, 2001; *Higashi et al.*, 2004; *Huber et al.*, 2003, 2007; *Moussard et al.*, 2006a; *Nakagawa et al.*, 2005b; *Reysenbach et al.*, 2000; *Takai et al.*, 2005]. The first members of this ubiquitous group were isolated in 2001 and since then have been cultured from vent fields around the world [*Campbell et al.*, 2001; *Takai et al.*, 2005]. They display wide metabolic and phylogenetic diversity, but in general most are hydrogen- and sulfur-oxidizing chemolithoautotrophs, including both mesophiles and thermophiles [*Campbell et al.*, 2006]. Epsilon-proteobacteria of the genus *Arcobacter* were isolated from coastal waters produce filamentous sulfur when grown under microaerophilic conditions using sulfide as the electron donor [*Sievert et al.*, 2007; *Taylor and Wirsen*, 1997; *Taylor et al.*, 1999; *Wirsen et al.*, 2002]. The sulfur end product of their metabolism is morphologically similar to the "snow blower" flocculent material observed at 9°N EPR, the CoAxial segment, and elsewhere following volcanic eruptions suggesting that these organisms are responsible for the production of this material [*Moussard et al.*, 2006a; *Sievert et al.*, 2007; *Taylor and Wirsen*, 1997; *Taylor et al.*, 1999; *Wirsen et al.*, 2002]. With the potential of the epsilon-proteobacteria to serve as a source of primary productivity in hydrothermal vents and their widespread distribution and dominance in all deep-sea vent habitats, they likely serve as important members of the subseafloor biosphere [*Alain et al.*, 2002; *Campbell et al.*, 2006; *Huber et al.*, 2003, 2007; *Sievert et al.*, 2007; *Taylor et al.*, 1999].

A group of high-temperature bacteria with a physiology similar to the epsilon-proteobacteria is the order Aquificales. Until recently, the Aquificales had only been isolated from terrestrial and shallow hydrothermal springs. Since 1998, they have been commonly isolated from or detected in diffuse hydrothermal fluids from sites in the Atlantic and Pacific Oceans [*Huber et al.*, 2003; *L'Haridon et al.*, 2006; *Reysenbach et al.*, 2002; *Takai et al.*, 2004a]. Some members of this group are strict chemolithoautotrophs and microaerophiles while others use various sulfur species and nitrate as their terminal electron acceptor [*L'Haridon et al.*, 2006; *Reysenbach et al.*, 2002]. All members are thermophilic or hyperthermophilic, allowing them to occupy higher temperature niches compared to epsilon-proteobacteria. These thermophilic chemolithoautotrophs likely represent important primary producers in warm subseafloor habitats.

A common group of archaea associated with diffuse hydrothermal fluids is the methanogens, which are strict anaerobes and generally use H_2 and CO_2 to produce CH_4 in hydrothermal environments. Before the first characterization of a methanogen from deep-sea hydrothermal vents, it was postulated that some of the volatiles in hot vent fluids, including CH_4, H_2, and CO_2, may be generated by microbes in anaerobic warm environments [*Baross et al.*, 1982]. Soon after, the hyperthermophile *Methanocaldococcus jannaschii* (formerly *Methanococcus jannaschii*) was isolated and characterized from a chimney from the EPR [*Jones et al.*, 1983]. The authors noted that because of its rapid autotrophic growth rate at high temperatures, *M. jannaschii* could contribute to CH_4 production and potentially serve as a primary producer in deep-sea hydrothermal vent habitats. Further reports suggest this genus is ubiquitous in deep-sea vents [*Holden et al.*, 1998; *Huber et al.*, 2002; *Jeanthon et al.*, 1998, 1999a, 1999b; *Jones et al.*, 1989; *Mehta and Baross*, 2006; *Summit and Baross*, 1998; *Takai et al.*, 2004a, 2004b]. Surveys for the methanogen-specific gene methyl coenzyme M reductase (*mcrA*) from various hydrothermal settings indicates tremendous diversity within this group [*Dhillon et al.*, 2005; *Nercessian et al.*, 2005; *Teske et al.*, 2003]. Beyond the direct detection of methanogens in diffuse fluids, there are also chemical indicators for the ubiquity of this group in hydrothermal vent fluids. For example, methane concentrations from hydrothermal fluids collected following the 1991 9°N EPR and 1993 CoAxial segment volcanic eruptions show more CH_4 in low-temperature fluids than would be expected by mixing of end-member hydrothermal fluids and seawater alone [*Von Damm and Lilley*, 2004]. It was suggested that this excess CH_4 was produced by subseafloor methanogens. Likewise, it was suggested that the anomalously high CH_4 concentrations found in hydrothermal fluids at the Endeavour segment and at Axial Seamount were due to methanogenesis [*Butterfield et al.*, 2004; *Lilley et al.*, 1993]. Because many methanogens are obligate chemolithoautotrophs, depend solely on geochemically derived energy and carbon sources, and grow at high temperatures under strictly anaerobic conditions, the methanogens are important members of the warm subseafloor microbial community.

Although not commonly detected in diffuse fluids at unsedimented ridges, anaerobic methane oxidizers and sulfate reducers are especially important at sediment-covered basalt hydrothermal systems. Anaerobic methane oxidation is carried out by a group of archaea closely related to methanogens. Metabolically, the process is the reversal of methanogenesis, using the same set of enzymes, and methanotrophic communities are often closely associated spatially with sulfate-reducing bacteria [*Valentine*, 2007]. Hydrothermally active sediments from Guaymas Basin harbor anaerobic methane oxidizers from the so-called ANME-1 and ANME-2 phylogenetic groups of Archaea [*Kallmeyer and Boetius*, 2004; *Teske et al.*, 2002]. Rates of sulfate reduction and methane oxidation were measured between 5° and 200°C [*Kallmeyer and Boetius*, 2004]. Sulfate reduction and methane oxidation rates were highest at 95°–100°C and 35°–60°C, respectively, and methane oxidation generally accounted for less than 5% of sulfate reduction [*Kallmeyer and Boetius*, 2004]. However, it indicates that methane oxidation, both aerobic and anaerobic, must be considered when developing biogeochemical models of diffuse vent systems.

The first sulfate reducer isolated from a deep-sea hydrothermal vent was the hyperthermophilic archaeon *Archaeoglobus profundus* [*Burggraf et al.*, 1990]. Members of the

Archaeoglobales grow both autotrophically and heterotrophically and reduce compounds such as sulfate, nitrate, and iron [*Kashefi et al.*, 2002; *Stetter et al.*, 1987; *Vorholt et al.*, 1995]. Sulfate reduction rates were measured in sediment core samples collected from Guaymas Basin, and maximum subsurface sulfate reduction rates consistently occurred at 70°–90°C [*Elsgaard et al.*, 1994; *Jørgensen et al.*, 1992; *Kallmeyer and Boetius*, 2004; *Weber and Jorgensen*, 2002]. The rates of sulfate reduction in these sediments were comparable to those measured in coastal environments by moderate temperature microorganisms. Members of the Archaeoglobales were also detected in ridge flank fluids from ODP Hole 1026B on the Juan de Fuca Ridge and may be important members of ridge flank subseafloor microbial communities [*Cowen et al.*, 2003; *Nakagawa et al.*, 2006a]. Sulfate reducers within the Bacteria have also been detected from Guaymas Basin [*Jeanthon et al.*, 2002], Axial Seamount [*Huber et al.*, 2003], and the EPR [*Alazard et al.*, 2003; *Elsgaard et al.*, 1995], including both mesophilic and thermophilic members of the *Desulfovibrio* and *Thermodesulfobacterium* genera. In addition, targeted functional gene analysis for dissimilatory sulfite reductase (*dsrAB*) suggests a much larger diversity of sulfate reducers than those presently in culture, as well as indicating a global distribution at deep-sea hydrothermal vents [*Dhillon et al.*, 2003; *Jeanthon et al.*, 2002; *Nakagawa et al.*, 2004b; *Nercessian et al.*, 2005; *Teske et al.*, 2003].

The final group of organisms commonly found in diffuse hydrothermal fluids is the Thermococcales, which are all hyperthermophilic archaea and obligate heterotrophs. They include the genera *Thermococcus*, *Pyrococcus*, and *Palaeococcus*. Because they grow well in artificial media, they are one of the more thoroughly studied groups of cultured organisms from diffuse vents. They are anaerobes that generally either reduce elemental sulfur to sulfide or protons to H_2 in the absence of sulfur [*Fiala and Stetter*, 1986; *Takai et al.*, 2000; *Zillig et al.*, 1983]. Thermococcales have also been detected in hydrothermal vent environments by using molecular techniques [*Huber et al.*, 2006a; *Nercessian et al.*, 2003; *Reysenbach et al.*, 2000; *Takai et al.*, 2004a]. Because of their relative ease of culturing, they serve as highly useful indicator organisms in diffuse flow systems [*Holland et al.*, 2004]. Using growth kinetics, hydrolase protein patterns, restriction fragment length polymorphism analyses of the 16S-23S rRNA intergenic spacer region, and 16S rRNA phylogenetics, it was shown that the Thermococcales isolated from diffuse hydrothermal fluids differed from those isolated from sulfide deposits and from site to site [*Holden et al.*, 2001; *Huber et al.*, 2006a; *Summit and Baross*, 2001]. Cultivation studies indicate that *Thermococcus* and *Pyrococcus* (as well as *Desulfurococcus*) have a high tolerance to O_2 exposure and also show long-term survival at temperatures below their minimal growth temperature, a characteristic thought to be important for their dispersal and wide distribution [*Jannasch et al.*, 1992; *Pledger and Baross*, 1989].

3.3. *Life Strategies: Syntrophy and Biofilm Formation*

Some diffuse vent and subseafloor microorganisms are not likely to grow in complete isolation from one another and may actually require each other's presence for optimal growth. Syntrophy, where two or more organisms combine their metabolic capabilities to carry out some process that neither can do alone, may be an important life strategy for organisms living in the subseafloor. A syntrophic relationship was found with a sulfur-reducing hyperthermophilic heterotroph that was able to grow without sulfur when grown with hyperthermophilic methanogens [*Bonch-Osmolovskaya and Stetter*, 1991]. Without sulfur and with the methanogen, the heterotroph formed H_2 instead of H_2S as a metabolite. The methanogen then consumed the H_2, which it requires for growth, and prevented H_2 inhibition of growth for the heterotroph. Similarly, others have found that *Methanocaldococcus* grows in the absence of exogenous sources of H_2 and CO_2 when cocultured with *Thermotoga*, which produces H_2 and CO_2 as metabolites [*Muralidharan et al.*, 1997]. Cell aggregates were observed when the two organisms were physically separated by a membrane barrier, and the methanogens did not grow, suggesting cell contact is necessary for the transfer of H_2 and CO_2.

The ability to attach to mineral surfaces and produce biofilms may allow organisms to interact with one another, and many subseafloor microbes possess the ability to attach to mineral surfaces and produce biofilms [*Pysz et al.*, 2004]. In almost all cases, biofilm-forming microbes produce extracellular polysaccharides (EPS) that form the matrix of the biofilm [*Jefferson*, 2004]. Attached cells within a biofilm allow interactions between different functional groups of microbes, and these cells may have a competitive advantage under flow conditions due to both the replenishment of nutrients and removal of waste products [*Jefferson*, 2004]. A number of isolates from deep-sea hydrothermal vents have been shown to produce exopolysaccharide and form biofilms [*Pysz et al.*, 2004; *Rinker and Kelly*, 1996], and two examples are shown in Figure 2. *Palaeococcus* spp. strain GR-1, a hyperthermophilic archaeon isolated from a hydrothermal plume following a deep-sea eruption, generated intracellular organic films and thick biofilms in liquid culture (Figure 2a) [*Summit*, 2000]. *Desulfurobacterium* spp. strain Ax99-59 is a thermophilic chemolithoautotroph isolated from a diffuse vent at Axial Seamount [*Huber et al.*, 2003], and it produces visible EPS and colonizes glass and mineral surfaces under liquid

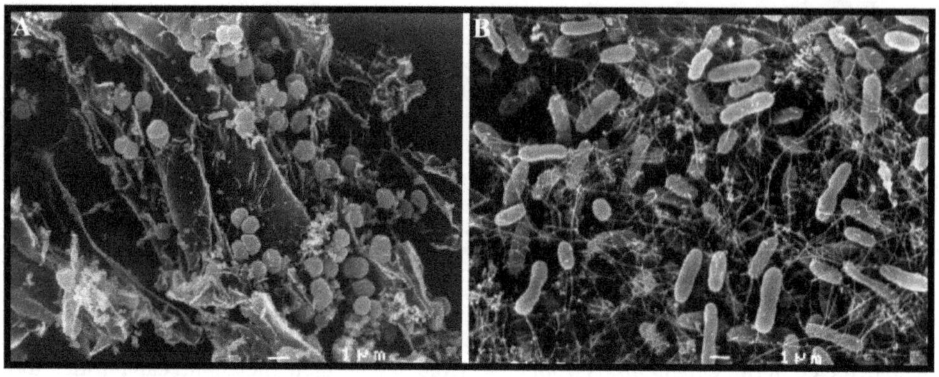

Figure 2. Scanning electron micrographs of two subseafloor isolates, including (a) hyperthermophilic archaeon GR1, isolated from an event plume associated with the 1996 North Gorda eruption; (b) thermophilic bacterium Ax99-59, isolated from a diffuse vent at Axial Seamount. In both cases, cells are embedded in a biofilm produced in unstirred culture. Scale bar is 1 μm. Image credits: (a) Melanie Holland and (b) Julie Huber.

growth conditions (Figure 2b). Cocultures of *Thermococcus* spp. and *Thermotoga* spp. show extensive exopolysaccharide production as well as biofilm growth on the walls of culturing flasks [*Rinker and Kelly*, 2000], and the hyperthermophilic sulfate reducer *Archaeoglobus fulgidus* produces a larger amount of biofilm when exposed to environmental stress, suggesting that biofilms also serve as a survival strategy [*Lapaglia and Hartzell*, 1997]. Aerobic, heterotrophic, mesophilic *Alteromonas* spp. that secrete polysaccharides have also been cultured from deep-sea vents [*Raguénès et al.*, 1996; *Vincent et al.*, 1994]. In addition, thermophilic, aerobic, and facultatively anaerobic *Bacillus* spp. were isolated from shallow vents, and the authors hypothesize that their production of exopolysaccharides may play a role in protecting the cells from stress [*Nicolaus et al.*, 2002, 2004]. These extracellular products may also serve as a source of organic carbon for heterotrophs. It has been suggested that autotrophic thermophiles and hyperthermophiles at deep-sea hydrothermal vents are a source of organic carbon for the heterotrophic populations of anaerobic microbes in the subseafloor, but there is no evidence to support this idea. However, in epilithic biofilms in an oligotrophic alpine system, it was shown that heterotrophic bacteria could use extracellular products from phototrophs for growth, thus allowing the heterotrophic population to live in a habitat that was otherwise unfavorable for growth [*McFeters et al.*, 1984].

3.4. Targeted Gene Analyses and Metagenomics

With recent progress in molecular methods, it is now possible to determine not only the diversity of microorganisms in diffuse fluids, but also the overall genetic potential of microbial communities and to link this potential to specific gene function and expression. Targeted primers for specific genes have revealed a large diversity of genes involved in nitrogen fixation, methanogenesis, sulfate reduction, and carbon fixation at deep-sea hydrothermal vents [*Campbell et al.*, 2003; *Campbell and Cary*, 2004; *Dhillon et al.*, 2003, 2005; *Mehta et al.*, 2003; *Nakagawa et al.*, 2004a, 2004b; *Nercessian et al.*, 2005; *Teske et al.*, 2003]. One particularly powerful tool is looking at the community metagenome, defined as the collective genome in a community [*Allen and Banfield*, 2005]. The application of metagenomics to microbial ecology is relatively new, but the field is progressing rapidly and analyses have been successfully applied in diverse marine environments including the open and coastal ocean [*DeLong et al.*, 2006; *Rusch et al.*, 2007], whale falls [*Tringe et al.*, 2005], the symbiont of a hydrothermal vent bivalve [*Newton et al.*, 2007], and uncultivated hydrothermal vent archaea [*Moussard et al.*, 2006b, 2006c]. The strength of this approach is that it allows one to access the genomes of uncultured microbes, therefore providing insight into community functional potential and metabolic capacity. In addition, the emerging field of environmental proteomics allows one to look at which genes are actually being expressed as proteins in a community, and this technology was recently applied to the simple endosymbiont community of the *Riftia pachyptila* tubeworm to study the metabolism of this uncultured consortia [*Markert et al.*, 2007]. Using these tools in combination with geochemical measurements, one can begin to assess the community adaptation to the habitat of interest by examining, for example, genes or processes indicative of an anaerobic thermophilic subseafloor microbial population. These analyses, in turn, can then guide cultivation efforts to gain further insight into the metabolism and physiology of hydrothermal vent microorganisms, as illustrated by the recent success in the isolation of vent thermoacidophiles and a hyperthermophilic nitrogen fixer [*Mehta and Baross*, 2006; *Reysenbach et al.*, 2006].

3.5. Biosignatures

Beyond genetic and geochemical biosignatures, there are other indicators of microbial activity in diffuse fluids, and the concept of microbial activity within the subseafloor at deep-sea hydrothermal vents originated soon after their discovery [*Corliss et al.*, 1979]. Based on high cell counts in discharging fluids (10^8–10^9 cells mL^{-1}), it was suggested that there must be a large microbial population living at depth in the basaltic rocks. *Karl et al.* [1980] used Adenosine Triphosphate (ATP) measurements as a proxy for biomass to estimate that there was 100–250 μg of microbial cell carbon per liter of vent fluid from the Galapagos. This is compared to 10 μg of total particulate organic carbon per liter of deep-sea water [*Karl et al.*, 1980]. There are other biosignatures of subseafloor activity as well. For example, Lang *et al.* [2006] measured higher dissolved organic carbon (DOC) concentrations in diffuse fluids from Axial Seamount relative to background seawater, and found that these high DOC concentrations were correlated with high cell counts. Because high-temperature fluids had depleted DOC relative to seawater, this is strong evidence of microbial biomass production in the subseafloor at diffuse vents [*Lang et al.*, 2006]. Another study used natural remnant magnetization as a proxy for microbial activity in the subseafloor [*Carlut et al.*, 2007]. By incubating sulfate-reducing bacteria with basalt, the authors found that young basalt lost part of its magnetization signal due to cycling of iron, suggesting that microorganisms may be involved in early basalt alteration. Phospholipid fatty acid (PLFA) analysis is another important biomarker that has been used successfully on sulfide chimneys, hydrothermally influenced sediments, and ridge flank crustal fluids [*Cowen et al.*, 2003; *Hedrick et al.*, 1992; *Schrenk et al.*, 2003; *Summit et al.*, 2000]. PLFA analysis can reveal differences in microbial community biomass, composition, and dynamics without culturing.

4. MODELING DIFFUSE FLUID MICROORGANISMS: A PROSPECTUS

4.1. Fluid Flux Estimates

Estimating vent-field scale diffuse fluid discharge is necessary to begin modeling the broader effect of microorganisms on the entire system. However, this has proven to be technically challenging because diffuse flow is spread over a wide area and both the temperature anomalies and the fluid velocities are small and difficult to measure [*Pruis and Johnson*, 2004]. Diffuse fluid heat flux and fluid flow have been estimated through direct measurements of diffuse venting on the seafloor and by inference through total heat flux and chemistry measurements for vent field hydrothermal plumes. Many studies, using both direct measurements and hydrothermal plume inferences, noted that heat flux is modulated by tidal effects [*Pruis and Johnson*, 2004; *Schultz et al.*, 1992; *Tivey et al.*, 2002].

Diffuse hydrothermal fluid discharge has been measured directly using electromagnetic induction and other devices to determine flow velocities, and arrays of temperature sensors have been used to calculate heat fluxes [*Pruis and Johnson*, 2004; *Schultz et al.*, 1992, 1996; *Tivey et al.*, 2002]. Vertical flow velocities and heat fluxes ranged from 0.07 to 0.15 m s^{-1} and 2 to 4 MW m^{-2}, respectively, on a sulfide-hosted diffuse vent at the Main Endeavour Field [*Schultz et al.*, 1992] to 1 to 4 m d^{-1} and 260 W m^{-2}, respectively, on a basalt-hosted diffuse vent at Axial Seamount [*Pruis and Johnson*, 2004]. This large difference in heat fluxes is likely due to the larger volume of hydrothermal fluid flowing through the sulfide. The Endeavour site study suggested that heat flux from diffuse hydrothermal fluid discharge may exceed that from high-temperature focused fluid discharge by a factor of 5 [*Schultz et al.*, 1992].

Heat flux and fluid flow from diffuse and focused venting is more commonly estimated using temperature sensors on human-occupied submersibles, autonomous undersea vehicles, and Conductivity-Temperature-Depth-Sensor (CTD) hydrocasts coupled with chemistry measurements, current meter mooring readings, and camera tows to estimate the total area of diffuse venting [*Dumont and Murrell*, 2005; *Lavelle and Wetzler*, 1999; *Rona and Trivett*, 1992; *Rudnicki and Elderfield*, 1992; *Veirs et al.*, 2006]. It is assumed that near-bottom plumes (<50 m above the seafloor) generally represent diffuse hydrothermal venting, whereas neutrally buoyant plumes 100–200 m above the seafloor are due to focused hydrothermal discharge [*Lavelle et al.*, 2001; *Rona and Trivett*, 1992], although these two fluids are not so easily segregated and each plume can be a mixture of the two sources [*Lavelle et al.*, 2001; *Veirs et al.*, 2006]. At the Main Endeavour Field, it was estimated that 85% of the flux from diffuse plumes is entrained by focused point-source plumes [*Veirs et al.*, 2006]. Total field-scale heat fluxes due to diffuse fluid discharge were estimated to be 15 to 75 MW at the ASHES vent field on Axial Seamount [*Rona and Trivett*, 1992], 250 to 470 MW at the TAG site [*Rudnicki and Elderfield*, 1992], 534 MW at the Cleft site on the Juan de Fuca Ridge [*Baker et al.*, 1993], 12 and 125 MW at two sites on the southern Juan de Fuca Ridge [*Trivett and Williams*, 1994], and 300 MW at the Main Endeavour Field [*Veirs et al.*, 2006]. These values can then be used to calculate the total volume of diffuse hydrothermal fluid that exits a vent-scale system.

The proportion of diffuse hydrothermal fluid output to focused high-temperature output varies significantly. In some plume studies, diffuse fluid heat fluxes were estimated to comprise more than 90% of the total heat flux [*Rona and Trivett*, 1992; *Schultz et al.*, 1992], whereas other studies estimated that 50 to 65% of the heat flux was from diffuse fluids [*Baker et al.*, 1993; *Rudnicki and Elderfield*, 1992; *Veirs et al.*, 2006] or that the contribution was less than 50% of the total [*Lavelle and Wetzler*, 1999; *Lavelle et al.*, 2001]. Variations in the contribution of diffuse fluid heat flux relative to the total are likely to be due in part to recent volcanic activity because diffuse fluid venting is generally more spatially extensive in the period immediately following an eruption [*Chadwick et al.*, 2001; *Shank et al.*, 1998]. These variations are also due to inherent differences in vent fields, the methods used to make the measurements, and the assumptions behind the calculations. Having the ability to estimate the vent-field scale output of diffuse hydrothermal fluid is an important component of developing first-order flow models of the systemwide impact of microorganisms within these fluids in the subseafloor and should be further developed.

4.2. Insights From the Ocean Drilling Program

At unsedimented mid-ocean ridges, we are still largely limited to examining surface expressions of basalt-hosted subseafloor processes via diffuse fluids. Currently, drilling into basalt is best accomplished in regions with sediment cover, including sedimented ridges and ridge flanks, but important questions related to microbial distribution, diversity, survival, and transport can be addressed using drilling in these environments. Middle Valley, a sediment-covered spreading center at the northern end of the Juan de Fuca Ridge, was drilled in 1993 and 1998 (Legs 139 and 169) to examine geological, geophysical, geochemical, and biological processes, including microbial community biomass estimates in the hydrothermally heated sediments [*Cragg and Parkes*, 1994; *Cragg et al.*, 2000; *Party*, 1998; *Summit et al.*, 2000].

Drilling projects on ridge flanks likewise are important for understanding microbial processes in the subseafloor of these environments. There is evidence for the existence of subseafloor ridge flank microbial communities in 3.5 Ma basement fluids from the Juan de Fuca ridge flank [*Cowen et al.*, 2003; *Cowen*, 2004; *Huber et al.*, 2006b; *Nakagawa et al.*, 2006a], as well as in flank fluids near the EPR [*Ehrhardt et al.*, 2007]. Studies on the Juan de Fuca ridge flank have benefited from the development of borehole observatories Circulation Obviation Retrofit Kit (CORKs) that allow for in situ experimentation and monitoring of subseafloor processes downhole [*Jannasch et al.*, 2003]. The development of this important technology is ongoing and will serve as a critical tool for building models of microbial processes in the subseafloor. For example, on the Juan de Fuca ridge flank, fluids recharge through a large volcanic outcrop and discharge through a smaller outcrop located 50 km away [*Fisher et al.*, 2003]. A series of holes drilled between the two sites would allow the monitoring and tracking of the flow of seawater in the crust as it moves through the crustal aquifer. Tracer experiments and isotope measurements would constrain seawater travel time in the basement, and this information coupled with measurements of the geochemistry and microbiology at each hole would foster the modeling of subseafloor processes. In this way, modeling of the subseafloor environment could go from first-order flow models to reactive transport models since the end-member fluids and intermediate fluids are accessible.

4.3. In Situ Monitoring

For nearly two decades, instruments have been deployed at various vent sites that recorded different parameters continuously for time spans ranging from weeks to months. However, these instruments generally required on-site recovery, the data were not available until recovery, and the different instruments were not necessarily integrated with one another. Access to real-time data from deep-sea vent sites began in 1993 with the use of the U.S. Navy's Sound Surveillance System (SOSUS) to monitor seismic activity and detected eruptions in the northeastern Pacific Ocean along the Gorda and Juan de Fuca ridges [*Bohnenstiehl et al.*, 2004; *Cowen and Baker*, 1998; *Dziak et al.*, 1995; *Dziak and Fox*, 1999]. New technologies are under development that will integrate different types of seafloor observations and then transmit the data either by buoy-satellite transmissions [e.g., NOAA New Millennium Observatory (NEMO), www.pmel.noaa.gov/vents/nemo/index.html] or directly through fiber-optic cables that also provide power to the instruments on the seafloor (e.g., NEPTUNE, www.neptune.washington.edu; ESONET, www.oceanlab.abdn.ac.uk/research/esonet.shtml). These observatories will significantly foster the development of flow models for diffuse vents by providing real-time measurements of fluid temperature, chemistry, and discharge over a prescribed area of diffuse hydrothermal venting. Microbial sampling of diffuse fluids with these monitoring measurements will address how the concentrations of various metabolic types of organisms vary with fluid temperature, chemistry, and discharge for modeling the impact of diffuse vent microorganisms along mid-ocean ridges and flanks.

4.4. Laboratory Experiments

The characterization of new microorganisms typically is performed using pure cultures and optimal growth conditions that are not necessarily representative of what the organisms experience in nature. Therefore, in order to better model the growth of diffuse vent organisms in nature, future laboratory experiments need to consider environmentally relevant growth conditions (e.g., nutrient limitation, mixed cultures, growth in a porous solid matrix, hydrostatic pressure) using organisms from the site of interest. For example, recent cultivation experiments with hyperthermophilic vent archaea combined stress factors such as extremes of pH, temperature, and pressure along with high sulfide and heavy metal concentrations, and results indicate that these organisms require a constant substrate and energy supply to survive under these extreme conditions [*Edgcomb et al.*, 2004, 2007; *Lloyd et al.*, 2005]. Furthermore, modeling the impact of microorganisms on diffuse fluid chemistry will require determining the growth yield from specific nutrients and metabolites. The theory is that an individual cell will consume or produce a compound at a constant rate and will use or make a constant amount of that compound per cell doubling for a given growth condition [*Adams et al.*, 2001; *Feinberg and Holden*, 2006]. During the growth of a cell culture, there is a linear relationship between the concentration of cells and the compound of interest and this can be used to calculate the amount of compound consumed or produced per cell doubling; then this value can be divided by the doubling time of the culture to calculate the rate of compound consumption or production. For example, the rate of phenylacetate production (a specific product of phenylalanine metabolism) by the hyperthermophilic archaeon *Pyrococcus furiosus* varied with growth conditions, and production rates were significantly higher when the media contained both peptides and elemental sulfur relative to cultures lacking either peptides or sulfur [*Adams et al.*, 2001]. Peptide metabolism in *P. furiosus* was not affected by the presence of carbohydrate (maltose), but it was demonstrated that growth on peptides was to a large extent dependent on sulfur availability. Similarly, there was a linear relationship between cell and Fe(II) concentrations in growth studies with the hyperthermophilic archaeon *Pyrobaculum aerophilum* [*Feinberg and Holden*, 2006]. Interestingly, the amount of ferrous iron produced per cell doubling was the same for growth on soluble (Fe(III) citrate) and insoluble (Fe(III) oxide hydroxide) forms of iron even though the growth rates and maximum cell concentrations for cultures grown under these conditions were significantly different. This suggests that the same number of electrons need to be passed to a dissimilatory terminal electron acceptor for each cell doubling regardless of the type of acceptor when conditions are otherwise optimal.

For modeling purposes, these types of experiments should be run under conditions that more closely resemble those found in nature. This could involve growth under nutrient limiting conditions, varying pH or oxidation-reduction potential, or in the presence of other organisms (competitive and beneficial) with the goal of determining how much of a compound is consumed or produced per cell doubling. For the purpose of developing a first-order fluid flux model, these studies would be coupled with field observations to determine the abundance of certain metabolic types of microorganisms in diffuse hydrothermal fluids. Assuming steady-state biomass levels within the basalt-hosted seafloor over relatively short time scales (e.g., days to weeks), the number of cells washed out of a diffuse vent per unit time would be equal to the number of cells produced within the subseafloor over that same period. The heat flux and fluid flow estimates described above would be used to estimate the total volume of diffuse fluid discharged over a given area, perhaps as large as a vent field. Therefore, the product of diffuse fluid volume discharged per unit time, cell concentration of a specific metabolic type of organism (e.g., methanogens), and the amount of compound produced or consumed per cell produced (e.g., CH_4 amount per methanogen doubling) would yield an estimate of the total biogenic input of that compound per unit time per unit area.

In the field, organisms of a specific metabolic type in diffuse hydrothermal fluids can be quantified by using targeted culturing [e.g., most probable number (MPN) estimates], molecular approaches (e.g., quantitative polymerase chain reaction), and microscopic techniques [e.g., fluorescence in situ hybridization (FISH)]. These measurements can then be merged with chemical data from the same sample and compared with the conditions used in the laboratory to better constrain and model the growth of these organisms in the laboratory.

5. SUMMARY

Ocean crust covers more than 60% of the Earth's surface, and the mixing of hydrothermal fluids with seawater in the crust creates abundant energy sources that can sustain microorganisms. There is a growing body of evidence that an indigenous subseafloor microbial community exists in ocean crust, but this globally distributed habitat remains relatively unexplored and poorly constrained. Diffuse fluids are common features along mid-ocean ridges and flanks, and these fluids provide one of the best points of access to the subseafloor. However, our knowledge of what microbes are present and how they are distributed in this dynamic geochemical environment over time and space is fragmentary. The development of fluid flow models of diffuse hydrothermal fluids and the organisms found within them will provide first-order estimates of what microbial processes are occurring within the subseafloor. As described in this chapter, the metabolic diversity of the microorganisms found in diffuse fluids is wide ranging. These organisms have specialized growth requirements that set them apart from other metabolic types of organisms; therefore, they may be useful as a tracer of subseafloor environmental conditions. This information, coupled with laboratory analyses of these organisms and field studies of the physical and chemical conditions found at diffuse hydrothermal vents, should provide a clearer picture of life within hydrothermally influenced regions of the subseafloor.

REFERENCES

Adams, M. W. W., et al. (2001), Key role for sulfur in peptide metabolism and in regulation of three hydrogenases in the hyperthermophilic Archaeon *Pyrococcus furiosus*, *J. Bacteriol.*, *183*, 716–724.

Alain, K., J. Querellou, F. Lesongeur, P. Pignet, P. Crassous, G. Raguenes, V. Cueff, and M.-A. Cambon-Bonavita (2002), *Caminibacter hydrogeniphilus* gen. nov., sp. nov., a novel thermophilic, hydrogen-oxidizing bacterium isolated from an East Pacific Rise hydrothermal vent, *Int. J. Syst. Evol. Microbiol.*, *52*, 1317–1323.

Alazard, D., S. Dukan, A. Urios, F. Verhe, N. Bouabida, F. Morel, P. Thomas, J.-L. Garcia, and B. Ollivier (2003), *Desulfovibrio hydrothermalis* sp. nov., a novel sulfate-reducing bacterium isolated from hydrothermal vents, *Int. J. Syst. Evol. Microbiol.*, *53*, 173–178.

Allen, E. E., and J. F. Banfield (2005), Community genomics in microbial ecology and evolution, *Nat. Rev. Microbiol.*, *3*, 489–498.

Alt, J. C., and W. Bach (2003), Alteration of oceanic crust: subsurface rock-water interactions, in *Energy and Mass Transfer in Marine Hydrothermal Systems*, edited by P. E. Halbach, et al., pp. 7–27, Dahlem University Press, Berlin.

Bach, W., and K. J. Edwards (2003), Iron and sulfide oxidation within basalt ocean crust: Implications for chemolithoautotrophic microbial biomass production, *Geochim. Cosmochim. Acta*, *67*, 3871–3887.

Baker, E. T., G. J. Massoth, S. L. Walker, and R. W. Embley (1993), A method for quantitatively estimating diffuse and discrete hydrothermal discharge, *Earth Planet. Sci. Lett.*, *118*, 235–249.

Baker, E. T., G. L. Massoth, R. A. Feely, G. A. Cannon, and R. E. Thomson (1998), The rise and fall of the CoAxial hydrothermal site, 1993–1996, *J. Geophys. Res.*, *103*, 9791–9806.

Baross, J. A., M. D. Lilley, and L. I. Gordon (1982), Is the CH_4, H_2, and CO venting from submarine hydrothermal systems produced by thermophilic bacteria? *Nature*, *298*, 366–368.

Bohnenstiehl, D. R., R. P. Dziak, M. Tolstoy, C. Fox, and M. Fowler (2004), Temporal and spatial history of the 1999–2000 Endeavour Segment seismic series, Juan de Fuca Ridge, *Geochem. Geophys. Geosyst.*, *10.1029/2004GC000735*.

Bonch-Osmolovskaya, E. A., and K. O. Stetter (1991), Interspecies hydrogen transfer in cocultures of thermophilic *Archaea*, *System. Appl. Microbiol.*, *14*, 205–208.

Burggraf, S., H. W. Jannasch, B. Nicolaus, and K. O. Stetter (1990), *Archaeoglobus profundus* sp. nov., represents a new species within the sulfate-reducing archaebacteria, *System. Appl. Microbiol.*, *13*, 24–28.

Butterfield, D. A., and G. J. Massoth (1994), Geochemistry of North Cleft segment vent fluids: Temporal changes in chlorinity and their possible relation to recent volcanism, *J. Geophys. Res.*, *99*(B3), 4951–4968.

Butterfield, D. A., et al. (1997), Seafloor eruptions and evolution of hydrothermal fluid chemistry, *Philos. Trans. R. Soc. Lond. A*, *355*, 369–386.

Butterfield, D. A., K. K. Roe, M. D. Lilley, J. A. Huber, J. A. Baross, R. W. Embley, and G. J. Massoth (2004), Mixing, reaction, and microbial activity in the sub-seafloor revealed by temporal and spatial variation in diffuse flow vents at Axial Volcano, in *The Subseafloor Biosphere at Mid-Ocean Ridges*, *Geophys. Monogr. Ser.*, vol. 144, edited by W. S. D. Wilcock, et al., pp. 269–289, AGU, Washington, D.C.

Campbell, B., C. Jeanthon, J. E. Kostka, G. W. I. Luther, and S. C. Cary (2001), Growth and phylogenetic properties of novel bacteria belonging to the epsilon subdivision of the *Proteobacteria* enriched from *Alvinella pompejana* and deep-sea hydrothermal vents, *Appl. Environ. Microbiol.*, *67*, 4566–4572.

Campbell, B. J., and S. C. Cary (2004), Abundance of reverse tricarboxylic acid cycles genes in free-living microorganisms at deep-sea hydrothermal vents, *Appl. Environ. Microbiol.*, *70*, 6282–6289.

Campbell, B. J., J. L. Stein, and S. C. Cary (2003), Evidence of chemolithoautrophy in the bacterial community associated with *Alvinella pompejana*, a hydrothermal vent polychaete, *Appl. Environ. Microbiol.*, *69*, 5070–5078.

Campbell, B. J., A. S. Engel, M. L. Porter, and K. Takai (2006), The versatile ε-proteobacteria: Key players in sulphidic habitats, *Nat. Rev. Microbiol.*, *4*, 458–468.

Carlut, J., H. Horen, and D. Janots (2007), Impact of micro-organisms activity on the natural remanent magnetization of the young oceanic crust, *Earth Planet. Sci. Lett.*, *253*, 497–506.

Chadwick, W. W., Jr., D. S. Scheirer, R. W. Embley, and H. P. Johnson (2001), High-resolution bathymetric surveys using scanning sonars: Lava flow morphology, hydrothermal vents, and geologic structure at recent eruption sites on the Juan de Fuca Ridge, *J. Geophys. Res.*, *106*(B8), 16,075–16,099.

Cooper, M. J., H. Elderfield, and A. Schultz (2000), Diffuse hydrothermal fluids from Lucky Strike hydrothermal vent field: Evidence for a shallow conductively heated system, *J. Geophys. Res.*, *105*(B8), 19,369–19,375.

Corliss, J. B., et al. (1979), Submarine thermal springs on the Galapagos rift, *Science*, *203*, 1073–1083.

Corre, E., A.-L. Reysenbach, and D. Prieur (2001), ε-Proteobacterial diversity from a deep-sea hydrothermal vent on the Mid-Atlantic Ridge, *FEMS Microbiol. Lett.*, *205*, 329–335.

Cowen, J. P. (2004), The microbial biosphere of sediment-buried oceanic basement, *Res. Microbiol.*, *155*, 497–506.

Cowen, J. P., and E. T. Baker (1998), The 1996 Gorda Ridge event detection and response activities: historical perspective and future implications, *Deep-Sea Res. II*, *45*, 2593–2512.

Cowen, J. P., R. Shackleford, D. McGee, P. Lam, E. T. Baker, and E. Olson (1999), Microbial biomass in hydrothermal plumes associated with the 1998 Axial Volcano eruption, *Geophys. Res. Lett.*, *26*, 3637–3640.

Cowen, J. P., S. J. Giovannoni, F. Kenig, H. P. Johnson, D. A. Butterfield, M. S. Rappe, M. Hutnak, and P. Lam (2003), Fluids from aging ocean crust that support microbial life, *Science*, *299*, 120–123.

Cragg, B. A., and R. J. Parkes (1994), Bacterial profiles in hydrothermally active deep sediment layers from Middle Valley (NE Pacific), Sites 857 and 858, in *Proc. Ocean Drill. Program, Sci. Results*, edited by M. J. Mottl, et al., pp. 509–516, Texas A&M University, Ocean Drilling Program, College Station, TX.

Cragg, B. A., M. Summit, and R. J. Parkes (2000), Bacterial profiles in a sulfide mound (Site 1035) and an area of active fluid venting (Site 1036) in hot hydrothermal sediments from Middle Valley (Northeast Pacific), in *Proceedings Ocean Drill. Program, Sci. Results*, edited by R. A. Zierenberg, et al., pp. 1–18, Texas A&M University, Ocean Drilling Program, College Station, TX.

DeLong, E. F., et al. (2006), Community genomics among stratified microbial assemblages in the ocean's interior, *Science*, *311*, 496–503.

Dhillon, A., A. Teske, J. Dillon, D. A. Stahl, and M. L. Sogin (2003), Molecular characterization of sulfate-reducing bacteria in the Guaymas Basin, *Appl. Environ. Microbiol.*, *69*, 2765–2772.

Dhillon, A., M. Lever, K. G. Lloyd, D. B. Albert, M. L. Sogin, and A. Teske (2005), Methanogen diversity evidenced by molecular characterization of methyl coenzyme M reductase A (mcrA) genes in hydrothermal sediments of the Guaymas Basin, *Appl. Environ. Microbiol.*, *71*, 4592–4601.

Dumont, M. G., and J. C. Murrell (2005), Stable isotope probing—linking microbial identity to function, *Nat. Rev. Microbiol.*, *3*, 499–504.

Dziak, R. P., C. G. Fox, and A. E. Schreiner (1995), The June-July 1993 seismo-acoustic event at CoAxial segment, Juan de Fuca Ridge: Evidence for a lateral dike injection, *Geophys. Res. Lett.*, *22*, 135–138.

Dziak, R. P., and C. G. Fox (1999), The January 1998 earthquake swarm at Axial Volcano, Juan de Fuca Ridge: Hydroacoustic evidence of seafloor volcanic activity, *Geophys. Res. Lett.*, *26*, 3429–3432.

Edgcomb, V. P., S. J. Molyneaux, M. A. Saito, K. Lloyd, S. Böer, C. O. Wirsen, M. S. Atkins, and A. Teske (2004), Sulfide ameliorates metal toxicity for deep-sea hydrothermal vent archaea, *Appl. Environ. Microbiol.*, *70*, 2551–2555.

Edgcomb, V. P., S. J. Molyneaux, S. Böer, C. O. Wirsen, M. Saito, M. S. Atkins, K. Lloyd, and A. Teske (2007), Survival and growth of two heterotrophic hydrothermal vent archaea, *Pyrococcus* strain GB-D and *Thermococcus fumicolans*, under low pH and high sulfide concentrations in combination with high temperature and pressure regimes, *Extremophiles*, *11*, 329–342.

Edmond, J. M., C. Measures, R. E. McDuff, L. H. Chan, R. Collier, B. Grant, L. I. Gordon, and J. B. Corliss (1979), Ridge crest hydrothermal activity and the balances of the major and minor elements in the ocean: The Galapagos data, *Earth Planet. Sci. Lett.*, *46*, 1–18.

Edmonds, H. N., and J. M. Edmond (1995), A three-component mixing model for ridge-crest hydrothermal fluids, *Earth Planet. Sci. Lett.*, *134*, 53–67.

Ehrhardt, C. J., R. M. Haymon, M. G. Lamontagne, and P. A. Holden (2007), Evidence for hydrothermal Archaea within the basaltic flanks of the East Pacific Rise, *Environ. Microbiol.*, *9*, 900–912.

Elsgaard, L., M. F. Isaksen, B. B. Jorgensen, A.-M. Alayse, and H. W. Jannasch (1994), Microbial sulfate reduction in deep-sea sediments at the Guaymas Basin hydrothermal vent area: Influence of temperature and substrates, *Geochim. Cosmochim. Acta*, *58*, 3335–3343.

Elsgaard, L., J. Guezennec, N. Benbouzid-Rollet, and D. Prieur (1995), Mesophilic sulfate-reducing bacteria from three deep-sea hydrothermal vent sites, *Oceanol. Acta*, *18*, 95–104.

Feinberg, L. F., and J. F. Holden (2006), Characterization of dissimilatory Fe(III) versus NO_3^- reduction in the hyperthermophilic archaeon *Pyrobaculum aerophilum*, *J. Bacteriol.*, *188*, 525–531.

Fiala, G., and K. O. Stetter (1986), *Pyrococcus furiosus* sp. nov, represents a novel genus of marine heterotrophic archaeobacteria growing optimally at 100°C, *Arch. Microbiol.*, *145*, 56–61.

Fisher, A. T. (1998), Permeability within basaltic oceanic crust, *Rev. Geophys.*, *36*, 143–182.

Fisher, A. T. (2003), Geophysical constraints on hydrothermal circulation, in *Energy and Mass Transfer in Marine Hydrothermal Systems*, edited by P. E. Halbach, et al., pp. 29–52, Dahlem University Press, Berlin.

Fisher, A. T., et al. (2003), Hydrothermal recharge and discharge across 50 km guided by seamounts on a young ridge flank, *Nature*, *421*, 618–621.

Gouze, P., and A. Coudrain-Ribstein (2000), Modeling permeability changes caused by hydrothermal circulation, *Nat. Resour. Res.*, *9*, 157–166.

Hedrick, D. B., R. D. Pledger, D. C. White, and J. A. Baross (1992), In situ microbial ecology of hydrothermal vent sediments, *FEMS Microbiol. Ecol.*, *101*, 1–10.

Higashi, Y., M. Sunamura, K. Kitamura, K. Nakamura, Y. Kurusu, J. Ishibashi, T. Urabe, and A. Maruyama (2004), Microbial diversity in hydrothermal surface to subsurface environments of Suiyo Seamount, Izu-Bonin Arc, using a catheter-type in situ growth chamber, *FEMS Microbiol. Ecol.*, *47*, 327–336.

Holden, J. F., and M. W. W. Adams (2003), Microbe-metal interactions in marine hydrothermal environments, *Curr. Opin. Chem. Biol.*, *7*, 160–165.

Holden, J. F., M. Summit, and J. A. Baross (1998), Thermophilic and hyperthermophilic microorganisms in 3–30°C hydrothermal fluids following a deep-sea volcanic eruption, *FEMS Microbiol. Ecol.*, *25*, 33–41.

Holden, J. F., K. Takai, M. Summit, S. Bolton, J. Zyskowski, and J. A. Baross (2001), Diversity among three novel groups of hyperthermophilic deep-sea *Thermococcus* species from three sites in the northeastern Pacific Ocean, *FEMS Microbiol. Ecol.*, *36*, 51–60.

Holland, M. E., J. A. Baross, and J. F. Holden (2004), Illuminating subseafloor ecosystems using microbial tracers, in *The Subseafloor Biosphere at Mid-Ocean Ridges, Geophys. Monogr. Ser.*, vol. 144, edited by W. S. D. Wilcock, et al., pp. 291–303, AGU, Washington, D.C.

Huber, J. A., D. A. Butterfield, and J. A. Baross (2002), Temporal changes in archaeal diversity and chemistry in a mid-ocean ridge subseafloor habitat, *Appl. Environ. Microbiol.*, *68*, 1585–1594.

Huber, J. A., D. A. Butterfield, and J. A. Baross (2003), Bacterial diversity in a subseafloor habitat following a deep-sea volcanic eruption, *FEMS Microbiol. Ecol.*, *43*, 393–409.

Huber, J. A., D. A. Butterfield, and J. A. Baross (2006a), Diversity and distribution of subseafloor Thermococcales populations in diffuse hydrothermal vents at an active deep-sea volcano in the northeast Pacific Ocean, *J. Geophys. Res.*, *111*, G04016, doi:10.1029/2005JG000097.

Huber, J. A., H. P. Johnson, D. A. Butterfield, and J. A. Baross (2006b), Microbial life in ridge flank crustal fluids, *Environ. Microbiol.*, *8*, 88–99.

Huber, J. A., D. B. Mark Welch, H. G. Morrison, S. M. Huse, P. R. Neal, D. A. Butterfield, and M. L. Sogin (2007), Microbial population structures in the deep marine biosphere, *Science*, *318*, 97–100.

Jacobson, R. S. (1992), Impact of crustal evolution on changes of the seismic properties of the uppermost ocean crust, *Rev. Geophys.*, *30*, 23–42.

James, R. H., and H. Elderfield (1996), Chemistry of ore-forming fluids and mineral formation rates in an active hydrothermal sulfide deposit on the Mid-Atlantic Ridge, *Geology*, *24*, 1147–1150.

Jannasch, H. W., and M. J. Mottl (1985), Geomicrobiology of deep-sea hydrothermal vents, *Science*, *229*, 717–725.

Jannasch, H. W., C. O. Wirsen, S. J. Molyneaux, and T. A. Langworthy (1992), Comparative physiological studies on hyperthermophilic archaea isolated from deep-sea hot vents with emphasis on *Pyrococcus* strain GB-D, *Appl. Environ. Microbiol.*, *58*, 3472–3481.

Jannasch, H. W., E. E. Davis, M. Kastner, J. D. Mooris, T. L. Pettigrew, J. N. Plant, E. A. Solomon, H. W. Villinger, and C. G. Wheat (2003), CORK II: Long-Term monitoring of fluid chemistry, fluxes, and hydrology in instrumented boreholes at the Costa Rica subduction zone, in *Proc. Ocean Drill. Program, Initial Rep.*, edited by J. D. Morris, et al., pp. 1–36, Ocean Drilling Program, College Station, TX.

Jeanthon, C., S. L'Haridon, A. L. Resyenbach, M. Vernet, P. Messner, U. B. Sleytr, and D. Prieur (1998), *Methanococcus infernus* sp. nov., a novel hyperthermophilic lithotrophic methanogen isolated from a deep-sea hydrothermal vent, *Int. J. Syst. Bacteriol.*, *48*, 913–919.

Jeanthon, C., S. L'Haridon, N. Pradel, and D. Prieur (1999a), Rapid identification of hyperthermophilic methanococci isolated from deep-sea hydrothermal vents, *Int. J. Syst. Bacteriol.*, *49*, 591–594.

Jeanthon, C., S. L'Haridon, A. L. Reysenbach, E. Corre, M. Vernet, P. Messner, U. B. Sleytr, and D. Prieur (1999b), *Methanococcus vulcanius* sp. nov., a novel hyperthermophilic methanogen isolated from East Pacific Rise, and identification of *Methanococcus* sp. DSM 4213T as *Methanococcus fervens* sp. nov., *Int. J. Syst. Evol. Microbiol.*, *49*, 583–589.

Jeanthon, C., S. L'Haridon, V. Cueff, A. Banta, A.-L. Reysenbach, and D. Prieur (2002), *Thermodesulfobacterium hydrogeniphilum* sp. nov., a thermophilic, chemoautotrophic, sulfate-reducing bacterium isolated from a deep-sea hydrothermal vent at Guaymas Basin, and emendation of the genus *Thermodesulfobacterium*, *Int. J. Syst. Evol. Microbiol.*, *52*, 765–772.

Jefferson, K. K. (2004), What drives bacteria to produce a biofilm? *FEMS Microbiol. Lett.*, *236*, 163–173.

Johnson, H. P., and M. J. Pruis (2003), Fluxes of fluid and heat from the oceanic crustal reservoir, *Earth Planet. Sci. Lett.*, *216*, 565–574.

Johnson, K. S., J. J. Childress, R. R. Hessler, C. M. Sakamoto-Arnold, and C. L. Beehler (1988), Chemical and biological interactions in the Rose Garden hydrothermal vent field, Galapagos spreading center, *Deep-Sea Res.*, *35*, 1723–1744.

Jones, W. J., J. A. Leigh, F. Mayer, C. R. Woese, and R. S. Wolfe (1983), *Methanococcus jannaschii* sp. nov., an extremely thermophilic methanogen from a submarine hydrothermal vent, *Arch. Microbiol.*, *136*, 254–261.

Jones, W. J., C. E. Stugard, and H. W. Jannasch (1989), Comparison of thermophilic methanogens from submarine hydrothermal vents, *Arch. Microbiol.*, *151*, 314–318.

Jørgensen, B. B., M. F. Isaksen, and H. W. Jannasch (1992), Bacterial sulfate reduction above 100°C in deep-sea hydrothermal vent sediments, *Science*, *258*, 1756–1757.

Juniper, S. K., and B. M. Tebo (1995), Microbe-metal interactions and mineral deposition at hydrothermal vents, in *The Microbiology of Deep-Sea Hydrothermal Vents*, edited by D. M. Karl, pp. 219–253, CRC Press, Boca Raton.

Juniper, S. K., P. Martineu, J. Sarrazin, and Y. Gélanis (1995), Microbial-mineral floc associated with nascent hydrothermal activity on CoAxial Segment, Juan de Fuca Ridge, *Geophys. Res. Lett.*, *22*, 179–182.

Juniper, S. K., D. F. Bird, M. Summit, M. P. Vong, and E. T. Baker (1998), Bacterial and viral abundances in hydrothermal event plumes over northern Gorda Ridge, *Deep-Sea Res. II*, *45*, 2739–2749.

Kallmeyer, J., and A. Boetius (2004), Effects of temperature and pressure on sulfate reduction and anaerobic oxidation of methane in hydrothermal sediments of Guaymas Basin, *Appl. Environ. Microbiol.*, *70*, 1231–1233.

Karl, D. M., C. O. Wirsen, and H. W. Jannasch (1980), Deep-sea primary production at the Galapagos hydrothermal vents, *Science*, *207*, 1345–1347.

Kashefi, K., J. M. Tor, D. E. Holmes, C. V. Gaw Van Praagh, A.-L. Reysenbach, and D. R. Lovley (2002), *Geoglobus ahangari* gen. nov., sp. nov., a novel hyperthermophilic archaeon capable of oxidizing organic acids and growing autotrophically on hydrogen with Fe(III) serving as the sole electron acceptor, *Int. J. Syst. Evol. Microbiol.*, *52*, 719–728.

Koschinsky, A., R. Seifert, P. Halbach, M. Bau, S. Brasse, L. M. De Carvalho, and N. M. Fonseca (2002), Geochemistry of diffuse low-temperature hydrothermal fluids in the North Fiji basin, *Geochim. Cosmochim. Acta*, *66*, 1409–1427.

Lang, S. Q., D. A. Butterfield, M. D. Lilley, H. Paul Johnson, and J. I. Hedges (2006), Dissolved organic carbon in ridge-axis and ridge-flank hydrothermal systems, *Geochim. Cosmochim. Acta*, *70*, 3830–3842.

Lapaglia, C., and P. L. Hartzell (1997), Stress-induced production of biofilm in the hyperthermophile *Archaeoglobus fulgidus*, *Appl. Environ. Microbiol.*, *63*, 3158–3163.

Lavelle, J. W., and M. A. Wetzler (1999), Diffuse venting and background contributions to chemical anomalies in a neutrally buoyant ocean hydrothermal plume, *J. Geophys. Res.*, *104*, 3201–3209.

Lavelle, W., M. A. Wetzler, E. T. Baker, and R. W. Embley (2001), Prospecting for hydrothermal vents using moored current meters and temperature data: Axial Volcano on the Juan de Fuca Ridge, Northeast Pacific, *J. Phys. Oceanogr.*, *31*, 827–838.

L'Haridon, S., et al. (2006), *Desulfurobacterium atlanticum* sp. nov., *Desulfurobacterium pacificum* sp. nov. and *Thermovibrio guaymasensis* sp. nov., three thermophilic members of the *Desulfurobacteriaceae* fam. nov., a deep branching lineage within the Bacteria, *Int J. Syst. Evol. Microbiol.*, *56*, 2843–2852.

Lilley, M. D., J. A. Baross, and L. I. Gordon (1984), Reduced gases and bacteria in hydrothermal fluids: The Galapagos spreading center and 21°N East Pacific Rise, in *Hydrothermal Processes at Seafloor Spreading Centers*, edited by P. A. Rona, et al., pp. 411–449, Plenum Press, New York.

Lilley, M. D., D. A. Butterfield, E. J. Olson, J. E. Lupton, S. A. Macko, and R. E. McDuff (1993), Anomalous CH_4 and NH_4^+ concentrations at an unsedimented mid-ocean-ridge hydrothermal system, *Nature*, *364*, 45–47.

Lloyd, K. G., V. P. Edgcomb, S. J. Molyneaux, S. Boer, C. O. Wirsen, M. S. Atkins, and A. Teske (2005), Effects of dissolved sulfide, pH, and temperature on growth and survival of marine hyperthermophilic archaea, *Appl. Environ. Microbiol.*, *71*, 6383–6387.

Markert, S., et al. (2007), Physiological proteomics of the uncultured endosymbiont of *Riftia pachyptila*, *Science*, *315*, 247–250.

McCollom, T. M. (2005), Energetic constraints on subsurface biomass production within igneous ocean crust, *Geochim. Cosmochim. Acta*, *69*, A200.

McFeters, G. A., et al. (1984), Biofilm development and its consequences: Group report, in *Microbial Adhesion and Aggregation*, edited by K. C. Marshall, pp. 109–124, Springer-Verlag, Berlin.

Mehta, M., D. A. Butterfield, and J. A. Baross (2003), Phylogenetic diversity of nitrogenase (*nifH*) genes in deep-sea and hydrothermal vent environments of the Juan de Fuca Ridge, *Appl. Environ. Microbiol.*, *69*, 960–970.

Mehta, M. P., and J. A. Baross (2006), Nitrogen fixation at 92°C by a hydrothermal vent archaeon, *Science*, *314*, 1783–1786.

Moussard, H., E. Corre, M.-A. Cambon-Bonavita, Y. Fouquet, and C. Jeanthon (2006a), Novel uncultured *Epsilonproteobacteria* dominate a filamentous sulphur mat from the 13°N hydrothermal vent field, East Pacific Rise, *FEMS Microbiol. Ecol.*, *58*, 449–463.

Moussard, H., G. Henneke, D. Moreira, V. Jouffe, P. Lopez-Garcia, and C. Jeanthon (2006b), Thermophilic lifestyle for an uncultured archaeon from hydrothermal vents: Evidence from environmental genomics, *Appl. Environ. Microbiol.*, *72*, 2268–2271.

Moussard, H., D. Moreira, M.-A. Cambon-Bonavita, P. Lopez-Garcia, and C. Jeanthon (2006c), Uncultured archaea in a hydrothermal microbial assemblage: Phylogenetic diversity and characterization of a genome fragment from a euryarchaeote, *FEMS Microbiol. Ecol.*, *57*, 452–469.

Muralidharan, V., K. D. Rinker, I. S. Hirsh, E. J. Bouwer, and R. M. Kelly (1997), Hydrogen transfer between methanogens and fermentative heterotrophs in hyperthermophilic cocultures, *Biotechnol. Bioeng.*, *56*, 268–278.

Nakagawa, S., et al. (2005a), Variability in microbial community and venting chemistry in a sediment-hosted back-arc hydrothermal system: Impacts of subseafloor phase-separation, *FEMS Microbiol. Ecol.*, *54*, 141–155.

Nakagawa, S., K. Takai, F. Inagaki, H. Hirayama, T. Nunoura, K. Horikoshi, and Y. Sako (2005b), Distribution, phylogenetic diversity and physiological characteristics of epsilon-*Proteobacteria* in a deep-sea hydrothermal field, *Environ. Microbiol.*, *7*, 1619–1632.

Nakagawa, S., et al. (2006a), Microbial community in black rust exposed to hot ridge flank crustal fluids, *Appl. Environ. Microbiol.*, *72*, 6789–6799.

Nakagawa, T., J. Ishibashi, A. Maruyama, T. Yamanaka, Y. Morimoto, H. Kimura, T. Urabe, and M. Fukui (2004a), Analysis of dissimilatory sulfite reductase and 16S rRNA gene fragments from deep-sea hydrothermal sites of the Suiyo Seamount, Izu-Bonin Arc, Western Pacific, *Appl. Environ. Microbiol.*, *70*, 393–403.

Nakagawa, T., S. Nakagawa, F. Inagaki, K. Takai, and K. Horikoshi (2004b), Phylogenetic diversity of sulfate-reducing prokaryotes in active deep-sea hydrothermal vent chimney structures, *FEMS Microbiol. Lett.*, *232*, 145–152.

Nakagawa, T., K. Takai, Y. Suzuki, H. Hirayama, U. Konno, U. Tsunogai, and K. Horikoshi (2006b), Geomicrobiological exploration and characterization of a novel deep-sea hydrothermal system at the TOTO caldera in the Mariana Volcanic Arc, *Environ. Microbiol.*, *8*, 37–49.

Nercessian, O., A.-L. Reysenbach, D. Prieur, and C. Jeanthon (2003), Archaeal diversity associated with *in situ* samplers deployed on hydrothermal vents on the East Pacific Rise (13°C), *Environ. Microbiol.*, *5*, 492–502.

Nercessian, O., N. Bienvenu, D. Moreira, D. Prieur, and C. Jeanthon (2005), Diversity of functional genes of methanogens, methanotrophs and sulfate reducers in deep-sea hydrothermal environments, *Environ. Microbiol.*, *7*, 118–132.

Newton, I. L. G., et al. (2007), The *Calyptogena magnifica* chemoautotrophic symbiont genome, *Science*, *315*, 998–1000.

Nicolaus, B., L. Lama, A. Panico, V. S. Moriello, I. Romano, and A. Gambacorta (2002), Production and characterization of exopolysaccharides excreted by thermophilic bacteria from shallow, marine hydrothermal vents of Flegrean Ares (Italy), *System. Appl. Microbiol.*, *25*, 319–325.

Nicolaus, B., V. S. Moriello, L. Lama, A. Poli, and A. Gambacorta (2004), Polysaccharides from extremophilic microorganisms, *Origins Life Evol. Biosphere*, *34*, 159–169.

Party, S. S. (1998), Investigation of hydrothermal circulation and genesis of massive sulfide deposits at sediment-covered spreading centers at Middle Valley and Escanaba Trough, pp. 7–16, Texas A&M University, Ocean Drilling Program, College Station.

Pledger, R. D., and J. A. Baross (1989), Characterization of an extremely thermophilic archaebacterium isolated from a black smoker polychaete (*Paralvinella* sp.) at the Juan de Fuca Ridge, *System. Appl. Microbiol.*, *12*, 249–256.

Pruis, M. J., and H. P. Johnson (2004), Tapping into the subseafloor: examining diffuse flow and temperature from an active seamount on the Juan de Fuca Ridge, *Earth Planet. Sci. Lett.*, *217*, 379–388.

Pysz, M. A., C. I. Montero, S. R. Chhabra, R. M. Kelly, and K. D. Rinker (2004), Significance of polysaccharides in microbial physiology and the ecology of hydrothermal vent environments, in *The Subseafloor Biosphere at Mid-Ocean Ridges*, *Geophys, Monogr. Ser.*, vol. 144, edited by W. S. D. Wilcock, et al., pp. 213–226, AGU, Washington, D.C.

Raguénès, G. H. C., P. Pignet, G. Gauthier, A. Peres, R. Christen, H. Rougeaux, G. Barbier, and J. Guezennec (1996), Description of a new polymer-secreting bacterium from a deep-sea hydrothermal vent, *Alteromonas macleodii* subsp. *fijiensis*, and preliminary characterization of the polymer, *Appl. Environ. Microbiol.*, *62*, 67–73.

Reysenbach, A.-L., K. Longnecker, and J. Kirshtein (2000), Novel bacterial and archaeal lineages from an in situ growth chamber deployed at a Mid-Atlantic Ridge hydrothermal vent, *Appl. Environ. Microbiol.*, *66*, 3798–3806.

Reysenbach, A.-L., D. Götz, A. Banta, C. Jeanthon, and Y. Fouquet (2002), Expanding the distribution of the *Aquificales* to the deep-sea vents on Mid-Atlantic Ridge and Central Indian Ridge, *Cah. Biol. Mar.*, *43*, 425–428.

Reysenbach, A.-L., Y. Liu, A. B. Banta, T. J. Beveridge, J. D. Kirshtein, S. Schouten, M. K. Tivey, K. L. Von Damm, and M. A. Voytek (2006), A ubiquitous thermoacidophilic archaeon from deep-sea hydrothermal vents, *Nature*, *442*, 444–447.

Rinker, K. D., and R. M. Kelly (1996), Growth physiology of the hyperthermophilic archaeon *Thermococcus litoralis*: Development of a sulfur-free defined medium, characterization of an exopolysaccharide, and evidence of biofilm formation, *Appl. Environ. Microbiol.*, *62*, 4478–4485.

Rinker, K. D., and R. M. Kelly (2000), Effect of carbon and nitrogen sources on growth dynamics and expolysaccharide production for the hyperthermophilic archaeon *Thermococcus litoralis* and bacterium *Thermotoga maritima*, *Biotechnol. Bioeng.*, *69*, 537–547.

Rona, P. A., and D. A. Trivett (1992), Discrete and diffuse heat transfer at ASHES vent field, *Earth Planet. Sci. Lett.*, *109*, 57–71.

Rudnicki, M. D., and H. Elderfield (1992), Theory applied to the Mid-Atlantic Ridge hydrothermal plumes: the finite difference approach, *J. Volcanol. Geotherm. Res.*, *50*, 163–174.

Rusch, D. B., et al. (2007), The Sorcerer II Global Ocean Sampling Expedition: Northwest Atlantic through Eastern Tropical Pacific, *PLoS Biol.*, *5*, e77.

Schrenk, M. O., D. S. Kelley, J. R. Delaney, and J. A. Baross (2003), Incidence and diversity of microorganisms within the walls of an active deep-sea sulfide chimney, *Appl. Environ. Microbiol.*, *69*, 3580–3592.

Schultz, A., J. R. Delaney, and R. E. McDuff (1992), On the partitioning of heat flux between diffuse and point source seafloor venting, *J. Geophys. Res.*, *97*(B9), 12,299–212,314.

Schultz, A., P. Dickson, and H. Elderfield (1996), Temporal variation in diffuse hydrothermal flow at TAG, *Geophys. Res. Lett.*, *23*, 3471–3474.

Shank, T. M., D. J. Fornari, K. L. Von Damm, M. D. Lilley, R. M. Haymon, and R. A. Lutz (1998), Temporal and spatial patterns of biological community development at nascent deep-sea hydrothermal vents (9 degrees 50′ N, East Pacific Rise), *Deep-Sea Res. II*, *45*, 465–515.

Sievert, S. M., E. B. A. Wieringa, C. O. Wirsen, and C. D. Taylor (2007), Growth and mechanism of filamentous-sulfur formation by *Candidatus* Arcobacter sulfidicus in opposing oxygen-sulfide gradients, *Environ. Microbiol.*, *9*, 271–276.

Sogin, M. L., H. G. Morrison, J. A. Huber, D. Mark Welch, S. M. Huse, P. R. Neal, J. M. Arrieta, and G. J. Herndl (2006), Microbial diversity in the deep sea and the underexplored "rare biosphere," *Proc. Natl. Acad. Sci. U.S.A.*, *103*, 12,115–112, 120.

Stetter, K. O., G. Lauerer, M. Thomm, and A. Neuner (1987), Isolation of extremely thermophilic sulfate reducers: Evidence for a novel branch of Archaebacteria, *Science*, *236*, 822–824.

Summit, M. (2000), Ecology, physiology, and phylogeny of subseafloor thermophiles from mid-ocean ridge environments, PhD thesis, 195 pp, University of Washington, Seattle.

Summit, M., and J. A. Baross (1998), Thermophilic subseafloor microorganisms from the 1996 North Gorda Eruption, *Deep-Sea Res. II*, *45*, 2751–2766.

Summit, M., and J. A. Baross (2001), A novel microbial habitat in the mid-ocean ridge subseafloor, *Proc. Natl. Acad. Sci. U.S.A.*, *98*, 2158–2163.

Summit, M., A. Peacock, D. Ringleberg, D. C. White, and J. A. Baross (2000), Phospholipid fatty acid-derived microbial biomass and community dynamics in hot, hydrothermally influenced sediments from Middle Valley, Juan de Fuca Ridge, in *Proc. Ocean Drill. Program, Sci. Results*, edited by R. A. Zierenberg, et al., pp. 1–19, Texas A&M University, Ocean Drilling Program, College Station.

Takai, K., and K. Horikoshi (1999), Genetic diversity of archaea in deep-sea hydrothermal vent environments, *Genetics*, *152*, 1285–1297.

Takai, K., A. Sugai, T. Itoh, and K. Horikoshi (2000), *Palaeococcus ferrophilus* gen. nov., sp. nov., a barophilic, hyperthermophilic archaeon from a deep sea hydrothermal vent chimney, *Int. J. Syst. Evol. Microbiol.*, *50*, 489–500.

Takai, K., T. Gamo, U. Tsunogai, N. Nakayama, H. Hirayama, K. H. Nealson, and K. Horikoshi (2004a), Geochemical and microbiological evidence for a hydrogen-based, hyperthermophilic subsurface lithoautotrophic microbial ecosystem (HyperSLiME) beneath an active deep-sea hydrothermal field, *Extremophiles*, *8*, 269–282.

Takai, K., K. H. Nealson, and K. Horikoshi (2004b), *Methanotorris formicicus* sp. nov., a novel extremely thermophilic, methane-producing archaeon isolated from a black smoker chimney in the Central Indian Ridge, *Int. J. Syst. Evol. Microbiol.*, *54*, 1095–1100.

Takai, K., et al. (2005), Enzymatic and genetic characterization of carbon and energy metabolisms by deep-sea hydrothermal chemolithautorophic isolates of *Epsilonproteobacteria*, *Appl. Environ. Microbiol.*, *71*, 7310–7320.

Taylor, C. D., and C. O. Wirsen (1997), Microbiology and ecology of filamentous sulfur formation, *Science*, *277*, 1483–1485.

Taylor, C. D., C. O. Wirsen, and F. Gaill (1999), Rapid microbial production of filamentous sulfur mats at hydrothermal vents, *Appl. Environ. Microbiol.*, *65*, 2253–2255.

Teske, A., K.-U. Hinrichs, V. Edgcomb, A. de Vera Gomiz, D. Kysela, S. P. Sylva, M. L. Sogin, and H. W. Jannasch (2002), Microbial diversity of hydrothermal sediments in the Guaymas Basin: Evidence for anaerobic methanotrophic communities, *Appl. Environ. Microbiol.*, *68*, 1994–2007.

Teske, A., A. Dhillon, and M. L. Sogin (2003), Genomic markers of ancient anaerobic microbial pathways: Sulfate reduction, methanogenesis, and methane oxidation, *Biol. Bull.*, *204*, 186–191.

Tivey, M. K., A. M. Bradley, T. M. Joyce, and D. Kadko (2002), Insights into tide-related variability at seafloor hydrothermal vents from time-series temperature measurements, *Earth Planet. Sci. Lett.*, *202*, 693–707.

Tringe, S. G., et al. (2005), Comparative metagenomics of microbial communities, *Science*, *308*, 554–557.

Trivett, D. A., and A. J. Williams III (1994), Effluent from diffuse hydrothermal venting 2. Measurement of plumes from diffuse hydrothermal vents at the southern Juan de Fuca Ridge, *J. Geophys. Res.*, *99*(C9), 18,417–418,432.

Valentine, D. L. (2007), Adaptations to energy stress dictate the ecology and evolution of the Archaea, *Nat. Rev. Microbiol.*, *5*, 316–323.

Veirs, S. R., R. E. McDuff, and F. R. Stahr (2006), Magnitude and variance of near-bottom horizontal heat flux at the Main Endeavour hydrothermal vent field, *Geochem. Geophys. Geosyst.*, *7*(Q02004).

Vincent, P., P. Pignet, F. Talmont, L. Bozzi, B. Fournet, J. Guezennec, C. Jeanthon, and D. Prieur (1994), Production and characterization of an exopolysaccharide excreted by a deep-sea hydrothermal vent bacterium isolated from the polychaete annelid *Alvinella pompejana*, *Appl. Environ. Microbiol.*, *60*, 4134–4141.

Von Damm, K. L., and M. D. Lilley (2004), Diffuse flow hydrothermal fluids from 9°50′ N East Pacific Rise: Origin, evolution, and biogeochemical controls, in *The Subseafloor Biosphere at Mid-Ocean Ridges*, *Geophys. Monogr. Ser.*, vol. 144, edited by W. S. D. Wilcock, et al., pp. 245–268, AGU, Washington, D. C.

Vorholt, J., J. Kunow, K. O. Stetter, and R. K. Thauer (1995), Enzymes and coenzymes of the carbon monoxide dehydrogenase pathway for autotrophic CO_2 fixation in *Archaeoglobus lithotrophicus* and the lack of carbon monoxide dehydrogenase in the heterotrophic *A. profundus*, *Arch. Microbiol.*, *163*, 112–118.

Weber, A., and B. B. Jorgensen (2002), Bacterial sulfate reduction in hydrothermal sediments of the Guaymas Basin, Gulf of California, Mexico, *Deep-Sea Res. Part I*, *49*, 827–841.

Winn, C. D., D. M. Karl, and G. J. Massoth (1986), Microorganisms in deep-sea hydrothermal plumes, *Nature*, *320*, 744–746.

Wirsen, C. O., S. M. Sievert, C. M. Cavanaugh, S. J. Molyneaux, A. Ahmad, L. T. Taylor, E. F. DeLong, and C. D. Taylor (2002), Characterization of an autotrophic sulfide-oxidizing marine *Arcobacter* sp. that produces filamentous sulfur, *Appl. Environ. Microbiol.*, *68*, 316–325.

Woese, C. R., O. Kandler, and M. L. Wheelis (1990), Towards a natural system of organisms: Proposal for the domains Archaea, Bacteria, and Eucarya, *Proc. Natl. Acad. Sci.*, *87*, 4576–4579.

Zillig, W., I. Holz, D. Janekovic, and W. Schäfer (1983), The archaebacterium *Thermococcus celer* represents a novel genus within the thermophilic branch of the archaebacteria, *System. Appl. Microbiol.*, *4*, 88–94.

J. A. Huber, Josephine Bay Paul Center, Marine Biological Laboratory, 7 MBL Street, Woods Hole, Massachusetts, 02543 USA. (jhuber@mbl.edu)

J. F. Holden, Department of Microbiology, 639 N. Pleasant St., University of Massachusetts, Amherst, Massachusetts, 01003 USA. (jholden@microbio.umass.edu)

Magma-To-Microbe Networks In The Context Of Sulfide Hosted Microbial Ecosystems

Matthew O. Schrenk

Geophysical Laboratory, Carnegie Institution of Washington, Washington, D. C., USA

James F. Holden

Department of Microbiology, University of Massachusetts, Amherst, Massachusetts, USA

John A. Baross

School of Oceanography, University of Washington, Seattle, Washington, USA

Hydrothermal vent sulfides are diverse and dynamic habitats for the growth and activities of microorganisms. Many well-known strains of thermophilic and hyperthermophilic microflora, and all of those capable of growth at the highest temperatures known to date have been obtained from vent chimney environments. The microbial diversity of chimney habitats mirrors this environmental complexity, spanning nearly the entire breadth of the archaeal and bacterial domains. Over the past 25 years, significant strides have been made in assessing the microbial diversity of chimney environments and measuring the relationships between microorganisms and their geologic settings. We review the microbiology of vent sulfide deposits in terms of habitat characteristics and suggest several avenues for future research to make progress in elucidating feedbacks between microbial ecology and geochemistry. However, this will ultimately require the development of reactive transport models that couple fluid flow, heat transfer, solute transport, geochemical reaction, and microbiological processes. The purpose of this review is to discuss microbial interactions with solids, fluids, and volatile phases within deep-sea hydrothermal sulfide deposits and provide a conceptual framework for future modeling efforts.

1. INTRODUCTION

Hydrothermal fluid circulation is widespread globally along the mid-ocean ridges and is driven by shallow to deep magmatic heat sources present beneath the ridge axis. Cold, oxic seawater is drawn through cracks and pores in the seafloor, reacts with hot rock and magma-derived compounds, and then exits the seafloor as hot, buoyant, metal- and volatile-rich fluids [*Alt*, 1995]. Upon contact with seawater, the dissolved metals and sulfide within the hydrothermal fluid precipitate forming the "smoke" that is characteristic of high-temperature deep-sea venting. Rapid heating of seawater, coupled with the high calcium and occasionally barium contents of the vent fluids, leads to the precipitation of sulfate minerals that form a porous barrier to mixing between

Magma to Microbe: Modeling Hydrothermal Processes at Ocean Spreading Centers
Geophysical Monograph Series 178

10.1029/178GM12

seawater and the hydrothermal fluids and serve as a nucleus for further mineral precipitation [*Goldfarb et al.*, 1983; *Haymon*, 1983]. This permeable shell is the beginning of what is known as a "chimney" or conduit for the emission of hot hydrothermal fluids into the deep sea. Eventually, parts of the interior and exterior of these structures become inhabited by microorganisms and macrofauna [*Juniper and Sarrazin*, 1995; *Sarrazin et al.*, 1999, 2002; *Pagé et al.*, 2004; *McCliment et al.*, 2006]. As high-temperature hydrothermal chimneys continue to grow and expand, they can develop less permeable barriers within the structures that affect fluid circulation pathways [*Haymon*, 1983; *Tivey*, 1995b; *Tivey et al.*, 1999] and incorporate organisms and/or remains of organisms growing on their exterior into hotter regions of the structure [*Tivey and Delaney*, 1986; *Cook and Stakes*, 1995; *Kristall et al.*, 2006].

Seafloor sulfide deposits are accessible natural laboratories for studying and quantifying the effects of magmatic and nonmagmatic processes on microbial communities and can be used to evaluate constraints (e.g., chemical disequilibria, temperature limits, and nutrient concentrations) upon microbial growth and habitability in deep-sea hydrothermal environments. They also serve as access points to the high-temperature environment above mid-ocean ridge spreading centers [*Deming and Baross*, 1993]. Understanding the microbe-mineral interactions within these deposits may be critical to evaluating the extent of habitable space within subseafloor environments and the impact of microorganisms within rock-hosted deep-sea ecosystems on global biogeochemical cycles [*Baross et al.*, 2004; *Edwards et al.*, 2005]. However, this will ultimately require the development of reactive transport models that couple fluid flow, heat transfer, solute transport, geochemical reaction, and microbiological processes [*Tivey*, 2004]. The purpose of this review is to discuss microbial interactions with solids, fluids, and volatile phases within deep-sea hydrothermal sulfide deposits, and to provide a conceptual framework for future modeling efforts.

2. DIVERSITY OF SULFIDE-HOSTED MICROBIAL HABITATS

2.1. Growth and Evolution of Sulfide Deposits on the Seafloor

Due to the reliance of chimney growth on a complex interplay of chemical and physical processes, the types of hydrothermal edifices that have been observed in the deep-sea are as diverse as the settings in which they are found. They range from the large, long-lived mounds of the slow-spreading Mid-Atlantic Ridge, to the small, ephemeral chimneys of the fast-spreading East Pacific Rise (EPR), and the tall, steep-sided structures observed on the Juan de Fuca Ridge [Figure 1; *Hannington et al.*, 1995; *Tivey*, 1995b; *Kelley et al.*, 2002]. Numerous physical, chemical, and possibly biological factors influence the ultimate structure and fate of a hydrothermal sulfide deposit and have been reviewed in several previous publications [*Hannington et al.*, 1995; *Tivey*, 1995b, 2004].

The development of hydrothermal chimney structures at deep-sea vents can be divided into three general stages: initial precipitation of a permeable, diffusive shell (stage I), mineralogical maturation and solidification (stage II), and senescence of the structure as venting ceases [*Goldfarb et al.*, 1983; *Haymon*, 1983; *Tivey*, 1995b; *Tivey et al.*, 1999]. Although a great deal of progress has been made in understanding the physical and chemical conditions that dictate the mineralogy of chimney structures [*Hannington and Scott*, 1988; *Tivey and McDuff*, 1990; *Hannington et al.*, 1995; *Tivey*, 1995a], significant gaps remain in our detailed knowledge of the processes involved in their growth and evolution into the diverse morphologies that have been observed. Factors that influence the morphology of these structures include the spatial extent and duration of hydrothermal discharge at a particular site as well as vent fluid chemistry, temperature, and flow velocity.

The growth and evolution of high-temperature (>300°C), "black smoker" chimney structures have been studied most extensively [*Goldfarb et al.*, 1983; *Haymon*, 1983; *Tivey*, 1995b]. The early stages of black smoker chimney growth are characterized by the emplacement of porous anhydrite ($CaSO_4$) walls that contain fine inclusions of sulfide minerals. Once in place, the anhydrite walls act to partially isolate hydrothermal fluid from seawater, resulting in stage II growth, which includes deposition of an inner lining of Cu-Fe sulfides, the precipitation of sulfides and anhydrite within pore spaces, and consequently a decrease in wall permeability.

A second type of high-temperature discharge results in the formation of "diffuser" or "beehive" chimneys, that have been observed at the Snakepit site of the Mid-Atlantic Ridge (MAR) and the Cleft Segment of the Juan de Fuca Ridge (JdFR). These structures form from similar temperature fluids (~350°C) to the black smoker chimneys, but lack a well-defined central conduit and exhibit a bulbous morphology. Their morphology is believed to be the consequence of slower, less-focused fluid flow [*Fouquet et al.*, 1993; *Koski et al.*, 1994]. Because fluids emanate from both the base and sides of these chimneys, the structures grow both vertically and laterally.

Many vent deposits are associated with lower temperature (100°–300°C) hydrothermal emissions [*Hannington et al.*, 1995]. The structures that result from this type of hydrothermal discharge lack major central conduits, and instead vent

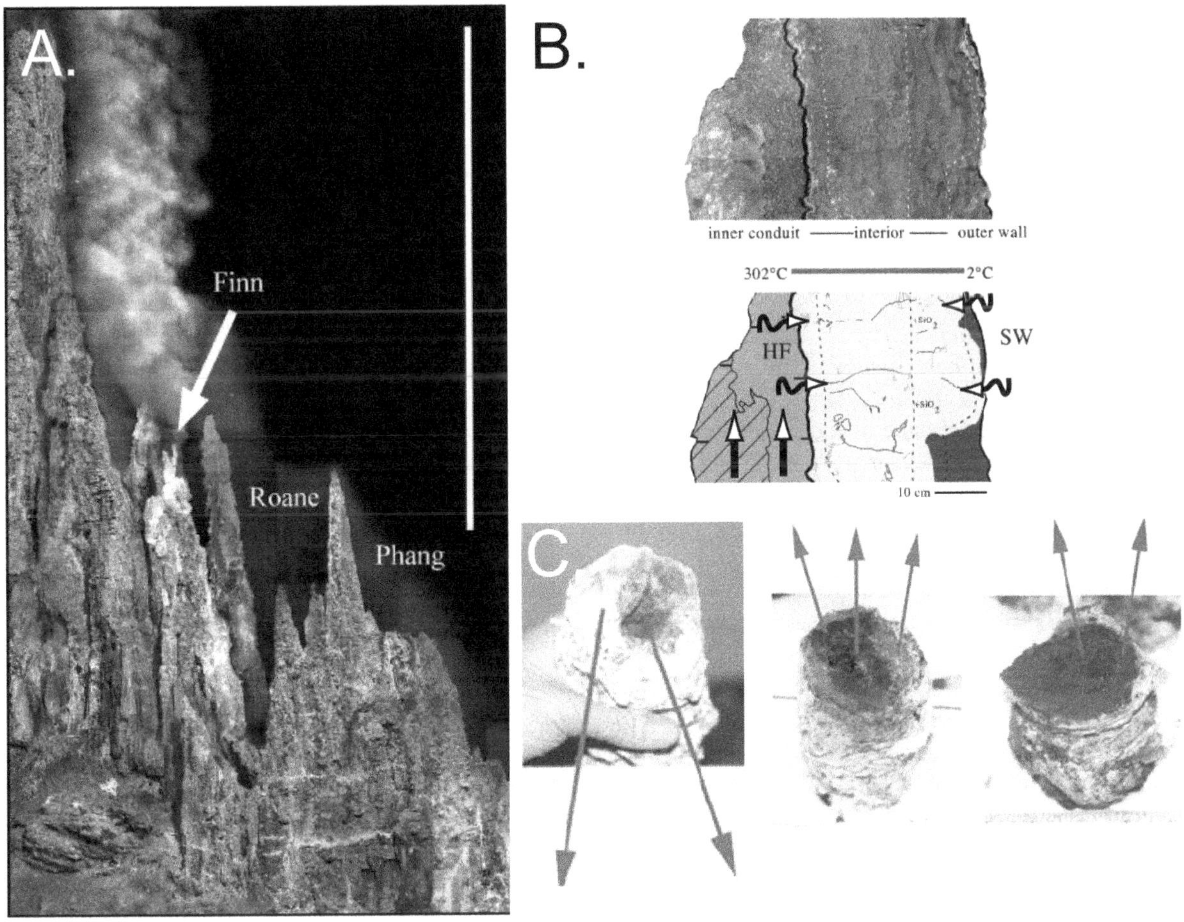

Figure 1. Examples of hydrothermal vent chimney habitats found at the deep-sea vents. (a) The Mothra Hydrothermal Vent Field from the Endeavour Segment of the JdFR contains numerous hydrothermal chimney structures in various stages of venting [*Kelley et al.*, 2001]. Three large chimneys were recovered from this vent field as part of the Edifice Rex Sulfide Recovery Project [*Delaney et al.*, 2001]. Scale bar is 13 m. (b) *Schrenk et al.* [2003] studied the microbial communities within the walls of a large black smoker chimney showing mineral zonation across the wall bound by >300°C hydrothermal fluids and seawater. (c) In contrast, a small (8–10 cm), white smoker chimney collected from the East Pacfic Rise, vented diffusely and did not show distinct mineral zonations [*Kormas et al.*, 2006]. Panels (a) and (b) were taken from *Schrenk et al.* [2003], reproduced with permission from *Applied and Environmental Microbiology*, American Society for Microbiology. Panel (c) was reproduced from *Kormas et al.* [2006], *Environmental Microbiology*, with permission from Blackwell Publishing Ltd., Oxford, UK.

from a series of interconnected channels. In many cases, sphalerite (Zn, Fe)S is the dominant matrix material, presumably as a result of lower temperatures. In other cases, barite- and silica-rich spires have been observed, especially along the JdFR. These structures form from a porous substrate of barite and amorphous silica, with subsequent precipitation of Zn and Cu sulfides [*Kristall et al.*, 2006].

Persistent hydrothermal activity at slow-spreading ridges over many hundreds of years can result in the coalescence of multiple hydrothermal deposits into large mounds such as those seen at the trans-Atlantic geotraverse (TAG) Hydrothermal Field of the MAR [*Hannington et al.*, 1995]. The sulfide mounds serve to restrict and sustain fluid flow pathways, reducing the loss of metals and directing the venting of the hydrothermal plume. However, access to the interior of these complex deposits is limited and requires the use of proxies such as vent fluid chemistry and relict hydrothermal sequences to understand subsurface structure and mineralogy [*Hannington et al.*, 1995].

A final stage in the history of a seafloor vent deposit is its eventual cessation of hydrothermal discharge and collapse. The time it takes to reach this stage can range from years to decades on the JdFR and up to centuries at the TAG site [*Hannington et al.*, 1995; *Lalou et al.*, 1995]. However, venting eventually wanes and the structures become subject to low-temperature weathering processes. These extinct structures continue to influence deep-sea microbiological activity as they represent a substantial long-term reservoir of reduced chemical energy that is a direct product of the hydrothermal venting [*Juniper and Tebo*, 1995; *Bach and Edwards*, 2003; *Edwards et al.*, 2003a].

2.2. Diversity of Potential Sulfide-Microbial Habitats: Thermal and Chemical Gradients

The most striking feature of the sulfide-microbial habitat is the stark gradient in temperature and chemistry that exists between internal and external portions of the vent structures. These gradients are most drastic within the walls of large, black smoker type chimneys where a central conduit venting >300°C fluids forms one boundary, and seawater bounds the exterior surface (Figure 1b). These gradients can exist at scales ranging from <1 cm to >0.5 m. Models of mixing between end-member hydrothermal fluids and seawater have been used to characterize the stability fields for mineral precipitation and dissolution across the walls of active chimney structures [*Tivey and McDuff*, 1990; *Tivey*, 1995a, 1995b]. These models demonstrate the sensitivity of thermal and chemical gradients within the chimney walls to characteristics of the environment, in particular to those which influence mass transport such as rock porosity and permeability [*Tivey*, 1995a, 1995b]. They predict that different mineral assemblages within the sulfide deposits are indicative of a gradient of temperature regimes with extensive portions sustained at high temperatures. The thermal and chemical gradients of black smoker walls are impacted by both advection and diffusive transfer of heat and mass.

Another type of chimney microbial habitat are hydrothermal flanges, ledges that protrude from the sides of active seafloor vent structures, and trap high-temperature (up to 375°C) fluids beneath them. Although flanges are not ubiquitous in all sites where chimneys occur, they play a substantial role in the growth and evolution of the sulfide structures at the Endeavour Segment vent fields along the JdFR [*Tivey and Delaney*, 1986]. Fluid flow from flanges occurs diffusely through the upper surfaces and sides, focused through small spires, and overflows from the sides of the trapped pool. Flanges are convenient systems to experimentally study heat and mass transfer due to the accessibility of end-member fluids, temperature measurements, and the mineralogical matrix. The thermal gradient across a flange has been modeled and demonstrates that both porosity and the thermal conductivity of the mineral assemblage can be significant variables in calculating the temperature profile [*Woods and Delaney*, 1992; *Tivey*, 1995b]. Based purely on the perspective of heat flux, the upper fifth (4 cm) of a 20-cm-thick flange venting 350°C fluids is within the habitable range for microbial life (<110°C; *Tivey*, 1995b). Future iterations of these models will require integrating chemical processes including the diffusion and advection of mass as well as reaction rates [*Tivey*, 1995b].

Diffuser chimneys lack the well-defined central conduit of the black smoker chimneys. They are defined by an axial zone of high porosity that favors outward percolation of hot fluids from all surfaces through a porous network of interconnecting channels [Figure 1c; *Fouquet et al.*, 1993; *Hannington et al.*, 1995; *Kormas et al.*, 2006]. Assuming a network of sulfide minerals and pore spaces, *Kormas et al.* [2006] estimated a temperature profile of 60–240°C in exterior portions of a diffuser structure from 9°N EPR and 240°–350°C in its internal core (Figure 1c).

2.3. The Chimney Microniche: Theory Versus Observations

The thermal and chemical gradients, and mineralogical framework found within sulfide deposits are predicted to be the important determinants for the microbial ecology of such structures [*McCollom and Shock*, 1997; *Schrenk et al.*, 2003; *Shock and Holland*, 2004]. Temperature exerts a key, first-order control on microbial populations as cell physiology and biochemistry is adapted to specific temperature ranges. Microbial communities are known to inhabit discrete thermal conditions over the range of ~2° to 120°C, and members of each thermal class have been cultured from the deep-sea vent environment [*Holland and Baross*, 2003]. Although there is nothing, a priori, that prohibits microbial growth above this range, microbial growth and activities have not been conclusively demonstrated at temperatures above 121°C [*Kashefi and Lovley*, 2003; *Holden and Daniel*, 2004].

Temperature, in combination with chemical composition, also influences the free energy available for biosynthesis [*Shock and Holland*, 2004]. The interface between hot, reducing hydrothermal fluids and cold, oxic seawater sustains chemical disequilibria that provide metabolic energy for the organisms contained within the chimney walls. Models that consider batch mixing between seawater and vent fluid examined the Gibbs free energy available at different mixing ratios between end-member hydrothermal fluids and seawater and predict a successive gradient of microbial metabolisms as a function of temperature [*McCollom and Shock*, 1997]. Specifically, they predict that oxidative metabolisms such as Fe, Mn, CH_4, or S oxidation are favored at lower, mesophilic temperatures (<38°C), whereas anaerobic metabolisms such as methanogenesis and SO_4 or S° reduction are favored at higher temperatures [*McCollom and Shock*, 1997]. Although this model serves as a useful starting point to investigate potential bioenergetic processes, iterations on the batch mixing approach have refined our understanding of the subtleties involved in such calculations.

A more recent study considered the effects of variations in fluid composition, diffusion, and advective transport upon the environmental conditions within vent chimney walls [*Tivey*, 2004]. It was found that the boundary between oxic and anoxic conditions can vary from between 3° and 92°C depending on the inputs considered. Furthermore, most habitats within the chimney wall matrix are likely to be acidic (pH = 3–6), which stands in stark contrast to the physiology of a majority of microorganisms that have been isolated from vent chimney habitats [*Tivey*, 2004]. *Shock and Holland* [2004] modified the original *McCollom and Shock* [1997] model to include hydrogen oxidation. As a result, the previously observed sharp distinction between high-temperature, anaerobic metabolisms and low-temperature, aerobic metabolisms was far less obvious. Oxidative and reductive metabolic strategies were predicted at both 25° and 100°C. These predictions greatly expand the range of metabolisms possible within chimney microhabitats.

For all of these models, organic compounds may also be present as another source of energy as a product of autotrophic microbial metabolism, as a result of thermal degradation of organic material from the surface of the sulfide, or from abiotic synthesis that would then be available for heterotrophic metabolism [*Karl*, 1995; *Shock and Schulte*, 1998]. Indeed, the organic carbon concentration of black smokers can be up to 1% (wt%), although it is typically considerably lower [*Wirsen et al.*, 1993], and the near-bottom seawater around hydrothermal discharge is enriched in both particulate and dissolved organic carbon [*Corliss et al.*, 1981; *Karl*, 1995]. A significant implication of these models is that chemical disequilibria, and thus free energy for microbial growth, are available throughout the entire span of the chimney wall, even at temperatures up to 350°C [*McCollom and Shock*, 1997].

Bioenergetic models agree reasonably well with observations from hydrothermal environments but fail to consider several phenomena known to exist in natural microbial ecosystems. Feedback from rapid mineral precipitation and dissolution, and microbial processes are difficult to consider without data on reaction rates [*Tivey*, 1995a, 1995b, 2004; *McCollom and Shock*, 1997]. In addition, these models do not consider microbial processes, such as the formation of consortial relationships between closely spaced microbial cells that can collectively overcome certain exergonic metabolic barriers in a process known as syntrophy [*Orphan et al.*, 2001]. The availability and chemical speciation of both major (N, P) and minor (e.g., Ni, Mo, W) nutrients are also infrequently addressed [*Mehta and Baross*, 2006]. Furthermore, new isolates from the deep-sea vents can oxidize solid phase iron at low temperatures [*Edwards et al.*, 2003b], and others can reduce insoluble iron at temperatures up to 121°C [*Kashefi and Lovley*, 2003]. Microbial interactions with solid phases, either in their immobilization or their precipitation, may influence mass transport through the chimney walls. Because many of the organisms grow as biofilms (see section 3.3 and Figure 2), the exopolymers may also have a significant impact on porosity and thus diffusion and/or advection within the chimney matrix [*LaPaglia and Hartzell*, 1997; *Stewart*, 2003; *Pysz et al.*, 2004b].

Modeling the conditions found within sulfide deposits is further complicated by the extremely dynamic character of the environment. Over its lifespan, a given sulfide deposit is likely to experience a range of temperature and chemical conditions and be affected by short time-scale periodic processes (e.g., from tidal effects) to longer time-scale, episodic perturbations such as earthquake activity, fractures opening and sealing [*Schultz et al.*, 1996; *Delaney*, 1998; *Dziak and Johnson*, 2002; *Tivey et al.*, 2002]. In situ measurements within a hydrothermal vent structure on the Cleft Segment (JdFR) documented periodic perturbations in temperature of ~2°C within a sulfide structure related to tidally driven bottom currents that varied the temperature on the exterior of the deposit, whereas the highest temperatures were relatively stable and had no tidal signal at all [*Tivey et al.*, 2002]. Detailed assessment of large chimney structures recovered from the JdFR also suggests the episodic influx of seawater into the chimney walls along fractures that is then conductively heated to precipitate anhydrite [*Schrenk et al.*, 2003].

At a larger scale, using time-series observations, *Von Damm et al.* [1995a, 1995b] found abrupt changes (on a time scale of only 1 week) in the temperature and chemistry (increased H_2S content, low chlorinity) of hydrothermal fluids from the EPR (9°N) immediately after a volcanic eruption. Abrupt changes in the chemistry and volatile content of hydrothermal fluids from the Endeavour Segment of the JdFR were also observed [*Lilley et al.*, 2003]. The authors of these studies conclude that these changes were the result of volcanic processes in the subseafloor.

Efforts to monitor temperature and chemistry in high resolution are beginning to be coordinated with measurements of microbial growth and activity and are making progress in delineating habitat characteristics within vent deposits. More detailed studies such as coupling in situ measurements with mineralogical and petrographic analyses are necessary to further model the characteristics and the scale of biogeochemical processes occurring within the walls of active hydrothermal chimney structures (Figure 3).

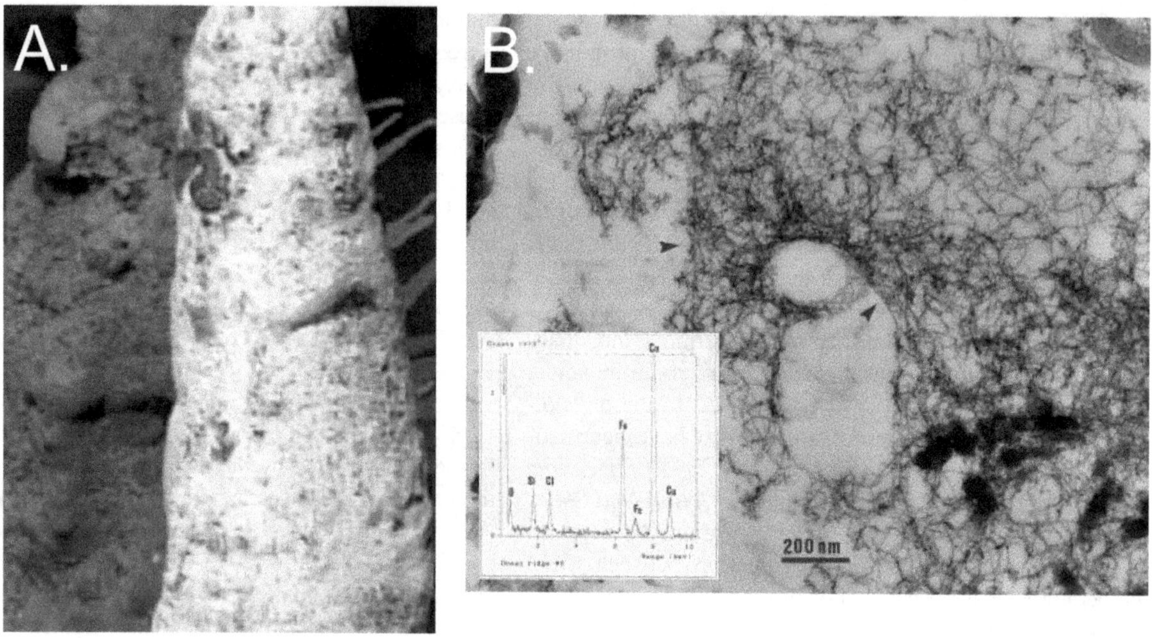

Figure 2. Microbial processes may influence mass transport in chimney habitats. (a) Filamentous bacterial "streamers" coat the exterior surface of a diffusely venting sulfide spire from the Central Indian Ridge [*Hoek et al.*, 2003]. (b) At a microscale, bacterial extracellular polymers may serve as nuclei for nanocrystalline mineral precipitation [*Fortin et al.*, 1998]. Energy dispersive spectroscopic analyses indicate that the extracellular polymers are extensively coated in fine iron oxide filaments (inset). Panel (a) was reproduced with permission from *Geobiology*, Blackwell Publishing Ltd., Oxford, UK, and panel (b) from *American Mineralogist*, Mineralogical Society of America, USA.

3. THE MICROBIOLOGY OF HYDROTHERMAL SULFIDE DEPOSITS

3.1. Forms of Microbial Metabolism

Sulfide deposits host microbial communities that use many different metabolic strategies (e.g., autotrophs and heterotrophs, aerobes and anaerobes), some of which have been obtained in laboratory cultures. Although these isolates represent only a small portion of the total microbial diversity present within sulfide deposits, they provide insight into adaptations to vent sulfide environments. Although many studies have examined the metabolic capabilities of mat-forming and endosymbiotic bacteria found on the outer surfaces of sulfide deposits [*Jannasch and Wirsen*, 1981; *Taylor et al.*, 1999], and those associated with low-temperature sulfide weathering [*Eberhard et al.*, 1995; *Rogers et al.*, 2003], this discussion will be limited to those microorganisms found within the interior of the deposits. Microbial metabolisms are differentiated on the basis of carbon source, electron donors and terminal electron acceptors, temperature, and chemical conditions such as pH and reduction potential.

The ability of some microorganisms in sulfide deposits to grow chemolithoautotrophically (i.e., "chemosynthetically") with CO_2 as the sole carbon source in the absence of sunlight and photosynthetically derived O_2 represents an avenue of long-standing interest in the study of life in deep-sea hydrothermal environments. The biochemical pathways of CO_2 assimilation often vary from those found in photosynthetic environments and from the symbiotic and mat-forming bacteria in vent environments and have important implications with regards to the natural history of autotrophy. Methanogens, which are all Archaea, combine H_2 and CO_2 to form CH_4, and are ubiquitous in sulfide deposits [*Harmsen et al.*, 1997a; *Takai and Horikoshi*, 1999; *Nercessian et al.*, 2003; *Schrenk et al.*, 2003]. Strains of methanogens isolated from deep-sea sulfide deposits include *Methanocaldococcus jannaschii* and *Methanopyrus kandleri* with growth at temperatures up to 92° and 110°C, respectively [*Jones et al.*, 1983; *Burggraf et al.*, 1991; *Mehta and Baross*, 2006]. These organisms assimilate CO_2 using an archaeal version of the acetyl-CoA pathway and several unique cofactors [*Ferry*, 1999]. Similarly, members of the Archaeoglobaceae (also Archaea) are commonly detected in sulfides [*Takai and Horikoshi*, 1999; *Reysenbach and Cady*, 2001] and the autotrophs within this family use a pathway similar to the methanogens for CO_2 assimilation but lack the final steps necessary to make CH_4 [*Vorholt et al.*, 1997].

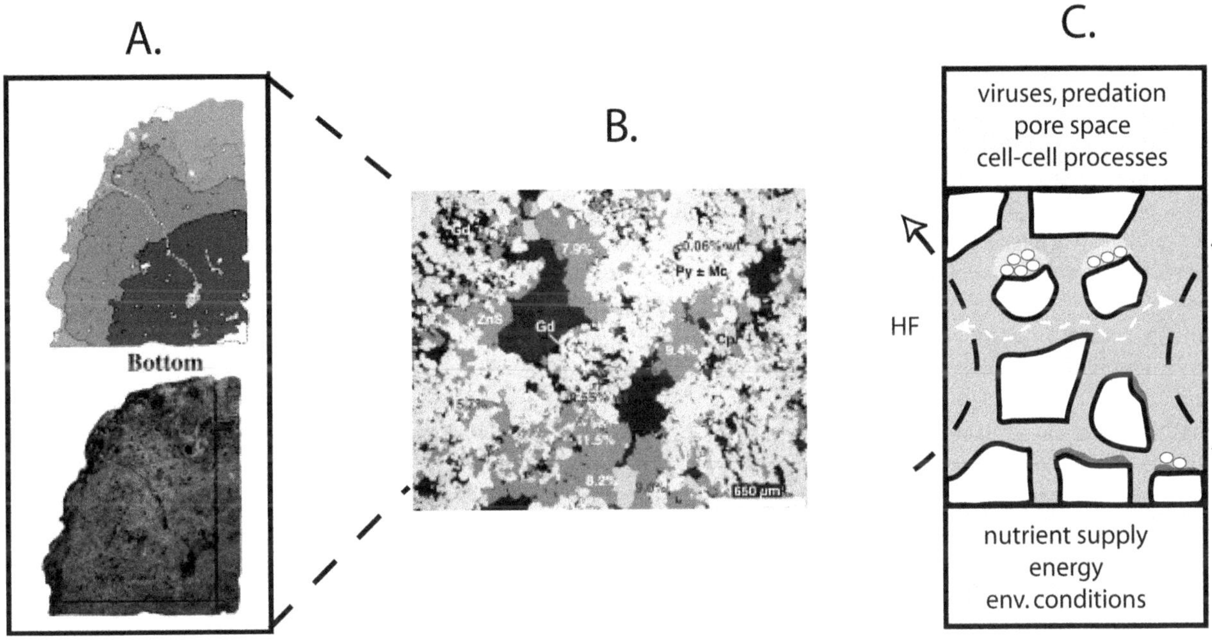

Figure 3. Detailed, nested studies can describe the sulfide-microbial habitat on the scale of vent fields, individual structures (a), and petrologic thin sections (b). Panel (a) is a horizontal cross section through the walls of a large sulfide chimney structure from the JdFR and corresponding mineralogical map. Panel (b) is a representative thin section from one of these mineralogical regions. However, consideration of microbial processes must take into account heterogeneity at the submillimeter scale (c), as well as time-integrated adaptive responses. Some of these considerations include "top-down" and "bottom-up" controlling factors, mineral surface topography and red-ox chemistry (dark gray), and the formation of exopolymeric biofilms (shown in light gray). Panels (a) and (b) were reproduced with permission from *Kristall et al.* [2006], AGU.

Autotrophic members of the Crenarchaeota that have been isolated from deep-sea hydrothermal sulfide material include *Ignicoccus pacificus* and *Pyrolobus fumarii* with maximum growth temperatures of 98° and 113°C, respectively [*Blöchl et al.*, 1997; *Huber et al.*, 2000]. *Pyrodictium abyssi* is a mixotroph that requires H_2 and CO_2 for growth as well as a source of organic carbon such as acetate or yeast extract [*Pley et al.*, 1991]. A survey of key CO_2 assimilation enzyme activities in *I. pacificus* and *P. abyssi* was unable to demonstrate the existence of any of the canonical CO_2 assimilation pathways, namely, the Calvin cycle, the reductive citric acid cycle, the acetyl-CoA pathway, and the 3-hydroxypropionate cycle [*Hügler et al.*, 2003]. Interestingly, low levels of ribulose-1,5-bisphosphate carboxylase activity (a key enzyme of the Calvin cycle) were measured in *P. abyssi* cell extracts, but the activity of another key enzyme of the cycle (ribulose-5-phosphate kinase) was not detected. Therefore, the CO_2 assimilation pathways in these Crenarchaeota are unknown and potentially utilize a novel autotrophic growth mechanism. Mesophilic and moderately thermophilic bacteria belonging to the ε-proteobacteria are also commonly detected in vent deposit material [*Campbell et al.*, 2006]. Two strains isolated from deep-sea vents were shown to use the reductive citric acid cycle for CO_2 assimilation and lacked the key enzyme activities of other CO_2 assimilation pathways [*Hügler et al.*, 2005].

Hyperthermophilic heterotrophs belonging to the Thermococcaceae are also commonly detected in vent deposit material [*Harmsen et al.*, 1997a; *Reysenbach et al.*, 2000; *Holden et al.*, 2001; *Takai et al.*, 2001; *Nercessian et al.*, 2003; *Schrenk et al.*, 2003]. The Thermococcaceae use either peptides or carbohydrates as their source of carbon and electrons [*Adams*, 1999], although the ability of *Pyrococcus furiosus* to use peptides required the addition of elemental sulfur [*Adams et al.*, 2001]. Heterotrophic metabolism at 95°C was readily measured in five of six small (<0.25 m diameter), newly formed black smokers from the Endeavour Segment but not in "sulfide microbial incubator" samples collected from a 9-month deployment in the interior of a massive sulfide deposit from the same site [*Kelley et al.*, 2004; J. F. Holden, unpublished data]. In contrast, H_2 consumption and chemolithoautotrophic growth was high at 95°C from the same incubator samples but not detected over a 3-day incubation period in the six black smoker samples. These data suggest that heterotroph activity is occurring near the source of macrofauna-derived organic compounds, whereas autotrophic activity is higher in the interior region of large sulfide deposits (J. F. Holden, unpublished data).

The availability of certain electron donors and acceptors is another key determinant of the kinds of organisms found within sulfide deposits. As noted above, CO_2 serves as the terminal electron acceptor for methanogens. Other electron acceptors used by organisms cultured from sulfide deposits include various sulfur species (e.g., SO_4^{2-}, $S_2O_3^{2-}$, S°), insoluble iron (III) oxyhydroxides, nitrate, and oxygen [*Jannasch*, 1995; *Holland and Baross*, 2003]. In marine sediments having a red-ox gradient, there is a spatial separation of microbial metabolisms coupled to these electron acceptors because the energy of various reactions is determined by the physical and chemical characteristics of different microniches. For example, although methanogens and sulfate reducers can occupy the same environments and grow commensurately, methanogens and dissimilatory iron reducers are generally segregated on the basis of the conditions present [*Lovely and Goodwin*, 1988; *Lovley et al.*, 1994]. Methanogens are not only very strict anaerobes but also require highly reduced conditions (<–330 mV) and higher H_2 concentrations for growth, whereas iron reducers tend to inhabit more oxidized environments and can grow with lower concentrations of H_2. In contrast to diffusion-dominated aquatic sediments, both advection and diffusion play a role in shaping the chemical environment of vent deposits. Energy-rich interfaces between hydrothermal fluids and seawater can be sustained and reaction byproducts removed, thereby enhancing the energetic potential of the microbial environment. It is important, however, to constrain both the energy flux through these dynamic systems, as well as quantifying the sources and sinks of reactants and products [*Holland and Baross*, 2003].

Chemical conditions may dictate which type of terminal electron acceptor process occurs in an organism capable of using various electron acceptors. For example, in the terrestrial hyperthermophiles *Pyrobaculum aerophilum* and *Pyrobaculum islandicum*, growth on soluble and insoluble forms of sulfur are preferred at lower reduction potentials and pHs, nitrate reduction is preferred at more neutral pHs and more oxidized conditions, and the reduction of soluble and insoluble forms of iron are favored at higher pH levels (pH = 8–9) and more oxidized conditions [*Feinberg et al.*, 2008]. Therefore, determining the spatial distribution and levels of specific metabolisms in sulfide deposits may provide an in situ tracer of the chemical conditions of that parcel of material.

Most of the microorganisms that have been cultured from hydrothermal sulfide deposits are neutrophiles (i.e., optimal growth near neutral pH levels). However, especially at higher growth temperatures, the prediction is that the pH of fluids within a sulfide deposit is more likely to be between pH 3 and 6 [*Tivey*, 2004]. The hyperthermophiles *P. fumarii* and *P. abyssi*, which were isolated from sulfide material, grow optimally at pH 5.5 [*Pley et al.*, 1991; *Blöchl et al.*, 1997]. The recently described *Aciduliprofundum boonei*, isolated from sulfide material, is a moderately thermophilic heterotroph capable of sulfur and iron reduction [*Reysenbach et al.*, 2006]. *A. boonei* and its close relatives are ubiquitous in hydrothermal systems based on environmental 16S ribosomal RNA (rRNA) sequence analyses, and the isolation of this organism elucidated the microbiology of one of the many common "unknown" sequences obtained from the vent environment (Table 5). Furthermore, *A. boonei* can grow between pH 3.3 and 5.8 with an optimum growth pH of 4.5, which validates earlier model predictions of acidic microhabitats within vent chimneys [*Tivey*, 2004]. The geologic context of microbial communities within sulfide deposits may help to constrain the conditions supporting microbial growth and activity (Figure 8).

3.2. Abundance and Diversity of Microorganisms Within Vent Sulfide Deposits

Although a systematic study of microbial populations within vent chimneys (akin to biogeographic studies of vent macrofauna) has yet to be compiled, the repeated observation of specific groups of microorganisms and their biomolecular signatures in deposits of different age and environmental settings provides a framework for future directed studies.

A number of studies have investigated the distribution of microbial populations within sulfide edifices across these thermal and chemical gradients [*Hedrick et al.*, 1992; *Harmsen et al.*, 1997a; *Takai et al.*, 2001; *Schrenk et al.*, 2003; *Kormas et al.*, 2006]. One of the first studies that examined changes in microbial populations within a sulfide deposit analyzed a large sulfide flange from the JdFR using intact membrane lipids as a proxy for microbial abundances and diversity. The flange was bounded by ~350°C fluid on its underside, which percolated up through the porous rock matrix to mix with seawater near the top of the structure [*Hedrick et al.*, 1992]. The lipid concentrations were highest in middle portions of the structure, but intact lipid molecules were recovered even from hot, internal portions of the flange. There was also a shift in the composition of the lipids from ester-linked (bacterial) to ether-linked (archaeal) lipids towards the warmer regions of the sulfide (Table 1). This study represented the first instance where microbial populations were placed within a physical-chemical context in situ at the deep-sea vents.

Culture-dependent and culture-independent methods were used to quantify and characterize the distribution of microorganisms within diffusive beehive hydrothermal chimneys

Table 1. Quantitative Assessments of Microbial Distributions Within Seafloor Sulfide Deposits

Sample Type	Fluid T, °C	Size, cm	Vent Site[a]	Horizon	Cells, g^{-1}	Comments	Arch	Bac	Ref.[b]
Sulfide flange							*lipids, µg/gdw*		
	350	several	JdFR	exterior		highest density	nd	4.50	1
				mid-exterior		on exterior	6.30	0.37	
				mid-interior		peak in middle	0.12	0.04	
				interior			0.17	0.01	
Very large chimney							*FISH[c], %*		
Wall	302	12–20	JdFR	exterior	1.9×10^6 (d)	highest density	33	53	2
				mid-exterior	1.7×10^8 (d)	on exterior	65	26	
				mid-interior	1.9×10^7 (d)	peak in middle	63	2	
				interior	2.1×10^5 (d)		23	0	
Large chimney							*qPCR/dot blot (%)*		3
Wall	>250	<10	MB	exterior orifice	3.2×10^7 (w)	highest density	30.5/5.3	67/95	
				interior-orifice	1.0×10^6 (w)	on exterior	20/7.4	78/93	
				exterior-wall	4.5×10^5 (w)		1.8/nd	98/100	
				mid-exterior	5.1×10^5 (w)		23.1/7.2	77/93	
				mid-interior	8.0×10^5 (w)		2.2/nd	95/100	
				vent conduit	4.5×10^5 (w)		2.8/2.8	99/97	
Small chimney									4
Wall	250	few cm	CIR	exterior	8.9×10^7 (w)	predominantly Eubacteria			
				interior	1.0×10^5 (w)	predominantly Archaea			
Beehiver diffuser chimney							*FISH[c], %*		5
Wall	<350	few cm	MAR	orifice	10^7 (d)	beehive diffuser	~50	~50	
				bottom	10^5 (d)		~50	~50	
				exterior	10^6 (d)		>60	<40	
				middle	10^5 (d)		<20	>80	
				Interior	10^5 (d)		<1	>99	
Small chimneys		few cm					*Clone library, %*		6
	300		SS	whole	10^4 (w)		3	97	
	~100		MK	whole	10^7 (w)		7	83	
	~180		MK	whole	10^7 (w)		4	96	
Small chimneys		few cm							7
Wall	<350		MAR	exterior	10^{10} (d)	highest density			
	<350		EPR	exterior	10^9 (d)	on exterior			
				middle	10^5 (d)				
				interior	Nd				
Large chimney							*Lipids µg/gdw*		1
Wall	>2	several	JdFR	exterior		predominantly Eubacteria	4.20	45.00	
				exterior			7.20	98.00	

[a]Vent sites: CIR, Central Indian Ridge; JdFR, Juan de Fuca Ridge; MAR, Mid-Atlantic Ridge; MB, Manus Basin; MK, Myojin Knoll; SS, Suiyo Seamount; EPR, East Pacific Rise.
[b]References: 1, *Hedrick* [1992]; 2, *Schrenk* [2003]; 3, *Takai* [2001]; 4, *Takai* [2004]; 5, *Harmsen* [1997b]; 6, Takai [1999]; 7, Chevaldonne [1997].
FISH[c], fluorescence in situ hybridization; nd, no data available (d), dry weight; (w), wet weight.

from the MAR [*Harmsen et al.*, 1997a, 1997b]. In this study, cell abundances were only slightly elevated in exterior portions of the structures relative to the interior, and both Eubacteria and Archaea contributed significantly to the microbial population structure [*Harmsen et al.*, 1997a, 1997b]. Molecular phylogenetic studies of black smoker chimneys from the Manus Basin and Okinawa Trough showed that there was substantial diversity represented by uncultivated phylotypes which included the deep-sea hydrothermal vent euryarchaeal group (DHVE) and other deeply branching lineages of Archaea [Figure 4b; *Takai and Horikoshi*, 1999; *Takai et al.*, 2001].

The microbial habitats of a large (~1 m diameter) black smoker chimney recovered from the Endeavour Segment of the JdFR were dissected using coregistered microbiological and mineralogical approaches [*Schrenk et al.*, 2003]. Cell concentrations ranged from extremely high (>10^9 cells/g) in mid-wall regions to <10^5 in hot internal regions (Table 1). In this study, the authors assumed that because the smoker was actively venting high-temperature fluid when recovered, the mineralogical record of the rock matrix could be used to infer the thermal and chemical conditions of the chimney wall [*Schrenk et al.*, 2003]. The relative percentage of Archaea within the populations increased from the exterior to the interior of the structure. Coupled phylogenetic analyses of 16S rRNA revealed that microbial diversity within the structure also varied. Archaeal phylotypes in external portions of the chimney resembled sequences from pelagic marine or deep-sea environments. Middle regions had the highest overall diversity and reflected contributions from both cooler, aerobic deep-sea sources and hydrothermally derived constituents. The innermost portions of the sulfide structure had predominantly uncultured phylotypes of Crenarchaeota, whose closest relatives are found in deep subsurface environments and other high-temperature environments [Table 2 and Figure 4b; *Schrenk et al.*, 2003].

Similar zonations were observed in the analysis of a small, ~250°C chimney from the Central Indian Ridge, in which *Thermococcus* and *Methanocaldococcus* species were abundant in internal (high-temperature) regions, and ε-proteobacteria were a significant component of the microbial community in external regions [Table 2 and Figure 4a; *Takai et al.*, 2004]. The authors of this study suggest that the predominance of methanogenic Archaea in this sample is related to the volatile-rich hydrothermal emissions associated with this site [*Takai et al., 2004*]. *Kormas et al.* [2006] also found microbiological differences between internal and external regions during subsampling of a diffuser chimney from the EPR. 16S rRNA sequences related to hyperthermophilic phylotypes within the Thermococcales and Archaeoglobales were found in internal portions of the vent structure, whereas bacterial sequences related to the ε-proteobacteria were most commonly found in the external regions (Table 2 and Figure 4). The authors point out the important artifacts that can be introduced during sample collection due to volume constriction of pore fluids, which can entrain seawater into the chimney walls and complicate phylogenetic interpretations [*Kormas et al.*, 2006].

Some conclusions can be made from these studies. Cell abundances and microbial diversity are commonly highest in middle regions of the sulfide edifices (Tables 1 and 2). Evidence of intact cells and cellular materials (e.g., membrane lipids) are evident in high-temperature internal portions of the chimney structures [*Takai et al.*, 2001; *Schrenk et al.*, 2003; *Kormas et al.*, 2006]. Whether this reflects an imprecise measurement of environmental conditions, a sampling artifact, or an incomplete understanding of microbial physiology remains to be addressed. From a microbiological perspective, these results emphasize the need for careful assessment of mineralogical context in addition to coordinated thermal and chemical measurements.

Comparison of sulfide-hosted microbial communities within different growth stages of hydrothermal chimneys reflects the temporal heterogeneity of microbial habitats within vent deposits. Early stages of chimney growth (stage I) as examined by *McCliment et al.* [2006] contain a relatively low diversity of largely autotrophic phylotypes. The "protochimneys" contained high proportions of Crenarchaeota related to the genus *Ignicoccus* as well as sequences related to its episymbiont *Nanoarchaeum equitans* [*McCliment et al.*, 2006]. Two distinct clades of Nanoarchaeota were noted in nascent versus mature chimney materials (Figure 5). Microbiological succession during the transition between stage I and stage II growth have not yet been examined in any great detail. However, a common observation is increased proportions of heterotrophic and fermentative organisms and

Figure 4. Phylogenetic diversity of microorganisms from deep-sea vent deposits. A substantial range of small subunit ribosomal RNA (signature) genes have been obtained from seafloor sulfide deposits, spanning the breadth of nearly the entire (a) eubacterial and (b) archaeal domains. Solid lines represent groups found in studies of deep-sea vent chimneys; dashed lines represent groups that have not yet been found. Trees were generated in the ARB software environment [*Ludwig et al.*, 2004] by pruning a bootstrapped maximum likelihood tree available from the Greengenes database [*DeSantis et al.*, 2006] while maintaining the tree topology. Scale bars represent the expected number of changes per nucleotide position.

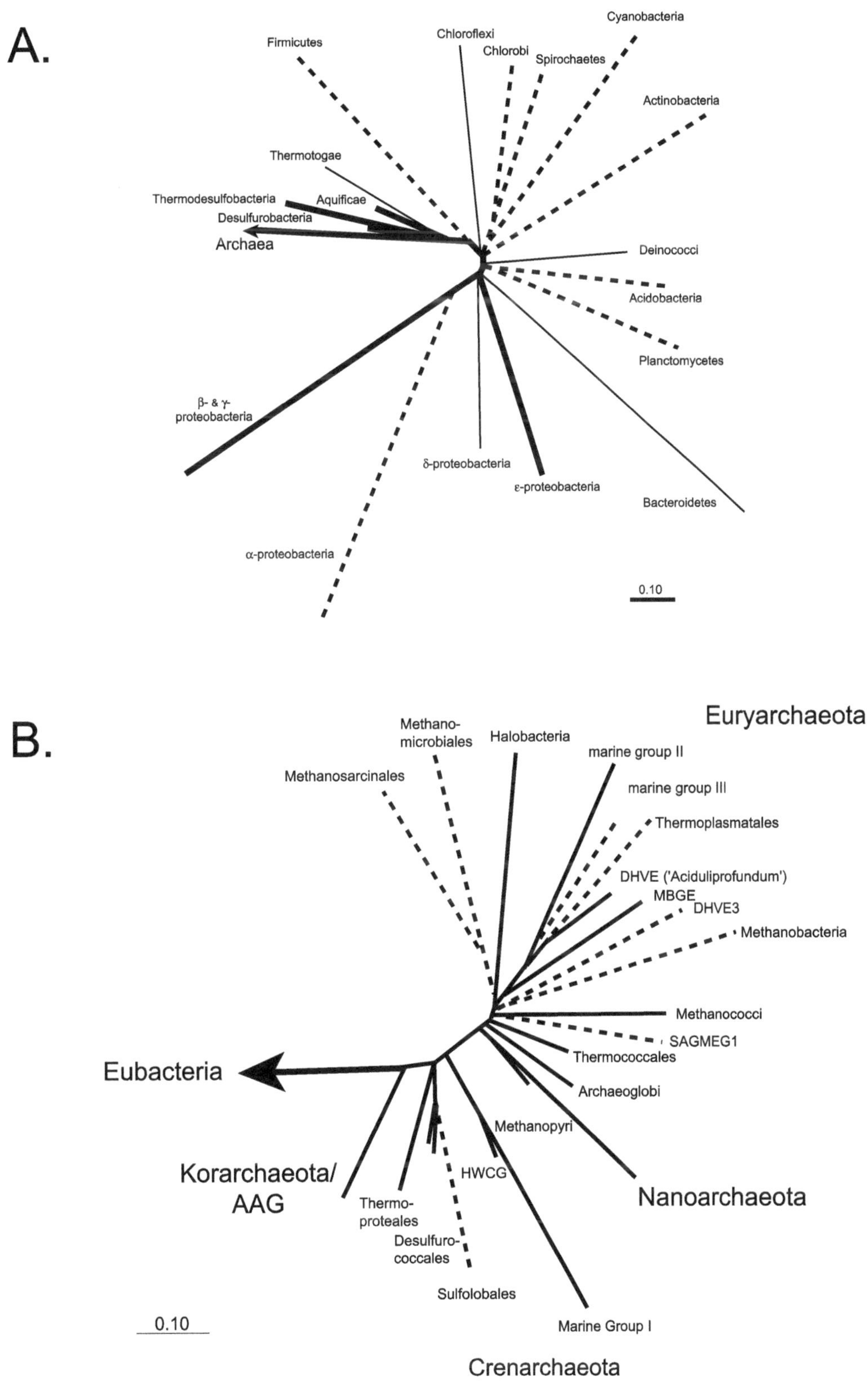
A.
Firmicutes
Chloroflexi
Chlorobi
Spirochaetes
Cyanobacteria
Actinobacteria
Thermotogae
Thermodesulfobacteria
Aquificae
Desulfurobacteria
Archaea
Deinococci
Acidobacteria
Planctomycetes
β- & γ-
proteobacteria
δ-proteobacteria
ε-proteobacteria
Bacteroidetes
α-proteobacteria
0.10
B.
Euryarchaeota
Methano-
microbiales
Halobacteria
marine group II
Methanosarcinales
marine group III
Thermoplasmatales
DHVE ('Aciduliprofundum')
MBGE
DHVE3
Methanobacteria
Methanococci
SAGMEG1
Thermococcales
Eubacteria
Archaeoglobi
Methanopyri
Korarchaeota/
AAG
HWCG
Nanoarchaeota
Thermo-
proteales
Desulfuro-
coccales
Sulfolobales
0.10
Marine Group I
Crenarchaeota

Table 2. Diversity of Microbial Populations Within Seafloor Sulfide Deposits

Type	Temp, °C	Site[a]	Horizon	Archaea	Eubacteria	Comments	Ref.[b]
Flange	350	JdFR				ether lipids increase, PLFA decrease towards interior	1
Wall			exterior			mostly PLFA	1
Wall	302	JdFR					2
			exterior	MGI Cren.; DHVE	γ-, ε-proteobacteria	high diversity, benthic and pelagic sequences	
			mid-exterior	Thermococcales, MGI Cren. Archaeoglobales, Korarchaeota	NT	high diversity, both hydrothermal and deep-sea sequences	
			mid-interior	HWCG	NT	low diversity	
			interior	ND	NT		
Wall	250	CIR					3
			exterior	Methanococcales			
			interior	Thermococcales			
Wall	diffuse	CIR	whole	DHVE, Thermococcales	ε-proteobacteria, Aquificales	acidic microenvironments within diffuse chimney	4
Wall	<350	MAR	whole	Thermococcales	Aquificales		5
Wall	diffuse	JdFR	exterior	DHVE(9), DHVE(3)	ε-proteobacteria	associated with *Paralvinella sulfincola*	6
Wall	300	SS	whole	AAG	ND	extensive diversity of novel sequences	7
	~100	MK	whole	MGI Cren., DHVE(2)	ND		
	~180	MK	whole	AAG, HWCG, MGI Cren. DHVE(I), DHVE(II)	ND		
Wall	<250	MB					8
			exterior orifice	DHVE, Thermococcus	NT	halophilic Archaea and Eubacteria near exterior	
			interior-orifice	Desulfurococcales, Thermococcales	NT	thermophiles near inside	
			exterior-wall	Thermococcales	NT		
			mid-interior	Halobacteriales	NT		
			vent conduit	Desulfurococcales	NT		
Wall	<253	JdFR	mound				
Flange	Gradient				γ-, ε-proteobacteria	presence of a cosmopolitan	
					γ-, ε-proteobacteria	group of Fe oxidizers	9
Wall	~100	MVA	outside	Desulfurococcales	γ-, ε-proteobacteria		
			inside	Thermococcales, DHVE	ε-proteobacteria		10
Wall	50–100	EPR	diffuser		*dsr* AB, *pmoA* gene	colonized by *Alvinella* sulfate reduction and methane oxidation potential	11
Wall	40–50	MAR	diffuser	*mcr* A gene[c]	NT	thermophilic methanogens	
	350	EPR	exterior	Thermococcales, Archaeoglobales Desulfurococcales	ε-proteobacteria, Aquificales		12
			interior	Thermococcales, Archaeoglobales	ε-proteobacteria, Aquificales		

Table 2. (continued)

Type	Temp, °C	Site[a]	Horizon	Archaea	Eubacteria	Comments	Ref.[b]
Wall	223	EPR	exterior	Desulfurococcales, Nanoarchaeota		abundance of Nanoarchaeota in nascent chimneys	13
	350	EPR	exterior	Desulfurococcales, Nanoarchaeota			
	98–203	EPR	exterior	Desulfurococcales, Nanoarchaeota			

[a]Vent sites: CIR, Central Indian Ridge; JdFR, Juan de Fuca Ridge; MB, Manus Basin; MK, Myojin Knoll; MVA, Marianas Volcanic Arc; SS, Suiyo Seamount; EPR, East Pacific Rise; MAR, Mid-Atlantic Ridge.

[b]References: 1, *Hedrick* [1993]; 2, *Schrenk* [2003]; 3, *Takai* [2004]; 4, *Hoek* [2003]; 5, *Harmsen* [1997a]; 6, *Page* [2004]; 7, *Takai* [1999]; 8, *Takai* [2001]; 9, *Rogers* [2003]; 10, *Nakagawa* [2006]; 11, *Nercessian* [2006]; 12, *Kormas* [2006]; 13, *McCliment* [2006].

[c]*mcrA* is a functional gene for methanogenesis; *dsrAB* for sulfate reduction, and *pmoA* for methane oxidation.

the occurrence of physiologically diverse species including aerobic and anaerobic thermophiles and mesophiles in more mineralogically mature, evolved chimneys [*Harmsen et al.*, 1997a; *Takai et al.*, 2001; *Schrenk et al.*, 2003]. These data presumably reflect an increase in niche diversity within the sulfide-microbial habitat.

In addition to the heterogeneity of microbial communities within individual structures, biogeographic studies of vent microorganisms represent a daunting challenge. Studies of both the phylogenetic and enzymatic profile of organisms from sulfide environments from the JdFR showed significant genotypic and phenotypic differences in microbial isolates of the same genus [*Holden et al.*, 2001; *Summit and Baross*, 2001]. A high level of phylogenetic diversity was found among Thermococcales isolates from six separate chimneys from the EPR using randomly amplified polymorphic DNA and 16S rRNA analyses [*Lepage et al.*, 2004]. These organisms are phylogenetically intermingled with isolates from terrestrial and shallow marine sites, suggesting that there is a high degree of dispersal among Thermococcales populations. Further efforts should be devoted to source tracking and comparison of genotypic and phenotypic diversity at local and global scales.

3.3. Microbial Biofilms

A biofilm is an extracellular layer of primarily exopolysaccharides (EPS) that coats and encapsulates a microbial population and serves as an adhesive to a solid surface. As part of the discovery of deep-sea hydrothermal vents, it was noted that large microbial mats formed in the vicinity of venting fluids use geochemical energy available from hydrothermal discharge (e.g., H_2S, Fe(II)) for primary production in lieu of photosynthesis [*Karl et al.*, 1980; *Jannasch and Wirsen*, 1981; *Jannasch*, 1985; *Wirsen et al.*, 1993].

Microscopic observations revealed dense microbial communities coating mineral surfaces in situ [*Baross and Deming*, 1985], and morphologically complex communities were observed coating sulfide mineral surfaces within the walls of deep-sea hydrothermal chimneys [Figure 6; *Juniper and Tebo*, 1995; *Harmsen et al.*, 1997a]. Samples of diffuse hydrothermal fluids and event plumes have also shown that subseafloor microbial populations are predisposed to associate with surfaces [*Summit*, 2000; *Huber et al.*, 2002; *Huber et al.*, 2003].

A few studies have documented the organisms associated with the initial colonization of materials in and around hydrothermal sites (Table 3). Incubation of both natural and artificial substrata within the "vent cap" and "smoker poker" devices placed atop smokers at Guaymas Basin in the mid-1980s found rapid and intense colonization of surfaces [*Karl et al.*, 1988]. Studies using assimilation of radiolabeled carbon, adenosine triphosphate (ATP) measurements, and cell counts found that microbial activity occurred optimally between 25° and 80°C (maximally at 45°C), and that there was no growth above 93°C [*Karl et al.*, 1988]. Other in situ experiments that sought to determine the upper temperature limit for life using solid substrates coupled with temperature and microbiological measurements found dense colonization of the moderate temperature, exterior regions of the sampling device but no evidence of microbial growth above 150°C [*Karl et al.*, 1988]. More recent studies have used the vent cap to study colonization processes and showed the predominance of ε-proteobacteria and Thermococcales during the early stages of colonization at temperatures between 20° and 70°C [*Reysenbach et al.*, 2000; *Longnecker and Reysenbach*, 2001; *Nercessian et al.*, 2003]. At higher temperatures, archaeal species related to *Ignicoccus* and *Nanoarcheum* spp. colonized mineral surfaces deployed in hydrothermal effluent within 92 h [*McCliment et al.*, 2006].

There is also laboratory evidence for high-temperature microbial attachment and biofilm formation. For example, *Pyrodictium* spp. that grow at temperatures up to 110°C produce microtubules between cells and thus form a network that most likely facilitates surface attachment [Figure 6; *Stetter et al.*, 1983]. The upper temperature limit for life was

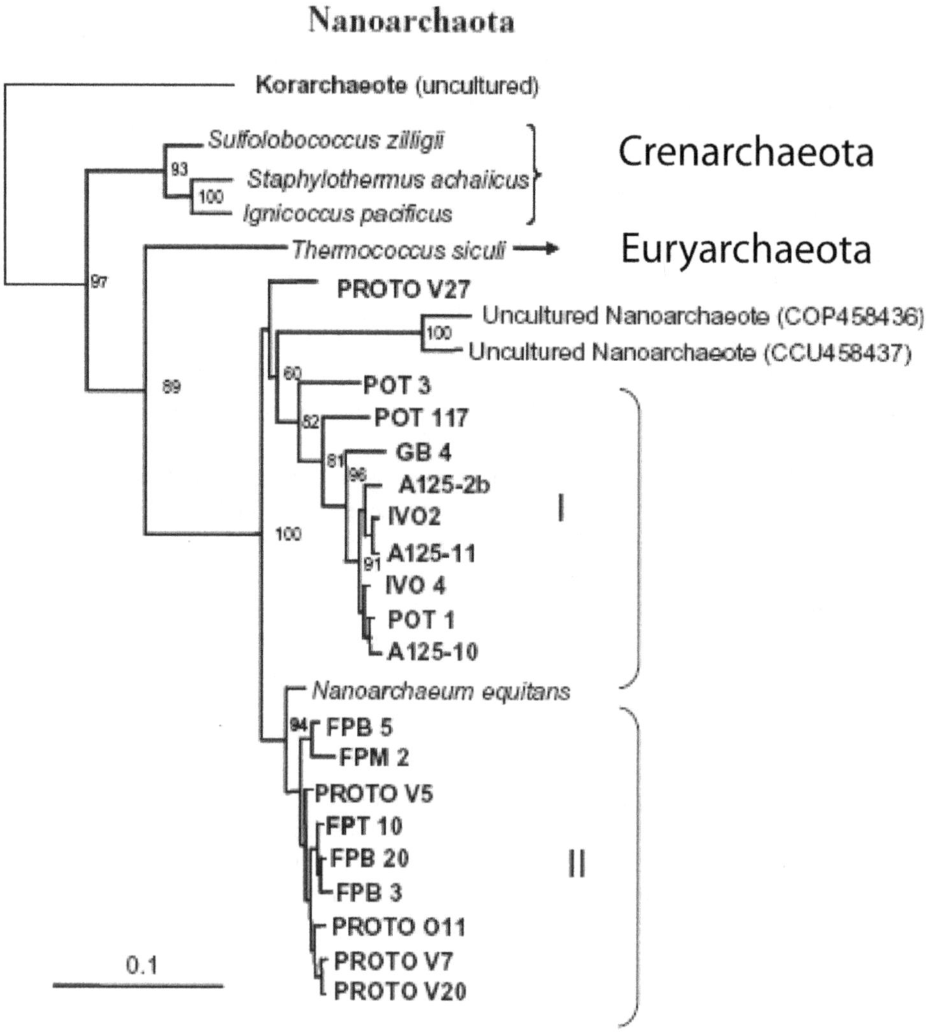

Figure 5. Careful studies placing the occurrence of sequence data into a geologic context are necessary to elucidate the relationships of these microorganisms to their environmental conditions. The study of *McCliment et al.* [2006] showed that distinct clades of Nanoarchaeota were specific to either mature (group I) or nascent (group II) chimney environments. Scale bar represents the expected number of changes per nucleotide position. Reproduced with permission from *Environmental Microbiology*, Blackwell Publishing Ltd., Oxford, UK.

extended to 120°C when a natural assemblage of organisms from a sulfide deposit was incubated under hydrostatic pressure in a solid gel matrix called Gelrite [*Deming and Baross*, 1986], and the highest known growth temperature of a pure culture organism was raised to 121°C when it was grown on insoluble iron [*Kashefi and Lovley*, 2003].

Studies on the hyperthermophilic archaeon *Archaeoglobus fulgidis* demonstrated the important role of biofilms for stress resistance and enhanced thermotolerance to superoptimal temperatures [Figure 7; *LaPaglia and Hartzell*, 1997]. In other studies, the hyperthermophiles *M. jannaschii* and *Thermotoga maritima* demonstrated increased growth rates when cultivated in coculture and were shown to form close cell-to-cell interactions between the species [*Muralidharan et al.*, 1997]. The hyperthermophilic archaeon *Thermococcus litoralis* formed biofilms when grown in continuous culture and analyses showed that mannose was the primary constituent of their extracellular polymeric substances [*Rinker and Kelly*, 1996]. Subsequently, gene expression and protein production studies related to biofilm formation at high temperatures showed numerous genes related to exopolysaccharide synthesis and the general stress response were up-regulated during biofilm formation [*Pysz et al.*, 2004a]. There remains much to be learned about the ecology of biofilms in high-

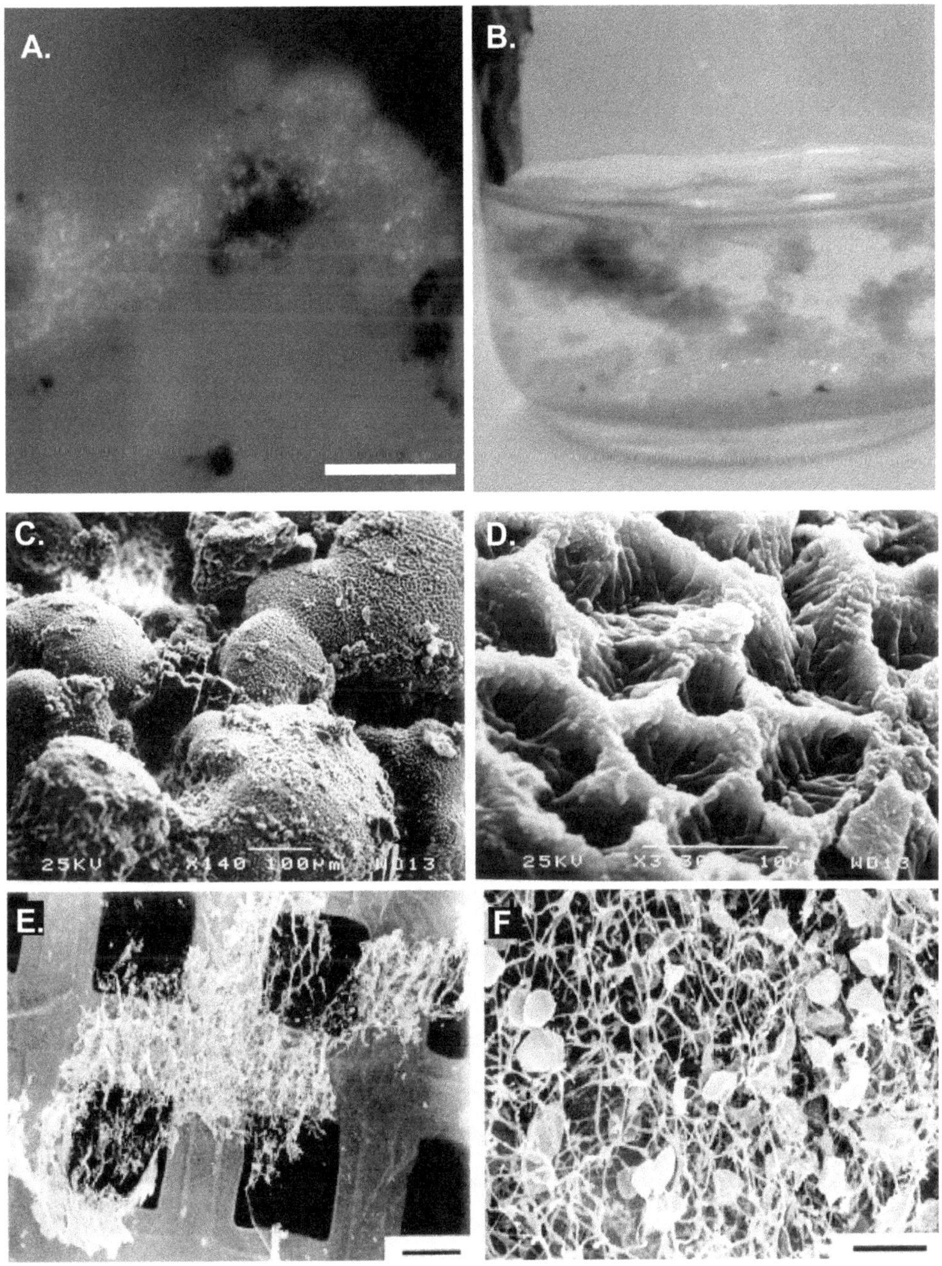

Figure 6. Montage of microbial biofilms formed by deep-sea hydrothermal vent microorganisms. (a) Microbial cells coating a sulfide mineral surface within the hydrothermal chimney Finn, a black smoker from the JdFR (shown in Figure 1). Scale bar is 20 μm. (b) Prolific biofilm material formed in a laboratory culture of a *Desulfurobacterium* sp. grown at 70°C. Each brown-colored aggregate contains millions of individual bacterial cells. Scanning electron micrograph (c) and close-up view (d) of bacterial colonies forming a regular three-dimensional pattern coating a sulfide mineral surface from the exterior of a hydrothermal vent chimney from the JdFR. (e, f) "Networked" biofilm formed by *P. abyssi* in laboratory culture [*Rieger et al.*, 1995]. *P. abyssi* was isolated from the MAR and is closely related to phylotypes commonly found at high temperature sulfide deposits. Scale bar is 20 μm (e) and 3 μm (f). Panels (e) and (f) were reproduced with permission from *Journal of Structural Biology*, Elsevier, Amsterdam, Netherlands.

Table 3. Microbial Colonization of Natural and Artificial Substrata Associated With Deep-Sea Sulfide Deposits

Type of Sample	Length of Deployment	*T*, °C	Vent[a]	Comments	Eubacteria	Archaea	Ref.[b]
Proto-chimney	92 h	140–162	EPR	natural chimney	NT	Desulfurococcales, Thermococcales	1
Mineral chamber	72 h	35–144	EPR	polished minerals	NT	Thermococcales, *Nanoarchaeota*	
	72 h	40–180	EPR		NT	Desulfurococcales, *Nanoarchaeota*	
	72 h	350–375	EPR		NT	ND	
In Situ Coln. System (ISCS)	3 d	<279	SS		ε-, γ-proteobacteria, Aquificales *dsr*AB related to *Thermodesulfo-bacterium*	Thermococcales, *Nanoarchaeota*	2
ISCS	4 d	>2	IN	high ε-proteo-bacteria diversity	ε-proteobacteria, Aquificales	Thermococcales, Methanococcales	3
	4 d	~2					
ISCS	7 d	365	CIR	predominantly Archaea	ND	Methanococcales	4
	3 d	250		predominantly Archaea	ND	Methanococcales	
	3 d	38		predominantly Eubacteria	*Aquificales*	Methanococcales	
	11 d	2		predominantly Eubacteria	ε-, γ-proteobacteria	Thermococcales, MGI Cren.	
	2 d	~2		predominantly Eubacteria	α-proteobacteria, other phyla	MGI Cren., Euryarchaeota	
In situ sampler	7 d	50–100	EPR	thermophilic metha-nogens	NT	*mcr*A	5
In situ sampler	7 d	50–100	EPR	predominantly Eubacteria	NT	Thermococcales, Methanopyrales MGI Cren.	6
In situ sampler	11 d	50–100		predominantly Eubacteria	NT	Thermococcales, Methanopyrales DHVEGII, Archaeoglobales	
In situ sampler	4 d	50–100		predominantly Eubacteria	NT	Thermococcus	
Vent Cap	5 d	30–70	MAR		ε-proteobacteria, Aquificales, *Desulfurobacterium*	Thermococcales, Archaeoglobales	7
Vent Cap	2 d	50–100	EPR		NT	Thermococcales, Desulfurococcales	6

Table 3. (Continued)

Type of Sample	Length of Deployment	T, °C	Vent[a]	Comments	Eubacteria	Archaea	Ref.[b]
Vent Cap	5 d	50–100		predominantly Eubacteria	NT	Thermococcales, Methanococcales, DHVEGI and II, MGI Cren. Methanopyrales, Korarchaeota	

[a]Vent sites: CIR, Central Indian Ridge; IN, Ihcya North; JdFR, Juan de Fuca Ridge; MB, Manus Basin; MK, Myojin Knoll; SS, Suiyo Seamount; MAR, Mid-Atlantic Ridge; EPR, East Pacific Rise.

[b]References: 1, *McCliment* [2006]; 2, *Nakagawa* [2004]; 3, *Nakagawa* [2005]; 4, *Takai* [2004]; 5, *Nercessian* [2006]; 6, *Nercessian* [2003]; 7, *Reysenbach* [2000]; *Nercessian* [2003].

ND, not detected; NT, not tested.

temperature environments, but they certainly play a role in the functioning and activity of natural ecosystems and may, given the integration between biofilm formation and stress responses, expand the limits to microbial growth and survival beyond what is known from planktonic cell cultures.

4. MODELING SULFIDE DEPOSIT HABITATS

4.1. What are the Scales of Geobiological Processes?

A significant challenge in interpreting the microbial ecology of rock-hosted habitats is in understanding the spatial and temporal scales of the processes involved. This challenge is accentuated within the diverse and dynamic milieu of the hydrothermal vent chimney. Most assessments of microbial distributions and diversity within the chimney habitat are at the bulk scale, meaning that in order to quantify or characterize the microbiological community, a relatively large-sized sample (~0.5 to 10 g), in comparison to the microhabitat, needs to be obtained (Table 4). It is not clear whether this large a sample allows one to accurately interpret meaningful biological phenomena that may be occurring at much smaller spatial scales (Figure 3). The potential for cross-contamination between mineralogical zones also amplifies the potential for misinterpretation [*Kormas et al.*, 2006].

Furthermore, the dynamic nature of fluid flow through sulfide deposits has yet to be fully incorporated into sulfide

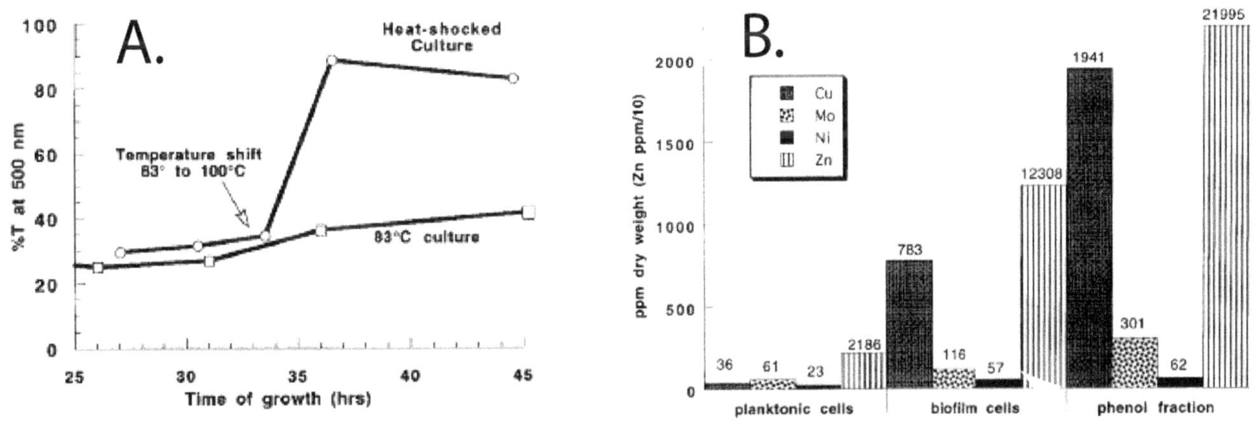

Figure 7. Protective effects of biofilm formation by vent microorganisms. (a) The euryarchaeon *A. fulgidis* demonstrated increased biofilm formation (measured as exoploysaccharide production) as a response to heat treatment relative to an untreated sample [*LaPaglia and Hartzell*, 1997] The y-axis depicts increased transmittance of the culture medium at 500-nm wavelength due to the binding of the dye Congo Red to extracellular polysaccharides. (b) *A. fulgidis* biofilms also bound trace metals from dilute solution which were enriched in phenol-insoluble (protein) fractions and could subsequently be used as a source of trace elements for other cultures [*LaPaglia and Hartzell*, 1997]. Reproduced with permission from *Applied and Environmental Microbiology*, American Society for Microbiology.

biogeochemical models. In addition to the stark thermal and chemical gradients within vent deposits, organisms within seafloor sulfide deposits may be impacted by both episodic and periodic perturbations that may influence both the microniche characteristics and community structures. An interesting question is how the dynamic environment shapes the community structure and activities of rock-hosted organisms. Could biofilms be a mechanism to deal with varying energetic fluxes and environmental stresses?

4.2. Do Products of Microbial Processes Impact Geochemistry?

Microorganisms can potentially impact the geochemistry and structure of sulfide deposits in several ways. The most obvious would be their interactions with solid phases. Some microorganisms can utilize solid iron oxides for growth [*Kashefi et al.*, 2002], whereas others serve as either passive or active nuclei for mineral deposition [Figure 2; *Juniper and Tebo*, 1995; *Fortin et al.*, 1998]. In addition to mineralogical effects, the role that the organisms play in extracting and shaping the surface of solid materials needs to be explored. Metal sulfide surfaces are effective at adsorbing organic molecules and trace nutrients depending on the ionic composition of the fluids [*Bebié and Schoonen*, 1999; *Bebié and Schoonen*, 2000]. They may also retain a record of varying reduction potentials and impact the metabolism of surface-associated cells. Such a process may have been recorded in the deposition of a layer of elemental sulfur noted in a study of a chimney from the Pito Seamount [*Verati et al.*, 1999] and may also be significant in the frequent use of iron (III) as an electron acceptor by hyperthermophilic organisms [*Kashefi et al.*, 2004].

The growth of organisms in exopolymer-encased biofilms, likewise, may have a number of effects [*Pysz et al.*, 2004b]. The polymers may play a role in clogging and affect the fluid fluxes through the structures. They may also promote synergy between organisms that may not be able to live independently [*Rinker and Kelly*, 1996; *Johnson et al.*, 2006b]. Additionally, quantitative surveys of microbial "functional" genes, such as *dsrAB* for sulfate reduction, *mcrA* for methanogenesis, and *pmoA* for aerobic methane oxidation, may also prove useful in linking microbial diversity to geochemical effects in situ [*Nercessian et al.*, 2005].

4.3. Are Sulfide Deposits Analogs of a High-Temperature Subseafloor Biosphere?

The presence of microorganisms, especially chemolithoautotrophic anaerobes, within sulfide deposits fed by hot, reduced, energy-rich hydrothermal fluids led in part to the idea that vast regions of the subseafloor could be inhabited by microorganisms that live independent of sunlight [*Jannasch*, 1985; *Jannasch and Mottl*, 1985; *Straube et al.*, 1990; *Whit-*

Table 4. Microbiological Methods Used to Characterize Hydrothermal Vent Microbial Communities

Method	Sample Size, g	Benefits	Limitations
Enumeration of specific taxa			
General cell abundance	0.5–10.0	high throughput	nonspecific
FISH; CARD-FISH	0.5–10.0	localization of taxa	requires knowledge of target
qPCR	0.5–10.0	localization of taxa	requires knowledge of target
MPN analyses	0.1–1.0	physiological/biochemical detail	difficulty in homogenizing sample
Enumeration of specific functions			
qPCR	0.5–>10.0	localization of functions	requires high-quality sample
CARD-FISH	0.5–10.0	localization of functions	labor-intensive
Enrichment Culturing	0.1–1.0	physiological/biochemical detail	culturing bias
Characterization of diversity			
tRFLP	0.5–>10.0	high throughput	limited resolution
Cloning and sequencing	0.5–>10.0	high resolution	labor-intensive, limited resolution
Metagenomics	>10	high resolution, links identity to function	resource- and sample-intensive
Phospholipid analyses	0.5–10.0	independent assessment	requires biochemical standards

FISH, fluorescence in situ hybridization; qPCR, quantitative polymerase chain reaction; MPN, most probable number enrichment analysis; CARD-FISH, catalyzed antibody reporter detection FISH; tRFLP, terminal restriction length polymorphism analysis.

man et al., 1998]. It has been suggested that up to 20% of the Earth's total biomass is present within the subseafloor [*Whitman et al.*, 1998], whereas another study proposed that the amount of biomass within the subsurface biosphere could exceed that of photosynthetically supported life on the Earth's surface [*Gold*, 1992]. The existence of microbial life based on geothermal energy rather than solar energy also has significant implications for life on the early Earth as well as for potential life on other planets and moons [*Corliss et al.*, 1981; *Jannasch*, 1985; *Bach et al.*, 2006]. Improved sampling methods since the discovery of vents, and a better understanding of both the geophysics and geochemistry of subseafloor environments, have allowed careful comparisons of geographic and temporal variations in vent fluids [*Von Damm*, 1990, 1995a, 1995b; *Lilley et al.*, 2003]. These studies have allowed hydrothermal fluxes to be accurately incorporated into global biogeochemical budgets [*Elderfield and Schultz*, 1996]. Unfortunately, a microbiological understanding of high-temperature environments at deep sea vents has not progressed at a similar pace.

The first reports of bacterial abundances in high-temperature vent fluids reported cell concentrations of $\sim 10^5$/mL [*Corliss et al.*, 1979; *Baross et al.*, 1982]. Subsequent studies using diverse methodologies have yielded variable and conflicting interpretations of the detection of life in these high-temperature hydrothermal fluids [*Karl et al.*, 1988; *Deming and Baross*, 1993]. These discrepancies led to a microbiology-oriented cruise to the Main Endeavour Field of the JdFR in 1988 to carefully obtain high-quality samples from numerous high-temperature orifices in a single vent field. Results from this expedition confirmed the observations of microorganisms in high-temperature fluids and documented an association between vent temperatures and particulate DNA (pDNA) concentrations [*Straube et al.*, 1990]. Surprisingly, the highest temperature vents contained the most pDNA. Based on the cumulative observations up to that point—that the high-temperature chimneys were conduits to subseafloor fluids, and that gradients within the subsurface were also within the chimney walls—black smoker chimneys and sulfide deposits became referred to as "windows to the subsurface biosphere" [*Deming and Baross*, 1993].

These observations were complicated by three factors: (1) it has not been possible to obtain completely undiluted samples of hydrothermal fluids, (2) there is generally very low biomass within the hot fluids making most microbiological analyses difficult, and (3) the chimney walls and diffuse peripheral environments contain extremely prolific biomass. More analyses, coupling detailed microbiological studies of chimney-hosted populations with those of vent fluids, need to be conducted to reconcile the consistent observation of microbial cells in hot hydrothermal fluids with the results of laboratory-based experiments. It appears though that a percentage of the microbial cells at vents can survive at least short-term exposures to temperatures well beyond their known limits to growth [*Deming and Baross*, 1993; *Takai et al.*, 2004; *Mitsuzawa et al.*, 2005]. In light of these complications, and the recognition that a far greater amount of heat flux and hydrothermal flow occurs away from the high-temperature orifices in oceanic crust up to 65 Ma [*Schultz et al.*, 1992; *Wheat and Mottl*, 2000], attention has turned to using other types of samples to characterize the subseafloor biosphere [*Huber et al.*, 2002, 2003; *Baross et al.*, 2004]. Diffuse fluids and event plumes associated with intense magmatic activity are the two types of samples most commonly used to access the subseafloor [*Huber et al.*, this volume; *Delaney*, 1998; *Holden et al.*, 1998; *Huber et al.*, 2002, 2003]. These samples appear to host microbial ecosystems that differ, to some extent, from those hosted within the chimney walls [*Holden et al.*, 2001; *Summit and Baross*, 2001]. In other instances, deep-sea drilling and related technologies have been used to obtain hydrothermally influenced samples from the deep subsurface [*Summit et al.*, 2000; *Johnson et al.*, 2006a]. Although these approaches all appear to sample subseafloor ecosystems, directly accessing subsurface microbial populations, and thus confirming their distributions and activities in situ, remains an enigma.

The walls of high-temperature chimneys remain a useful tool to study microbial populations in the deep-sea vent environment (Figure 8). Gradients within the subseafloor that delimit the extent of the subsurface biosphere share broad similarities with the steeper gradients within the walls of hydrothermal chimney structures [*Deming and Baross*, 1993; *McCollom and Shock*, 1997; *Delaney et al.*, 1998]. They are considerably more accessible than samples of the subseafloor at mid-ocean ridges, and permit experimentation and observation at numerous scales. The walls of high-temperature hydrothermal chimneys are particularity well suited for the study of microbial adaptation to high temperatures. They facilitate the open exchange of mass (including microorganisms) between the high-temperature subsurface and more moderate conditions, they contain a preponderance of energy resources, and liquid water persists within these systems at temperatures well above the known upper limits to life. Learning about the microbial populations that inhabit the energy-rich gradients of chimney walls and the processes controlling their distribution will help us explore the more enigmatic subseafloor.

4.4. What Data Are We Missing to Advance the Models?

One of the primary challenges in deep-sea hydrothermal systems, as in most natural ecosystems, is to understand the significance and activities of uncultured microorganisms.

These studies should be buoyed by the recent successful isolation of a ubiquitous archaeal group through integration of microbiological data with habitat characteristics [Table 5; *Reysenbach et al.*, 2006]. The ε-proteobacteria (Figure 4a) are another group with cosmopolitan distribution and the properties of cultured species from this group have recently been reviewed [*Campbell et al.*, 2006]. Because cultured representative ε-proteobacteria have a large habitat range and diverse physiological characteristics, future efforts should target the activities and rates of biogeochemical processes modulated by this group [*Taylor et al.*, 1999]. This is likely to be challenging because recent analyses of the diversity of ε-proteobacteria from diffuse flow vents using new molecular techniques indicate orders-of-magnitude greater diversity than thought to exist based on data from clone library sequences [*Huber et al.*, 2007]. Likewise, despite advancements in understanding their physiology in the laboratory, the roles of Crenarchaeota inhabiting high-temperature portions of the system remain to be elucidated. Cultured representatives of this group oxidize hydrogen coupled to the reduction of Fe(III) or intermediate sulfur compounds. Where the overlap of these compounds occurs and how widespread these groups are needs to be quantified and incorporated into models.

Several studies of environmental samples from vent deposits have found nucleotide sequences that root phylogenetically near the base of the tree of life, but all that is known about their source organisms comes from their 16S rRNA genes (Figure 4b and Table 5). One such example are the Korarchaeota, which were first found in Yellowstone National Park and sustained at low abundances within a hydrogen-rich chemostat [Table 5 and Figure 4b; *Burggraf et al.*, 1997; *Auchtung et al.*, 2006]. They have since been found in other hot hydrothermal environments including sulfide chimneys, shallow marine sites, and colonization devices. They frequently co-occur with Eubacteria of the Aquificales group and hyperthermophilic Crenarchaeota. A recent study reported a high level of endemicity (living in a defined geographic area) within this clade [*Auchtung et al.*, 2006].

A second avenue of research to make headway in modeling the microbiology of sulfide deposits is an accurate assessment of rates and activities of microorganisms at environmentally relevant conditions [*Houghton et al.*, 2007]. We need to understand both the relevance of the organisms we have in culture and the impact of their activities on the physical and chemical environments present at the vents [Figure 8; *Reysenbach and Shock*, 2002]. Most cultures are designed to net maximal yields, and the conditions that yield these optimal growth rates are extrapolated to environmental relevance based on natural selection. It is unknown whether this is actually the case. Another development to advance our models of vent geomicrobiology is an improvement in the identification and measurement of biosignatures of mi-

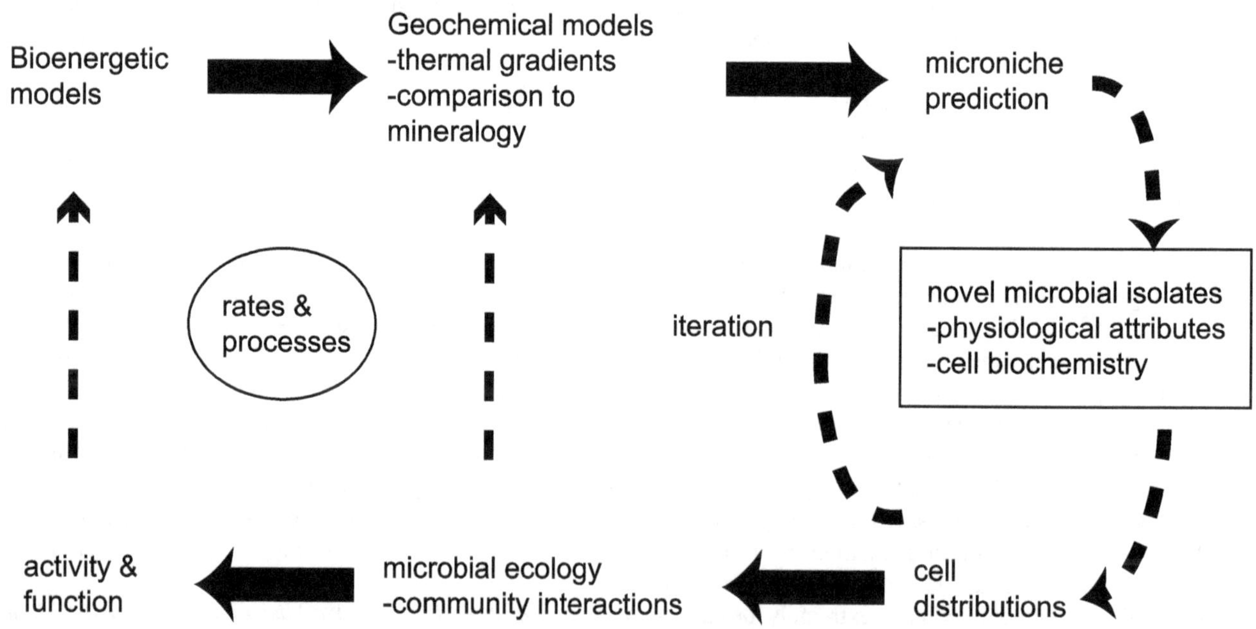

Figure 8. Flowchart depicting the utility of predictive bioenergetic and geochemical models for the isolation of new microorganisms. Microbial isolates can then be used to study cell physiology and biochemistry in the laboratory and query the activities and abundance of organisms within natural systems. Careful comparison and iteration of abiological and biological models can be used to ascertain rates of microbial processes and their impact on the geochemical environment.

Table 5. Key Taxonomic Groups Without Cultured Representatives From Deep-Sea Vent Deposits

Taxonomic Designation	Locations Found	Habitat Characteristics	Ref.[a]
Deep-sea hydrothermal vent euryarchaeal group (DHVE)[b]			
ELSC, EPR, MK, CIR, MB, GB, MVA[c]	chimney exterior	acidic pH organic materials	1, 2
Marine benthic group E (MBGE)			
MK, JdFR, MAR	sulfide chimneys, hydrothermal sediments	benthic environments	2–4
Hot water crenarchaeal group (HWCG)			
JdFR, Japanese Gold Mine	hot interior regions of black smokers	high temperatures low diversity	4
Korarchaeota			
EPR, JdFR, CIR, GB	chimney surface, vent cap nascent chimney	Fe, S, H_2 intermittent O_2 exposure	5
Ancient archaeal group (AAG)			
MK, JdFR	small hydrothermal chimneys	unknown	6
Marine group II Euryarchaeota			
CIR	seawater	abundant in pelagic ocean	7

[a]References: 1, *Reysenbach et al.* [2006]; 2, *Takai* [2004]; 3, *Nercessian* [2005]; 4, *Schrenk* [2003]; 5, *Auchtung* [2006]; 6, *Takai* [1999]; 7, *Takai* [2004].

[b]Recently isolated by *Reysenbach et al.* [2006].

[c]ELSC, Eastern Lau Spreading Center; EPR, East Pacific Rise; MK, Myojin Knoll; CIR, Central Indian Ridge; GB, Guaymas Basin; MVA, Mariana Volcanic Arc; MAR, Mid-Atlantic Ridge.

crobial activity. We need multiple means of assessing the presence of microorganisms, especially when many of the populations are at low abundance, inaccessible, or largely unknown. The ever-increasing database of genomic sequence data will help to develop this effort and allow identification of the genes that are expressed in situ using either proteomics- or transcriptomics-based approaches.

5. SUMMARY AND CONCLUSIONS

Hydrothermal chimneys remain an excellent option to determine the relationships between magmatic and microbial processes at high temperatures, although much remains to be discovered regarding the geochemistry, evolution, and microbiology of these systems. Their relative accessibility also makes them an ideal natural laboratory for studying microbial processes associated with deep-sea hydrothermal vent discharge and the relationship between microorganisms, hydrothermal fluids, and mineral deposits. Understanding these interactions may be critical to evaluating (1) the extent of habitable space in subseafloor environments, (2) the upper growth temperature for microorganisms, and (3) the impact of microorganisms within rock-hosted deep-sea ecosystems on global biogeochemical cycles [*Gold*, 1992; *Baross et al.*, 2004; *Wilcock et al.*, 2004].

Acknowledgments. We thank the organizers of the Ridge Theoretical Institute for creating a forum that contributed to the ideas presented in this paper. We thank J. Huber for helpful discussions and constructive reviews and W. Brazelton for assistance in tree construction. This work was prepared with support from the NASA Astrobiology Institute through a Postdoctoral Fellowship to M.O.S

REFERENCES

Adams, M. W. W. (1999), The biochemical diversity of life near and above 100°C in marine environments, *J. Appl. Microbiol.*, *85*, 108S–117S.

Adams, M. W. W., et al. (2001), Key role for sulfur in peptide metabolism and in regulation of three hydrogenases in the hyperthermophilic archaeon *Pyrococcus furiosus*, *J. Bacteriol.*, *183*(2), 716–724.

Alt, J. C. (1995), Subseafloor processes in mid-ocean ridge hydrothermal systems, in *Seafloor Hydrothermal Systems: Physical, Chemical, Biological, and Geological Interactions*, *Geophys. Monogr. Ser.*, vol. 91, edited by S. E. Humphries, et al., pp. 85–114, AGU, Washington, D. C.

Auchtung, T. A., C. D. Takacs-Vesbach, et al. (2006), 16S rRNA phylogenetic investigation of the candidate division "*Korarchaeota*," *Appl. Environ. Microbiol.*, *72*(7), 5077–5082.

Bach, W., and K. J. Edwards (2003), Iron and sulfide oxidation within the basaltic oceanic crust: Implication for chemolithoautotrophic microbial biomass production, *Geochim. Cosmochim. Acta*, *67*(20), 3871–3887.

Bach, W., K. J. Edwards, J. M. Hayes, S. Sievert, J. A. Huber, and M. L. Sogin (2006), Energy in the dark: Fuel for life in the deep ocean and beyond, *Eos Trans. AGU*, *87*(7), 73.

Baross, J. A., and J. W. Deming (1985), The role of bacteria in the ecology of black-smoker environments, *Biol. Soc. Wash. Bull.*, *6*, 355–371.

Baross, J. A., et al. (1982), Is the CH_4, H_2, and CO venting from submarine hydrothermal systems produced by thermophilic bacteria?, *Nature*, *298*, 366–368.

Baross, J. A., W. S. D. Wilcock, D. S. Kelley, E. F. DeLong and S. C. Cary (2004), The subsurface biosphere at mid-ocean ridges: Issues and challenges, in *The Subseafloor Biosphere at Mid-ocean Ridges*, *Geophys. Monogr. Ser.*, vol. 144, edited by W. S. D. Wilcock, et al., pp. 1–12, AGU, Washington, D. C.

Bebié, J., and M. A. A. Schoonen (1999), Pyrite and phosphate in anoxia and an origin-of-life hypothesis, *Earth Planet. Sci. Lett.*, *171*, 1–5.

Bebié, J., and M. A. A. Schoonen (2000), Pyrite surface interaction with selected organic aqueous species under anoxic conditions, *Geochem. Trans. Pap.*, *8*.

Blöchl, E., et al. (1997), *Pyrolobus fumarii*, gen. and sp. nov. represents a novel group of archaea, extending the upper temperature limit to life to 113°C, *Extremophiles*, *1*, 14–21.

Burggraf, S., K. O. Stetter, et al. (1991), *Methanopyrus kandleri*: An archaeal methanogen unrelated to all other known methanogens, *System. Appl. Microbiol.*, *14*, 346–351.

Burggraf, S., et al. (1997), A pivotal Archaea group, *Nature*, *385*, 780.

Campbell, B. J., A. S. Engel, et al. (2006), The versatile ε-proteobacteria: Key players in sulphidic habitats, *Nature*, *4*, 458–468.

Cook, T. L., and D. S. Stakes (1995), Biogeological mineralization in deep-sea hydrothermal deposits, *Science*, *267*, 1975–1979.

Corliss, J. B., J. Dymond, et al. (1979). Submarine thermal springs on the Galapagos Rift, *Science*, *203*(4385), 1073–1083.

Corliss, J. B., J. A. Baross, et al. (1981), A hypothesis concerning the relationship between submarine hot springs and the origin of life on Earth, *Oceanol. Acta*, Suppl 4., 59–69.

Delaney, J. R., D.S. Kelley, M.D. Lilley, D.A. Butterfield, J.A. Baross, W.S.D. Wilcock, R.W. Embley, and M. Summit (1998), The quantum event of oceanic crustal accretion: Impacts of diking at mid-ocean ridges, *Science*, *281*, 222–230.

Delaney, J. R., D. S. Kelley, E. A. Mathez, D. R. Yoerger, J. Baross, M. O. Schrenk, M. K. Tivey, J. Kaye, and V. Robigou (2001), "Edifice Rex" sulfide recovery project: Analysis of a submarine hydrothermal microbial habitat, *EoS Trans.*, *82*(6), 67–73.

Deming, J. W., and J. A. Baross (1986), Solid medium for culturing black smoker bacteria at temperatures to 120°C, *Appl. Environ. Microbiol.*, *51*(2), 238–243.

Deming, J. W., and J. A. Baross (1993), Deep-sea smokers: Windows to a subsurface biosphere?, *Geochim. Cosmochim. Acta*, *57*, 3219–3230.

DeSantis, T. Z., P. Hugenholtz, et al. (2006), Greengenes, a chimera-checked 16S rRNA gene database and workbench compatible with ARB, *Appl. Environ. Microbiol.*, *72*, 5069–5072.

Dziak, R. P., and H. P. Johnson (2002), Stirring the oceanic incubator, *Science*, *296*, 1406–1407.

Eberhard, C., C. O. Wirsen, et al. (1995), Oxidation of polymetal sulfides by chemolithoautotrophic bacteria from deep-sea hydrothermal vents, *Geomicrobiol. J.*, *13*, 145–164.

Edwards, K. J., T. M. McCollom, et al. (2003a), Seafloor bioalteration of sulfide minerals: Results from in situ incubation studies, *Geochim. Cosmochim. Acta*, *67*(15), 2843–2856.

Edwards, K. J., D. R. Rogers, et al. (2003b), Isolation and characterization of novel psychrophilic, neutrophilic, Fe-oxidizing, chemolithoautotrophic α- and γ-proteobacteria from the deep sea, *Appl. Environ. Microbiol.*, *69*, 2906–2913.

Edwards, K. J., W. Bach, et al. (2005), Geomicrobiology in oceanography: Microbe-mineral interactions at and below the seafloor, *Trends Microbiol.*, *13*(9), 449–456.

Elderfield, H., and A. Schultz (1996), Mid-ocean ridge hydrothermal fluxes and the chemical composition of the ocean, *Annu. Rev. Earth Planet. Sci.*, *24*, 191–224.

Feinberg, L. F., R. Srikanth, et al. (2008), Constraints on anaerobic respiration in the hyperthermophilic Archaea *Pyrobaculum islandicum* and *Pyrobaculum aerophilum*, *Appl. Environ. Microbiol.*, *74*, 396–402.

Ferry, J. G. (1999), Enzymology of one-carbon metabolism in methanogenic pathways, *FEMS Microbiol. Rev.*, *23*(1), 13–38.

Fortin, D., F. G. Ferris, et al. (1998), Formation of Fe-silicates and Fe-oxides on bacterial surfaces in samples collected near hydrothermal vents on the Southern Explorer Ridge in northeast Pacific Ocean, *Am. Mineral.*, *83*, 1399–1408.

Fouquet, Y., A. Wafik, et al. (1993), Tectonic setting, mineralogical and geochemical zonation in the Snake Pit sulfide deposit, *Econ. Geol.*, *88*, 2048–2036.

Gold, T. (1992), The deep, hot biosphere, *Proc. Natl. Acad. Sci. U. S. A.*, *89*, 6045–6049.

Goldfarb, M. S., D. R. Converse, et al. (1983), The genesis of hot spring deposits on the East Pacific Rise, 21°N. *Econ. Geol. Monogr.*, *5*, 184–197.

Hannington, M. D., and S. D. Scott (1988), Mineralogy and geochemistry of a hydrothermal silica-sulfide-sulfate spire in the caldera of Axial Seamount, Juan de Fuca Ridge, *Can. Mineral.*, *26*, 603–625.

Hannington, M. D., et al. (1995), Physical and chemical processes of seafloor mineralization at mid-ocean ridges, in *Seafloor Hydrothermal Systems: Physical, Chemical, Biological, and Geological Interactions*, *Geophys. Monogr. Ser.*, vol. 91, edited by S. E. Humphris, et al., pp. 115–157, AGU, Washington, D. C.

Harmsen, H. J. M., D. Prieur, et al. (1997a), Distribution of microorganisms in deep-sea hydrothermal vent chimneys investigated by whole-cell hybridization and enrichment culture of thermophilic subpopulations, *Appl. Environ. Microbiol.*, *63*, 2876–2883.

Harmsen, H. J. M., D. Prieur, et al. (1997b), Group-specific 16S rRNA-targeted oligonucleotide probes to identify thermophilic bacteria in marine hydrothermal vents, *Appl. Environ. Microbiol.*, *63*(10), 4061–4068.

Haymon, R. M. (1983), Growth history of hydrothermal black smoker chimneys, *Nature*, *301*, 695–698.

Hedrick, D. B., R. D. Pledger, et al. (1992), In-situ microbial ecology of hydrothermal vent sediments, *FEMS Microbiol. Ecol.*, *101*, 1–10.

Hoek, J., A. Banta, et al. (2003), Microbial diversity of a sulphide spire located in the Edmond deep-sea hydrothermal vent field on the Central Indian Ridge, *Geobiology*, *1*, 119–127.

Holden, J. F., and R. M. Daniel (2004), Upper temperature limit of life based on hyperthermophile culture experiments and field observations, in *The Subseafloor Biosphere at Mid-ocean Ridges*, *Geophys. Monogr. Ser.*, vol. 144, edited by W. S. D. Wilcock, et al., pp. 13–24, AGU, Washington, D. C.

Holden, J. F., M. Summit, et al. (1998), Thermophilic and hyperthermophilic microorganisms in 3–30°C fluids following a deep-sea volcanic eruption, *FEMS Microbiol. Ecol.*, *25*, 33–41.

Holden, J. F., K. Takai, et al. (2001), Diversity among three novel groups of hyperthermophilic deep-sea *Thermococcus* species from three sites in the northeastern Pacific Ocean, *FEMS Microbiol. Ecol.*, *36*, 51–60.

Holland, M. E., and J. A. Baross (2003), Limits to life in hydrothermal systems, in *Energy and Mass Transfer in Marine Hydrothermal Systems*, edited by P. E. Halbach, V. Tunnicliffe, and J. R. Hein, pp. 235–248, Dahlem University Press, Berlin.

Houghton, J. L., W. E. Seyfried Jr., et al. (2007), Continuous enrichment culturing of thermophiles under sulfate and nitrate-reducing conditions and at deep-sea hydrostatic pressures, *Extremophiles*, *11*, 371–382.

Huber, H., S. Burggraf, et al. (2000), *Ignicoccus* gen. nov., a novel genus of hyperthermophilic, chemolithoautotrophic *Archaea*, represented by two new species, *Ignicoccus islandicus* sp. nov. and *Ignicoccus pacificus* sp. nov., *Int. J. System. Evol. Microbiol.*, *50*, 2093–2100.

Huber, J. A., D. A. Butterfield, et al. (2002), Temporal changes in archaeal diversity and chemistry in a mid-ocean ridge subseafloor habitat, *Appl. Environ. Microbiol.*, *68*, 1585–1594.

Huber, J. A., D. A. Butterfield, et al. (2003), Bacterial diversity in a subseafloor habitat following a deep-sea volcanic eruption, *FEMS Microbiol. Ecol.*, *43*, 393–409.

Huber, J. A., D. B. M. Welch, et al. (2007), Microbial population structures in the deep marine biosphere, *Science*, *318*, 97–100.

Hügler, M., H. Huber, et al. (2003), Autotrophic CO_2 fixation pathways in archaea (Crenarcheota), *Arch. Microbiol.*, *179*, 160–173.

Hügler, M., C. O. Wirsen, et al. (2005), Evidence for autotrophic CO_2 fixation via the reductive tricarboxylic acid cycle by members of the epsilon subdivision of proteobacteria, *J. Bacteriol.*, *187*(9), 3020–3027.

Jannasch, H. W. (1985), The chemosynthetic support of life and the microbial diversity at deep-sea hydrothermal vents, *Proc. R. Soc. Lond., Ser. B, Biol. Sci.*, *225*(1240), 277–297.

Jannasch, H. W. (1995), Microbial interactions with hydrothermal fluids, in *Seafloor Hydrothermal Systems: Physical, Chemical, Biological, and Geological Interactions*, *Geophys. Monogr. Ser.*, vol. 91, edited by S. E. Humphries, et al., pp. 273–296, AGU, Washington, D. C.

Jannasch, H. W., and M. J. Mottl (1985), Geomicrobiology of deep-sea hydrothermal vents, *Science*, *229*(4715), 717–725.

Jannasch, H. W., and C. O. Wirsen (1981), Morphological survey of microbial mats near deep-sea thermal vents, *Appl. Environ. Microbiol.*, *41*, 528–538.

Johnson, H. P., J. A. Baross, et al. (2006a), On sampling the oceanic upper crustal reservoir, *Geofluids*, *6*, 251–271.

Johnson, M. R., S. B. Conners, et al. (2006b), The *Thermotoga maritima* phenotye is impacted by syntrophic interaction with *Methanococus jannaschii* in hyperthermophilic coculture, *Appl. Environ. Microbiol.*, *72*(1), 811–818.

Jones, W. J., J. A. Leigh, et al. (1983), *Methanococcus jannaschii* sp. nov., an extremely thermophilic methanogen from a submarine hydrothermal vent, *Arch. Microbiol.*, 136, 254–261.

Juniper, S. K., and J. Sarrazin (1995), Interaction of vent biota and hydrothermal deposits: Present evidence and future experimentation, in *Seafloor Hydrothermal Systems: Physical, Chemical, Biological, and Geological Interactions*, *Geophys. Monogr. Ser.*, vol. 91, edited by S. E. Humphris, et al., pp. 178–193, AGU.

Juniper, S. K., and B. M. Tebo (1995), Microbe-metal interactions and mineral deposition at hydrothermal vents, in *The Microbiology of Deep-Sea Hydrothermal Vents*, edited by D. S. Karl, pp. 219–253, CRC Press, Boca Raton, FL.

Karl, D. M. (1995), Ecology of free-living hydrothermal vent microbial communities, in *The Microbiology of Deep-Sea Hydrothermal Vents*, edited by D. M. Karl, pp. 35–124, CRC Press, Boca Raton, FL.

Karl, D. M., G. T. Taylor, et al. (1988), A microbiological study of Guaymas Basin high temperature hydrothermal vents, *Deep-Sea Res.*, *35*(5), 777–791.

Karl, D. M., C. O. Wirsen, et al. (1980), Deep-sea primary productivity at the Galapagos hydrothermal vents, *Science*, *207*, 1345–1347.

Kashefi, K., and D. R. Lovley (2003), Extending the upper temperature limit for life, *Science*, *301*, 934.

Kashefi, K., D. E. Holmes, et al. (2002), Use of Fe(III) as an electron acceptor to recover previously uncultured hyperthermophiles: Isolation and characterization of *Geothermobacterium ferrireducens* gen. nov., sp. nov., *Appl. Environ. Microbiol.*, *68*(4), 1735–1742.

Kashefi, K., D. E. Holmes, et al. (2004), Potential importance of dissimilatory Fe(III)-reducing microorganisms in hot sedimentary environments, in *The Subseafloor Biosphere at Mid-Ocean Ridges*, *Geophys. Monogr. Ser.*, vol. 144, edited by W. S. D. Wilcock, et al., pp. 199–121, AGU, Washington, D.C.

Kelley, D., et al. (2004), Towards determining the upper temperature limits to life on Earth: An in-situ sulfide-microbial incubator, *Eos. Trans. AGU*, *85*(47): Fall Meet. Suppl., Abstract OS42A-05.

Kelley, D. S., J. A. Baross, et al. (2002), Volcanoes, fluids, and life at mid-ocean ridge spreading centers, *Annu. Rev. Earth Planet. Sci.*, *30*, 385–491.

Kelley, D. S., J.R. Delaney, and D.R. Yoerger (2001), Geology and venting characteristics of the Mothra Hydrothermal field, Endeavour Segment, Juan de Fuca Ridge." *Geology*, *29*, 959–962.

Kormas, K. A., M. K. Tivey, et al. (2006), Bacterial and archaeal phylotypes associated with distinct mineralogical layers of a white smoker spire from a deep-sea hydrothermal vent site (9°N, East Pacific Rise), *Environ. Microbiol.*, *8*(5), 909–920.

Koski, R. A., I. R. Jonasson, D. C. Kadko, V. K. Smith, and F. L. Wong (1994), Compositions, growth mechanisms, and temporal relations of hydrothermal sulfide-sulfate-silica chimneys at the northern Cleft Segment, Juan de Fuca Ridge, *J. Geophys. Res.*, *99*, 4813–4832.

Kristall, B., D. S. Kelley, M. D. Hannington, and J. R. Delaney (2006), Growth history of a diffusely venting sulfide structure from the Juan de Fuca Ridge: A petrological and geochemical study, *Geochem., Geophys., Geosyst.*, *7*, Q07001, doi:10.1029/2005GC001166.

Lalou, C., J.-L. Reyss, E. Brichet, P. A. Rona, and G. Thompson (1995), Hydrothermal activity on a 10^5-year scale at a slow-spreading ridge, TAG hydrothermal field, Mid-Atlantic Ridge 26°N, *J. Geophys. Res.*, *100*, 17,855–17,862.

LaPaglia, C., and P. L. Hartzell (1997), Stress-induced production of biofilm in the hyperthermophile *Archaeoglobus fulgidis*, *Appl. Environ. Microbiol.*, *63*(8), 3158–3163.

Lepage, E., E. Marguet, et al. (2004), Molecular diversity of new *Thermococcales* isolates from a single area of hydrothermal deep-sea vents as Revealed by randomly amplified polymorphic DNA fingerprinting and 16S rRNA gene sequence analysis, *Appl. Environ. Microbiol.*, *70*(3), 1277–1286.

Lilley, M. D., D. A. Butterfield, et al. (2003), Magmatic events can produce rapid changes in hydrothermal vent chemistry, *Nature*, *422*, 878–881.

Longnecker, K., and A.-L. Reysenbach (2001), Expansion of the geographic distribution of a novel lineage of ε-proteobacteria to a hydrothermal vent site on the southern East Pacific Rise, *FEMS Microbiol. Ecol.*, *35*, 287–293.

Lovely, D. R., and S. Goodwin (1988), Hydrogen concentrations as an indicator of the predominant terminal electron accepting process in aquatic sediments, *Geochim. Cosmochim. Acta*, *52*, 11–18.

Lovley, D. R., F. H. Chapelle, et al. (1994), Use of dissolved H_2 concentrations to determine distribution of microbially catalyzed redox reactions in anoxic groundwater, *Environ. Sci. Technol.*, *28*(7), 1205–1210.

Ludwig, W., O. Strunk, et al. (2004), ARB: A software environment for sequence data, *Nucleic Acids Res.*, *32*, 1363–1371.

McCliment, E. A., K. M. Voglesonger, et al. (2006), Colonization of nascent, deep-sea hydrothermal vents by a novel Archaeal and Nanoarchaeal assemblage, *Environ. Microbiol.*, *8*(1), 114–125.

McCollom, T. M., and E. L. Shock (1997), Geochemical constraints on chemolithoautotrophic metabolism by microorganisms in seafloor hydrothermal systems, *Geochim. Cosmochim. Acta*, *61*, 4375–4391.

Mehta, M. P., and J. A. Baross (2006), Nitrogen fixation at 92°C by a hydrothermal vent archaeon, *Science*, *314*, 1783–1786.

Mitsuzawa, S., S. Deguchi, et al. (2005), Flow-type apparatus for studying thermotolerance of hyperthermophiles under conditions simulating hydrothermal vent circulation, *Deep-Sea Res. I*, *52*(6), 1085–1092.

Muralidharan, V., K. D. Rinker, et al. (1997), Hydrogen transfer between methanogens and fermentative heterotrophs in hyperthermophilic cocultures, *Biotechnol. Bioeng.*, *56*(3), 268–278.

Nercessian, O., A.-L. Reysenbach, et al. (2003), Archaeal diversity associated with in situ samplers deployed on hydrothermal vents on the East Pacific Rise (13°N), *Environ. Microbiol.*, *5*(6), 492–502.

Nercessian, O., N. Bienvenu, et al. (2005), Diversity of functional genes of methanogens, methanotrophs and sulfate reducers in deep-sea hydrothermal environments, *Environ. Microbiol.*, *7*(1), 118–132.

Orphan, V. J., C. H. House, et al. (2001), Methane-consuming Archaea revealed by directly-coupled isotopic and phylogenetic analysis, *Science*, *293*, 484–487.

Pagé, A., S. K. Juniper, et al. (2004), Microbial diversity associated with a *Paralvinella sulfincola* tube and adjacent substratum on an active deep-sea vent chimney, *Geobiology*, 2, 225–238.

Pley, U., J. Schipka, et al. (1991), *Pyrodictium abyssi* sp. nov. represents a novel heterotrophic marine archaeal hyperthermophile growing at 110°C, *System. Appl. Microbiol.*, 14, 245–253.

Pysz, M. A., S. B. Conners, et al. (2004a), Transcriptional analysis of biofilm formation processes in the anaerobic, hyperthermophilic bacterium *Thermotoga maritima*, *Appl. Environ. Microbiol.*, *70*(10), 6098–6112.

Pysz, M. A., C. I. Montero, S. R. Chhabra, R. M. Kelly, and K. D. Rinker (2004b), Significance of polysaccharides in microbial physiology and the ecology of hydrothermal vent environments, in *The Subseafloor Biosphere at Mid-ocean Ridges*, *Geophys. Monogr.*, vol. 144, edited by W. S. D. Wilcock, et al., pp. 213–226, AGU, Washington, D. C.

Reysenbach, A.-L., and S. L. Cady (2001), Microbiology of ancient and modern hydrothermal systems, *Trends Microbiol.*, *9*, 79–86.

Reysenbach, A.-L., and E. Shock (2002), Merging genomes with geochemistry in hydrothermal ecosystems, *Science*, *296*, 1077–1082.

Reysenbach, A.-L., K. Longnecker, et al. (2000), Novel bacterial and archaeal lineages from an in situ growth chamber deployed at a Mid-Atlantic Ridge hydrothermal vent, *Appl. Environ. Microbiol.*, *66*(9), 3798–3806.

Reysenbach, A.-L., Y. Liu, et al. (2006), A ubiquitous thermoacidophilic archaeon from deep-sea hydrothermal vents, *Nature*, *442*, 444–447.

Rieger, G., R. Rachel, et al. (1995), Ultrastructure of the hyperthermophilic archaeon *Pyrodictium abyssi*, *J. Struct. Biol.*, *115*, 78–87.

Rieger, G., K. Müller, et al. (1997), Cultivation of hyperthermophilic archaea in cellulose capillaries resulting in improved preservation of fine structures, *Arch. Microbiol.*, *168*, 373–379.

Rinker, K. D., and R. M. Kelly (1996), Growth physiology of the hyperthermophilic archaeon *Thermococcus litoralis*: Development of a sulfur-free defined medium, characterization of an exopolysaccharide, and evidence of biofilm formation, *Appl. Environ. Microbiol.*, *62*(12), 4478–4485.

Rogers, D. R., C. M. Santelli, et al. (2003), Geomicrobiology of deep-sea deposits: Estimating community diversity from low-temperature seafloor rocks and minerals, *Geobiology*, *1*, 109–117.

Sarrazin, J., S. K. Juniper, et al. (1999), Physical and chemical factors influencing species distribution on hydrothermal sulfide edifices of the Juan de Fuca Ridge, northeast Pacific, *Mar. Ecol. Prog. Ser.*, *190*, 89–112.

Sarrazin, J., C. Levesque, et al. (2002), Mosaic community dynamics on Juan de Fuca Ridge sulfide edifices: Substratum, temperature and implications for trophic structure, *Cahr. Biol. Mar.*, *43*, 275–279.

Schrenk, M. O., D. S. Kelley, et al. (2003), Incidence and diversity of microorganisms within the walls of an active deep-sea sulfide chimney, *Appl. Environ. Microbiol.*, *69*(6), 3580–3592.

Schultz, A., J. R. Delaney, et al. (1992), On the partitioning of heat flux between diffuse and point source seafloor venting, *J. Geophys. Res.*, *97*, 12299–12314.

Schultz, A., P. Dickson, et al. (1996), Temporal variations in diffuse hydrothermal flow at TAG, *Geophys. Res. Lett.*, *23*, 3471–3474.

Shock, E. L., and M. E. Holland (2004), Geochemical energy sources that support the subsurface biosphere, in *The Subseafloor Biosphere at Mid-ocean Ridges*, *Geophys. Monogr. Ser.*, vol. 144, edited by W. S. D. Wilcock, et al., pp. 153–165, AGU, Washington, D. C.

Shock, E. L., and M. D. Schulte (1998), Organic synthesis during fluid mixing in hydrothermal systems, *J. Geophys. Res.*, *103*(E12), 28,513–28,527.

Stetter, K. O., H. König, et al. (1983), *Pyrodictium* gen. nov., a new genus of submarine disc-shaped sulfur-reducing archaebacteria growing optimally at 105°C, *System. Appl. Microbiol.*, *4*, 535–551.

Stewart, P. S. (2003), Diffusion in biofilms, *J. Bacteriol.*, *185*(5), 1485–1491.

Straube, W. L., J. W. Deming, et al. (1990), Particulate DNA in smoker fluids: Evidence for existence of microbial populations in hot hydrothermal systems, *Appl. Environ. Microbiol.*, *56*(5), 1440–1447.

Summit, M. (2000), Ecology, physiology, and phylogeny of subseafloor thermophiles from mid-ocean ridge environments, School of Oceanography, Seattle, University of Washington.

Summit, M., and J. A. Baross (2001), A novel microbial habitat in the mid-ocean ridge subseafloor, *Proc. Natl. Acad. Sci. U. S. A.*, *98*(5), 2158–2163.

Summit, M., A. Peacock, et al. (2000), Phoshpholipid fatty-acid derived microbial biomass and community dynamics in hot, hydrothermally influenced sediments from the Middle Valley, Juan de Fuca Ridge, in *Ocean Drill. Program, Sci. Results*, edited by R. A. Zierenberg, Fouquet, Y., Miller, D.J., Normark, W.R, *169*: 1–19.

Takai, K., and K. Horikoshi (1999), Genetic diversity of Archaea in deep-sea hydrothermal vent environments, *Genetics*, *152*, 1285–1297.

Takai, K., T. Komatsu, et al. (2001), Distribution of Archaea in a black smoker chimney structure, *Appl. Environ. Microbiol.*, *67*, 3618–3629.

Takai, K., T. Gamo, et al. (2004), Geochemical and microbiological evidence for a hydrogen-based, hyperthermophilic subsurface lithoautotrophic microbial ecosystem (HyperSLiME) beneath an active deep-sea hydrothermal field, *Extremophiles*, *8*, 269–282.

Taylor, C. D., C. O. Wirsen, et al. (1999), Rapid microbial production of filamentous sulfur mats at hydrothermal vents, *Appl. Environ. Microbiol.*, *65*, 2253–2255.

Tivey, M. K. (1995a), The influence of hydrothermal fluid composition and advection rates on black smoker chimney mineralogy: Insights from modeling transport and reaction, *Geochim. Cosmochim. Acta*, *59*, 1933–1949.

Tivey, M. K. (1995b), Modeling chimney growth and associated fluid flow at seafloor hydrothermal vent sites, in *Seafloor Hydrothermal Systems: Physical, Chemical, Biological, and Geological Interactions*, *Geophys. Monogr. Ser.*, vol. 91, edited by S. E. Humphries, et al., pp. 158–177, AGU, Washington, D. C.

Tivey, M. K. (2004), Environmental conditions within active seafloor vent structures: Sensitivity to vent fluid composition and fluid flow, in *The Subseafloor Biosphere at Mid-ocean Ridges*, *Geophys. Monogr. Ser.*, vol. 144, edited by W. S. D. Wilcock, et al., pp. 137–152, AGU, Washington, D. C.

Tivey, M. K., and J. R. Delaney (1986), Growth of large sulfide structures on the Endeavour Segment of the Juan de Fuca Ridge, *Earth Planet. Sci. Lett.*, *77*, 303–317.

Tivey, M. K., and R. E. McDuff (1990), Mineral precipitation in the walls of black smoker chimneys: A quantitative model of transport and chemical reaction, *J. Geophys. Res.*, *95*, 12617–12637.

Tivey, M. K., A. M. Bradley, et al. (2002), Insights into tide-related variability at seafloor hydrothermal vents from time-series temperature measurements, *Earth Planet. Sci. Lett.*, *202*, 693–707.

Tivey, M. K., D. S. Stakes, et al. (1999), A model for growth of steep-sided vent structures on the Endeavour Segment of the Juan de Fuca Ridge: Results of the petrologic and geochemical study, *J. Geophys. Res.*, *102*, 22859–22883.

Verati, C., P. d. Donato, et al. (1999), Evidence of bacterial activity from micrometer-scale layer analyses of black-smoker sulfide structures (Pito Seamount Site, Easter microplate), *Chem. Geol.*, *158*, 257–269.

Von Damm, K. L. (1990), Seafloor hydrothermal activity: Black smoker chemistry and chimneys, *Annu. Rev. Earth Planet. Sci.*, *18*, 173–204.

Von Damm, K. L. (1995a). Controls on the chemistry and temporal variability of seafloor hydrothermal fluids, in *Seafloor Hydrothermal Systems: Physical, Chemical, Biological, and Geological Interactions, Geophys. Monogr.*, vol. 91, S. E. Humphries, R. A. Zierenberg, S. Mullineaux, and R. E. Thomson, pp. 222–247, American Geophysical Union, Washington, D. C.

Von Damm, K. L., S. E. Oosting, et al. (1995b), Evolution of East Pacific Rise hydrothermal vent fluids following a volcanic eruption, *Nature*, *375*, 47–50.

Vorholt, J. A., D. Hafenbradl, et al. (1997), Pathways of autotrophic CO_2 fixation and of dissimilatory nitrate reduction to N_2O in *Ferroglobus placidus*, *Arch. Microbiol.*, *167*, 19–23.

Wheat, C. G., and M. J. Mottl (2000), Composition of pore and spring waters from Baby Bare: Global implications of geochemical fluxes from a ridge flank hydrothermal system, *Geochim. Cosmochim. Acta*, *64*, 629–642.

Whitman, W. B., D. C. Coleman, et al. (1998), Prokaryotes: The unseen majority, *Proc. Natl. Acad. Sci. U. S. A.*, *95*, 6578–6583.

Wilcock, W. S. D., E. F. DeLong, D. S. Kelley, J. A. Baross, S. C. Cary (Eds.) (2004), *The Subsurface Biosphere at Mid-Ocean Ridges*, AGU, Washington, D. C.

Wirsen, C. O., H. W. Jannasch, et al. (1993), Chemosynthetic microbial activity at Mid-Atlantic Ridge hydrothermal vent sites, *J. Geophys. Res.*, *98*, 9693–9703.

Woods, A. W., and J. R. Delaney (1992), The heat and fluid transfer associated with the flanges on hydrothermal venting structures, *Earth Planet. Sci. Lett.*, *112*, 117–129.

Processes and Interactions in Macrofaunal Assemblages at Hydrothermal Vents: A Modeling Perspective

Katriona Shea,[1] Anna Metaxas,[2] Curtis R. Young,[3] and Charles R. Fisher[4]

Our understanding of the biological assemblages at hydrothermal vents is growing rapidly, in part facilitated by the coupling of experimental and observational methods with theoretical modeling efforts. We review theoretical approaches in four main areas and link them to empirical studies of the biology of hydrothermal vent systems. First, we describe models of dispersal of vent organisms within and among vent sites. These models fall broadly into two categories: those based on larval biology and currents that predict dispersal capabilities from a source, and those using genetic data to address historical genetic links between populations. Then we discuss models for individual and population growth at a location and give examples from work in a biologically similar deep-sea environment: cold seeps. Third, we address how these two aspects can be integrated using models that couple dispersal and demography. Last, we summarize existing modeling approaches for community succession in other types of habitats, and outline the conceptual models and the sorts of data that currently exist for hydrothermal vent communities. Throughout, we also examine the use of models in improving experimental design.

1. INTRODUCTION

The geological setting of hydrothermal vents presents the biological assemblages that inhabit them with several important challenges. First, the physical and chemical characteristics of vent habitats limit their accessibility to relatively few specifically adapted species, and often limit them to a particular type of microhabitat. Second, these habitats can be geologically unstable on annual to decadal timescales and chemically unstable on subannual scales. Individual habitats are subject to destruction by paving, chimney collapse or cessation of hydrothermal flow, as well as other less extreme forms of disturbance such as changes in the chemistry, temperature, or flow paths of the local source of vent fluid. Third, vent sites can be separated by hundreds to thousands of kilometers of habitat that is inhospitable to vent organisms. The difficulties faced by organisms in this extreme, unstable, and patchy habitat are also faced by the researchers attempting to understand their dispersal, establishment, population growth, and the biological communities that result. Observational and experimental studies that would be relatively simple in terrestrial or even shallow-water marine systems are logistically challenging or even impossible in these extreme deep sea-habitats. The use of models as a tool to augment empirical efforts has a strong appeal.

[1]Ecology Program and Department of Biology, The Pennsylvania State University, University Park, Pennsylvania, USA.

[2]Department of Oceanography, Dalhousie University, Halifax, Nova Scotia, Canada.

[3]Department of Organismic and Evolutionary Biology, Harvard University, Cambridge, Massachusetts, USA.

[4]Department of Biology, The Pennsylvania State University, University Park, Pennsylvania, USA.

Magma to Microbe: Modeling Hydrothermal Processes at Ocean Spreading Centers
Geophysical Monograph Series 178

10.1029/178GM13

Although modeling of the physical and chemical environment at deep-sea hydrothermal vents is relatively advanced, modeling of biological systems in these locations is in its initial stages. In part, this is the result of a lack of sufficient biological information for many types of models. Although it can be argued that models are difficult, if not impossible, to formulate without adequate biological information, models can, in fact, significantly advance our understanding of these deep-sea systems. For example, by developing alternative models based on our best understanding of the systems, and using insights from other marine systems, it may be possible to assess the likelihood that certain mechanisms and processes explain observed patterns. Where alternative models differ in predictions, we may be able to design studies to assess which model is more plausible. In addition, models can be used to develop or refine the experimental methods used to assess such alternatives. Once a mechanistic understanding is reached, models can also be used to identify important parameters on which to focus more detailed empirical efforts.

The dynamics of deep-sea hydrothermal vent populations and communities are driven by a combination of local population growth and community development within a suitable patch, and the dispersal of organisms between patches. The existence of the patches themselves in the generally nutrient-limited deep sea is a result of microbial primary productivity. Within a patch, spatial and temporal differences in the macrobiological communities are tightly tied to chemical and thermal properties of diffuse vent effluent. Models addressing chemical transformations that occur in hot vent fluid and sulfide structures, and the microbial populations that contribute to these transformations, are discussed elsewhere in this volume [*Huber and Holden*, this volume; *McCollom*, this volume; *Schrenk et al.*, this volume]. In this paper, we discuss models that focus on four main aspects of the macrobiological communities of these systems. For these models, microbial chemoautotrophic primary production is tied to vent chemistry, although details of chemical transformations that determine the properties of diffuse vent fluid are not considered. First, we discuss models of dispersal, both in general and specifically between suitable habitat patches, including the resultant genetics of the postsettlement population. These spatial aspects have been receiving increased experimental and theoretical attention in recent years, and there is a rich literature on similar processes in terrestrial and shallow marine systems. Second, we consider models of population growth and of energy flow and allocation within individuals at a location, which are only now becoming possible because of our increased understanding of the basic biology of some key vent species. Third, we present a suite of increasingly complex models that incorporate both the regional and local processes affecting populations at a variety of spatial and temporal scales. Increasingly complex models that link demography and dispersal are being developed in terrestrial systems [see, e.g., *Neubert and Caswell*, 2000; *Easterling et al.*, 2000; *Jongejans et al.*, 2008; *Skarpaas and Shea*, 2007], and clearly many of these can be usefully applied to address vent population and community dynamics. Several ecological studies and long-term observations over the past 15 years have described patterns of community change over time and have begun to explore potential factors that regulate these patterns. In the fourth section, we provide an overview of observations and experiments, and of conceptual models for community succession at hydrothermal vents. Throughout this work, we also address ways in which models can be used to assist in directing research, or in designing empirical studies.

2. MODELS OF DISPERSAL

Because of the patchy and ephemeral nature of vents and the high endemism of vent biota, successful colonization of newly formed vent sites can usually only be achieved by long-distance larval dispersal. Dispersal is defined as the movement of diaspores away from their source (sensu *Nathan* [2001]). A diaspore is the unit of dispersal and may be a whole organism (at any life stage) or part of an organism (e.g., a piece of a modular organism). Because most vent organisms disperse during their larval stage (either through evolutionary constraints or because adults are sessile), we will focus on these early life-history dispersal stages. However, we note that adults of some groups (e.g., fish, crabs, shrimp) are quite motile and may also have limited dispersal potential.

There are three main sequential components to dispersal, which relate to the source of the diaspores, the movement of the diaspores, and the cessation of movement at a destination [*Isard and Gage*, 2001]. Most vent larvae are effectively free-drifting because they are very small, with relatively little ability to propel themselves over significant distances by swimming. Successful movement between suitable hydrothermal vent habitat patches is a complex process, and depends on interactions between physical flows, particularly deep-sea currents, hydrothermal plume dynamics, and diaspore biology (e.g., larval physiology, buoyancy, and life span), as well as on the availability of suitable habitat once the larvae are ready to settle. The two main approaches that have been used to better understand the mechanisms of animal dispersal between vent sites arise from focusing on either the movement of all diaspores away from a source (be they successful or not), or on successful movement between a source and a particular destination.

2.1 Models of Movement From a Source

In models that focus on the movement process, combination of estimates of biological parameters (such as larval abundance, larval development/survival times, fecundity, recruit to adult ratio) and of point estimates of hydrodynamics (used to constrain current speeds and directions) are used to predict the physical displacement of larvae from the time of release until the end of competence (the last stage in larval development before settlement). This estimated displacement is independent of the availability of habitat suitable for settlement at that time; many diaspores disperse without ever reaching suitable habitat and thus are lost. In non-vent marine ecosystems, larval mortality during the dispersal phase of life history can be as high as 90% [*Rumrill*, 1990]. For the tubeworm *Riftia pachyptila*, *Marsh et al.* [2001] estimated that 61% of larvae would be lost during dispersal through advection off the spreading axis to unsuitable habitats. Even for larvae that are retained within the axis, suitable habitat is limited to areas with direct exposure to venting fluid, and consequently only a small percentage of those larvae will successfully settle and colonize.

Different conceptual models of larval dispersal potential have been proposed for spatial scales ranging from between vents within a vent field (10 m–1 km scale), to between vent fields (1–100 km scale) to between ridge segments (100–1000 km scales). At the smallest scale, larval abundance in the water column within a vent field has been measured in a handful of studies on the East Pacific Rise (EPR) and the Juan de Fuca Ridge (JdFR). At all locations, larvae of vent species were more abundant within 5 m of the bottom than at heights >20 m above the bottom [*Mullineaux et al.*, 1995; *Kim and Mullineaux*, 1998; *Metaxas*, 2004; *Mullineaux et al.*, 2005]. Abundance may [*Mullineaux et al.*, 2005] or may not [*Kim and Mullineaux*, 1998; *Metaxas*, 2004] decrease with distance from the vents up to ~10 km, and the observed pattern has been attributed to hydrodynamics [*Metaxas*, 2004]. These studies suggest that larval dispersal of many vent species likely occurs near bottom, where the mean direction of advective currents is typically along the axial valley [*Mullineaux and France*, 1995; *Marsh et al.*, 2001; *Thompson et al.*, 2003].

The intermediate spatial scale (10–100 km) appears to provide a realistic range for along-axis larval dispersal potential within one larval cohort/generation, at least for tubeworms and alvinellid polychaetes. Based on the energy reserves of the egg and the energy requirements for respiration, larvae of the tubeworm *R. pachyptila* can potentially survive in the water column for 38 d [*Marsh et al.*, 2001]. Based on models that used simulated releases of batches of tubeworm larvae from a vent at 9°50′N on the EPR and point measures of currents collected for the area, *Marsh et al.* [2001] suggested that tubeworm larvae can disperse approximately 90 km along axis within 20 days after release.

In addition to modeling actual dispersal, related models can be used to design optimal strategies for sampling propagules or recruits. For example, in cases where the number of logistically possible sampling stations may be limited, the location of such samples will be critical in providing the most relevant information. Models have been used to address such issues for plant seeds dispersing by wind [*Skarpaas et al.*, 2005; *Bullock et al.*, 2006]. Using a simulation model, a range of possible seed-trap arrangements was tested against a range of plausible dispersal kernels, for scenarios of isotropic and anisotropic diaspore dispersal [*Skarpaas et al.*, 2005]. The trapping designs that consistently recaptured the underlying "true" dispersal kernel distributions were the ones most likely to return useful information on dispersal patterns. Given that some commonly used spatial sampling arrangements performed very poorly in these simulations, assessment of the effectiveness of different sampling designs before sample collection will enable more productive use of limited resources and submersible time.

Using alvinellid polychaetes as an example organism, *Chevaldonné et al.* [1997] utilized a propagule flux model (PFM) to define the range of potential dispersal among all known vent fields between 14°N and 7°N on the EPR. The PFM calculates the number of migrants (*Nm*) that are potentially exchanged in one generation between two randomly chosen populations. *Nm* depends on the number of eggs produced per generation in the source population (i.e., produced propagules) and the probability that a dispersing propagule will encounter an active vent while still alive, and it has to equal at least 1 to prevent genetic drift and speciation between populations. Not surprisingly, *Nm* decreased logarithmically with increasing duration of dispersal, and fell below 1 after 8 days. Based on the current regime, this duration corresponds to a traveled distance of 8–40 km. Given the spatial separation of known vent fields in this area, the range of potential dispersal is short and only possible within, but not between, sectors [*Chevaldonné et al.*, 1997].

An alternative conceptual model proposed that mesoscale flows, generated by the buoyant plume, can influence larval dispersal on 100 km scales [*Mullineaux and France*, 1995]. Mesoscale vortices can form in the plume, which are eventually shed downstream. The model proposed that if larvae become entrained and concentrated in the plume, they can be transported in patches in the shed vortex. *Kim et al.* [1994] used a passive tracer to confirm that plume models can be used to estimate larval flux by entrainment. *Kim et al.* [1994] estimate that 100–1000 larvae are entrained per hour and their models suggest that larval concentration in the spreading plume is 2 orders of magnitude lower than near the bottom.

Given the low probability of larval dispersal between active vent fields suggested by existing models, most larvae within a vent field must be produced locally to sustain populations on ecological time scales [*Metaxas*, 2004; *Van Dover and Lutz*, 2004]. *Metaxas* [2004] measured larval supply at Axial Volcano on the JdFR, using passive larval collectors, and found little spatial variability among vents. She proposed that a homogeneous larval pool existed within the entire vent field, which is formed in large part by retention of locally produced larvae whose dispersal is constrained by local near-bottom hydrodynamics.

2.2. Models of Movement Between a Source and a Destination

Models that focus on the links between a source and a destination are only concerned with realized dispersal (i.e., dispersal that also includes successful establishment). Studies utilizing genetic data focus on documenting genetic relationships within and between populations along a ridge axis to examine realized dispersal. Mutations in DNA sequences contain historical information about the demography of species and can be used to test demographic hypotheses. The spatial scale of larval movement determines the genetic structure of populations because adults of most vent invertebrates are sessile or have limited dispersal capabilities.

There is considerable interest in using genetic data to infer population connectivity in the shallow water marine environment. Initially, the lack of obvious physical barriers in the ocean led to the parsimonious expectation of broad larval dispersal and genetically well-mixed populations. In fact, self-recruitment and high local retention of larvae is the rule rather than the exception in many shallow water marine species [*Swearer et al.*, 2002; *Jones et al.*, 2005]. Distinct genetic subpopulations have been documented empirically for many marine species (e.g., reef fish [*Taylor and Hellberg*, 2003], Atlantic cod [*Pogson et al.*, 2001], corals [*Baums et al.*, 2005], and barnacles [*Sotka et al.*, 2004]) and likely arise from critical breaks in larval movement due to current patterns at the time of dispersal, even in species with relatively long larval duration. Models suggest that larval behavior may in part explain local retention [*Kingsford et al.*, 2002]. Biophysical models have been used to constrain the spatial extent of ecologically relevant larval dispersal [*James et al.*, 2002; *Cowen et al.*, 2006] and confirm that the interaction of currents and larval biology can lead to such genetically rich patterns.

The relationship between physical and geological processes (e.g., ocean circulation and birth/death of vents), genetic diversity within vent populations, genetic divergence between populations, and total metapopulation diversity is complex. Nevertheless, population geneticists have made considerable progress toward developing genetic models that incorporate population structure and the dynamics of extinction/recolonization processes. Substantial progress also has been made in the development of statistical methodologies to estimate parameters, test hypotheses, and determine model adequacy under structured population models [*Beerli and Felsenstein*, 2001; *Nielsen and Wakeley*, 2001; *Hey and Nielsen*, 2004; *Young et al.*, 2008]. The development of coalescent theory over the last 20 years has contributed immensely to this progress [*Hudson*, 1990; *Rosenberg and Nordborg*, 2002].

Coalescent theory focuses on modeling samples of DNA sequences rather than entire populations. Just as the birth and death of species is a branching process, the birth and death of individuals in a population is a branching process. Thus, the relationships among individuals in a sample can be represented by a genealogy composed of the branching order of individuals and ancestors, as well as the times that they first share a common ancestor (Figure 1a). Coalescent theory predicts the structure of sample genealogies under various demographic scenarios, where a species' demography might include population number and sizes, dispersal rates, and extinction/recolonization rates. The particular configuration of mutations in DNA sequences is the result of mutations that randomly accumulate along this genealogy. Therefore, coalescent theory links DNA sequence data to model parameters of interest through the stochastic ancestral and mutation processes. Using this theory, we can extract historical information about demography from genetic data.

Genetic models generally operate on a time scale of N generations, where N is the total number of individuals in a population. Because many factors (e.g., rapid population fluctuations or a high variance in reproductive success) affect the time scale, geneticists refer to N as an effective population size. In many cases, this time scale might be millions of years [e.g., *Johnson et al.*, 2006]. Coalescent-based genetic estimates of dispersal are averages over thousands to millions of generations, unlike the ecological models previously discussed that generally focus on dispersal over one or a few generations. In addition to the vastly different temporal scales of the direct and indirect approaches, genetic homogeneity of populations is achieved if populations exchange as few as tens of migrants per generation [*Wright*, 1931]. Therefore, although population connectivity in the ecological sense can be quite low, populations may appear genetically homogenous [e.g., *Becker et al.*, 2007]. In addition, species with strongly leptokurtic dispersal kernels (i.e., with a probability distribution of dispersal distances that has

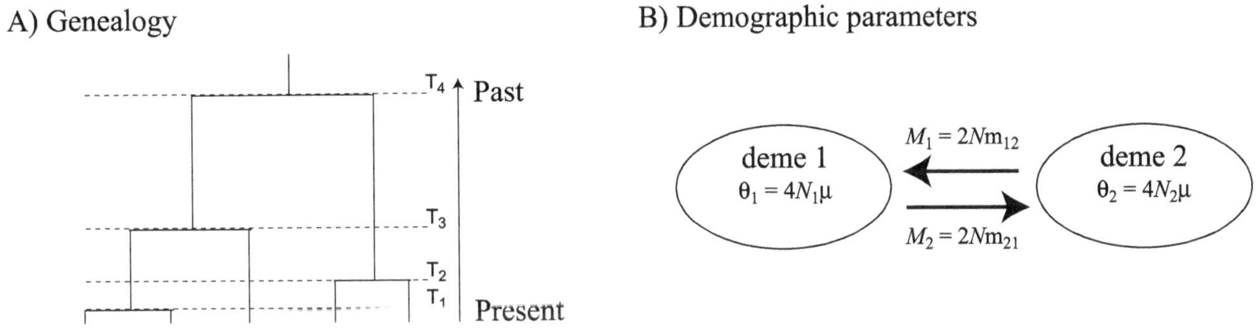

Figure 1. (a) Genealogy composed of branching order and times of ancestor events (T_n). Coalescent theory models genealogies as a reverse-time Markov death process on the number of ancestors in the sample. The waiting times are exponentially distributed and depend on the demographic parameters of the model. DNA sequence mutations are then scattered on the genealogy in proportion to the branch lengths. See *Hudson* [1990] for a detailed treatment of coalescent theory. (b) Demographic parameters in a simple two-deme genetic model.

fatter tails and narrower peaks than the normally distributed equivalent) will generally exhibit genetic differentiation indicative of high dispersal, even if average dispersal distances are considerably lower [*Kinlan and Gaines*, 2003]. Genetic- and invasion-based estimates of dispersal distances for marine taxa often substantially exceed direct estimates of dispersal distances [*Kinlan and Gaines*, 2003]. Genetic estimates can also be influenced by infrequent or sporadic dispersal events due to the small amount of genetic exchange necessary to homogenize populations [*Slatkin*, 1987]. For these reasons, ecological and genetic estimates of dispersal offer complementary information about dispersal processes.

Models of genetic variation in structured (i.e., nonrandomly mating) populations have been extensively studied [*Donnelly and Tavare*, 1995]. A structured population consists of a number of demes of size N, where a deme is a local, randomly mating population. N is estimated from genetic data, which cannot distinguish the effect of mutation rate from that of population size on levels of genetic diversity. Therefore, the population size of deme i is represented by the composite parameter, $\theta_i = 4N_i\mu$, where μ is the per-gene per-generation mutation rate. The number of immigrants entering deme i is $M_i = 2Nm_{ij}$, where m_{ij} is the proportion of immigrants in deme i arriving from deme j in each generation (Figure 1b). We can compare population sizes among demes by comparing estimates of θ_i, because we assume that the mutation rate is the same among populations. Alternately, we can use prior estimates of mutation rates to determine the absolute sizes of populations. To measure dispersal, we are interested in comparing estimates of M_i among different populations.

One fundamental test of dispersal involves the discrimination between island and stepping-stone models of dispersal (Figures 2a and 2b). The island model assumes that immigrants arriving in any deme are drawn from a well-mixed larval pool (i.e., M_i is equal for all demes). In other words, immigrants are equally likely to arrive from any of the demes in the metapopulation so that dispersal occurs on about the same spatial scale as the entire metapopulation. Alternatively, migration might only occur between adjacent demes, such that demes act as stepping-stones for migrants (i.e., M_i is appreciable only for adjacent demes). In this case, dispersal occurs on a much smaller spatial scale than that of the metapopulation, and a pattern of isolation by distance arises [*Wright*, 1943]. The stepping-stone model predicts that genetic divergence between populations (measured by F_{ST}) will be correlated with geographical distance among populations, whereas the island model predicts no such correlation. The likelihood of rejecting the island model depends on several factors including the spatial distribution of demes comprising the metapopulation, the mutation rate of the genetic loci under consideration, the migration rates among demes, the spatial scale of sampling (i.e., 2–5 times the average dispersal distance of the species), the overall magnitude of genetic divergence relative to sampling error in the study, and critically, whether the metapopulation has reached an equilibrium between gene flow and genetic drift [*Slatkin*, 1993; *Palumbi*, 2003].

The island model has been tested in several vent endemic species [*Van Dover et al.*, 2002; *Vrijenhoek*, 1997]. Some vent species show evidence for stepping-stone dispersal. The genetic population structure of the tubeworm *R. pachyptila* [*Black et al.*, 1994], the amphipod *Ventiella sulfuris* [*France et al.*, 1992], and the mussel *Bathymodiolus thermophilus* (*Won et al.* [2003]; but see *Craddock et al.* [1995]) along the EPR fit the stepping-stone model of dis-

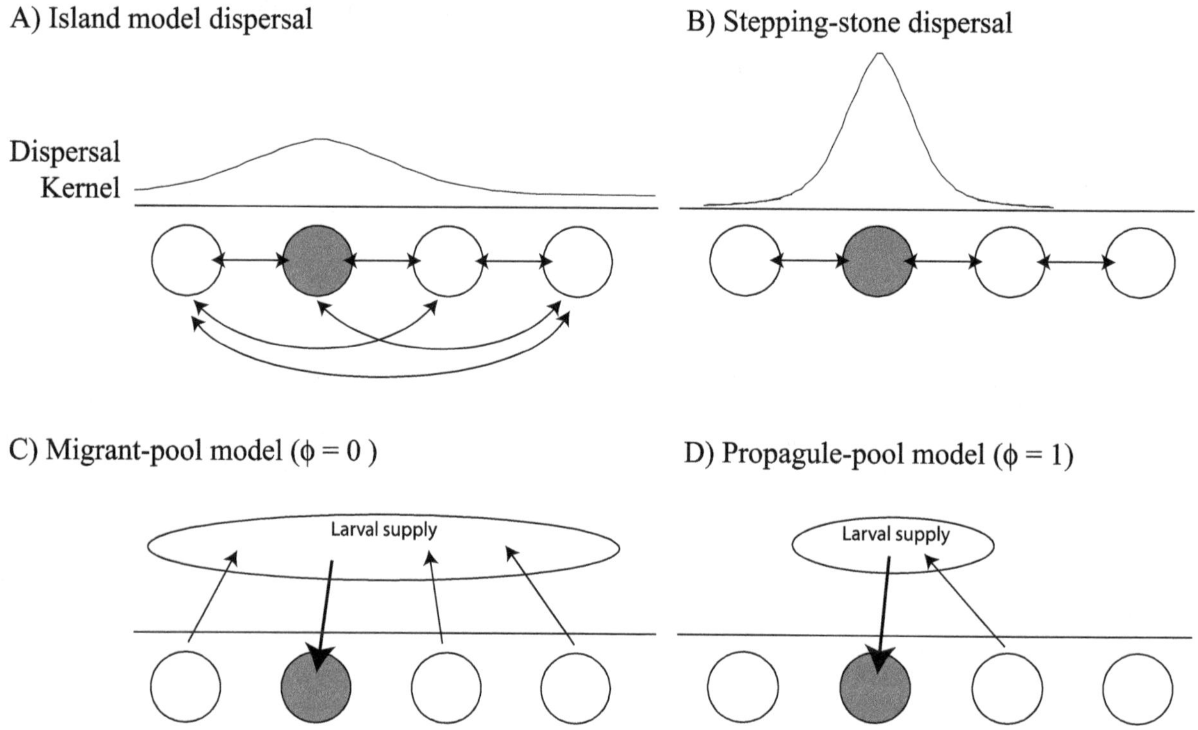

Figure 2. Genetic models of dispersal (a and b) and colonization (c and d). (a) The island model assumes that all demes (circles) exchange migrants. The arrows in the figure denote this connectivity. The hypothetical dispersal kernel in the figure depicts the distribution of larvae produced by the gray deme. (b) The stepping-stone model assumes that the only appreciable migration occurs between adjacent demes. The arrows in the figure denote this limited connectivity. The hypothetical dispersal kernel in the figure depicts the distribution of larvae produced by the gray deme. (c) The migrant-pool model assumes that colonists arriving in the gray deme are drawn from the entire metapopulation, similar to the assumption in the island model that immigrants can originate in any deme. (d) The propagule-pool model assumes that colonists arriving in the gray deme are drawn from a single deme in the metapopulation.

persal. However, other vent invertebrates appear to follow the island model, such as the clam *Calyptogena magnifica* [*Karl et al.*, 1996], as well as two species of archaeogastropods [*Craddock et al.*, 1997] on the northern EPR. The horizontal distance represented in these studies far exceeds our current estimates of the dispersal range of a single larva based on the ecological models discussed in the previous section. However, for these species the genetic data suggest that colonization on these scales is either not limited by larval dispersal, or more likely, one or more of the issues discussed above (e.g., statistical power due to sampling or whether the metapopulation has achieved drift-migration equilibrium) affects these species. Extinction and recolonization can also affect the degree of genetic divergence among populations [*Slatkin*, 1977; *Whitlock and McCauley*, 1990; *Pannell and Charlesworth*, 2000]. These processes may well affect estimates of migration rates between populations as well as alter the likelihood of rejecting the island model of dispersal.

Two fundamental genetic models [*Slatkin*, 1977] of extinction/recolonization processes exist that differ in how populations are colonized, although others have been examined in the literature [*Pannell and Charlesworth*, 2000; *Wakeley*, 2004]. These two models assume a constant number of occupied demes in the metapopulation. Under this assumption, a deme is immediately recolonized after an extinction event that occurs with some probability, e, per generation. The migrant-pool model (Figure 2c), assumes that k colonists are randomly sampled from the entire metapopulation, whereas in the propagule pool model (Figure 2d), colonists are randomly sampled from a single deme. *Whitlock and McCauley* [1990] generalized these two models by introducing a parameter, ϕ, describing the probability that two genes in the colonists come from the same deme. Assuming an infinite number of demes in the metapopulation and island-model migration, they determined that population divergence (measured by F_{ST}) will be decreased under recurrent extinction if:

$$k > \frac{2Nm}{1-\phi} + \frac{1}{2}, \quad (1)$$

where $\phi = 0$ and $\phi = 1$ correspond to the migrant- and propagule-pool models, respectively. If larvae from more than a single deme colonize new demes (i.e., $\phi > 0$) and k is larger than approximately twice the number of migrants arriving in the deme per generation, then population divergence will be reduced by recurrent extinction. Low population divergence has been observed over broad geographic regions in some vent taxa [*Craddock et al.*, 1995, 1997; *Won et al.*, 2003; *Hurtado et al.*, 2004; *Johnson et al.*, 2006; *Young et al.*, 2008]. The high rate of extinction/recolonization at hydrothermal vents suggests that, if the above conditions apply, we should expect less genetic differentiation among populations of vent taxa than that predicted by restricted migration alone. Consequently, applying models that ignore the effects of extinction and recolonization will overestimate dispersal ability. However, we currently know very little about the ratio of the number of immigrants to the numbers of individuals colonizing new vents, whether colonists arrive from multiple sources, or even the rates of extinction of vent communities.

Genetic studies conducted thus far in hydrothermal vent systems use coalescent models that do not explicitly include extinction/recolonization dynamics, because the statistical methodology to fit these models has not yet been developed. However, results from structured models have been carefully interpreted in light of theoretical results from extinction/recolonization models [e.g., *Won et al.*, 2003; *Johnson et al.*, 2006; *Young et al.*, 2008]. Future work will undoubtedly benefit from the ability of these newer models to distinguish the effects of dispersal from those of extinction/recolonization dynamics [*Wakeley and Aliacar*, 2001]. Clearly, ecological estimates of dispersal distances, comparisons of recruitment rates in new and established sites, and information on the frequency of habitat turnover in these systems will help clarify the roles of these processes in shaping the distribution of genetic variation in these systems.

Often, bathymetric features, such as the Easter Microplate or transform faults, are associated with significant genetic divergence of vent species, although these features affect species dispersal to varying degrees. In these instances, the application of genetic and ocean circulation models has clarified the nature of these boundaries and helped to point to mechanisms responsible for reduced dispersal or species isolation. *B. thermophilus* populations north and south of the Easter Microplate are genetically isolated from each other [*Won et al.*, 2003]. Significant cross axis currents produced by the inflated bathymetry of the Easter Microplate were hypothesized to limit dispersal in this mussel species [*Won et al.*, 2003]. In contrast, there is no significant genetic differentiation among populations of either the worm that lives commensally in the mantle cavity of this mussel or among populations of the Pompeii worm *Alvinella pompejana* separated by the Easter Microplate [*Hurtado et al.*, 2004]. The tubeworms, *R. pachyptila* and *Tevnia jerichonana* also seem to traverse the Easter Microplate region more easily than *B. thermophilus*, although significant genetic divergence is observed for these species [*Hurtado et al.*, 2004].

Similar patterns have been observed for some species in the northeast Pacific. A striking example is that of the limpet *Lepetodrilus*, where gene flow was so restricted that speciation has occurred across the Blanco transform fault, resulting in *Lepetodrilus fucensis* occurring to the north and *L. gordensis* to the south of the fault [*Johnson et al.*, 2006]. Although the authors of this study have suggested that major bathymetric barriers can inhibit dispersal, the effect of barriers of similar magnitude is inconsistent, and thus the mechanism of inhibition remains unclear. Recently, deep-ocean circulation was explicitly implicated in the genetic subdivision of the tubeworm *Ridgeia piscesae* across the Blanco transform fault [*Young et al.*, 2008]. These authors proposed that bathymetry and the predominantly southern direction of ocean currents facilitate greater migration (and consequently gene flow) southwards than northwards across the fault (i.e., $M_{GOR} > M_{JDF}$). Interestingly, the genetic diversity of both the limpets and the tubeworms is higher on the Gorda than the Juan de Fuca Ridge (i.e., $\theta_{GOR} > \theta_{JDF}$), possibly because populations on the heavily sedimented Gorda Ridge experience fewer extinction events. However, given the complex relationship between demographic parameters and metapopulation effective size, this hypothesis should be tested by multidisciplinary studies.

The application of genetic models to the study of demography in hydrothermal vent systems has significantly increased our understanding of realized dispersal in these organisms. This progress has benefited greatly from the development of statistical methodologies that allow tests of model predictions and even direct tests between simple models in a likelihood or Bayesian framework. Although statistical methodologies have not yet been developed for some of the more complex genetic models, these methods will undoubtedly become commonplace in the near future. Bayesian inference is particularly promising in that genetic data can be coupled with prior information from ecological estimates of the number of colonists (k), and the frequency of habitat turnover in these systems (e).

3. MODELS OF INDIVIDUAL AND POPULATION GROWTH FOR A SINGLE SPECIES

Modeling growth of individuals and populations at a location can be achieved by using a variety of different growth models. This may be particularly informative if applied to keystone species or ecosystem engineers that provide key habitat and alter nutrient flows to improve habitat suitability for additional species. For example, matrix models [*Caswell*, 2001] have been developed to describe and project the dynamics of a key species at a location. One such study focuses on the numerically dominant limpet *L. fucensis* on the JdFR (Kelly and Metaxas, in preparation). Life history parameters that are incorporated in these population models, such as growth, fecundity, and mortality, are quantified from empirical data collected at vents in Axial Volcano (Kelly and Metaxas, in preparation). The frequent use of matrix models in the terrestrial literature [e.g., *Shea et al.*, 2005] and shallow marine literature [e.g., *Crouse et al.*, 1987] suggests that such applications will likely become more common. Integral projection models, which include continuous demographic variables (such as size of individuals), have also recently been developed [*Easterling et al.*, 2000; *Ellner and Rees*, 2006].

One extremely promising approach is the use of dynamic energy budget (DEB) models, which model energy flow and allocation within individuals [*Kooijman*, 1993, 2000]. Originally developed for toxicological studies, they have proven useful over a range of situations (e.g., Daphnia growth, insect development *Kooijman* [2000]) and at a variety of organizational levels. Excitingly, they allow metabolic processes to be scaled up through individual growth patterns to population and community level dynamics [*Nisbet at al.*, 2000]. These models can be used to improve our understanding of the basic biology of foundation species in vent ecosystems, such as the tubeworms in the Eastern Pacific or the pandemic bathymodiolid mussels. In addition, these models can be used to quantify the role of these species (which also are dominant in terms of biomass) in energy and nutrient cycling, both within the vent ecosystems and in the export potential to the surrounding deep sea.

The laws of thermodynamics apply to all living organisms, and the principle of conservation of energy means that we can model an individual by accounting for all its energetic inputs and outputs. DEB models examine how the energy acquired by an individual is allocated to maintenance, growth and development, and reproduction (Figures 3 and 4). Such models are dynamic, and comprise a series of coupled differential equations linking different processes within the individual, which can be investigated either analytically or numerically. In the models, energy is assimilated and utilized by the individual. Where there are alternative hypotheses regarding the order of allocation to different requirements, alternative models can be developed. For example, in one plausible scenario (Figure 4a), maintenance demands are met first and then the remaining energy is allocated proportionally to growth or to reproductive processes. In another scenario (based on the κ-rule model of *Kooijman* [2000]; Figure 4b), a set proportion of energy is allocated to growth and maintenance (with maintenance requirements being met first), while the remaining proportion is allocated to reproductive development or output. In the latter case, energy flux is modeled so that reproduction may continue even when growth has ceased, while in the former case, both processes co-occur. The true state of affairs would be difficult to assess directly, but differences in model predictions (e.g., predicted size at maturity in different energy flow regimes) can then be used to distinguish between such alternatives and hence to improve our mechanistic understanding. Once developed, the DEB models can be integrated across size and age classes and applied to questions of population and community level dynamics, for example, to estimate the primary and net exported productivity of the communities in these deep-sea environments.

This broad type of approach has been used successfully for tubeworm aggregations at cold seeps [*Cordes et al.*, 2003, 2005]. First, the model quantified the levels of sulfide supply necessary to meet the physiological requirements of the aggregations, by incorporating growth, recruitment and metabolic rates, symbiont metabolism, substrate concentrations, diffusion gradients, and kinetics [*Cordes et al.*, 2003]. The models were then used to examine several alternative mechanisms for the production and supply of sulfide to the interstices of the buried posterior portions of worms [*Cordes et al.*, 2005]. Surprisingly, the models indicated that the natural sulfide production in, and fluxes from, the sediment were not sufficient to sustain large tubeworm aggregates to the ages observed at hydrocarbon seeps. However, when sulfate release (sulfate is one end-product of chemoautotrophic sulfide oxidation by the tubeworm symbionts) through the tubeworm's posterior rootlike extensions into hydrocarbon-rich sediments was invoked, sufficient sulfide was produced to allow the model aggregations to persist for centuries. As a result, sulfate release by the tubeworm roots was investigated and has now been confirmed empirically [*Dattagupta et al.*, 2006]. For any model, sensitivity analyses indicate those parameters with the greatest influence on the system, and can direct the collection of data which otherwise may be overlooked. Importantly, sensitivity analyses of these models guided some of the empirical data collection (such as root length), which were subsequently used to better refine the models.

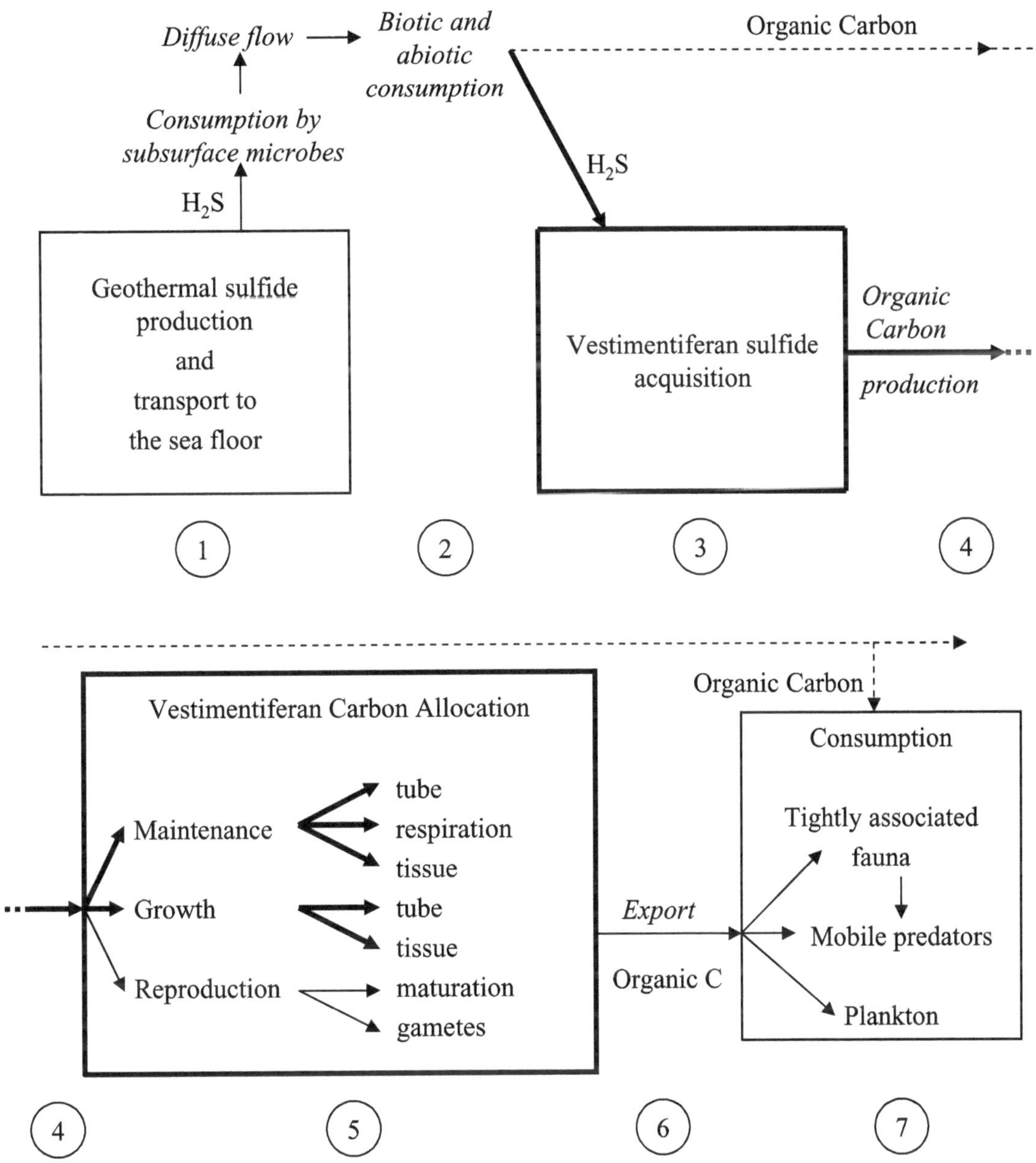

Figure 3. Modeling framework for the study of EPR vestimentiferan communities.

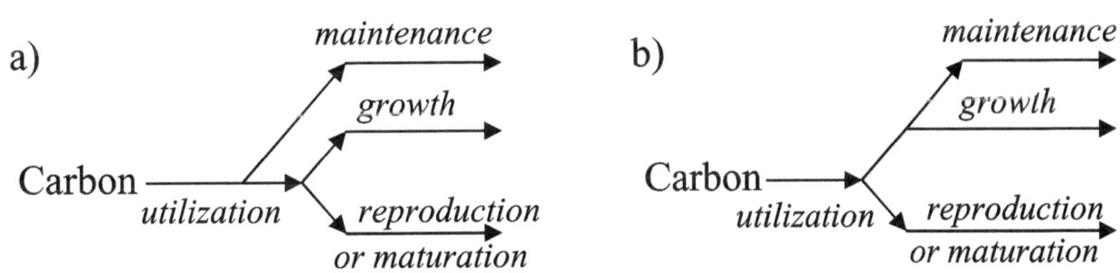

Figure 4. Examples of alternative models for box 5 in Figure 3.

The application of models such as those described above to individuals, populations, and communities at spreading centers would be an obvious and exciting next step. This will require some empirical advances; vent communities exist in a much more spatially and temporally complex environment than seep communities and will require finer scale data collection. However, continuing advances in our ability to survey, monitor, and sample appropriate variables on appropriate scales will enable the new generation of vent models.

4. COUPLED DEMOGRAPHIC AND SPATIAL MODELS

To understand and describe the dynamics of species and communities at landscape scales, models for movement between suitable habitat patches and for growth of individuals, populations, or communities at a location must be combined. There are several axes along which model complexity can be varied: the spectrum from nonspatial, through pseudo-spatial to explicitly spatial models; the shift from single species models, through models with two or a few species, to models that consider whole communities; and the increase in demographic considerations from simple documentation of presence or absence, through models of abundance, to models that include complex demography. Even when considering one species at a time, a fairly complex matrix of possible combinations of levels of detail is possible [e.g., *Jongejans et al.*, 2008]. In the next two sections, we review a few of the most relevant model types, and span these axes, but do not attempt to be exhaustive.

The simplest models that include both demographic and dispersal elements are metapopulation models [*Levins*, 1969; *Hanski*, 1982], commonly used to model the dynamics of patchy terrestrial and marine systems. For a single vent species, the simplest metapopulation models would involve a pseudo-spatial (i.e., space is considered, in that habitat is divided into suitable and unsuitable patches, but exactly where those patches are located is not defined) network of interconnected patches and movement between patches formulated in a relatively simple form. The patches are either occupied by the focal species or not, patches go extinct at a rate determined by geological activity and are colonized at a rate determined by a simple dispersal parameter. Clearly, this approach is highly simplistic, but it incorporates the essential features of local dynamics and between-patch dispersal that underlies more complex models. Patch-occupancy models can be applied where patch states are usually defined by the presence or absence of the species (not its abundance). Documenting species abundances is more complicated, but certainly feasible, and is relatively straightforward for simpler communities. Metapopulation models with specified distances between patches have also been developed [*Hanski et al.*, 1996]. However, metapopulation models of the underlying variation in the patches themselves are less common [e.g., *Nee and May*, 1992].

A metapopulation model (GENEFLUX) has been used to describe the population dynamics of the alvinellid polychaete *A. pompejana* in the vent fields between 14°N and 7°N on the EPR [*Jollivet et al.*, 1999]. The model incorporated interconnected subroutines of population structure, propagule flux (PFM, as in *Chevaldonné et al.* [1997]), genetic drift, and vent site spatial and temporal dynamics. The goal was to examine the effect of vent instability (extinction and activation, modeled as relocation) on long-term (thousands of generations) population structure of alvinellids. Population genetic structure (allele frequency) was modeled by using a series of juvenile and adult mortality-affected "boxes," each simulating a single cohort, with multiple adult cohorts contributing to a single migrant pool per generation. Genetic drift was modeled as the probability distribution of the frequency of a particular allele in every set of migrants, given migrant abundance and initial allele frequency. Randomly selected vents were "relocated" (one vent would shut down and another start up) every 40 generations at locations constrained by one of four modes, each representing a different conceptual model for vent site propagation. The model showed that vent "relocation" can affect metapopulation size, migrant fluxes and thus gene flow, allele fixation and genetic drift, and extinction [*Jollivet et al.*, 1999]. The authors concluded that, within hydrothermally active areas on a ridge crest, local populations are strongly interconnected, whereas between sectors recolonization cannot theoretically balance extinction by propagule dispersal, unless spatial redistribution of hydrothermally active zones over time is also considered.

Spatial detail can also be combined with models of a species' abundance using reaction-diffusion models [*Fisher*, 1937; *Skellam*, 1951; *Shigesada and Kawasaki*, 1997; *Okubo and Levin*, 2001], although analytical versions assume homogeneous space. The number of individuals, n, at location x at time t depends on the growth of an unstructured population with intrinsic growth rate, r, carrying capacity, K, and with movement defined by the diffusion coefficient, D:

$$\frac{\partial n(x, t+1)}{\partial t} = rn\left(1 - \frac{n}{K}\right) + D\frac{\partial^2 n(x, t)}{dx^2} \tag{2}$$

Such models ignore demographic complexity, and require the addition of asymmetry (advection-diffusion models) to model anisotropic movement in currents.

Integrodifference equation models for the spread of structured populations (based on matrix population models) [*Neubert and Caswell*, 2000; *Neubert and Parker*, 2004] have recently been developed. These models allow certain life history stages to disperse, while others remain stationary, providing a considerable improvement over demographically unstructured models. This is particularly important in the deep sea, where sessile adults may be long-lived with respect to the dispersive larval stage. In these models, the number of individuals, n, at location x at time t depends on the growth of the population as defined by a population projection matrix B_n and a dispersal kernel, K, of movement over space.

$$n(x, t+1) = \int_{-\infty}^{\infty} [K(x, y) \circ B_n(y)]\, n(y, t)\, dy \qquad (3)$$

Additional biological and spatial complexity can be incorporated into models, but usually at the cost of reduction in analytical rigor for simulation approaches. Although cellular automata and individual-based models can incorporate great amounts of biological and spatial information, this must be done numerically, and hence they can also be very time-consuming. However, given the interactions of the vent fauna with the chemical and physical environments at vent systems, and their extreme spatial heterogeneity, such complexity is warranted. These models are also heuristically more accessible than extremely complex analytic models. Models of disturbance, in particular, lend themselves well to individual-based modeling approaches [e.g., *Roxburgh et al.*, 2004], where different aspects of disturbance (frequency, extent, duration, timing, and intensity) can be addressed for any number of species.

5. MODELS OF COMMUNITY SUCCESSION

Scaling up to multiple species can be computationally challenging. If only species richness is of interest, models based on the theory of island biogeography [*MacArthur and Wilson*, 1967] are pertinent. Under the assumption that patch persistence is greater than the generation time of the organisms concerned, models of species richness based on patch size and isolation can be developed; smaller, isolated patches are expected to have fewer species. Lower-than-expected species richness would point to recent disturbances.

Metacommunity models shift the focus from a simple description of species richness to explicitly documenting which species are present. An important contribution in this area used a multispecies metapopulation modeling approach to address observed patterns of species diversity at vent systems [*Neubert et al.*, 2006]. This metacommunity model addressed facilitation-driven succession, where the state is no longer the presence or absence of a single species, but an inventory of all species in a patch. For a single species, two patch types are of interest: occupied or unoccupied. For two species, four patch types are pertinent: empty patches, patches with the first or the second species only, or patches with both. Clearly, as the number of species increases, the number of possible states increases rapidly, and this makes modeling the transitions in diverse community computationally challenging. *Neubert et al.* [2006] approached this problem by using a successional hierarchy to simplify the state space enough to make it tractable.

Matrix models, similar to those used for modeling population growth described above, have also been used to model succession. Whereas in population models the states are ages or life history stages of individuals, in successional models they represent different states of the community, and transitions between them represent probabilities of switching between community types. Most are first-order Markov models and thus assume that only the current state affects the transition probabilities, and that prior history is unimportant. Such models have a relatively long pedigree in terrestrial ecology [*Waggoner and Stephens*, 1970; *Horn*, 1975; *Usher*, 1979; *Cooke*, 1980; *McAuliffe*, 1988] and can incorporate the three main conceptual models of succession: facilitation, inhibition, tolerance [*Gotelli*, 2001], in which late successional species are facilitated by, inhibited by, or indifferent to earlier successional species, respectively. Some of the most detailed and thorough of these models have been developed for shallow marine systems [*Tanner et al.*, 1994, 1996; *Wootton*, 2001; *Hill et al.*, 2002, 2004]. Models of community succession are still relatively rare for deep-sea systems, and to date have been predominately conceptual. However, because empirical data are relatively abundant and successional processes are critical, the development of relevant successional models is likely to be rapid and productive.

The geological instability of hydrothermal vents on ecological time scales (decades) can cause the complete destruction of the assemblages that occupy them by repaving the substratum with fresh lava flows. Recovery (or reestablishment, depending on the extent of destruction) of the disturbed assemblages depends on: (1) distance to neighboring populations operating as larval sources, (2) successful colonization, and (3) intra- and interspecific interactions. Despite the presumed frequency of occurrence of these events, there have only been three opportunities when both the disturbance (eruption) and the subsequent primary community succession were recorded: at the 9–10°N sector of the EPR in 1991 [*Haymon et al.*, 1993; *Shank et al.*, 1998] and in

2006 (*Tolstoy et al.* [2006]; ongoing biological studies), and at CoAxial Seamount on the JdFR in 1993 [*Embley et al.*, 1995; *Tunnicliffe et al.*, 1997]. Thus, our measures of primary succession from the onset of colonization of the newly created substrate at these habitats are limited. In contrast, we are rapidly increasing our understanding of processes and regulating factors involved in secondary succession following less destructive changes, such as changes in vent effluent chemistry and small-scale changes in hydrothermal flow patterns.

Repaving events occur at spatial scales of 100s m^{-2} to <10 km^{-2}, and well within the range of larval dispersal from neighboring populations [*Chevaldonné et al.*, 1997; *Marsh et al.*, 2001]. In 1991, several newly formed hydrothermal vents were discovered between 9°45′N and 9°52′N on the EPR [*Haymon et al.*, 1993]. The number of colonizing species visible in time-lapse imagery increased incrementally over the recovery period, and was equal to 73% of the regional species pool within 5 years after the eruption [*Shank et al.*, 1998]. Because potential larval sources for colonizers were at distances ranging from 30 to 1450 m [*Shank et al.*, 1998], the successional sequence was most likely not limited by larval availability. At the CoAxial segment of the JdFR, the tubeworm *R. piscesea* and the polychaete *Paralvinella pandorae* had colonized the newly formed Floc Site within 1 year of the eruption and perhaps as early as 7 months after eruption [*Tunnicliffe et al.*, 1997]. The number of mega- and macrofaunal species present 2 years after the eruption was >75% of the regional species pool [*Tunnicliffe et al.*, 1997]. In contrast to 9°50′N, the nearest known extant vents that could have provided a source population were 15–18 km from the eruption site at CoAxial Segment [*Tunnicliffe et al.*, 1997]. These studies suggest that, given the spatial scales of repaving events, larval availability does not limit recolonization of hydrothermal vent habitats, even after complete destruction of the assemblage.

Although sufficient data for quantitative models of colonization and succession are not available for vent communities, robust conceptual models have been developed. Successful colonization during primary and secondary succession depends both on fluid chemistry and local biological interactions. In established communities (i.e., in the absence of a repaving event), rates and patterns of colonization have been measured experimentally at vents in the Galapagos Spreading Center (GSC), EPR, and JdFR by deploying panels or blocks of natural substrate (either slate or basalt), or artificial aggregations of tubeworms over 1- to 5-year periods. Colonization rate is initially (in the first month) slow, but increases over several months to a year, leading to spatially variable patterns in abundance among vents [*Van Dover et al.*, 1988; *Kelly et al.*, 2007]. This spatial variability can decrease over periods >1 year, as the communities become more similar. Abundance and diversity of colonizers typically decreases with increasing distance from the vents [*Van Dover et al.*, 1988; *Govenar and Fisher*, 2007; *Kelly et al.*, 2007], but colonizers can be found outside the range of adults of the same species [*Mullineaux et al.*, 1998; *Govenar and Fisher*, 2007]. The physicochemical characteristics of the vent fluid can influence these colonization patterns [*Mullineaux et al.*, 1998], particularly by broadly constraining colonization through physiological tolerance. However, it appears that biological processes play a significant role in fine-tuning these patterns within the imposed bounds [*Kelly and Metaxas*, 2006; *Kelly et al.*, 2007].

Biological processes (both positive and negative interactions) have been shown to influence patterns of colonization at vent sites near 9°50′N on the EPR [*Mullineaux et al.*, 2003], and Axial Volcano and Endeavour on the JdFR [*Kelly et al.*, 2007]. At the EPR, the fauna at hydrothermal vents show a distinct biological zonation with respect to fluid flux [*Hessler et al.*, 1985]. *Mullineaux et al.* [2003] found that on the periphery of areas of active vent fluid flux (dominated by suspension feeders), positive biological interactions influenced colonization, such as facilitation of sessile colonists by conspecifics. In contrast, where diffuse fluid flux was greatest (and tubeworms dominated), negative interactions, such as bulldozing of colonists by mobile species and space limitation by the settlement of sessile species, influenced colonization. At JdFR, although faunal zonation relative to the gradient in fluid flux is not as distinct as at the EPR, particular assemblages are associated with certain physicochemical conditions [*Sarrazin at al.*, 1999]. On the JdFR, both positive (e.g., gregarious settlement for the polychaete *Amphisamytha galapagensis*) and negative (e.g., early recruitment mortality for the numerically dominant gastropod *L. fucensis*) interactions influenced colonization [*Kelly et al.*, 2007].

Positive biological interactions during colonization also appear to be important drivers of the patterns of primary succession examined at 9°50′N. After the 1991 eruption, community succession was directed, and associated with changes in fluid chemistry and flow [*Shank et al.*, 1998]. In the first year, dense microbial material was either expelled from the orifices or formed a dense mat on the substratum. *T. jerichonana* had colonized within 12 months, and *R. pachyptila* within 32 months, mostly in areas with existing dense populations of *Tevnia* [*Shank et al.*, 1998]. *Mullineaux et al.* [2000] proposed a conceptual facilitation model to explain the successional pattern of tubeworms at this site. Using colonization blocks early in the successional sequence (3–4 years after the eruption), they found that *Tevnia* colonized in the absence of other species, but *R. pachyptila* and *O. alvinae* (another tubeworm) did not colonize in the absence of *Tevnia*. However, during a later successional stage in the same area (7–9 years after the eruption), *Riftia* colonized blocks in the absence of *Tevnia* [*Hunt et al.*, 2004]. *Hunt et al.* [2004] proposed that facilitation by *Tevnia* may only be relevant in the early stages of community succession, when the more abundant *Tevnia* may be providing a stronger cue. Such positive interactions have been shown to be critical in terrestrial succession [*Chapin et al.*, 1994], in coastal waters [*Harris et al.*, 1984], and in intertidal communities [e.g., *Turner*, 1983]. They are highly likely to be key in other vent communities as well.

The conceptual model for succession in established communities of sulfide edifices at the JdFR proposes that the chemical environment is the main driver, and biological processes are only important at certain stages [*Sarrazin and Juniper*, 1999]. The assemblages that inhabit sulfide edifices are extremely variable on spatial scales of tens of centimeters to meters and on annual temporal scales [*Sarrazin et al.*, 1997; 1999; *Govenar et al.*, 2002]. Each of the five distinct assemblages (I–V) that were described was associated with a particular temperature/fluid flow regime and, accordingly, assigned to a stage in the successional sequence of the community [*Sarrazin et al.*, 1997; 1999]. Observations over a 4-year period indicated that transitions between any pair of assemblages are possible [*Sarrazin et al.*, 1997], and a dynamic succession model has been proposed to describe these transitions [*Sarrazin and Juniper*, 1999]. Based on their model, most transitions occur as a result of changes in the flow patterns (increases or decreases) and fluid chemistry, but two transitions are biologically facilitated (by the insulation from high temperatures in the fluid by a thermotolerant species, and the increase in space available for colonization arising from tubeworm growth).

6. CONCLUSION

A wide variety of models are used in terrestrial and shallow marine systems. Their use in deep-sea hydrothermal vent systems is less common, possibly because a lack of biological information has hampered the development of such models. In this review, we have outlined the state of model development for understanding dispersal, local growth, spatial dynamics, and community succession; in some areas models are fairly detailed and quantitative, in others they are still at the conceptual stage. At the same time, we have tried to indicate where significant progress can be made with relatively little effort, by drawing on studies from other systems.

One of the major challenges facing biologists in studying this system is its relative inaccessibility. To address this, we have also outlined ways in which theoretical and empirical methods can be coupled to advance the study of vent systems more rapidly, rather than simply continuing to collect empirical information before developing detailed models. For example, we can use models to optimize or refine our experimental and sampling designs. Or, related models may be used to summarize alternative hypotheses, pinpointing where their predictions differ in empirically meaningful ways, and thus help us focus on the studies that would provide the most information on poorly understood mechanisms or processes. Some models may even be used to confirm our intuitions, or to make predictions where we have absolutely no intuition. Integrating local demographic and community processes with geophysically and hydrodynamically driven spatial dynamics will be critical in these spatially complex and highly variable environments. As we discover more about these fascinating systems, models offer considerable potential to accelerate our learning.

Acknowledgments. We thank Iliana Baums for helpful discussions and the Ridge Theoretical Institute for stimulating this paper. K. Shea acknowledges NSF grants DEB-0315860 and DEB-0614065. A. Metaxas was funded by a Discovery grant from NSERC (Canada). C. R. Young acknowledges NSF grant OCE-0453901. C. R. Fisher acknowledges NSF grants OCE-0117050 and OCE-0116823.

REFERENCES

Baums, I. B., M. W. Miller, and M. E. Hellberg (2005), Regionally isolated populations of an imperiled Caribbean coral, *Acropora palmate*, *Mol. Ecol.*, *14*, 1377–1390.

Becker, B. J., L. A. Levin, F. J. Fodrie, and P. A. McMillan (2007), Complex larval connectivity patterns among marine invertebrate populations, *Proc. Natl. Acad. Sci. U. S. A.*, *104*, 3267–3272.

Beerli, P., and J. Felsentein (2001), Maximum likelihood estimation of a migration matrix and effective population sizes in n subpopulations by using a coalescent approach, *Proc. Natl. Acad. Sci. U. S. A.*, *98*, 4563–4568.

Black, M. B., R. A. Lutz, and R. C. Vrijenhoek (1994), Gene flow among vestimentiferan tube worm (*Riftia pachyptila*) populations from hydrothermal vents of the eastern Pacific, *Mar. Biol.*, *120*, 33–39.

Bullock, J. M., K. Shea, and O. Skarpaas (2006), Measuring plant dispersal: A guide to field methods and experimental design, *Plant Ecol.*, *186*, 217–234.

Caswell, H. (2001), *Matrix Population Models: Construction, Analysis, and Interpretation*, Sinauer Associates, Sunderland, Mass.

Chapin, F. S., L. R. Walker, C. L. Fastie, and L. C. Sharman (1994), Mechanisms of primary succession following deglaciation at Glacier Bay, Alaska, *Ecol. Mon.*, *64*, 149–175.

Chevaldonné, P., D. Jollivet, A. Vangriesheim, and D. Desbruyères (1997), Hydrothermal-vent alvinellid polychaete dispersal in the eastern Pacific. 1. Influence of vent site distribution, bottom currents, and biological patterns, *Limnol. Oceanogr.*, *42*, 67–80.

Cooke, D. (1980), A Markov chain model of plant succession, in *The Mathematical Theory of the Dynamics of Biological Populations II*, edited by R. W. Hiorns and D. Cooke, pp. 231–247, Academic Press, Inc., London.

Cordes, E. E., D. C. Bergquist, K. Shea, and C. R. Fisher (2003), Hydrogen sulphide demand of long-lived vestimentiferan tube worm aggregations modifies the chemical environment at deep-sea hydrocarbon seeps, *Ecol. Lett.*, *6*, 1–9.

Cordes, E. E., M. A. Arthur, K. Shea, R. S. Arvidson, and C. R. Fisher (2005), Modeling the mutualistic interactions between tubeworms and microbial consortia, *PLoS Biol.*, *3*, e77.

Cowen, R. K., C. B. Paris, and A. Srinivasan (2006), Scaling of connectivity in marine populations, *Science*, *311*, 522–527.

Craddock, C., W. R. Hoeh, R. A. Lutz, and R. C. Vrijenhoek (1995), Extensive gene flow among mytilid (*Bathymodiolus thermophilus*) populations from hydrothermal vents of the eastern Pacific, *Mar. Biol.*, *124*, 137–146.

Craddock, C., R. A. Lutz, and R. C. Vrijenhoek (1997), Patterns of dispersal and larval development of archaeogastropod limpets at hydrothermal vents in the eastern Pacific, *J. Exp. Mar. Biol. Ecol.*, *210*, 37–51.

Crouse, D. T., L. B. Crowder, and H. Caswell (1987), A stage-based population model for loggerhead sea turtles and implications for conservation, *Ecology*, *68*, 1412–1423,

Dattagupta, S., L. L. Miles, M. S. Barnabei, and C. R. Fisher (2006), The hydrocarbon seep tubeworm *Lamellibrachia luymesi* primarily eliminates sulfate and hydrogben ions across its roots to conserve energy and ensure sulfide supply, *J. Exp. Biol.*, 209, 3795–3805.

Donnelly, P., and S. Tavare (1995), Coalescents and genealogical structure under neutrality, *Annu. Rev. Genet.*, *29*, 401–421.

Easterling, M. R., S. P. Ellner, and P. M. Dixon (2000), Size-specific sensitivity: Applying a new structured population model, *Ecology*, *81*, 694–708.

Ellner, S. P., and M. Rees (2006), Integral projection models for species with complex demography, *Am. Nat.*, *167*, 410–428.

Embley, R., W. Chadwick Jr., I. Jonasson, D. Butterfield, and E. Baker (1995), Initial results of the rapid response to the 1993 CoAxial event: Relationships between hydrothermal and volcanic processes, *Geophys. Res. Lett.*, *22*(2), 143–146.

Fisher, R. A. (1937), The wave of advance of advantageous genes, *Ann. Eugen. London*, *37*, 355–369.

France, S. C., R. R. Hessler, and R. C. Vrijenhoek (1992), Genetic differentiation between spatially-disjunct populations of the deep-sea, hydrothermal vent-endemic amphipod *Ventiella sulfuris*, *Mar. Biol.*, *114*, 551–559.

Gotelli, N. J. (2001), *A Primer of Ecology*, Sinauer Associates, Sunderland, Mass, 265 pp.

Govenar, B., and C. R. Fisher (2007), Experimental evidence of habitat provision by the siboglinid polychaete *Riftia pachyptila*, *Mar. Ecol.*, *28*, 3–14.

Govenar, B., D. C. Bergquist, I. A. Urcuyo, J. T. Eckner, and C. R. Fisher (2002), Three *Ridgeia piscesae* assemblages from a single Juan de Fuca Ridge sulfide edifice: Structurally different and functionally similar, *Cah. Biol. Mar.*, *43*, 247–252.

Hanski, I. (1982), Dynamics of regional distribution: The core and satellite species hypothesis, *Oikos*, *38*, 210–221.

Hanski, I., A. Moilanen, T. Pakkala, and M. Kuussaari (1996), The quantitative incidence function model and persistence of an endangered butterfly metapopulation, *Conserv. Biol.*, *10*, 578–590.

Harris, L. G., A. W. Ebeling, D. R. Laur, and R. J. Rowley (1984), Community recovery after storm damage: A case of facilitation in primary succession, *Science*, *224*, 1336–1338.

Haymon, R. M., D. J. Fornari, K. L. Von Damm, M. D. Lilley, M. R. Perfit, J. M. Edmond, W. C. Shanks, R. A. Lutz, J. M. Grebmeier, S. Carbotte, D. Wright, E. McLaughlin, M. Smith, N. Beedle, and E. Olson (1993), Volcanic eruption of the mid-ocean ridge along the East Pacific Rise crest at 9°45.52′N: Direct submersible observations of sea floor phenomena associated with an eruption event in April 2001, *Earth Planet. Sci. Lett.*, *119*, 85–101.

Hessler, R. R., W. M. Smithey, M. A. Boudrias, C. H. Keller, R. A. Lutz, and J. J. Childress (1985), Spatial and temporal variation of giant clams, tube worms and mussels at deep-sea hydrothermal vents, *Bull. Biol. Soc. Wash.*, *6*, 411–428.

Hey, J., and R. Nielsen (2004), Multilocus methods for estimating population sizes, migration rates and divergence time, with applications to the divergence of *Drosophila pseudoobscura* and *D. persimilis*, *Genetics*, *167*, 747–760.

Hill, M. F., J. D. Witman, and H. Caswell (2002), Spatio-temporal variation in Markov chain models of subtidal community succession, *Ecol. Lett.*, *5*, 665–675.

Hill, M. F., J. D. Witman, and H. Caswell (2004), Markov chain analysis of succession in a rocky subtidal community, *Am. Nat.*, *164*, E46–E61.

Horn, H. S. (1975), Markovian properties of forest succession, in *Ecology and Evolution of Communities*, edited by M. L. Cody and J. M. Diamond, pp. 196–211, Belknap Press, Cambridge, Mass.

Huber, J. A., and J. F. Holden (2008), Modeling the impact of diffuse vent microorganisms along mid-ocean ridges and flanks, this volume.

Hudson, R. R. (1990), Gene genealogies and the coalescent process, *Oxford Surv. Evol. Biol.*, *7*, 1–44.

Hunt, H. L., A. Metaxas, R. M. Jennings, K. M. Halanych, and L. S. Mullineaux (2004), Testing biological control of colonization by vestimentiferan tubeworms at deep-sea hydrothermal vents (East Pacific Rise, 9°50′N), *Deep-Sea Res. I*, *51*, 225–234.

Hurtado, L. A., R. A. Lutz, and R. C. Vrijenhoek (2004), Distinct patterns of genetic differentiation among annelids of eastern Pacific hydrothermal vents, *Mol. Ecol.*, *13*, 2603–2615.

Isard, S. A., and S. H. Gage (2001), Flow of life in the atmosphere: An airscape approach to invasive organisms, Michigan State University Press, East Lansing, Mich., 240 pp.

James, M. K., P. R. Armsworth, L. B. Mason, and L. Bode (2002), The structure of reef fish metapopulations: Modelling larval dispersal and retention patterns, *Proc. R. Soc. Lond. B*, *269*, 2079–2086.

Johnson, S. B., C. R. Young, W. J. Jones, A. R. Warén, and R. C. Vrijenhoek (2006), Migration, isolation, and speciation of hydrothermal vent limpets (Gastropoda: Lepitodrilidae) across the Blanco transform fault, *Biol. Bull.*, *210*, 140–157.

Jollivet, D., P. Chevaldonné, and B. Planque (1999), Hydrothermal-vent alvinellid polychaete dispersal in the eastern Pacific. 2. A metapopulation model based on habitat types, *Evolution*, *53*, 1128–1142.

Jones, G. P., S. Planes, and S. R. Thorrold (2005), Coral reef fish larvae settle close to home, *Curr. Biol.*, *15*, 1314–1318.

Jongejans, E., O. Skarpaas, and K. Shea (2008), Dispersal and spread models for conservation and control management, *Perspect. Plant Ecol. Evol. System.*, in press.

Karl, S. A., S. J. Schutz, D. Desbruyères, R. A. Lutz, and R. C. Vrijenhoek (1996), Molecular analysis of gene flow in the hydrothermal-vent clam *Clayptogena magnifica*, *Mol. Mar. Biol. Biotechnol.*, *5*, 193–202.

Kelly, N., and A. Metaxas (2006), Recruitment patterns of invertebrates at anhydrite hydrothermal vents on the Juan de Fuca Ridge, NE Pacific, *Cah. Biol. Mar.*, *47*, 435–441.

Kelly, N., A. Metaxas, and D. Butterfield (2007), Spatial and temporal patterns in colonization of deep-sea hydrothermal vent invertebrates on the Juan de Fuca Ridge, NE Pacific, *Aquatic Biology*, 1:1–16.

Kim, S. L., and L. S. Mullineaux (1998), Distribution and near-bottom transport of larvae and other plankton at hydrothermal vents, *Deep-Sea Res. II*, *45*, 423–440.

Kim, S., L. Mullineaux, and K. Helfrich (1994), Larval dispersal via entrainment into hydrothermal vent plumes, *J. Geophys. Res.*, *99*(C6), 12655–12665.

Kingsford, M. J., J. M. Leis, A. Shanks, K. C. Lindeman, S. G. Morgan, and J. Pineda (2002), Sensory environments, larval abilities and local self-recruitment, *Bull. Mar. Sci.*, *70*, 309–340.

Kinlan, B. P., and S. D. Gaines (2003), Propagule dispersal in marine and terrestrial environments: A community perspective, *Ecology*, *84*, 2007–2020.

Kooijman, S. A. L. M. (1993), *Dynamic Energy Budgets in Biological Systems: Theory and Applications in Ecotoxicology*, Cambridge University Press, Cambridge, United Kingdom, 350 pp.

Kooijman, S. A. L. M. (2000), *Dynamic Energy Mass Budgets in Biological Systems*, Cambridge University Press, Cambridge, United Kingdom, 424 pp.

Levins, R. (1969), Effect of random variations of different types on population growth, *Proc. Natl. Acad. Sci. U. S. A.*, *62*, 1061.

MacArthur, R. H., and E. O. Wilson (1967), *The Theory of Island Biogeography*, Princeton University Press, Princeton, NJ.

Marsh, A. G., L. S. Mullineaux, C. M. Young, and D. T. Manahan (2001), Larval dispersal potential of the tubeworm *Riftia pachyptila* at deep-sea hydrothermal vents, *Nature*, *411*, 77–80.

McAuliffe, J. R. (1988), Markovian dynamics of simple and complex desert plant communities, *Am. Nat.*, *131*, 459–490.

McCollum, T. M., Observational, experimental, and theoretical constraints on carbon cycling in mid-ocean ridge hydrothermal systems, this volume.

Metaxas, A. (2004), Spatial and temporal patterns in larval supply at hydrothermal vents on the northwest Pacific Ocean, *Limnol. Oceanogr.*, *49*, 1949–1956.

Mullineaux, L. S., and S. C. France (1995), Dispersal mechanisms of deep-sea hydrothermal vent fauna, in *Seafloor Hydrothermal Systems: Physical, Chemical, Biological and Geological Interactions*, *Geophys. Monogr.*, vol. 91, edited by S. E. Humphris, R. A. Zierenberg, L. S. Mullineaux, and R. E. Thompson, pp. 408–424, American Geophysical Union, Washington, D. C.

Mullineaux, L. S., P. H. Wiebe, and E. T. Baker (1995), Larvae of benthic invertebrates in hydrothermal vent plumes over Juan de Fuca Ridge, *Mar. Biol.*, *122*, 585–596.

Mullineaux, L. S., S. W. Mills, and E. Goldman (1998), Recruitment variation during a pilot colonization study of hydrothermal vents, *Deep-Sea Res. II*, *45*, 441–464.

Mullineaux, L. S., C. R. Fisher, C. H. Peterson, and S. W. Schaeffer (2000), Tubeworm succession at hydrothermal vents: Use of biogenic cues to reduce habitat selection error? *Oecologia*, *123*, 275–284.

Mullineaux, L. S., C. H. Peterson, F. Micheli, and S. W. Mills (2003), Successional mechanism varies along a gradient in hydrothermal fluid flux at deep-sea vents, *Ecol. Monogr.*, *73*, 523–542.

Mullineaux, L. S., S. W. Mills, A. K. Sweetman, A. H. Boudreau, A. Metaxas, and H. L. Hunt (2005), Vertical, lateral and temporal structure in larval distributions at hydrothermal vents, *Mar. Ecol. Prog. Ser.*, *293*, 1–16.

Nathan, R. (2001), Dispersal biogeography, in *Encyclopedia of Biodiversity*, edited by S. A. Levin, pp. 127–152, Academic Press, San Diego, Calif.

Nee, S., and R. M. May (1992), Dynamics of metapopulations—Habitat destruction and competitive coexistence, *J. Anim. Ecol.*, *61*, 37–40.

Neubert, M. G., and H. Caswell (2000), Demography and dispersal: Calculation and sensitivity analysis of invasion speed for structured populations, *Ecology*, *81*, 1613–1628.

Neubert, M. G., and I. M. Parker (2004), Projecting rates of spread for invasive species, *Risk Anal.*, *24*, 817–831.

Neubert, M. G., L. S. Mullineaux, and M. F. Hill (2006), A metapopulation approach to interpreting diversity at deep-sea hydrothermal vents, in *Marine Metapopulations*, edited by J. P. Kritzer and P. F. Sale, pp. 321–350, Elsevier Academic Press, Boston, Mass.

Nielsen, R., and J. Wakeley (2001), Distinguishing migration from isolation: A Markov chain Monte Carlo approach, *Genetics*, *158*, 885–896.

Nisbet, R. M., E. B. Muller, L. Lika, and S. A. L. M. Kooijman (2000), From molecules to ecosystems through Dynamic Energy Budget models, *J. Anim. Ecol.*, *69*, 913–926.

Okubo, A., and S. A. Levin (2001), *Diffusion and Ecological Problems: Modern Perspectives*, Springer, New York, 467 pp.

Palumbi, S. R. (2003), Population genetics, demographic connectivity, and the design of marine reserves, *Ecol. Appl.*, *13*, S146–S158.

Pannell, J. R., and B. Charlesworth (2000), Effects of metapopulation processes on measures of genetic diversity, *Philos. Trans. R. Soc. Lond. B*, *355*, 1851–1864.

Pogson, G. H., C. T. Taggart, K. A. Mesa, and R. G. Boutilier (2001), Isolation by distance in the Atlantic cod, *Gadus morhua*, at large and small geographic scales, *Evolution*, *55*, 131–146.

Rosenberg, N. A., and M. Nordborg (2002), Genealogical trees, coalescent theory and the analysis of genetic polymorphisms, *Nat. Rev. Genet.*, *3*, 380–390.

Roxburgh, S. H., K. Shea, and J. Bastow Wilson (2004), The intermediate disturbance hypothesis: Patch dynamics and mechanisms of species coexistence, *Ecology*, *85*, 359–371.

Rumrill, S. S. (1990), Natural mortality of marine invertebrate larvae, *Ophelia*, *32*, 163–198.

Sarrazin, J., and S. K. Juniper (1999), Biological characteristics of a hydrothermal edifice mosaic community, *Mar. Ecol. Prog. Ser.*, *185*, 1–19.

Sarrazin, J., V. Robigou, S. K. Juniper, and J. R. Delaney (1997), Biological and geological dynamics over four years on a high-temp sulfide structure at the Juan de Fuca Ridge hydrothermal observatory, *Mar. Ecol. Prog. Ser.*, *153*, 5–24.

Sarrazin, J., S. K. Juniper, G. Massoth, and P. Legendre (1999), Physical and chemical factors influencing species distributions on hydrothermal sulfide edifices of the Juan de Fuca Ridge, northeast Pacific, *Mar. Ecol. Prog. Ser.*, *190*, 89–112.

Schrenk. M. O., J. F. Holden, and J. A. Baross (2008), Magma-to-microbe networks in the context of sulfide hosted microbial ecosystems, this volume.

Shank, T. M., D. J. Fornari, K. L. Von Damm, M. D. Lilley, R. M. Haymon, and R. A. Lutz (1998), Temporal and spatial patterns of biological community development at nascent deep-sea hydrothermal vents, *Deep-Sea Res. II*, *45*, 465–515.

Shea, K., D. Kelly, A. W. Sheppard, and T. L. Woodburn (2005), Context-dependent biological control of an invasive thistle, *Ecology*, *86*, 3174–3181.

Shigesada, N., and K. Kawasaki (1997), *Biological Invasions: Theory and Practice*, Oxford University Press, Oxford, 205 pp.

Skarpaas, O., and K. Shea (2007), Dispersal patterns, dispersal mechanisms and invasion wave speeds for *Carduus* thistles. *Am. Nat.*, *170*(3), 421–430.

Skarpaas, O., K. Shea, and J. M. Bullock (2005), Optimising dispersal study design by Monte Carlo simulation, *J. Appl. Ecol.*, *42*, 731–739.

Skellam, J. G. (1951), Random dispersal in theoretical populations, *Biometrika*, *38*, 433–435.

Slatkin, M. (1977), Gene flow and genetic drift in a species subject to frequent local extinctions, *Theor. Popul. Biol.*, *12*, 253–262.

Slatkin, M. (1987), Gene flow and the geographic structure of natural populations, *Science*, *236*, 787–792.

Slatkin, M. (1993), Isolation by distance in equilibrium and non-equilibrium populations, *Evolution*, *47*, 264–279.

Sotka, E. E., J. P. Wares, J. A. Barth, R. K. Grosberg, and S. R. Palumbi (2004), Strong genetic clines and geographical variation in gene flow in the rocky intertidal barnacle *Balanus glandula*, *Mol. Ecol.*, *13*, 2143–2156.

Swearer, S. E., J. S. Shima, M. E. Hellberg, S. R. Thorrold, G. P. Jones, D. R. Robertson, S. G. Morgan, K. A. Selkoe, G. M. Ruiz, and R. R. Warner (2002), Evidence of self-recruitment in demersal marine populations, *Bull. Mar. Sci.*, *70*, 251–271.

Tanner, J. E., T. P. Hughes, and J. H. Connell (1994), Species coexistence, keystone species, and succession: A sensitivity analysis, *Ecology*, *75*, 2204–2219.

Tanner, J. E., T. P. Hughes, and J. H. Connell (1996), The role of history in community dynamics: A modelling approach, *Ecology*, *77*, 108–117.

Taylor, M. S., and M. E. Hellberg (2003), Genetic evidence for local retention of pelagic larvae in a Caribbean reef fish, *Science*, *299*, 107–109.

Thompson, R. E., S. F. Mihály, A. B. Rabinovich, R. E. McDuff, S. R. Veirs, and F. R. Stahr (2003), Constrained circulation at Endeavour ridge facilitates colonization by vent larvae, *Nature*, *424*, 545–549.

Tolstoy, M., J. P. Cowen, E. T. Baker, D. J. Fornari, K. H. Rubin, T. M. Shank, F. Waldhauser, D. R. Bohnenstiehl, D. W. Forsyth, R. C. Holmes, B. Love, M. R. Perfit, R. T. Weekly, S. A. Soule, and B. Glazer (2006), A sea-floor spreading event captured by seismometers, *Science*, *314*, 1920–1922.

Tunnicliffe, V., R. W. Embley, J. F. Holden, D. A. Butterfield, B. J. Massoth, and S. K. Juniper (1997), Biological colonization of new hydrothermal vents following an eruption on Juan de Fuca Ridge, *Deep-Sea Res. I*, *44*, 1627–1644.

Turner, T. (1983), Facilitation as a successional mechanism in a rocky intertidal community, *Am. Nat.*, 121, 729–738.

Usher, M. B. (1979), Markovian approaches to ecological succession, *J. Anim. Ecol.*, *48*, 413–426.

Van Dover, C. L., and R. Lutz (2004), Experimental ecology at deep-sea hydrothermal vents: A perspective, *J. Exp. Mar. Biol. Ecol.*, *300*, 273–307.

Van Dover, C. L., C. J. Berg Jr., and R. D. Turner (1988), Recruitment of marine invertebrates to hard substrates at deep-sea hydrothermal vents on the East Pacific Rise and Galapagos spreading center, *Deep-Sea Res. A*, *35*, 1833–1849.

Van Dover, C. L., C. R. German, K. G. Speer, L. M. Parson, and R. C. Vrijenhoek (2002), Evolution and biogeography of deep-sea vent and seep invertebrates, *Science*, *295*, 1253–1257.

Vrijenhoek, R. C. (1997), Gene flow and genetic diversity in naturally fragmented metapopulations of deep-sea hydrothermal vent animals, *J. Hered*, *88*, 285–293.

Waggoner, P. E., and G. R. Stephens (1970), Transition probabilities for a forest, *Nature*, *225*, 1160–1161.

Wakeley, J. (2004), Metapopulation models for historical inference, *Mol. Ecol.*, *13*, 865–875.

Wakeley, J., and N. Aliacar (2001), Gene genealogies in a metapopulation, *Genetics*, *159*, 893–905.

Whitlock, M. C., and D. E. McCauley (1990), Some population genetic consequences of colony formation and extinction: Genetic correlations within founding groups, *Evolution*, *44*, 1717–1724.

Won, Y., C. R. Young, R. A. Lutz, and R. C. Vrijenhoek (2003), Dispersal barriers and isolation among deep-sea mussel populations (Mytilidae: *Bathymodiolus*) from eastern Pacific hydrothermal vents, *Mol. Ecol.*, *12*, 169–184.

Wootton, J. T. (2001), Causes of species diversity differences: A comparative analysis of Markov models, *Ecol. Lett.*, *4*, 46–56.

Wright, S. (1931), Evolution in Mendelian populations, *Genetics*, *16*, 97–159.

Wright, S. (1943), Isolation by distance, *Genetics*, *28*, 114–138.

Young, C. R., S. Fujio, and R. C. Vrijenhoek (2008), Directional dispersal between mid-ocean ridges: Deep-ocean circulation and gene flow in *Ridgeia piscesae*, *Mol. Ecol.*, *17*, 1718–1731.

Katriona Shea, Department of Biology, 208 Mueller Laboratory, The Pennsylvania State University, University Park, Pennsylvania 16802, USA. (K-shea@psu.edu)

The Role of Seafloor Hydrothermal Systems in the Evolution of Seawater Composition During the Phanerozoic

Lee R. Kump

Department of Geosciences, The Pennsylvania State University, University Park, Pennsylvania, USA

Seafloor hydrothermal systems clearly have played an important role in controlling the composition of seawater over geologic time. However, controversy exists concerning their role as drivers of secular variability in major element composition of seawater during the Phanerozoic. The history of Phanerozoic changes in calcium (Ca^{2+}) and magnesium (Mg^{2+}) concentration has been reconstructed through analysis of fluid inclusions in evaporites, and this history is inconsistent with global geochemical cycling models that link both volcanic CO_2 release and Mg^{2+} uptake/Ca^{2+} release to inferred rates of seafloor production. These models generate little variability in Mg^{2+} concentrations because of compensating effects. Some models that recreate the observed trends either do not conserve alkalinity or do not link seafloor spreading, inferred from past variations in continental flooding presumed to reflect changing mid-ocean-ridge volumes, to volcanic CO_2 production. The most comprehensive models reproduce the variations in seawater Mg^{2+}/Ca^{2+} through time quite well, but not because this ratio is tied to variations in seafloor spreading rate, and despite considerable mismatches to the Mg^{2+} record. This cacophony arises in part from our lack of quantitative understanding of fundamental relationships between heat flow, seafloor production rates, sea level, hydrothermal circulation rates, volcanic CO_2 release rates, and ocean chemical changes. Nevertheless, the current fluid inclusion data seem to indicate that ocean composition has an "attractor" that drives Ca^{2+} and Mg^{2+} toward equilibration with seafloor hydrothermal mineral assemblages [20 m*m* (millimolal) Ca^{2+} and 0 m*m* Mg^{2+}], thwarted by other processes (weathering and riverine inputs, biogenic mineral precipitation) that drive the system away from this attractor.

1. INTRODUCTION

The evolution of the chemistry of the ocean has been a topic of great interest for many decades. Recurring themes include calculations about the age of Earth based on the notion of accumulating salt [*Joly*, 1923], the question of equilibrium vs. kinetic control of seawater chemistry [*Broecker*, 1971; *Holland*, 1972; *Rubey*, 1951; *Sillen*, 1961], and evolving thoughts about how variations in continental weathering, riverine delivery, and sediment deposition might alter ocean composition on geologic time scales [*Mackenzie and Garrels*, 1966].

A relative latecomer to the debate has been the role of seafloor hydrothermal fluids in determining the composition of the ocean. Before the discovery of black smokers, efforts to balance the inputs and outputs of the elements for the ocean (presuming steady state) often came up short. The most notorious of these

Magma to Microbe: Modeling Hydrothermal Processes at Ocean Spreading Centers
Geophysical Monograph Series 178

10.1029/178GM14

was the "Mg problem," best highlighted by the classic work of *Drever* [1974]. In this work, Drever showed that the input of Mg from rivers greatly exceeds the output of Mg from the ocean associated with the deposition of Mg-containing marine carbonates (calcite and dolomite) and authigenic clays. The discovery of seafloor hydrothermal systems debauching hot waters with chemical compositions quite distinct from seawater immediately raised the possibility that interaction with hot oceanic crust served as a sink for some elements and a source of others. The fact that these waters were essentially depleted of Mg seemed to once and for all solve the enigma of the missing Mg sink [e.g., *Wolery and Sleep*, 1976]. More recent analyses of the Mg budget highlight the significant uncertainties associated with off-axis hydrothermal and low-temperature alteration of seafloor basalt [*Wolery and Sleep*, 1988; *Elderfield and Schultz*, 1996]. These uncertainties affect the crustal Ca balance, which shows excess Ca that could reflect the cumulative effects of Ca release from basalt during hydrothermal alteration at both high and low temperatures [*de Villiers*, 1998].

One thing seems certain: over billion year time scales the influence of seafloor hydrothermal systems on ocean chemistry has diminished. *Veizer et al.* [1982] developed the notion that the Precambrian ocean was a "mantle-dominated" ocean because of higher seafloor production rates in the distant geologic past. The best demonstration of mantle dominance was the Sr isotopic composition of seawater, recorded in marine carbonates, which evolved from the mantle end-member to a mix of continental and mantle inputs over the course of geologic time (Figure 1). High hydrothermal Fe fluxes have also been invoked to explain the abundance of "banded iron formations" in the Archean and Paleoproterozoic [*Isley*, 1997], and these fluxes were likely strongly enhanced by the generally more reduced nature of hydrothermal fluids at that time given the likelihood that the oceans were essentially devoid of sulfate, a major oxidant in modern hydrothermal systems [*Kump and Seyfried*, 2005]. What little sulfate entered seafloor hydrothermal systems would have been reduced to sulfide and deposited in vent chimneys. Gradually, with time, heat flow, and thus, presumably, fluxes from seafloor hydrothermal systems diminished. Also with the growth of continents, chemical weathering fluxes increased, so the ocean's $^{87}Sr/^{86}Sr$ deviated from the mantle evolution line, reflecting a balance between these two end-member compositions.

The focus of research on Precambrian ocean chemistry has been redox evolution: the progressive oxygenation of the deep sea, and associated changes in sulfur and nitrogen cycling. The general notion is that the Archean ocean was oxygen free except for "oxygen oases," certain regions of the surface ocean where cyanobacterial blooms may have supported spatially and temporally limited buildups of molecular oxygen in an otherwise reducing ocean/atmosphere system [*Kasting*, 1991]. There is a strong evidence in the mass-independent fractionation of sulfur isotopes preserved in pyrites for oxygenation of the atmosphere at about 2.4–2.5 Ga [*Bekker et al.*, 2004], and this likely led to deep-ocean oxygenation. However, oxidative weathering would have provided a new source of sulfate to the ocean. In a somewhat counterintuitive way, this may have promoted (by at least 1.8 Ga) a deep ocean enriched in hydrogen sulfide through the activity of sulfate-reducing bacteria [*Canfield*, 1998]. The consequences of a euxinic ocean during the Proterozoic have been pursued by a number of investigators [*Anbar and Knoll*, 2002; *Poulton et al.*, 2004; *Shen et al.*, 2003].

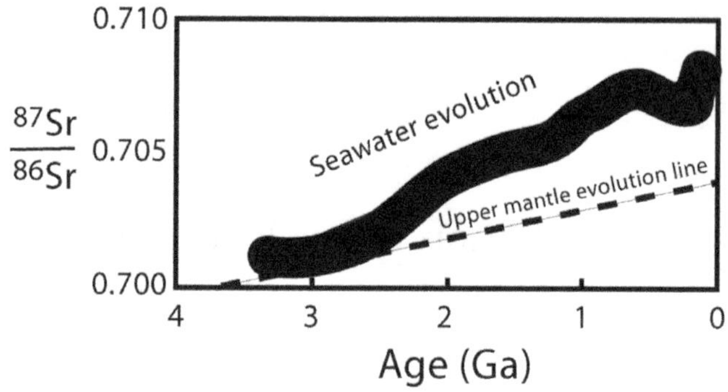

Figure 1. Schematic of the evolution of Sr isotopic composition of seawater as recorded in marine carbonate rocks through geologic time [after *Veizer and Compston*, 1982], demonstrating the "mantle dominance" of Archean seawater and a shift to crustal weathering dominance in the Proterozoic and Phanerozoic.

Consideration of changes in ocean chemistry during Phanerozoic time (the last 542 million years) has focused on cyclical variations in seafloor hydrothermal activity tied to the Wilson cycle of supercontinent assembly, breakup, dispersal, and reassembly. This paper reviews the evolving thoughts concerning the role of seafloor hydrothermal fluxes in driving ocean chemistry on this timescale. It is motivated by recent analyses of fluid inclusions from evaporites that have been interpreted to demonstrate substantial variations in the major element composition of seawater during the Phanerozoic, and controversy over whether these fluctuations could be driven by variable rates of seafloor production and hydrothermal activity. Arguments linking the two often have been advanced without comprehensive consideration of the intricate couplings among the element cycles and between them and climate. Models that link the cycles generate little variability in major element (especially Ca^{2+} and Mg^{2+}) composition driven by variations in seafloor hydrothermal activity. The reasons why, and implications for our understanding of these cycles, are explored below.

2. MODELING THE MG AND CA CHEMISTRY OF THE OCEANS

The notion that variation in the rate of circulation of seawater through mid-ocean ridges might drive changes in ocean chemistry was first quantified with the so-called BLAG model [*Berner et al.*, 1983]. In BLAG, the Mg^{2+} content of seawater, for example, depended on inputs from weathering of Mg-containing silicates and dolomite and outputs from dolomite formation (considered by BLAG to be unimportant from mid-Cretaceous to present) and uptake at mid-ocean ridges (presumed to be balanced stoichiometrically by release of Ca^{2+}). BLAG was a fully coupled model, so the weathering flux of Mg^{2+} was linked to the consumption of carbon dioxide from the atmosphere and transfer of bicarbonate (alkalinity) to the ocean. Thus, any conclusions one would draw from the results of BLAG concerning changes in the Mg^{2+} or Ca^{2+} content of seawater would be consistent with the strict constraints imposed by alkalinity balance in the ocean (a constraint that cannot be violated on time scales longer than the alkalinity adjustment time of the ocean, approximately 10 ka; *Broecker and Peng* [1987]).

One of the key attributes of BLAG, its predecessor model by *Walker et al.* [1981], and many subsequent models, is the parameterized sensitivity of chemical weathering rates to climate, and thus, through the greenhouse effect, to changes in atmospheric pCO_2. The first models ignored the organic carbon subcycle, thus changes in atmospheric CO_2 were driven solely by slight imbalances in the input of carbon dioxide from volcanism and output via silicate-rock weathering. Because volcanic activity was presumed to be linearly dependent on seafloor production rate, an increase in the global average seafloor spreading rate was modeled to drive a proportional increase in the volcanic CO_2 release rate, silicate weathering rate, and finally, $CaCO_3$ deposition rate [*Berner et al.*, 1983].

To assess the implications of varying seafloor-spreading rates on ocean chemistry, *Berner et al.* [1983] made the further assumption that seafloor hydrothermal fluxes would be linearly proportional to seafloor spreading rates. More specifically, the uptake of Mg was proportional to the seafloor production rate and the concentration of Mg^{2+} ($[Mg^{2+}]$) of seawater, and the release rate of Ca^{2+} was equal to that of Mg^{2+} uptake. Finally, various reconstructions of spreading rates in the geologic past were used as forcing functions of the model (together with reconstructed changes in land area available for weathering). The result of this modeling effort was "postdictions" of the evolution of ocean chemistry, specifically changes in Ca^{2+} and Mg^{2+} concentrations over the last 100 Ma (Figure 2). According to BLAG, oceanic $[Mg^{2+}]$ rose in the Late Cretaceous and then fell through the Cenozoic, whereas $[Ca^{2+}]$ displayed a mirror-image pattern, as one might expect from a system that was dominated by seafloor hydrothermal exchange of Mg^{2+} for Ca^{2+}.

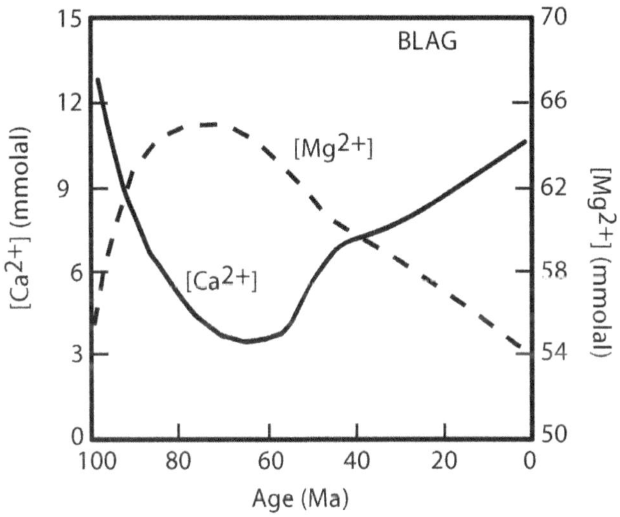

Figure 2. Calculated variations in calcium and magnesium seawater concentrations over the last 100 Ma with a model that couples volcanic CO_2 release and hydrothermal uptake of magnesium and release of calcium to seafloor spreading rates (the so-called BLAG model; *Berner et al.*, 1983).

3. RECONSTRUCTING ANCIENT SEAWATER CHEMISTRY FROM BRINE INCLUSIONS IN SALT DEPOSITS

Subsequent to BLAG there has been a considerable amount of modeling by Berner and his colleagues (the generations of GEOCARB models), and by others, aimed both at refining the estimates of ancient levels of atmospheric CO_2 and O_2 and at improving our understanding of the processes that drive long-term evolution in ocean and atmospheric chemistry [e.g., *Arvidson et al.*, 2006; *Berner*, 2006; *Berner and Kothalava*, 2001; *Godderis et al.*, 2007; *Wallmann*, 2001]. Little attention has been paid, until recently, to improving the calculation of secular changes in major element (e.g., Ca^{2+} and Mg^{2+}) composition of seawater. The most significant breakthrough in the past 20 years of research on ancient-ocean chemistry has instead been the analysis of brine inclusions in salt and their interpretation in terms of changes in ocean chemistry through geologic time. Two groups have worked in parallel to establish these trends: the "Harvard" group led by H. D. Holland [e.g., *Horita et al.*, 2002, and references therein] and the "Johns Hopkins/Binghamton" group led by L. Hardie, T. Lowenstein, and R. Demicco (see references in *Lowenstein et al.* [2003]). These workers extract minute quantities of evaporated seawater that have been retained and preserved in salt crystals. Analyses of the composition of these fluids are then corrected for inferred evaporation paths and mineral precipitation sequences so that ancient seawater compositions can be estimated.

The latest results from both groups are quite consistent, especially in terms of mirror-image variations in Ca and Mg concentrations (Figure 3) not unlike those of BLAG. The early Paleozoic and mid-Mesozoic were times of high $[Ca^{2+}]$ and low $[Mg^{2+}]$. These results are consistent with the finding that marine carbonates deposited during this time were largely originally calcite, a mineral that is kinetically favored over its polymorph aragonite when $[Mg^{2+}]/[Ca^{2+}]$ ratios are low [e.g., *Stanley and Hardie*, 1998]. In contrast, the Late Paleozoic/Early Mesozoic and much of the Cenozoic were times of high $[Mg^{2+}]/[Ca^{2+}]$, with abundant evidence for primary aragonite.

Two interpretations of these trends have been proposed:

1. They represent variations in the rates of Mg uptake and Ca release during high-temperature hydrothermal circulation of seawater through mid-ocean ridges [e.g., *Spencer and Hardie*, 1990; *Demicco et al.*, 2005]. For the Cretaceous to Recent, these rates are estimated based on measured variations in seafloor production rates. For earlier times, the extent of flooding of the continents is used as a proxy for sea level, itself an assumed proxy for seafloor spreading (production) rates [*Gaffin*, 1987]. Times of maximal continental flooding (e.g., the early Paleozoic and mid-Mesozoic) were times of high sea level caused by broad, shallow mid-ocean ridges created by fast seafloor spreading. Fast seafloor spreading supported high rates of Ca release and Mg uptake at mid-ocean ridges, and thus low ratios of $[Mg^{2+}]/[Ca^{2+}]$ in the ocean. Intervals during which the continents were maximally exposed were times of low sea level, slow seafloor spreading, and low rates of cation exchange leading to high ratios of $[Mg^{2+}]/[Ca^{2+}]$.

2. They represent variations in the rates of dolomitization of preexisting limestones or contemporaneously deposited limestones [*Holland*, 2005]. Times of maximum continental flooding are times when shallow, restricted marine basins should be abundant; these types of environments are presumed to be ideal for the development of dolomitizing basinal fluids.

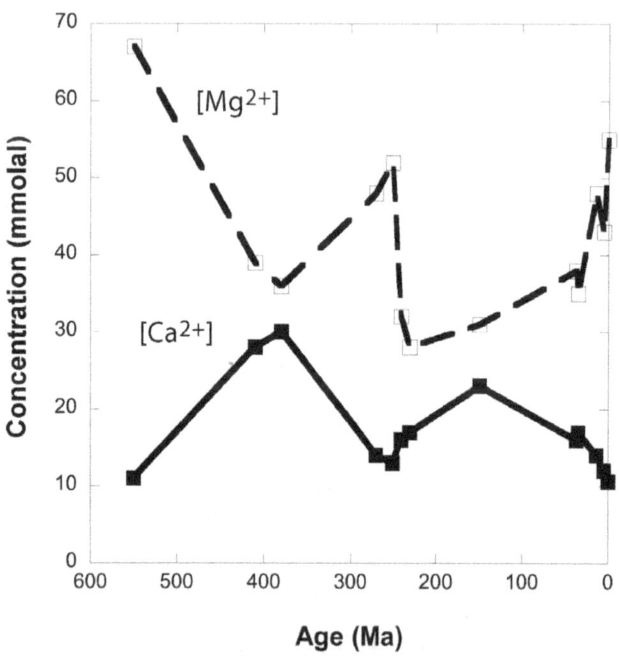

Figure 3. Evolution of seawater concentration of Mg^{2+} and Ca^{2+} reconstructed from the analysis of fluid inclusions in marine evaporites [*Horita et al.*, 2002].

The seafloor hydrothermal interpretation was proffered first, and provided what seemed to be a self-consistent explanation for secular variations in seawater chemistry, biogenic mineralogy, and sea level. However, any celebrations were short-lived; the analysis of seafloor production rates through time by *Rowley* [2002] concluded that the present-day distribution of ages and areas of seafloor is entirely consistent with a constant rate of production of seafloor through time (over the last 150 Ma or so) coupled to age-independent seafloor destruction rates. Indeed, it was the work of *Rowley* [2002] that prompted the second explanation for inverse covariation of $[Mg^{2+}]$ and $[Ca^{2+}]$ through the Phanerozoic. Moreover, a reanalysis of eustatic sea level variations by *Miller et al.* [2005] downscaled the amplitude of Mesozoic-Cenozoic variations from 200 to 300 m to much closer to that explainable by glacioeustasy (100–200 m). If Rowley is correct [cf. *Demicco*, 2004], and seafloor spreading rates have been essentially invariant in time, then variations in hydrothermal fluxes cannot be invoked to explain the pattern of variation of $[Mg^{2+}]/[Ca^{2+}]$.

4. RECONCILING MODELS AND PROXIES OF $[MG^{2+}]$ AND $[CA^{2+}]$ VARIATION

Although much-improved models exist, the BLAG model is the simplest expression of an important concept: secular variations in seafloor spreading rates drive changes in volcanic CO_2 release that lead to climate change on long time scales. Moreover, these variations are also reflected in changes in the Mg/Ca ratio of seawater. Curiously, the BLAG model calculations of oceanic $[Mg^{2+}]$ and $[Ca^{2+}]$ bear little resemblance to the time histories derived from the study of brine inclusions. The data clearly indicate that

$[Mg^{2+}]$ has increased and $[Ca^{2+}]$ has decreased from the mid-Cretaceous to Recent, whereas the BLAG calculation shows just the reverse. This is even more peculiar when one realizes that modeling done by the Hopkins/Binghamton group [e.g., *Demicco et al.*, 2005] reproduces the secular variations quite closely. How can two classes of numerical model, both driven by essentially the same presumed secular variation in seafloor spreading rates, generate such different results?

Even more baffling are the recent modeling predictions of $[Mg^{2+}]$ and $[Ca^{2+}]$ again by *Berner* [2004] that reproduce the observed trends from fluid inclusions quite closely. What is the key difference between BLAG of *Berner et al.* [1983] and *Berner* [2004]?

First, we must delve into the modeling of *Berner* [2004]. In this work, Berner utilized the fact that, over long time scales (>10^4 years), the alkalinity inputs to and outputs from the ocean must be balanced. When that constraint and the exchange of Ca for Mg at mid-ocean ridges are applied to the balance equations for Mg^{2+} and Ca^{2+}, an interesting result emerges: secular variations in Ca^{2+} must mirror those of Mg^{2+} except to the extent that imbalances of the sulfur cycle allow Ca to vary independently of Mg^{2+}:

$$\frac{d[\mathrm{Ca}^{2+}]}{dt} = -\frac{d[\mathrm{Mg}^{2+}]}{dt} + (\text{gypsum weathering} - \text{deposition}) + 0.7\,(\text{pyrite weathering} - \text{pyrite burial}).$$

Pyrite burial and weathering are involved because the acidity produced by pyrite oxidation is in part involved in Ca mineral weathering, whereas the alkalinity generated during sulfate reduction drives the precipitation of $CaCO_3$ (0.7 represents the fraction of the sulfuric acid generated that reacts with Ca-bearing minerals). *Berner* [2004] uses the C and S isotope modeling of *Berner* [2001] to provide the C and S fluxes necessary to drive calculations of Ca and Mg variation through time. These generate mirror-image trends that match observations quite well. The calculations represent an improvement over those of the Hopkins/Binghamton group because they explicitly conform to alkalinity mass balance.

However, both the *Berner* [2004] and the Hopkins/Binghamton group models lacked the explicit link between seafloor spreading and volcanic CO_2 emission of BLAG and GEOCARB models. In models where this link is included, an increase in seafloor spreading rate drives proportional increased hydrothermal uptake of Mg^{2+}, and $[Mg^{2+}]$ tends to fall, as in all other models. However, increased spreading drives proportionally higher weathering rates (i.e., Mg^{2+} input rates). Thus, the tendency toward lower $[Mg^{2+}]$ from increased hydrothermal uptake is countered by increased riverine Mg^{2+} delivery, and $[Mg^{2+}]$ remains relatively unchanged (as long as the weathering adjustment is fast relative to the Mg^{2+} residence time of the ocean). $[Ca^{2+}]$ variability is also damped, but for a different reason: increased spreading increases both the weathering and hydrothermal inputs, which would tend to increase $[Ca^{2+}]$. However, the ocean remains near saturation with respect to $CaCO_3$, so this excess Ca input is quickly countered by increased $CaCO_3$ burial.

This tendency toward damping is best illustrated in a highly simplified model. Here the simplified carbonate-silicate geochemical cycle model of *Payne and Kump* [2007] is modified to also calculate changes in Ca and Mg with time. The oceanic reservoir of Mg^{2+} changes through time according to:

$$\frac{d(\mathrm{Mg}^{2+})}{dt} = F_{\mathrm{wMgsil}} + F_{\mathrm{wdol}} - F_{\mathrm{hydrMg}}$$

where F_{wMgsil} and F_{wdol} are the weathering flux of Mg^{2+} from silicates and dolomite, respectively, each calculated by the model to depend on changes in atmospheric pCO_2 [*Payne and Kump*, 2007], and F_{hydrMg} is the rate of hydrothermal uptake of Mg^{2+}, expressed as:

$$F_{\mathrm{hydrMg}} = F^0_{\mathrm{hydrMg}} \cdot f_{\mathrm{SR}} \cdot \frac{[\mathrm{Mg}^{2+}]}{[\mathrm{Mg}^{2+}]^0}.$$

Here, F^0_{hydrMg} is the present-day steady state hydrothermal uptake rate of Mg^{2+}, 5.2×10^{18} mol/yr [*Berner et al.*, 1983], f_{SR} is the seafloor spreading rate according to *Gaffin* [1987] as presented by *Demicco et al.* [2005], and $[Mg^{2+}]^0$ is today's Mg^{2+} concentration [55 m*m* (millimolal)]. Similarly, the oceanic reservoir of Ca^{2+} changes through time according to:

$$\frac{d(\mathrm{Ca}^{2+})}{dt} = F_{\mathrm{wCasil}} + F_{\mathrm{wdol}} + F_{\mathrm{wcarb}} + F_{\mathrm{hydrCa}} - F_{\mathrm{bcarb}}$$

where F_{wCasil}, F_{wdol}, and F_{wcarb} are the weathering fluxes of Ca^{2+} from silicates, dolomites, and limestones, respectively, each calculated by the model to depend on changes in atmospheric pCO_2 [*Payne and Kump*, 2007], and F_{hydrCa} is the rate of hydrothermal uptake of Ca^{2+}, expressed as:

$$F_{\mathrm{hydrCa}} = F^0_{\mathrm{hydrCa}} \cdot f_{\mathrm{SR}} \cdot \frac{41 - [\mathrm{Ca}^{2+}]}{31}.$$

Here, F^0_{hydrCa} is the present-day steady-state hydrothermal uptake rate of Ca^{2+}, 5.2×10^{18} mol/yr [*Berner et al.*, 1983], and the numbers 31 and 41 are selected to (1) give a value of the quotient of 1 for the modern $[Ca^{2+}]$ of 10 m*m*, and (2) be consistent with the vent fluid composition chosen by *Demicco et al.* [2005] (41 m*m*; see below).

In the fully coupled model, changes in spreading rate drive changes in volcanic CO_2 outgassing and proportional changes in Mg uptake and Ca release from hydrothermal systems. This system behaves as described above: even with spreading rates 2.5× higher than present, the Ca^{2+} and Mg^{2+} contents of the ocean are essentially identical to today (Figure 4) and unchanging in time (this model does not include the sulfur cycle dependencies of *Berner* [2004]). Only with the dependence of CO_2 outgassing on seafloor spreading disabled do fluctuations in seafloor spreading drive variations in Ca^{2+} and Mg^{2+} that mimic the observed record (Figure 4).

What then explains the temporal variability of [Mg^{2+}] and [Ca^{2+}] in BLAG? There are two factors involved:

(1) *Berner et al.* [1983] initiated their model at 100 Ma far from steady state, so [Mg^{2+}] increased and [Ca^{2+}] decreased for the first 30 Ma of the simulation as the system sought steady state [*Kasting*, 1984]; and

(2) the quasi-steady state [Mg^{2+}], once achieved, declined from 100 Ma to the present because the sedimentary dolomite Mg reservoir shrunk, and so the weathering input of Mg^{2+}, presumed proportional to the reservoir size, diminished as the input of Ca^{2+} increased. The dolomite reservoir decreased (and limestone reservoir increased) over this interval because *Berner et al.* [1983] specified that no dolomite formation has occurred during this interval of time (a common presumption). If, instead, dolomite production and weathering were closely balanced, BLAG would dictate that [Mg^{2+}] was essentially invariant in time.

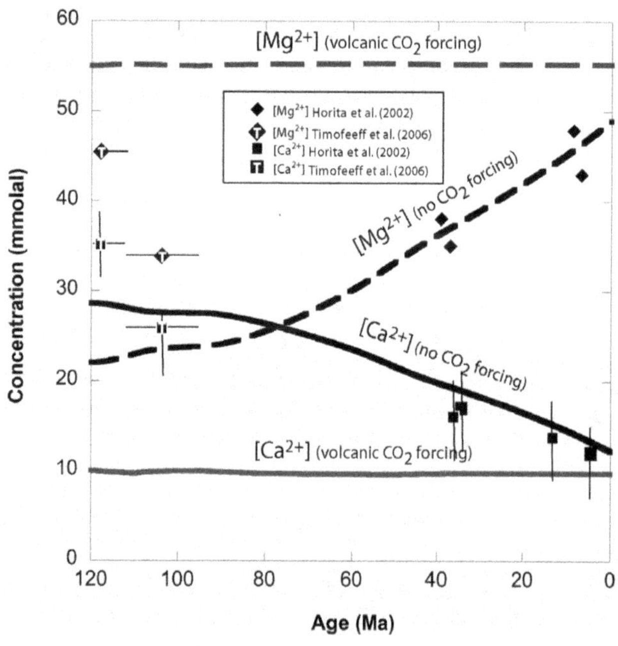

Figure 4. Modeled and observed variations of Mg^{2+} and Ca^{2+} from 120 Ma ago to the present. Two simulations are shown; one in which variations in seafloor spreading rate drive changes in atmospheric CO_2 as well as in Ca^{2+} and Mg^{2+}, and another in which the link to CO_2 is removed. Data from *Horita et al.* [2002] and *Timofeeff et al.* [2006].

The MAGic model of *Arvidson et al.* [2006] provides a state-of-the-art box model of the geochemical cycling of the elements, highly evolved from BLAG, fully coupled, and comprehensive, including factors like dolomite formation that were not part of BLAG. *Arvidson et al.* [2006] use MAGic to address important issues in ocean chemical evolution such as the impact of the evolution of pelagic carbonate production, the importance of dolomite formation and whether it occurs during early diagenesis or deep burial, and the impact of authigenic silicate mineral formation in the ocean (reverse weathering). In the standard MAGic simulations, spreading rate variations drive changes in volcanic CO_2 outgassing and mid-ocean ridge hydrothermal fluxes. Interestingly, although MAGic fits the fluid inclusion [Ca^{2+}] and Mg/Ca ratio quite closely, [Mg^{2+}] variations deviate from observed levels: whereas observed [Mg^{2+}] reaches a minimum at the end of the Paleozoic, MAGic reaches a maximum. Likewise, the Mesozoic trends are generally inverted. Similarly, the model of *Hansen and Wallmann* [2003] predicts little variation in [Mg^{2+}] from seafloor spreading variations because of compensating changes in riverine Mg input. The modeled Cenozoic rise in [Mg^{2+}] is rather the result of intensified carbonate weathering caused by increased exposure of carbonate terrains.

Where does this leave us? Models that incorporate a tight coupling between seafloor spreading, volcanism, climate, and weathering predict a less variable seawater composition (especially for Mg) than reconstructed from the brine inclusion record because of compensating effects. However, there is a compelling correspondence between the extent of continental flooding and the observed [Mg^{2+}] and [Ca^{2+}] of seawater. Either seawater chemistry is indeed driven by variations in seafloor hydrothermal fluxes reflected in inferred sea level fluctuations, and the seafloor spreading-volcanism-CO_2 link is much weaker than previously thought (i.e., the modeling of the Hopkins/Binghamton group is largely correct), or the chemistry of seawater instead has been driven primarily by other factors, for example, variations in rates of dolomitization modulated by secular changes in the extent of continental inundation [*Holland*, 2005] resulting from some factor other than seafloor spreading-driven mid-ocean ridge volume changes [*Rowley*, 2002]. Of course, this option also lessens the importance of the relationship between seafloor spreading and volcanic CO_2 emission anyway, because seafloor spreading rates do no vary. In either case, changes in atmospheric CO_2 driven by variations in volcanic CO_2 release not tied to seafloor spreading, and tectonic changes driving variations in the rate of CO_2 consumption during weathering, both important factors in subsequent GEOCARB modeling by Berner and colleagues [e.g., *Berner and Kothalava*, 2001], together with biotically driven changes in the organic carbon cycle, remain as likely climate drivers over Phanerozoic time scales. Indeed, sensitivity analysis of the most recent models, including GEOCARB and MAGic, produces atmospheric CO_2 curves that match the proxy record quite well while specifying constant seafloor spreading rates.

Perhaps the answer to the question of what drives coupled variations in [Mg^{2+} and Ca^{2+}] still resides in the evaporite brine inclusion record. The Phanerozoic data, recast in phase space, reveal an interesting pattern (Figure 5): [Mg^{2+}] and [Ca^{2+}] are inversely related when [Ca^{2+}] is <~20 m*m*, and positively related above this concentration. One possible explanation of this behavior is that seafloor hydrothermal activity is indeed driving ocean chemical composition toward an "attractor," a fluid equilibrated with hot basalt. This attractor would have an equilibrium [Ca^{2+}] of ~20 m*m* (based on Figure 5) and an equilibrium [Mg^{2+}] of 0. Interestingly, the observed [Ca^{2+}] of vent fluids, corrected for phase separation by normalizing to seawater salinity, is 20 m*m* [*Seyfried et al.*, 1991] (note that this is considerably smaller than the 41 mm used by *Demicco et al.* [2005]). Hypothetically, when [Ca^{2+}] in the oceans is <20 m*m*, seawater circulating through the ridges loses Mg^{2+} and accumulates Ca^{2+} with

$$\frac{d[\mathrm{Mg}^{2+}]}{d[\mathrm{Ca}^{2+}]} = -\frac{[\mathrm{Mg}^{2+}]}{20 - [\mathrm{Ca}^{2+}]},$$

or a slope with [Mg^{2+}] = 50 m*m* and [Ca^{2+}] = 10 m*m* of −5 (approximately the slope through this range of [Ca^{2+}] in Figure 5). When [Ca^{2+}] exceeds 20 m*m*, seafloor alteration acts as a sink for both Ca and Mg; the slope of this relationship for a [Mg^{2+}] =40 m*m* and [Ca^{2+}] = 30 m*m* is +4, again compatible with the observed slope of the relationship at these higher concentrations of Ca.

Of course, this mechanism is highly speculative, and requires considerably more investigation. If secular variations in seafloor hydrothermal activity need to be invoked to explain the Ca and Mg chemical evolution of the oceans, and if these variations tend to correlate with extent of continental flooding, then we need to address the mechanism that relates these phenomena [cf. *Rowley*, 2002; *Demicco*, 2004]. Alternatively, if the "attractor" idea has merit, then perhaps other variables such as rates of dolomitization [*Holland*, 2005] or rates of sulfur cycling [*Berner*, 2004] drive the system away from the attractor, and then hydrothermal activity (possibly relatively invariant) carries the system back toward the attractor.

In any event, the notion that the rate of volcanic CO_2 release scales proportionately to seafloor production rates is challenged by the observed variations in seawater chemistry through time, causing us to look elsewhere, for example, tectonically driven changes in metamorphic degassing and continental weatherability, biological innovations that affect carbon cycling, that is, the multitude of other factors that go into the current GEOCARB, MAGic, and related models. Progress on understanding seawater chemical evolution depends on fundamental breakthroughs in our understanding of the relationships between variations in seafloor production rate (if they exist), seafloor spreading rates, sea level, hydrothermal fluid fluxes and compositions, and volcanic CO_2 release rates.

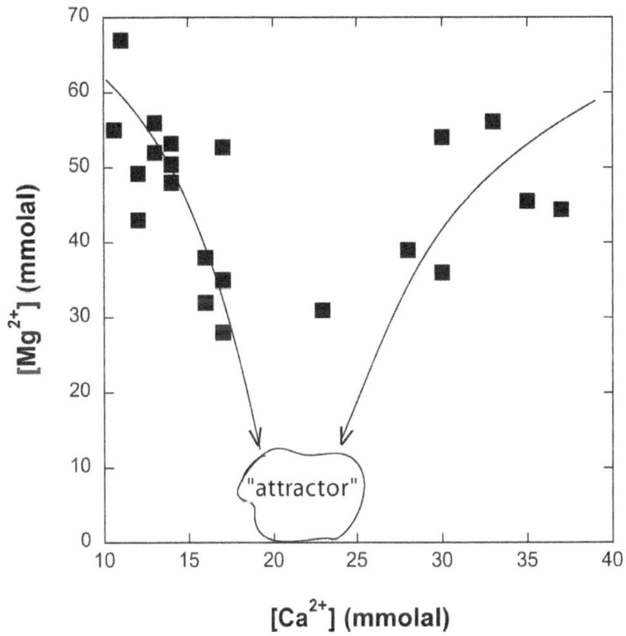

Figure 5. Relationship between Mg^{2+} and Ca^{2+} concentrations in Phanerozoic seawater (data from *Horita et al.* 2002; *Lowenstein et al.* 2003) showing an inferred "attractor" for seawater composition determined by the equilibrium composition of seafloor hydrothermal vent fluids.

Acknowledgments. The author thanks Fred Mackenzie and an anonymous reviewer for their helpful comments on an earlier draft of this manuscript. The NASA Astrobiology Institute and the Geobiology and Environmental Geochemistry Program at the U. S. National Science Foundation supported this work.

REFERENCES

Anbar, A. D., and A. H. Knoll (2002), Proterozoic ocean chemistry and evolution: A bioinorganic bridge?, *Science*, *297*, 1137–1142.

Arvidson, R. S., F. T. Mackenzie, and M. Guidry (2006), MAGic: A Phanerozoic model for the geochemical cycling of major rock-forming components, *Am. J. Sci.*, *306*, 135–190.

Bekker, A., H. D. Holland, P. L. Wang, D. Rumble, H. J. Stein, J. L. Hannah, L. L. Coetzee, and N. J. Beukes (2004), Dating the rise of atmospheric oxygen, *Nature*, *427*, 117–120.

Berner, R. A. (2001), Modeling atmospheric O_2 over Phanerozoic time, *Geochim. Cosmochim. Acta*, *65*, 685–694.

Berner, R. A. (2004), A model for calcium, magnesium and sulfate in seawater over Phanerozoic time, *Am. J. Sci.*, *304*, 438–453.

Berner, R. A. (2006), GEOCARBSULF: A combined model for Phanerozoic atmospheric O_2 and CO_2, *Geochim. Cosmochim. Acta*, *70*, 5653–5664.

Berner, R. A., and Z. Kothalava (2001), GEOCARB III: A revised model of atmospheric CO_2 over Phanerozoic time, *Am. J. Sci.*, *301*, 182–204.

Berner, R. A., A. C. Lasaga, and R. M. Garrels (1983), The carbonate-silicate geochemical cycle and its effect on atmospheric carbon dioxide over the past 100 million years, *Am. J. Sci.*, *283*, 641–683.

Broecker, W. S. (1971), A kinetic model for the chemical composition of sea water, *Quat. Res.*, *1*, 188–207.

Broecker, W. S., and T.-H. Peng (1987), The role of $CaCO_3$ compensation in the glacial to interglacial atmospheric CO_2 change, *Global Biogeochem. Cycles*, *1*, 15–29.

Canfield, D. E. (1998), A new model for Proterozoic ocean chemistry, *Nature*, *396*, 450–453.

Demicco, R. V. (2004), Modeling seafloor-spreading rates through time, *Geology*, *32*, 485–488.

Demicco, R. V., T. K. Lowenstein, L. A. Hardie, and R. J. Spencer (2005), Model of seawater composition for the Phanerozoic, *Geology*, *33*, 877–880.

de Villiers, S. (1998), Excess dissolved Ca in the deep ocean: A hydrothermal hypothesis, *Earth Planet. Sci. Lett.*, *164*, 627–641.

Drever, J. I. (1974), The magnesium problem, in *The Sea*, edited by E. D. Goldberg, pp. 337–357, Wiley Interscience, New York.

Elderfield, H., and A. Schultz (1996), Mid-ocean ridge hydrothermal fluxes and the chemical composition of the ocean, *Annu. Rev. Earth Planet. Sci.*, *24*, 191–224.

Gaffin, S. (1987), Ridge volume dependence on sea-floor generation rate and inversion using long-term sea level change, *Am. J. Sci.*, *287*, 596–611.

Godderis, Y., Y. Donnadieu, C. Dessert, B. Dupre, F. Fluteau, L. M. Francois, J. Meert, A. Nedelec, and G. Ramstein (2007), Coupled modeling of global carbon cycle and climate in the Neoproterozoic: links between Rodinia breakup and major glaciations, *C. R. Geosci.*, *339*, 212–222.

Hansen, K. W., and K. Wallmann (2003), Cretaceous and Cenozoic evolution of seawater composition, atmospheric O_2 and CO_2: A model perspective, *Am. J. Sci.*, *303*, 94–148.

Holland, H. D. (1972), The geologic history of sea water—an attempt to solve the problem, *Geochim. Cosmochim. Acta*, *36*, 637–651.

Holland, H. D. (2005), Sea level, sediments and the composition of seawater, *Am. J. Sci.*, *305*, 220–239.

Horita, J., H. Zimmerman, and H. D. Holland (2002), The chemical evolution of seawater during the Phanerozoic: Implications from the record of marine evaporites, *Geochim. Cosmochim. Acta*, *66*.

Isley, A. (1997), Hydrothermal plumes and the delivery of iron to banded iron formation, *J. Geol.*, *103*, 169–185.

Joly, J. (1923), The age of the Earth, *Sci. Mon.*, *16*, 205–216.

Kasting, J. F. (1984), The carbonate-silicate geochemical cycle and its effect on atmospheric carbon-dioxide over the past 100 million years, *Am. J. Sci.*, *284*, 1175–1182.

Kasting, J. F. (1991), Box models for the evolution of atmospheric oxygen—an update, *Global Planet. Change*, *97*, 125–131.

Kump, L. R., and W. E. Seyfried (2005), Hydrothermal Fe fluxes during the Precambrian: Effect of low oceanic sulfate concentrations and low hydrostatic pressure on the composition of black smokers, *Earth Planet. Sci. Lett.*, *235*, 654–662.

Lowenstein, T. K., L. A. Hardie, M. N. Timofeeff, and R. V. Demicco (2003), Secular variation in seawater chemistry and the origin of calcium chloride basinal brines, *Geology*, *31*, 857–860.

Mackenzie, F. T., and R. M. Garrels (1966), Chemical mass balance between rivers and oceans, *Am. J. Sci.*, *264*, 507–525.

Miller, K. G., M. A. Kominz, J. V. Browning, J. D. Wright, G. S. Mountain, M. E. Katz, P. J. Sugarman, B. S. Cramer, N. Christie-Blick, and S. F. Pekar (2005), The Phanerozoic record of global sea-level change, *Science*, *310*, 1293–1298.

Payne, J. L., and L. R. Kump (2007), Evidence for recurrent Early Triassic massive volcanism from quantitative interpretation of carbon isotope fluctuations, *Earth Planet. Sci. Lett.*, *256*, 264–277.

Poulton, S. W., P. W. Fralick, and D. E. Canfield (2004), The transition to a sulphidic ocean ~1.84 billion years ago, *Nature*, *431*, 173–178.

Rowley, D. B. (2002), Rate of plate creation and destruction: 180 Ma to present, *Geol. Soc. Am. Bull.*, *114*, 927–933.

Rubey, W. W. (1951), The geologic history of seawater—an attempt to state the problem, *Bull. Geol. Soc. Am.*, *62*, 1111–1148.

Seyfried, W. E., K. Ding, and M. E. Berndt (1991), Phase-Equilibria Constraints on the Chemistry of Hot-Spring Fluids at Midocean Ridges, *Geochim. Cosmochim. Acta*, *55*, 3559–3580.

Shen, Y., A. H. Knoll, and M. R. Walter (2003), Evidence for low sulphate and anoxia in a mid-Proterozoic marine basin, *Nature*, *423*, 632–635.

Sillen, L. G. (1961), The physical chemistry of sea water, in *Oceanography*, edited by M. Sears, pp. 549–581, American Association for the Advancement of Science, Washington, D. C.

Spencer, R. J., and L. A. Hardie (1990), Control of seawater composition by mixing of river waters and midocean ridge hydrothermal brines, in *Fluid-Mineral Interactions: A Tribute to H. P. Eugster*, edited by R. J. Spencer and L.-M. Chou, pp. 409–419, Geochemical Society Special Publications.

Timofeeff, M. N., T. K. Lowenstein, M. A. da Silva, and N. B. Harris (2006), Secular variation in the major-ion chemistry of seawater: Evidence from fluid inclusions in Cretaceous halites, *Geochim. Cosmochim. Acta*, *70*, 1977–1994.

Veizer, J., W. Compston, J. Hoefs, and H. Nielsen (1982), Mantle buffering of the early oceans, *Naturwissenschaften*, *69*, 173–180.

Walker, J. C. G., P. B. Hays, and J. F. Kasting (1981), A negative feedback mechanism for the long-term stabilization of Earth's surface temperature, *J. Geophys. Res.*, *86*, 9776–9782.

Wallmann, K. (2001), Controls on the Cretaceous and Cenozoic evolution of seawater composition, atmospheric CO_2 and climate, *Geochim. Cosmochim. Acta*, *65*, 3005–3025.

Wolery, T. J., and N. H. Sleep (1976), Hydrothermal circulation and geochemical flux at Mid-Ocean Ridges, *J. Geol.*, *84*, 249–275.

Wolery, T. J., and N. H. Sleep (1988), Interactions of geochemical cycles with the mantle, in *Chemical Cycles in the Evolution of the Earth*, edited by C. B. Gregor, et al., pp. 77–104, John Wiley and Sons, Inc., New York.

L.R. Kump, Department of Geosciences, Pennsylvania State University, 535 Deike Building, University Park, Pennsylvania 16827, USA. (lkump@psu.edu)

Index

www.ingramcontent.com/pod-product-compliance
Lightning Source LLC
LaVergne TN
LVHW082119200926
848484LV00008B/1980